Power Plant System Design

White Bluff Steam Electric Station Units #1 and #2 of Arkansas Power and Light Co., 1983. The plant consists of two coal-fired units, each rated at approximately 800 MW, complete with electrostatic precipitators and natural draft cooling towers. Steam conditions are 2400 psig, 1000 F superheat, and 1000 F reheat. Coal fuel is low-sulfur, western subbituminous, delivered by unit trains. (Courtesy of Arkansas Power and Light Co.)

Power Plant System Design

Kam W. Li

North Dakota State University, Fargo

A. Paul Priddy

Chas. T. Main, Inc., Engineers, Boston

John Wiley & Sons

New York   Chichester   Brisbane   Toronto   Singapore

TK
1191
L5
1985

Library of Congress Cataloging in Publication Data:

Li, Kam W.
  Power plant system design.

  Includes index.
  1. Electric power-plants–Design and construction.
I. Priddy, A. Paul.   II. Title.

TK1191.L5 1985   621.31′2132   84-22177
ISBN 0-471-88847-8

Printed and bound in the United States of America by Braun-Brumfield, Inc.

10 9 8 7 6

Preface

This textbook is the outgrowth of our consulting engineering work and teaching in electric power generation. It was written to meet the needs of mechanical engineering students and engineers. In the last 20 years, the changes in technology include substantial growth in unit size (from approximately 200 MW in the 1950s to 1100 MW in the 1970s) and the use of different steam conditions (from subcritical in the 1950s to supercritical conditions in the 1970s). In addition, the plant capital and fuel costs have escalated so rapidly that the plant system design has become a subject of increasing importance in the power industry.

The aim of this book is the design of optimum power plant systems. There are two basic concepts in power plant design that will be embodied in this book: component design and system design. The system generally consists of one or more components related to each other to perform one particular task. In power plants the system may be very simple, such as a section of steam pipe between the superheater and the high-pressure turbine, or very complex, such as the turbine cold-end system, which may consist of the turbine exhaust end, condenser, and cooling tower. This book will emphasize systems rather than components. The selection of components will be made in terms of its impacts on the system. However, basic knowledge of the components is a necessary ingredient for understanding the system.

Design is a decision-making process. The design process frequently results in a set of drawings, or a report that may include calculations and descriptions of equipment. In this textbook attention will be focused on the system design rather than the component design; on the thermal design rather than the mechanical design. When we write "thermal design" we mean that the calculations or decisions are based on the principles of thermodynamics, heat transfer, and fluid mechanics. The system design procedures will generate several optional solutions. Apparently, not all these solutions are equally acceptable. Some are better than others. The final decision as to which solution to use will be made by utilizing various simulation and optimization techniques.

This book serves as an introduction to power plant system design. Since the electric power generating system is complex, we do not intend to cover all aspects. Rather, attention is focused on the steam turbine, steam supply systems, condenser, and cooling tower, as well as their combined system. However, the design methodology introduced here is so general that it can be easily adapted to other system design problems.

The use of the digital computer in power plant design is another feature of this textbook. Several computer programs are introduced and may be obtained from us. These programs have been thoroughly verified and tested in a Boston consulting firm. The reason for including these programs is to provide students with an opportunity to use them for system design. Without them, students may have to spend a lot of time in design calculation and not have enough time to appreciate the effects of various design parameters. These computer programs may also serve as models for the further development of computer programs for power plant system design. However, the computer materials were presented in such a way that omitting them would not in any way disturb the continuity of the text.

The book is intended for use at the undergraduate and beginning graduate levels. It should provide sufficient materials, including homework problems, for one four-credit course in universities and colleges. The prerequisites are the first course of thermodynamics, heat transfer, and fluid mechanics. This book is also suitable as a reference for engineers in consulting engineering firms and in utility and manufacturing companies.

The subject matter included in this text is arranged to provide the instructor with a certain degree of flexibility in developing a particular engineering course. When the text is used in a system course (such as power plant system design or thermal system design in general), some background and component materials should be omitted. For this purpose it is suggested that Chapters 2, 5, and 6 be quickly reviewed or entirely omitted. When the text is used in a low-level course such as "Energy Conversion" or "Introduction to Power Plant Systems," the design materials presented in the text should be de-emphasized to some extent. In either case the instructor must select the material to be covered according to the background of the student and the purpose of the course.

During the preparation of this book students were foremost in the our minds. The objective was to develop in students an awareness and understanding of the relationship between the power plant system design and thermal science courses. Efforts were made to demonstrate by examples the use of the principles and working procedures in system design. The book has been tested for two years at North Dakota State University. In 1982 it was also used as a text for the short course "Power Plant System Simulation and Design Optimization" at the Center for Professional Advancement in New Brunswick, New Jersey. We appreciated very much the constructive criticisms both from the practicing engineers and university students.

No claim is made for complete originality of the text. We have been influenced by the excellent publications of many organizations and individuals, especially *Steam/Its Generation and Use* by Babcock & Wilcox, *Combustion, Fossil Power Systems* by Combustion Engineering, and those by General Electric and Westinghouse. We feel that these excellent publications should be acknowledged separately in addition to their being listed in the reference sections in the text.

We are indebted to Northern States Power Company (Minneapolis), Chas. T. Main, Inc. (Boston), and North Dakota State University for the assistance rendered in their professional development. We also thank North Dakota State's Department of Mechanical Engineering and Applied Mechanics for their support in preparing the manuscript and to Brenda Stotser and Debbie Coon for their typing.

Kam W. Li
A. Paul Priddy

Contents

Power Plant System Design

CHAPTER ONE

Introduction

1.1 INTRODUCTION

As civilization has evolved, energy consumption has increased. Table 1-1 shows a rapid increase in the recent decades of world energy consumption. For the last 80 years the consumption increased more than tenfold. Energy consumption is mainly in the form of coal, oil, natural gas, and hydroelectric and nuclear energy. The electric power generated by these natural resources is the most convenient form in which this energy can be used. In 1970 the total world generation of electricity was about 5×10^{12} kWh, which was about 10% of the world energy consumption in that year. If the average conversion efficiency was roughly one-third, then approximately 30% of the world energy consumption was devoted to the production of electricity. This percentage is expected to increase gradually in the years to come.

The demand for electrical energy is not constant. Figure 1-1 shows the typical daily load curve for a metropolitan area. The ratio of valley to peak (minimum daily output to maximum daily output) is between 0.5 to 0.8. If we include weekends, when the demand from the industries is greatly reduced, and consider the maximum annual demand, we will find an even smaller ratio of annual minimum to annual maximum. This relationship becomes apparent in the annual load duration curve shown in Fig. 1-2. Any point on the load duration curve will indicate the number of hours in a given period during which the given load and higher loads prevail. The hatched area can be interpreted as the electrical energy generated in the year by the network system, while the remaining area represents the unused capacity. The ratio of the hatched area to the total area is frequently called the annual system capacity factor, which is generally around 60%. It follows then that sufficient power generating units must be available to meet the highest load, but for most of the year some of these units will be at a standstill. To satisfy the load variation, the power industry must have different generating systems. The generating systems can be classified into three groups as presented in Table 1-2.

The peak-load generating units are for use during the load peaks. Accordingly, they can be started and shut down several times every day. The number of operation hours per year varies from a few hundred to about 2500 hours. Because of the ease in startup, these units are also used as standby or emergency units. These units are generally characterized by a low capital investment and high fuel costs.

The base-load generating units are operated at full load as long as possible during the year. They have high conversion efficiency and can generate electric power at the lowest cost. These units are generally characterized by high capital investment and low fuel cost. Because of the complexity in the generating system, the base-load

Table 1-1
World Energy Consumption [3]

Year	Annual Consumption (10^{18} J)	Annual Consumption (10^{15} Btu)	Average Annual Growth Rate (%)	Annual Consumption per Capita (10^9 J)	Annual Consumption per Capita (10^6 Btu)
1900	22	20.8	2.9	14	13.2
1925	45	42.6	2.2	23	21.8
1950	80	75.8	4.0	32	30.3
1960	118	111.9	5.5	40	37.9
1965	155	147.0	5.4	55	52.1
1970	201	190.5	5.3	58	54.9
1972	218	206.6	3.3	60	56.9
1974	233	220.8	3.1	60	57.2
1976 ·	264	250.4	6.5	65	61.8
1978	279	264.8	2.8	66	62.7
1980	289	274.1	1.7	66	62.3
1981	294	278.3	1.5	65	62.0

Note: The data in this table have been supplemented with additional data from the United Nations report, "World Energy Supplies, 1929 to 1981."

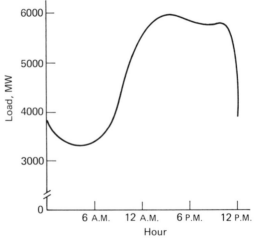

Figure 1-1. Typical daily load curve for a metropolitan area.

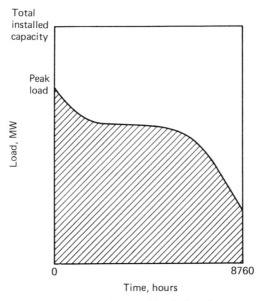

Figure 1-2.   Typical annual load duration curve.

generating units have a poor load change capability. In other words, they take more time to respond to load demand than the peak-load generating units.

The medium-load generating units are operated predominantly on weekdays and shut down at night and on the weekend. The number of operation hours per year varies between 2000 and 5000 hr. These units have operation and economic characteristics somewhere between those of base-load and peak-load generating units. The conversion efficiency is higher than that of peak-load plants, but lower than that of base-load plants.

Table 1-2
Classification of Electric Power Generation Units

1. Peak-load generating units
    - Gas turbine
    - Diesel engine
    - Hydropumped storage plant
    - Old simple steam turbine plant
2. Base-load generating units
    - Nuclear plant
    - High-performance steam turbine plant
    - Advanced combined gas and steam turbine plant
    - Hydroelectric plant
3. Medium-load generating units
    - Simple steam turbine plant
    - Old base-load plant
    - Combined gas and steam turbine plant

Electric power generation is a capital-intensive business. In 1978 it costs around $600 to $1000 to build one kilowatt of generating facility. For a 1000 MW plant, this means a capital investment of 600 million to 1 billion dollars. The cost of operating a power plant is equally expensive. For instance, a 1000 MW coal-fired power plant would consume fuel at a rate of 58.34 trillion Btu per year (capacity factor 70% at plant net heat rate 9513 Btu/kWh), or 2.03 million tons of coal per year (coal heating value 12851 Btu/lb). At the price of $0.7 per million Btu, the annual fuel cost for this 1000 MW plant would be $40.833 million. One percent of fuel saving resulting from improved engineering design would mean a savings of $408,330 every year. These capital-intensive characteristics and high operating costs make the engineering design of power plants extremely important and challenging.

Power plant design is a complex process. The general design objective is to produce a reliable facility that can generate electric power at the lowest cost and with acceptable impacts on the environment. The designer must take into account the progress in technology and the availability and economics of fuels. In the last 20 years, the changes in technology include exponential growth in unit size (from approximately 200 MW in 1960s to 1100 MW in 1970s) and the use of different inlet steam conditions (from subcritical conditions in the 1960s to supercritical conditions in the 1970s). Because of increasing fuel cost, plant efficiency has become a matter of greater significance.

Power plant construction is also a complicated and time-consuming process. In recent years it took 12 years or more to complete a nuclear generating unit and seven to eight years for a coal-fired generating unit. This long lead time is mainly due to regulatory processes and occasionally materials shortage. It follows therefore that load forecasting and long-range planning are especially important in the electric utility. There are also many other reasons for forecasting and long-range planning. One of these is the scheduling of financing by the utility company on a long-term basis. In general, system planning is carried out on a 10- to 20-year basis. Rates are affected by changes that may take place in the company's service area. Sales programs, devised to improve the company's position, are based on long-term forecasts.

Environmental impact of power plant operation is another concern to the plant designer. The magnitude of waste heat is significant. For every kilowatt of electricity generated, there are approximately two kilowatts of waste heat. The availability of cooling water is frequently considered to be one of the important factors for the plant site selection. The plant cooling system affects not only the plant performance but also the plant environment.

Air pollution problems associated with fossil-fuel power plant operation are serious. Of the air pollutants emitted from fossil-fuel power plants, SO_2 is the most serious one. Table 1-3 shows the effects of SO_2 on health by level of exposure. In general, illness from SO_2 may be acute (short-lived) or chronic (usually permanent and irreversible). Chronic bronchitis may develop in persons exposed to even moderate levels of SO_2 over many years—an exposure range of 100 to 350 $\mu g/m^3$ with associated particulate levels of 66 to 365 $\mu g/m^3$. In the past few years various methods of controlling SO_2 concentration levels have been proposed. These methods

Table 1-3
Some Observed Health Effects of SO$_2$ and Particulates [3][†]

Concentration of SO$_2$		Concentration of Particulates		
µg / m^3	Measured as	µg / m^3	Measured as	Possible Effect
1500	24-hr average			Increased mortality
715	24-hr mean	750	24-hr mean	Increased daily death rate; a sharp rise in illness rates among bronchitis over age 54
300–500	24-hr mean	Low	24-hr mean	Increased hospital admissions of elderly respiratory disease cases; increased absenteeism among older workers
600	24-hr mean	300	24-hr mean	Symptoms of chronic lung disease cases accentuated
105–265	Annual mean	185	Annual mean	Increased frequency of respiratory symptoms and lung disease
120	Annual mean	100	Annual mean	Increased frequency and severity of respiratory diseases among schoolchildren
115	Annual mean	160	Annual mean	Increased mortality from bronchitis and lung cancer

[†] Reprinted with permission from MIT Press.

seem to group themselves into categories ranging from fuel selection to intermittent control of emitted SO$_2$. More progress, including fluidized bed combustion, is expected in the near future.

1.2 COAL-FIRED POWER PLANTS

Coal-fired plants make up slightly over one-half of the electric power generation in the United States and in most other parts of the world. There are vast recoverable reserves of coal in the United States, estimated at over 5000 quads (one quad is 10^{15} Btu). Approximately 16 quads were consumed in the United States in 1981, 75% of which was in electricity generation.

Coals are generally classified by geological age, beginning with peat, then lignite, subbituminous, bituminous, semianthracite, and anthracite. There are meta-anthracite reserves in New England, but the ability to burn it in power plants has not been developed, because of its hardness.

The percentage of oxygen and volatile matter are greatest in peat and lignite, and decrease through subbituminous, to the lowest percentage in anthracite. Conversely, the percentage of fixed carbon is greatest in anthracite and decreases through the bituminous grades to the lowest percentage in lignite and peat.

The lignite and subbituminous coals appear more abundantly in the western area and the Gulf Coast areas. The bituminous coals are more abundant in the central and eastern areas. There is relatively little anthracite, which is found chiefly in eastern Pennsylvania.

Coal is usually hauled from the mines to the power plant by railroad. In some cases the coal may be shipped partly by river barge or ocean ship. If shipped by rail, each rail car will contain up to 100 tons, which will fuel a 220 MW plant at full load for approximately one hour.

Coal is usually prepared for shipment in one of several ways, such as by reducing the lump size to a maximum of approximately $1\frac{1}{4}$ inches. At the plant site, the coal is further crushed to approximately a $\frac{3}{4}$ inch maximum. In some cases the coal is washed at the mine site to reduce pyrites and sulphur.

Rail cars may be unloaded at the power plant, either by opening doors at the bottom of the cars, dumping into a hopper, or the cars may be turned upside-down in a rotary trunnion over the hopper. In some cases a mechanical shaking device must be applied to the car to facilitate the removal of coal from the bottom hopper doors. In colder climates thawing facilities must be provided if the coal is frozen in the cars.

From the unloading hopper the coal is conveyed by moving inclined rubber belts mounted on steel-frame galleries to a crusher tower and to the storage yard or to the elevated coal bunkers, or silos, above the pulverizer mills. Plant coal storage is usually divided into two parts; a small area for "ready" use to be used, if required, between coal deliveries, and a "dead" storage area sufficient to serve the plant for an extended period, possibly 90 days, in the event of a major stoppage of coal delivery. The "dead" storage coal must be compacted and coated with a bitumastic material to prevent erosion and possible spontaneous combustion.

Deciding on the location of the plant site also requires a comprehensive engineering study. The site should be large enough for sufficient coal storage and should be accessible to coal delivery. It should also be suitable for condenser cooling water, either from river, or water body, or by cooling towers.

The plant should be supplied with an elevated coal bunker above each pulverizer mill, each sized for several hours of storage. Coal will flow by gravity to the coal feeders at each mill. A pulverizer mill has a grinding capacity of up to 60 tons of coal per hour. An extra or spare mill is usually supplied with each boiler unit.

Coal "grindability" is an important factor in selecting coal for a plant, and also for selecting and sizing pulverizer mills. Grindability is an index of the coal's relative ease of being pulverized. The index is named the Hardgrove Index for the man who developed it. The index covers the most grindable at approximately 40 to over 100. Anthracite coals have the higher indices.

Steam generators designed to burn coal must be "tailored" for the specific coal analysis. The amount of ash and the analysis of the ash are very important. The softening temperature of the ash determines its tendency to slag (deposit on metal surfaces). Softening temperatures of 2600 F or above are more suitable for "dry-bottom" furnaces in which the large ash particles drop to the bottom of the furnace in a solid form (approximately 15–20% of the total ash) and the smaller particles are carried out in the combustion gases in dry form called "fly-ash" (approximately 80–85%). Coals with ash-softening temperatures of less than 2500 F may be more suitable for "wet-bottom" furnaces or "cyclone" furnaces, which will be described in greater detail in Chapter 5. The chemical composition of the ash is also very important. The corrosive qualities of the ash have caused boiler designers and operators many problems.

The suitable coals are usually selected by the utility company. However, the suitable coals may be more expensive and may require greater hauling distance. In most cases all local coal available in the required quantities should be used even if some coal qualities are objectionable.

The boiler designer can, to some extent, compensate for undesirable coal qualities. The geometry of the furnace and the size and placement of coal burners should be such as to prevent flame impingement on furnace walls. Superheater tubes near the top of the furnace section should be at a suitable distance above the topmost coal burners. Superheater and reheater tube sections should have adequate free spacing for gas passage so that slag will not be trapped between the tubes. All tubes in the gas passages should be "bare." The use of "finned" tubes will increase heat transfer but also form traps for slag and fouling. Economizer tubes in the rear boiler pass may be "finned" for boilers firing natural gas, but not for those using coal.

The generous application of soot blowers will help to remove the fouling of slag, which inevitably will occur. Soot blowers consist of jets of high-pressure air or steam, which is intended literally to blow the molten stag off the tubes and wall surfaces. Soot blowers are of two general types; the "wall" blowers, which penetrate the water walls at intervals and are operated by entering the furnace for a few feet and "blowing" in sequence, then withdrawn from the furnace for protection; another type of soot blower is the "long-retractable" units for blowing convection surface sections. These long-retractable blowers, on large, wide boilers may extend from each side to the midpoint of the boiler (as much as 24 or 25 ft). These blowers are withdrawn when not in use and rest on hangers with access platforms. When in use these blowers, with their air or steam nozzles, are traversed into the boiler for sequential blowing of superheater and reheater surfaces. The reason for sequential blowing is to minimize the air compressor and air storage tank size required, or the steam consumption demand. Soot-blower access platforms and furnace wall penetrations should be designed into the boiler initially, because it is either impossible, or very expensive, to discover later that more are needed. This very frequently has happened with coal-fired boilers. Boilers that were designed for oil or natural gas fuel can seldom be satisfactorily converted to use coal fuel. For example the furnaces are usually too small, the burners are not spaced properly, the tube spacing is too "tight," the tube surfaces may be "finned" in some places, and there is no space for soot-blower installation.

After the combustion gases pass through all the heat transfer surfaces, they must be passed through dust precipitators, and then possibly through "scrubbers" to remove sulphur compounds before the gas is discharged into the atmosphere. The dust collector consists of electrostatically charged plates or wires installed in a large boxlike structure. These devices are described in detail in Chapter 5.

Waste disposal is a major concern in coal-fired generating plants. Coal fly ash must be collected in electrostatic precipitators. Fly ash is ash that is carried in the combustion gases. Heavier ash will fall to the bottom of the boiler furnace, where it is collected in a hopper. Facilities must be provided to remove the dry fly ash from the precipitator hoppers, and the bottom ash from the boiler bottom hopper. The fly ash is usually removed by pneumatic conveyors to a storage silo system, from which it may be trucked away. In many cases the dry ash may be used in concrete construction. The boiler bottom ash may be water sluiced to a storage area, or the dewatering facility, from which it may be trucked to a landfill area.

Steam turbines for power generation have grown significantly in size in the last 30 years. Single-shaft units operating at the generator synchronous speed of 3600 rpm for 60 Hz generation, were limited in capacity to approximately 100,000 kW in 1950. Higher capacities were available at lower speeds. The largest earlier units were designed for 1200 and 1800 rpm. At lower speeds, such as 1800 rpm, the last stage diameter can be approximately twice that at 3600 rpm. There is a substantial economic advantage to the 3600 rpm unit, which has substantially reduced weight, stage diameters and, consequently, manufacturing cost. However, the principal effect of the higher speed has been that designers must now cope with higher pressures and temperatures.

The steam loading on the last row of turbine blades was limited earlier to 12,000 lb per hour and per square foot of exhaust annulus area. Later design and metallurgical improvements permitted an increase in allowable loading to 15,000 lb/hr-ft^2. This limit has remained in effect up to the present time.

Increasing the maximum capacity of the 3600 rpm turbine has been achieved over the years by: (1) doubling the capacity of the low-pressure section by using a "double-flow" exhaust (and even more than one double-flow exhaust section); (2) developing metallurgy and design to permit increasing the maximum length of the last stage blades from 23 to 25–26 in., then to 28–30 in., and finally to $33\frac{1}{2}$ in.; (3) increasing the throttle pressure from the 1450–1800 psig level to 2400 and 3500 psig; (4) increasing throttle steam temperature to 1000 F from the 900 to 950 F temperatures used more frequently in the 1930s and 1940s; (5) returning a portion of the steam after partial expansion to the boiler to be reheated and returned to the turbine. Reheating is a common practice now. The steam expands to approximately 500–700 psig, with a temperature of approximately 600 F, is reheated to 1000 F, and is returned to the turbine to continue its expansion. Such reheating increases the turbine efficiency and the maximum available capability for a specific last stage blade length.

At the present time the maximum capability that can be developed on a single shaft (tandem-compound) is approximately 1,100,000 kW. This is achieved by using three double-flow exhaust sections, each with maximum length last-row blades, a separate casing for the reheated steam intermediate-pressure section, and a separate casing for the high-pressure steam. Very few units of this type are in service, and most of these units were designed for lower capability at high efficiency.

Cross-compound (two shafts) turbine units have been employed occasionally to achieve greater capability and higher efficiency. Maximum capabilities have ranged from 500 to 1300 MW. Steam turbines are discussed further in Chapter 7.

Condensers are usually located below the turbine floor. Steam from each turbine exhaust section is discharged from the bottom. This arrangement usually requires two separate condenser shells, one below each exhaust section. This is done because the turbine must be supported by foundations between the two exhaust sections, and these foundations are made more economically and structurally sound if the condensing function is in two separate shells. In such an installation, the condenser tubes are perpendicular to the axis of the turbine, and access to the water boxes and tube installation is from one side of the unit.

Condensers have been designed for installation with the tubes parallel to the turbine axis, with the foundation structures "straddling" the longitudinal condenser. Such a condenser has tubes that are somewhat longer than in the perpendicular condenser.

In years prior to the 1970s, the large power plant units were nearly always located adjacent to available rivers, harbors, lakes, or on manmade lakes, so that cooling water could be pumped through the condensers from the body of water and discharged back to the source (with intake and discharge separated a sufficient distance to prevent recirculation of the heated water). This method of condensing water became known as the "once-through" arrangement.

The quality of condenser water required for a large turbine-generator could be around 400,000 to 500,000 gallons of water per minute (gpm). A power plant consisting of four 600 to 800 MW units would require 1,600,000 to 2,000,000 gpm of water.

The water used would be pumped through the condenser tubes and condense the steam, which would be exposed to the outside of the tubes. The water temperature would be increased by as much as 20 F. Environmentalists have claimed that variations in water temperature have adverse effects upon marine life. This has resulted in the application of cooling towers and other systems to many new power plants. (See Chapter 8 for a discussion of cooling towers.)

1.3 OTHER ELECTRIC POWER GENERATING SYSTEMS

While coal-fired plants constitute most of generating systems used in the United States and in most other parts of the world, other systems such as oil-fired and nuclear plants are available and frequently used. This section contains a brief description of several electric power generating systems.

Oil-Fired Power Plants

The use of fuel oil as a power plant fuel of prime importance began in the late 1950s when the price of crude oil dropped to around $2 per barrel. Prior to that time fuel oil was a relatively minor fuel. Light distillates were used in diesel engines, gas turbines, and smaller installations. The heavier grades of oils such as No. 6 were used for marine applications and emergency back-up for gas- and coal-fired power plants. Fuel oil did not play a major role in large power generation.

When the price of oil dropped, it became highly competitive with coal. The increased use of refined grades of gasoline for automobiles made available large amounts of all grades of residuals from the refining process. Beginning around 1960 many large central station coal-fired power plants were converted for oil firing. The conversions were feasible because the boiler internals, which were designed for coal, were more than ample for oil firing. The conversions consisted primarily of installing the oil burners, or gun, usually within the frame of the coal burners, and adding the necessary piping, pumps, tankage, and other equipment. In many cases the old coal-handling equipment was partially or completely dismantled, and over a period of years became difficult to reconvert to coal firing.

During the early and mid-1970s the oil-producing nations (OPEC) increased the price of crude oil, first to $12.00 per barrel, and later to $30.00 or higher. The oil price later receded to $25.00 per barrel.

In addition to the conversion of existing coal-fired plants to burn oil, many new plants were designed and built to burn only oil. The boiler designed for burning oil cannot be converted to burn coal without substantial derating. There are several reasons for this: (1) Furnace volumes and tube spacing can be "tighter" for oil firing. Coal firing would cause considerable "slagging" of surfaces; (2) Space for coal-handling facilities were not provided in the initial design of the oil-fired unit.

At the present time approximately 10% of the power generation in the United States is oil fired (down from 17% in 1974). It is expected that the percentage of oil fuel will continue to decline, because there is no long-range prospect of oil price declining, in comparison to coal. It is not expected that any new large oil-fired central stations will be built in the foreseeable future.

Fuel oil for power generation consists, basically, of two grades, both of which are by-products of the oil-refining process. No. 2 distillate oil is a light oil (specific gravity 0.8654 at 50 F) and usually contains 0.4 to 0.7% sulphur. It is used for gas turbines, diesels, and small industrial boilers. No. 6 residual oil is much heavier (specific gravity 0.9861 at 60 F) and contains up to 2.8% sulphur. No. 6 oil is the grade that generally has been used in large central station boilers. This latter grade of oil has a viscosity of up to 3000 SUS (Saybolt universal seconds) at 122 F, and must be heated to approximately 200 F to become free-flowing and suitable for atomizing in the burners.

Oil facilities at an oil-fired power plant must include storage tanks (with containing dikes), unloading pumps, and steam heaters. Storage tanks usually are sized to provide several weeks, or months, of storage capacity. Smaller tanks are usually installed near the boilers to provide flexibility in operation. These latter tanks are called "day-tanks."

Turbine system and other parts of oil-fired plant are similar to those in a coal-fired unit.

Natural Gas-Fired Plants

Natural gas-fired plants made up approximately 15% of the electric generation in the United States in 1981. Gas plants are located principally in the west southcentral portion of the United States. In 1970 the percentage of electricity generated by gas in the United States was around 24%. Since that time generation by gas firing has leveled off, while other forms of generation have increased. New central station gas-fired plants are not now permitted by government regulatory agencies.

Proven natural gas reserves vary from time to time. A figure of approximately 400 quads (10^{15} Btu) for 1980 has been estimated by the United States Geological Survey (USGS). This figure is relatively small as compared with over 5000 quads of proven coal reserves. A typical analysis of natural gas is

Methane CH_4	90%
Ethane C_2H_6	6%
Other hydrocarbons	2%
Nitrogen and CO_2	2%
Total	100%

Natural gas-fired plants are the simplest and cleanest of all fossil power plants. Except in the boiler design and operation, the plant system is similar to that of a coal-fired unit.

Nuclear Power Plants

After World War II there was a worldwide push for the development of nuclear energy. The United States government formed the Atomic Energy Commission (AEC) to promote the "peaceful use of the atom." A substantial amount of money and effort was expended by the U.S. government and private industry in the development of nuclear power. This was primarily a government effort, pushed by public pressure. Great strides were made in the development of nuclear power reactors in the subsequent years. In 1973, it was predicted by national forums that approximately 62% of the electric generation would be by nuclear power in 1985.

More than 70 nuclear power generating units are now in operation in the United States. These plants have operated remarkably well and have generated electric power at a lower cost than the average of other modes of generation.

There has, however, been a strong reversal of public pressure in the 1970s and the 1980s, and the further development of nuclear power has been essentially halted in the United States. The percentage of electricity provided by nuclear energy in 1985 in the United States will not be 62%, but more like the 12% that obtained for 1981.

The cost of those plants that were "trapped" in partial completion in the late 1970s and early 1980s was driven up substantially. After the OPEC oil embargo in 1973–1974, energy conservation, which was much needed, reduced the electricity demand and future forecasts. Licensing delays caused a substantial buildup of "interest during construction" at the high interest rates that prevailed during the period. The construction period of these units have extended to 12 to 14 years from the date of inception. The cost per kilowatt has increased from original forecasts of around $600 to as much as $2500 or more in some cases.

Nuclear reactors for power production are discussed in detail in Chapter 6. The production of steam in a nuclear reactor is by means of the reactor coolant fluid, which is water in most cases. In the boiling water reactor (BWR) saturated steam is generated directly in the reactor vessel at 800 to 1000 psig. Then the steam is piped from the reactor vessel to the steam turbine, and the slightly radioactive steam must be shielded to protect personnel from contamination. In another type of system, the pressurized water reactor (PWR), the reactor coolant is pumped through the reactor in a completely liquid state, and then passed to a separate steam generator vessel. In this vessel, the water flows through internal tubes and generates steam in uncontaminated water outside the tubes.

In either type of reactor system, the steam that is piped to the turbine is saturated. No superheat can be developed in the present-day reactor. A high level of cycle efficiency in fossil-fueled turbine units has evolved over the past 50 years by improved steam conditions such as increased pressure and higher superheating and reheating. The saturated steam conditions in the water-cooled reactor appear to be a reversal of this development. Improvements in nuclear turbine cycles have been made by means of reheating the steam in the intermediate stages of expansion in the turbine. This reheating is accomplished by use of the throttle saturated steam. Moisture separators are also used to reduce the moisture in the steam during the expansion cycle.

The low-pressure saturated steam of the nuclear cycles necessitates the use of turbines with large last-stage blade lengths in order to obtain suitable power output to match the economic reactor capacity. However, a blade length of 44 and 53 in. can be employed only in 1800 rpm machines, and in four-flow and six-flow exhaust sections. These are physically very large machines with capacities of 900 and 1100 MW.

Nuclear turbine cycles have a much higher heat rate than the high-pressure, high-temperature fossil cycles. A nuclear turbine heat rate at rated capacity is typically around 10,000 Btu/kWh, while the fossil heat rate under similar circumstances is probably around 8000 Btu/kWh. This difference is the result of the use of low-pressure saturated steam in the nuclear turbine cycle. The additional heat loss in the nuclear cycle must be dissipated in the condenser and cooling system. For this reason the condensers and cooling towers for the nuclear plant are approximately 40% larger than for the equivalent fossil plant. Offsetting the lower efficiency of the nuclear plant cycle is the fact that nuclear fuel cost is considerably less expensive than the cost of coal.

Spent fuel is a problem in nuclear power generation. Waste disposal and nuclear fuel reprocessing is a political issue in the United States. At the present time, the existing nuclear plants are storing their spent fuel rods in their station spent fuel tanks.

Gas Turbine and Combined Cycle Plants

The gas turbine power plant has secured a prominent place in electric utility systems. The plant is compact, relatively inexpensive, and can be constructed usually within two years. It is suitable for "peaking" service during daily periods of high load demand.

The plant occupies a very small space. If fueled with natural gas, no fuel storage is required. Water requirements are minimal, and the generator can be air cooled.

Gas turbine plants have been installed in vacant corners of utility substation sites. The plant has a low profile and is aesthetically compatible with the surrounding environment.

Gas turbines are environmentally more suitable than other fossil or nuclear plants. Environmentally harmful discharge gases, such as NO_x, can be reduced, if necessary, to allowable limits by steam or water injection. The plant is relatively free of noise, if equipped with the proper level of inlet silencers.

Gas turbine heat rates are somewhat higher than other fossil or nuclear plants, and the fuels required are more expensive than coal or nuclear fuel. Gas turbine fuels are natural gas, distillate oils, and in some oil-producing nations, crude oil is frequently used.

Gas turbine plants have the further advantage of quick starting. They can be "on the line" at rated load within 30 minutes. The units can be equipped with automatic controls that permit remote starting and operating. The sizes of single units of gas turbines are manufactured in many steps up to 100,000 kW and even larger.

Combined-cycle plants are an extension of gas turbine plants, having the addition of heat recovery boilers to utilize the gas turbine exhaust heat and a steam turbine generator.

The combined-cycle plant has most of the advantages of the simple cycle gas turbine, but with the added advantage of a heat rate that is lower than all other fossil-fired or nuclear plants.

Conversely, high fuel cost, which is the chief disadvantage of the gas turbine plant, also is the chief disadvantage of the combined-cycle plant. These plants, with their low heat rates and other advantages listed above, give promise of greater use with the further development of synthetic fuels, and possibly higher firing temperatures in the gas turbines.

The typical combined-cycle plant consists of two or more gas turbines, each with a heat recovery boiler serving steam to a single steam turbine generator. Plant capacities can be in steps from approximately 150,000 to 600,000 kW.

Hydroelectric and Pumped Storage Plants

Hydroelectric plants were among the first developed for electric and mechanical power. In 1920, approximately 41% of the electric generation in the United States was hydroelectric. Although hydroelectric generation has increased steadily from 16 billion kWh in 1920 to 258 billion kWh in 1981, its percentage of the total generation has decreased to 11% in 1981.

Hydroelectric generation utilizes the flow of natural rivers as they move from higher altitudes toward the oceans. Dams are built to channel the river flow through the hydraulic turbines with as much head as the natural gradient and terrain will feasibly permit.

Some hydroelectric plants are designed to utilize the actual stream flow. These plants are called "run-of-the-river" plants, and have little storage reservoir behind the dam. Such plants are usually located on substantial rivers with fairly uniform flow. Other hydro plants rely upon large storage reservoirs created by dams, which can store water during times of maximum flow, and release the water through the turbines at rates that are compatible with a desired schedule of generation.

The hydraulic "head" of a plant may be as low as a few feet, or as high as 1500 or more feet, depending upon the natural terrain. Output from a hydroelectric plant is a product of head and water flow rate. Therefore, low-head plants must use proportionately more water flow to obtain the same power output as a high-head plant.

At the present time nearly 50% of the hydroelectric generating capacity in the United States is in the Pacific Northwest. The Columbia River and its tributaries furnish the greater part of this capacity. No other area is outstanding in the United States. The Tennessee Valley system (TVA) has 8 to 10% of the U.S. capacity. The major hydroelectric developments at the present time are outside the United States, with many activities in the developing countries.

Pumped storage generation utilizes reversible pump-generation hydraulic units that alternately generate power when water flows from a higher level reservoir through the generating unit to a lower reservoir, then act as a pump to reverse the flow back to the upper reservoir. The pumped storage plants may be located adjacent to a river or other body of water, which could serve as the lower reservoir, with an upper reservoir nearby at a suitable natural elevation. Such an arrangement is economically feasible only if the natural contours of the upper reservoir sites are suitable for economical application.

Pumped storage operation does not, of course, produce a net output. Its overall efficiency is approximately 67%. That is, the generated output as a turbine–generator is approximately 67% of the electric input from the system to supply the unit when serving as a pump.

Pumped storage units are designed to serve peak power to an electric system and to restore the upper reservoir level during off-peak hours.

Diesel Engine Power Plants

Diesels are internal combustion reciprocating engines. They were developed primarily for marine propulsion and stationary service. The size of these units, in power output, is relatively small for power generation, but physically very large per unit of output, as compared to other types of generation.

In the early years of electric generation, where service areas were small and electric demand was low, diesel engine generators were prominent. Some plants consisted of units having a single 12 to 14 in. cylinder, rated at 50 to 100 bhp (brake horsepower), operating at speeds of less than 300 rpm. Some units were multicylinder with as many as 10 or more cylinders in line. Such a 10-cylinder unit would require a building space of about 3000 ft^2. Later refinements in design and materials permitted the increase in speeds to 1200 rpm or more.

The conversion of coal-fired steam locomotives to diesel engines was completed after World War II because of the low price of fuel oil during that period. The higher speed diesel units developed by that time were more economical than coal-fired reciprocating engines.

However, in electric power generation, the expansion of electric utility systems and the development of large economical central power stations using the steam turbine made the cumbersome diesel engine too small in output to serve economically. At the present time diesel generator power stations are still serving small municipal plants with power outputs up to approximately 4000 kW.

Diesel engines are relatively efficient and may have net heat rates less than 10,000 Btu/kWh. Fuels for diesels consist of most all grades of distillate and residual oils.

1.4 ENGINEER'S ROLES IN POWER GENERATION

Engineers are responsible for the siting, design, construction, and operation of power plants. The utility company engineering staff takes the lead in developing the load forecasts for the future. They may employ specialists in the field to develop the forecast, with the advice and support of the utility company staff. An electrical system load study may be prepared to determine the most feasible additions to transmission facilities and generation facilities. The final report would lead to selection of the general area location, size, and scheduling of generation installations. These forecasts are generally long range and are updated periodically.

Plant site studies are a continuing process for the utility company engineering staff, with periodic assistance from outside engineering firms. A site investigation is a comprehensive evaluation of prospective sites in terms of environmental impact and site development cost.

Engineering feasibility studies must be prepared to determine the type of generation to be installed. This usually involves conceptual plant arrangement drawings, showing fuel supply access, condensing water arrangements, transmission line access, construction cost estimates, construction schedules, cash flow, and estimates of fuel costs and operating costs of various alternatives. An architect-engineering firm is usually selected to perform the preceding site and feasibility studies. The construction cost estimates at this stage are "preliminary" and new revised estimates must be made from time to time.

After the site has been selected, and the necessary permits and financing have been obtained by the utility company, the actual plant design is begun. Again, an architect-engineer (A-E) firm is selected to perform this design service. The design of a steam power plant requires the services of a large number of engineers of many different disciplines. The actual number varies with the type and complexity of the plant. A nuclear plant design would require hundreds of engineers by the A-E firm alone. Table 1.4 shows in tabular form the different engineering disciplines employed by the various principals involved in the design of a coal-fired steam power plant.

The architect-engineers staff serves as application engineers, fitting together and shaping the many different components into a working system. The suppliers and manufacturers of the individual components each have an engineering staff to design their respective equipment.

Time is of the essence in power plant design and construction. The design function must take several courses simultaneously. One major engineering function that must commence in the early stages of the design process is the preparation of purchase specifications for the principal equipment. Time must be allowed for the equipment manufacturers to design their respective equipment and to furnish design information to the plant designers so that the latter can proceed with their processes.

Table 1-4
Engineering Involvement in Steam Power Plant Design

Principals	Consulting	Mechanical	Architects	Civil-hydraulic	Civil-concrete	Civil-structural	Electrical power	Electronic	Chemical water and waste	Environmental	Instrumentation and control
Utility company		X				X	X	X	X	X	X
Regulatory commissions	X	X					X			X	
Architect–engineer	X	X	X	X	X	X	X	X	X	X	X
Steam generator manufacturer	X	X						X			X
Turbine-generator manufacturer	X	X					X	X			X
Coal-handling equipment		X				X	X	X			X
Condenser manufacturer		X									
Ash-handling equipment manufacturer		X		X	X	X					X
Pump manufacturer		X		X							
Water treatment manufacturer									X	X	X
Waste treatment manufacturer									X	X	X
Fan manufacturer		X				X					
Pollution control manufacturer								X	X	X	X
Feedwater heater manufacturer		X									
Piping		X									
Valve manufacturer		X									
Electric switchgear manufacturer							X	X			
Instruments and controls		X						X			X
Electric motor manufacturer							X				

Manufacturer lead time is a major element in the scheduling process, and forms a part of the "critical path" through the engineering and construction schedule. The order of priority and the assigned engineering discipline for the preparation of the principal specifications is usually as follows.

Specification	Engineering Discipline
Turbine-generator	Mechanical and electrical
Steam generator	Mechanical
Condenser and auxiliaries	Mechanical
Feedwater heaters	Mechanical
Boiler feed pumps and drivers	Mechanical and electrical
Main transformers	Electrical
Electrical switchgear	Electrical

These purchase specifications may require revisions before final purchase orders are placed to accommodate adjustments required by system design studies or other developments.

Simultaneously with specification preparation should be the preparation of studies for several system designs. These studies probably would include:

Feedwater cycle arrangement, including number of heaters
Condenser and cooling tower optimization
Boiler feed pumps: number, sizing, and method of driving
Deaerating heater
Feedwater heater terminal differences, and drain cooler
    approaches
Boiler forced and induced draft fan drives
Major pipe sizing

The preparation of purchase specifications and system design studies require engineering input from engineers employed by various manufacturers.

The analysis of proposals received in response to the issuance of the purchase specifications for equipment for the plant is also a major function of the A-E engineering staff. This culminates in the preparation of purchase orders, and revising the initial specifications as required after the analysis of proposals.

The A-E begins preparation of plant arrangement drawings, showing equipment locations and building outlines as soon as preliminary equipment drawings are available from the principal manufacturers. From these a "Plot Plan" drawing is made of the orientation of the plant on the site, and locations are developed for soil test borings for foundations by the civil-concrete engineers. The civil-structural engineers begin their steel framing drawings. Architects begin preparation of building features, and the office and service buildings. Mechanical engineers begin preparation of piping diagrams. Electrical engineers begin preparation of their design of auxiliary power supply system. For these purposes frequent conferences are held with the utility company staff engineers and the various manufacturer's engineers.

Preparation of drawings for construction purposes constitutes the most time-consuming design effort by all engineering disciplines. Computerized design and drafting have been developed in recent years in order to save time, improve accuracy, and coordinate the efforts of those who are preparing construction drawings in the same general physical area.

The engineering and design of a coal-fired power plant unit of large capacity (500 to 800 MW) may require as many as 800,000 hours time of the A-E staff, which includes engineers, clerical, purchasing, expediting, and shop inspection staff. A considerable number of man-hours is expended by the utility company engineering staff and by many manufacturers' engineering staffs.

A quality assurance program must be established at the beginning of the design process and extended through construction to plant operation. During construction, engineers of the various disciplines serve as liaison between the A-E design staff and the construction staff. Engineers must check the accuracy of installation and perform quality control. The checking out and starting up of equipment is performed by manufacturer's engineers, with coordination by the utility company operating staff and the construction staff. While construction is in progress, the utility company operation management develops the plant operating and maintenance staff. This sometimes involves training sessions with manufacturer's schools.

1.5 OUTLINE OF THIS TEXT

As mentioned earlier, power plant system design is one of the important functions for engineers. The quality of system design affects the initial and operating cost as well as the system reliability.

The methodology of arriving at an optimal power plant design is complex, not only because of the arithmetic involved, but also because of many qualitative judgments that have to be introduced. Figure 1.3 shows a general system design flowchart.

The inputs to the system design include, along with the problem statement or specification, the component design information and constraints. The problem statement may specify a general goal of the system under consideration. Since the system consists of many components, the component information must be known and available in input form. The design constraints may be in the form of available space or in the form of environmental impacts.

The system design procedures will generate optional solutions (system designs). Apparently, not all these solutions are equally acceptable. Some will be better than others. In the next step, therefore, schemes must be developed to search for the best solution (the best system design). This is the function of the optimization and simulation procedures.

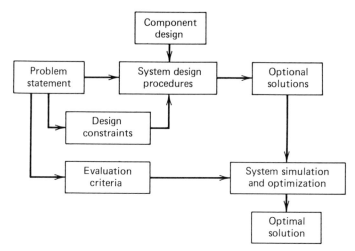

Figure 1-3. A general system design flowchart.

A system is a collection of components whose performance parameters are interrelated. System simulation is the calculation of operating variables for the system under various conditions. System simulation assumes knowledge of the performance characteristics of all components. Simulation is used when it is not possible or not economical to observe the real system. To search for the optimal design, system simulation must be exercised.

Optimization is the process of finding the conditions that give maximum or minimum value of a function. In the power plant system design, optimization generally means a search process for a system design that will have an optimal performance. More specifically, the optimal system is the system that will result directly or indirectly in the lowest production cost and at the same time will have an acceptable impact on the plant environment.

The goal of this textbook is to introduce power plant system design. Since the electric power generating system is complex, there is no intention to cover all aspects here. Rather, attention is focused on the boiler, steam turbine, condenser, cooling towers, as well as their combined system. However, the design methodology to be introduced is so general that it can be easily adapted to other system design problems.

Use of the digital computer in power plant designs is also emphasized in this book. Several computer programs are to be included. These programs have been thoroughly verified and tested, and provide the student with an opportunity to use them for system design. Without them, students may have to spend a lot of time in design calculation and not have enough time to appreciate the effects of various design parameters.

Chapter 2 presents a review of thermodynamic principles and their applications in power plant cycle analysis. This chapter covers a wide range of thermodynamic cycles some of which may not be implemented in today's practice. Students with previous training in this area may skip Chapter 2 for the power plant system design course.

Chapter 3 presents a brief review of economic evaluation methods. These materials are introduced first because they are inseparable from the system designs presented in the subsequent chapters. In fact, the design criterion is frequently an economic one in the power plant design. The readers of this text are assumed to have a first-course knowledge in engineering economy.

Chapter 4 presents the performance characteristics for various generating systems including nuclear, coal-fired, and gas turbine. It is intended to provide the readers with an overview of the utility network operation. This includes economic scheduling principles and the procedures by which the generating facilities are selected.

Chapters 5 through 9 cover the steam generator (both fossil and nuclear), steam turbine, condenser, and evaporative cooling tower. The objective of these chapters is to provide the readers with a knowledge of this equipment so that they can design the plant system intelligently. Particular emphasis is on the parameters that affect the component design as well as its operation performance.

Since system stimulation and optimization are involved in most system design, Chapters 10 and 11 present the fundamentals of this engineering subject. These include formulation of performance models for components and systems. Various optimization techniques are also introduced and presented in such a manner that they can be easily implemented in the computer.

In Chapter 12 the power plant cooling system, which consists of the low-pressure turbine exhaust end, condenser, and cooling towers, is used to demonstrate the system design procedures presented in the text. These include the cooling system configuration design and equipment proposal evaluation. Finally, Chapter 13 presents gas turbine plants, combined cycles, and cogenerations. Further discussion on system design is included.

The appendix includes a power plant heat balance and system design case study. These are prepared so to serve as a reference to engineering students or to those who have limited experience in power plant system design.

SELECTED REFERENCES

1. Babcock and Wilcox Company, *Steam/Its Generation and Use*, 38th edition, 1972.

2. J. G. Singer (editor), *Combustion, Fossil Power System*, Combustion Engineering, Inc. 1981.

3. P. G. Hill, *Power Generation*, MIT Press, 1977.

4. W. F. Stoecker, *Design of Thermal Systems*, McGraw-Hill, 1971.

CHAPTER TWO

Power Plant
Thermodynamic Cycles

The design of electric power generating system starts with the application of certain basic engineering principles called the first and second laws of thermodynamics. These thermodynamic laws provide a quantitative method of looking at the sequential processes at which the working substance, such as steam and air, performs various functions. When the working substance (or a system in general) in a given initial state goes through a series of changes in its thermodynamic properties and finally returns to the initial state, the working substance is said to undergo a thermodynamic cycle. An understanding of power plant thermodynamic cycles is essential in system designs.

This chapter presents a brief review of the thermodynamic principles and their application to various processes. In addition, the chapter covers the basic thermodynamic cycles frequently utilized in electric power generating systems. These include Rankine, Brayton, and other cycles. The purpose of this chapter is to introduce the engineering fundamentals needed for a power plant cycle analysis.

2.1 THERMODYNAMIC PRINCIPLES

The first and second laws of thermodynamics provide the fundamental relationships for a power plant cycle analysis. These laws are presented in equation form as follows.

The first law is

$$\Sigma \Delta m_i h_i - \Sigma \Delta m_e h_e + \Delta Q_{cv} - \Delta W_{cv} = \Delta (mu)_{cv} \qquad (2\text{-}1)$$

where

ΔQ_{cv} = heat transferred to control volume (cv)

ΔW_{cv} = work produced by control volume (cv)

$\Sigma \Delta m_i h_i$ = enthalpy (internal energy plus flow work) convected into the control volume by mass flow

$\Sigma \Delta m_e h_e$ = enthalpy (internal energy plus flow work) convected from the control volume by mass flow

$\Delta (mu)_{cv}$ = internal energy change in control volume (cv)

21

Equation (2-1) presents an energy balance for a control volume, indicating the net energy flow in and out of the control volume is equal to the change of the internal energy inside the control volume. For simplicity, the kinetic and potential energies are omitted from this equation. Similarly, the mass balance equation for a control volume can be written as

$$\Sigma \Delta m_i - \Sigma \Delta m_e = \Delta(m)_{cv} \qquad (2\text{-}2)$$

For a steady-state and steady-flow process both equations are further simplified by the fact that there is no mass and energy accumulation in the control volume. In equation form, they become

$$\Sigma \Delta m_i h_i - \Sigma \Delta m_e h_e + \Delta Q_{cv} - \Delta W_{cv} = 0 \qquad (2\text{-}3)$$

and

$$\Sigma \Delta m_i - \Sigma \Delta m_e = 0 \qquad (2\text{-}4)$$

Equation (2-1) applies equally well to a closed system where the working substance does not cross the system boundary. Thus the first-law equation for a closed system is

$$\Delta Q_{sys} - \Delta W_{sys} = \Delta(mu)_{sys} \qquad (2\text{-}5)$$

The second law is frequently expressed by the equation

$$\Sigma \Delta m_i s_i - \Sigma \Delta m_e s_e + \Sigma \frac{\Delta Q_{cv}}{T} + \Sigma \Delta \sigma = \Delta(ms)_{cv} \qquad (2\text{-}6)$$

where

$$\Sigma \frac{\Delta Q_{cv}}{T} = \text{entropy increase by heat transfer}$$

$\Sigma \Delta \sigma$ = entropy increase due to internal irreversibility (such as friction)

$\Sigma \Delta m_i s_i$ = entropy associated with the mass flow entering the control volume

$\Sigma \Delta m_e s_e$ = entropy associated with the mass flow leaving the control volume

$\Delta(ms)_{cv}$ = entropy change in the control volume

Equation (2-6) presents an entropy balance for a control volume. When the entropy induced by the internal irreversibility is taken into account, the net entropy flow in and out of the control volume must be equal to the change of entropy in the control volume. The induced entropy is intimately related to the lost work, the work that could not be realized in the process because of internal friction. In equation form the entropy induced by internal irreversibility is

$$\Delta\sigma = \frac{LW}{T} \qquad (2\text{-}7)$$

where

LW = lost work

For the steady-state, steady-flow process such as those frequently encountered in power plants, the entropy equation is simplified by setting the term $\Delta(ms)_{cv}$ to zero. That is,

$$\Sigma\Delta m_i s_i - \Sigma\Delta m_e s_e + \Sigma\frac{\Delta Q}{T} + \Sigma\Delta\sigma = 0 \qquad (2\text{-}8)$$

Equation (2-6) is also applicable to a closed system. In this case the terms associated with mass convection would vanish. Therefore, the second-law equation for a closed system is

$$\frac{\Delta Q_{sys}}{T} + \Sigma\Delta\sigma = \Delta(ms)_{sys} \qquad (2\text{-}9)$$

This equation is further simplified if the closed system undergoes a reversible process. In a reversible process the internally generated entropy becomes zero and the change of entropy in the system is directly related to the heat transfer. That is,

$$\Delta(ms)_{sys} = \frac{\Delta Q_{sys}}{T} \qquad (2\text{-}10)$$

The following examples illustrate the application of the thermodynamic principles to various processes and show the methods of determining the thermodynamic properties.

Turbine Process

For most practical cases the turbine process is assumed to be adiabatic, and no change in kinetic and potential energy occurs. Under steady-state and steady-flow conditions, the energy equation becomes

$$\Delta m_i h_i - \Delta m_e h_e = \Delta W_{cv}$$

or

$$w_t = \frac{\Delta W_{cv}}{\Delta m_i} = h_i - h_e \qquad (2\text{-}11)$$

This equation presents the turbine work by unit mass of working substance as the difference of enthalpy h_i at the turbine inlet and enthalpy h_e at the turbine outlet. Where the working substance is steam, the enthalpy h_i is determined by the steam temperature and pressure at the turbine inlet. The steam table and Mollier diagram are very useful for these calculations. To determine the enthalpy h_e, the turbine internal efficiency is frequently utilized. The turbine internal efficiency is defined as

$$\eta_t = \frac{h_i - h_e}{h_i - h_{es}} \tag{2-12}$$

It is the ratio of the actual enthalpy drop to the enthalpy drop that would occur in the corresponding adiabatic and reversible process. Figure 2-1 indicates these two enthalpy drops in an h-s diagram. The adiabatic and reversible process is seen as constant entropy process (frequently called isentropic process). In practice, he turbine internal efficiency is given, and the enthalpy h_e is calculated by using Eq. (2-12). Other thermodynamic properties at the turbine exhaust end can be easily determined with steam table.

Where the working substance in the turbine process is a gas (or air), Eq. (2-11) is further simplified and expressed as

$$w_t = C_p(T_i - T_e) \tag{2-13}$$

Equation (2-13) is based on the assumption that the gas behaves as an ideal gas, and the temperature change in the process is not substantial so that a constant specific heat (C_p) can be used. For this case the turbine internal efficiency becomes

$$\eta_t = \frac{T_i - T_e}{T_i - T_{es}} \tag{2-14}$$

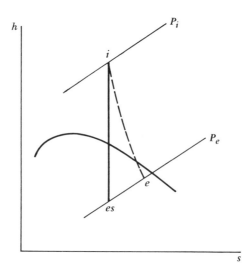

Figure 2-1. Irreversible expansion process in a steam turbine.

Equation (2-14) provides the relationship for the calculation of the gas exhaust temperature T_e. The temperature T_{es} is the temperature of the exhaust gas if the gas would expand isentropically in the turbine. This temperature is determined by

$$\frac{T_{es}}{T_i} = \left(\frac{P_e}{P_i}\right)^{(k-1)/k}$$

(2-15)

The constant k is the ratio of constant-pressure specific heat to constant-volume specific heat. For air the constant k is approximately 1.4.

Compressor or Pump Process

The compressor process is, in some sense, a reversed turbine process. Figure 2-2 shows a compression process and its corresponding isentropic process in a T-s diagram. The reason that the dotted line is used for the actual process is the uncertainty of actual compression in the compressor. This is the characteristics of all real processes that are irreversible to some extent. The compressor efficiency is defined by

$$\eta_c = \frac{T_i - T_{es}}{T_i - T_e}$$

(2-16)

and the compressor work for unit mass

$$w_c = C_p(T_i - T_e)$$

(2-17)

Equation (2-17) indicates a negative value for compressor work. In thermodynamics

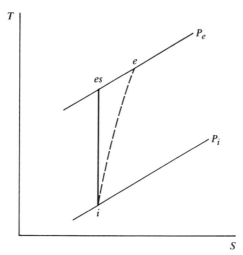

Figure 2-2. Irreversible compression process from state i to state e.

the negative sign only means the work received by the control volume (i.e., the compressor in this case).

The pumping process is similar to the compression process. Equation (2-11) is equally applicable. For convenience, the reversible pump work can be also calculated by the equation

$$w_p = -\int_i^e v\,dp \tag{2-18}$$

or

$$w_p = -v_i(P_e - P_i) \tag{2-19}$$

if the substance is incompressible and has constant specific volume in the process. To obtain the actual pump work per unit mass, the reversible pump work must be divided by the pump efficiency.

Heat Exchanger

Figure 2-3 shows a schematic heat exchanger diagram. With the heat exchanger as a control volume, the first-law equation is

$$\Delta m_h h_{hi} + \Delta m_c h_{ci} = \Delta m_h h_{he} + \Delta m_c h_{ce}$$

or

$$\Delta m_h(h_{hi} - h_{he}) = \Delta m_c(h_{ce} - h_{ci}) \tag{2-20}$$

The equation assumes negligible heat loss from the heat exchanger. Like the previous examples, the kinetic and potential energy are omitted from consideration.

The enthalpy change is generally determined by using thermodynamic tables. Where the substance can be treated as an ideal gas, the following equation is used as an approximation.

$$h_i - h_e = \int_i^e C_p\,dT \tag{2-21}$$

The condenser is one of the heat exchangers frequently encountered in power plants. In the condenser steam from the turbine exhaust end will return to the saturated liquid state and in so doing have latent heat released and transferred to the cooling water flowing inside the condenser tubes. Equation (2-20) is still valid for the

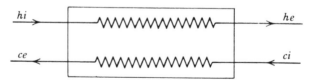

Figure 2-3.   Schematic diagram for a heat exchanger.

condenser process. The amount of heat removed is either equal to the enthalpy drop in steam

$$\Delta Q = \Delta m_h (h_{hi} - h_{he}) \tag{2-22}$$

or equal to the heat gained by the cooling water

$$\Delta Q = \Delta m_c (h_{ce} - h_{ci})$$

$$= \Delta m_c C_{pc} (T_{ce} - T_{ci}) \tag{2-23}$$

The boiler is another kind of heat exchanger. In a boiler fuel is combusted and its energy is released for steam generation. In other words, the energy gained by the steam (or water) must be equal to the amount of energy released by fuel. Taking the boiler efficiency η_b into account, the first-law equation is

$$\Delta Q_f = \frac{\Delta Q_s}{\eta_b} = \frac{\Delta m_s}{\eta_b} (h_{se} - h_{si}) \tag{2-24}$$

where

ΔQ_f = heat released by fuel

ΔQ_s = heat gained by steam

Throttling Process

The throttling process is one of the processes frequently encountered in electric power generating systems. When steam goes through a valve and experiences a pressure drop, the steam is said to undergo a throttling process. Because of no work produced and negligible heat loss from the valve, the first-law equation is for a steady-state, steady-flow process

$$h_i = h_e \tag{2-25}$$

The equation indicates the enthalpy at the valve inlet is equal to the enthalpy at the valve outlet. Therefore, the process is sometimes referred to as a constant enthalpy process. If the substance can be treated as an ideal gas, the throttling process is also a constant temperature process or isothermal process.

2.2 CARNOT CYCLE

In 1824, Sadi Carnot, a French engineer, presented for the first time the concepts of the cycle and reversible processes. Later these concepts proved to be the foundation of the second law of thermodynamics. The reversible cycle he proposed, often called the Carnot cycle, is the most efficient cycle that can operate between two constant temperature reservoirs. In fact, the Carnot cycle is still used as a comparison for other power-producing cycles.

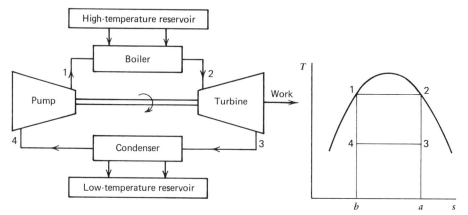

Figure 2-4. A Carnot engine utilizing a two-phase saturated vapor as the working substance.

The Carnot cycle consists of the following four processes as shown in Fig. 2-4.

1. The process 1-2 is reversible and isothermal. Heat is transferred from the high-temperature reservoir.

2. The process 2-3 is reversible and adiabatic. The working substance expands and has its temperature decreased to that of the low-temperature reservoir.

3. The process 3-4 is reversible and isothermal. Heat is transferred to the low-temperature reservoir.

4. The process 4-1 is reversible and adiabatic. The working substance is compressed and has its temperature increased back to that of the high-temperature reservoir.

To achieve the first reversible and isothermal process, the working substance must have the temperature infinitesimally lower than the high-temperature reservoir. Since the temperature of the reservoir remains constant, the temperature of the substance must be constant. If the working substance is a pure substance such as water, this isothermal process can be carried out in an evaporation process. The second process occurs in the turbine. Because there is no heat transfer, the process must be isentropic and has the same entropy at the turbine inlet and outlet. In this process steam expands reversibly and produces the useful work. In the third process, heat is transferred from the condensing steam to the low-temperature reservoir. When steam condensation takes place at a temperature infinitesimally higher than that of the low-temperature reservoir, the process is reversible and isothermal. The fourth process occurs in the pump. Similar to the second process, the entropy of the working substance will be the same both at the pump inlet and outlet. In this process the substance is compressed back to the initial state at the beginning of the cycle.

The Carnot cycle is an ideal cycle that could not be attained in practice. To realize a heat transfer process through an infinitesimal temperature difference, an infinite amount of time or an infinite amount of heat transfer surface would be

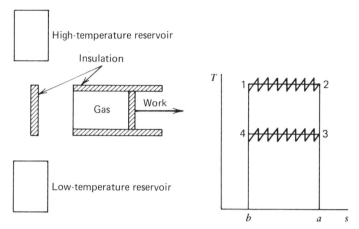

Figure 2-5. A Carnot engine utilizing a gas as the working substance.

required. To have a frictionless expansion or compression, the fluid would have to move at a velocity infinitesimally close to zero. Also shown in Fig. 2-4, both expansion and compression processes are in the wet steam region. These will undoubtedly present a difficult problem in design as well as operation.

It should be pointed out that the Carnot cycle can have many different working substances. When the working substance is a hot gas, the gas temperature will change as it receives or releases heat. To approximate an isothermal heat addition process, the method is to utilize a series of isentropic expansions and constant pressure reheats. As the number of processes in the series increases, the series will approach the isothermal process as shown in Fig. 2-5. Similarly, the isothermal heat rejection process can be approximated by a series of isentropic compressions and constant pressure intercoolings.

There are various possible arrangements of machinery for the Carnot cycle. Figure 2-4 shows a typical arrangement with a two-phase saturated vapor as the working substance. The Carnot cycle can be also devised so that it takes place entirely within a cylinder such as that in Fig. 2-5. In this arrangement a gas is used as the working substance.

The Carnot cycle has no counterpart in practice. Since the processes are reversible, the Carnot cycle offers maximum thermal efficiency attainable between two constant temperature reservoirs. The cycle thermal efficiency is generally defined as

$$\eta_{cy} = \frac{\text{work produced by the cycle}}{\text{heat supplied to the cycle}} \tag{2-26}$$

For the Carnot cycle the thermal efficiency becomes

$$\eta_{cy} = \frac{Q_h - Q_l}{Q_h} \tag{2-27}$$

or

$$\eta_{cy} = \frac{T_h - T_l}{T_h} = 1 - \frac{T_l}{T_h} \tag{2-28}$$

where T_h is the temperature of the heat source (i.e., the high-temperature reservoir) and T_l is the temperature of the heat sink (i.e., the low-temperature reservoir). Equation (2-28) clearly indicates that the Carnot cycle efficiency can be increased by either increasing the temperature T_h or lowering the temperature T_l. At this point it is appropriate to point out that when changes of kinetic and potential energy are neglected, heat transfer and work may be represented by various areas on the T-s diagram. Referring back to Fig. 2-4, we represent the heat transferred to the working substance by the area 1-2-a-b-1 and the heat transferred from the working substance by the area 3-a-b-4-3. From the first law the difference of these two areas, namely 1-2-3-4-1, must represent the work produced in the Carnot cycle. Using these areas, the Carnot cycle efficiency is

$$\eta_{cy} = \frac{\text{area 1-2-3-4-1}}{\text{area 1-2-}a\text{-}b\text{-1}} \tag{2-29}$$

In analyzing the Carnot and other cycles it is helpful to think of these areas in relation to the cycle efficiency. Any relative change in these areas will affect the cycle efficiency.

2.3 RANKINE CYCLE

The ideal cycle for a simple steam power plant is a modification of the Carnot cycle proposed by Professor William John Macquorn Rankine (1820–1872). The ideal cycle was later named as the Rankine cycle in his honor. The processes that comprise the cycle are shown in Fig. 2-6.

1. Constant pressure heat addition process in the boiler (1-2).
2. Reversible and adiabatic expansion process in the turbine (2-3).
3. Constant pressure heat rejection process in the condenser (3-4).
4. Reversible and adiabatic compression process in the pump (4-1).

Figure 2-6 indicates that the Rankine cycle is similar to the Carnot cycle with one exception in the condensation process. In the Rankine cycle the condensation process terminates at the saturated liquid state. Because of this, a simple liquid pump can replace the two-phase compressor. In operation the pump power is greatly reduced and only amounts to 1% or less of the turbine output.

It is evident in the T-s diagram that the Rankine cycle is less efficient than a Carnot cycle for the same maximum and minimum temperatures. The Rankine cycle work represented by the area 1-a-2-3-4-1 is smaller than the Carnot-cycle work

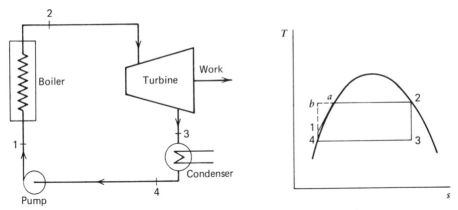

Figure 2-6. Simple steam power plant operating on the Rankine cycle.

represented by the area 1-b-2-3-4-1. In other words, the triangle area bounded by 1-a-b-1 is the loss of cycle work due to the irreversibility in the liquid heating. The lower efficiency of Rankine cycle is also evident in terms of the average temperature at which the working substance receives heat. In the Rankine cycle, the average temperature for the liquid heating process $(1 - a)$ is much lower than the temperature during evaporation; thus the average temperature for heat addition is lower than the maximum temperature.

EXAMPLE 2-1. An ideal Rankine cycle utilizes steam as the working substance. Determine the cycle efficiency for the following conditions:

Turbine inlet condition	300 psia (saturated vapor)
Condenser pressure	1 psia

Solution: It is very convenient and helpful to use the Mollier diagram for calculation of steam turbine work. First, we use the turbine inlet conditions (300 psia, saturated vapor) to determine state 2 as shown in Fig. 2-7. Since the turbine process is isentropic process in the ideal Rankine cycle, the process in the *h-s* diagram must be a vertical straight line. The interception between this line and the pressure curve (1 psia) will locate state 3, which represents the turbine exit conditions. In this case we read from the Mollier diagram.

$$h_2 = 1204 \text{ Btu/lb}$$

$$h_3 = 844 \text{ Btu/lb}$$

$$s_3 = s_2 = 1.5105 \text{ Btu/lb-R}$$

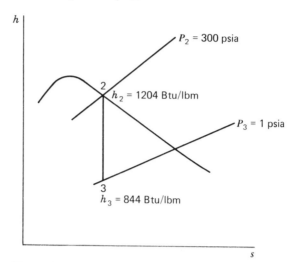

Figure 2-7. A turbine process in the Mollier diagram.

Considering the steam turbine as a control volume, we determine the turbine work by the first-law equation as below:

$$w_t = h_2 - h_3$$

$$= 1204 - 844 = 360 \text{ Btu/lb}$$

The pump work is calculated by

$$w_p = -\int_4^1 v\, dp$$

$$= -v_4(p_1 - p_4)$$

Substituting the steam and water properties into the above equation and omitting the negative for simplicity gives us

$$w_p = 0.01614(300 - 1)\frac{144}{778}$$

$$= 0.893 \text{ Btu/lb}$$

The net work is then

$$w_n = w_t - w_p$$

$$= 360 - 0.893 = 359.1 \text{ Btu/lb}$$

Considering the boiler as the control volume, the first law provides the equation for calculation of heat input to the ideal Rankine cycle. The equation is

$$q_h = h_2 - h_1$$

The enthalpy at state 1 is the water enthalpy at the pump outlet. This enthalpy is approximated by

$$h_1 = h_4 - w_p$$

$$= 69.7 + 0.89 = 70.6 \text{ Btu/lb}$$

Finally, we have the heat supplied as

$$q_h = 1204 - 70.6$$

$$= 1133.4 \text{ Btu/lb}$$

and the cycle efficiency

$$\eta_{cy} = \frac{w_{net}}{q_h} = \frac{359.1}{1133.4} = 0.317 \qquad \text{or} \qquad 31.7\%$$

There are several important parameters affecting the Rankine cycle efficiency. These include condenser pressure, steam conditions, use of reheating and regenerative process. First, let us consider the effect of condenser pressure on the Rankine cycle. Figure 2-8 indicates a Rankine cycle with two different condenser pressures in the *T-s* diagram. The one with the condenser pressure p_3 has the cycle work represented by the area 1-2-3-4-1 while the one with the lower pressure p_3' has the work by the area 1'-2-3'-4'-1'. Evidently, the work produced in the Rankine cycle can be increased by lowering the condenser pressure. Since the heat transferred to the steam is almost the same for both cases, the net result is an increase in cycle efficiency. However, it does not mean the condenser pressure should be reduced infinitely. As shown in Fig. 2-8, lowering the condenser pressure can cause an increase in the moisture content in the turbine exhaust end. These in turn will affect adversely the turbine internal efficiency and the erosion of turbine blades. Also, a low condenser pressure will result in an increase in condenser size and cooling water flow rate. In modern steam turbine design the moisture content in steam is usually limited to 15% or less.

Next we consider the effect of steam conditions on the Rankine cycle. Figure 2-9 indicates a Rankine cycle with two different steam temperatures at the turbine inlet. One is saturated steam while another is the superheated steam. The area bounded by 2-2'-3'-3-2 (shown by the crosshatching) is the increase in the cycle work when the steam temperature is increased from T_2 to T_2'. Increasing the steam temperature also results in an increase of heat supplied in the boiler. This increase is represented by

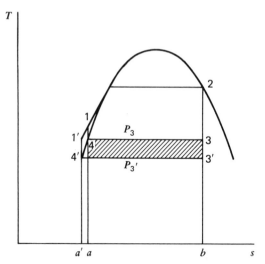

Figure 2-8. The effect of condenser pressure on the Rankine cycle efficiency.

the area 2-2'-b'-b-2. Since the ratio of increase in the cycle work to increase in the heat supply is greater than the ratio for the balance of the cycle, the net result is an increase in cycle efficiency. This is also evident from the fact that the average temperature at which heat is transferred to the steam is increased. Increasing the steam temperature not only improves the cycle efficiency, but also reduces the moisture content at the turbine exhaust end. In steam turbine design the maximum steam temperature is in the range of 1000 to 1100 F.

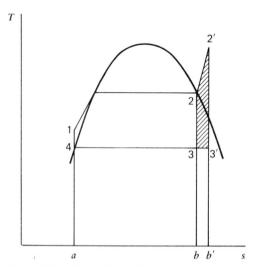

Figure 2-9. The effect of steam temperature on the Rankine cycle efficiency.

The steam pressure in the turbine inlet is important. To have a fair comparison, the maximum steam temperature and the condenser pressure are held constant as shown in Fig. 2-10. It is seen that as the steam pressure increases, the net work tends to remain unchanged (i.e., the single-crosshatching area is approximately equal to the double-crosshatching area). Since the heat rejected decreases by the area 3-*b*-*b'*-3'-3, the net result is an increase in cycle efficiency. This conclusion is also evident by determining the average temperature at which heat is supplied to the steam. While the thermodynamic analysis indicates that the Rankine cycle efficiency can be improved by increasing the steam pressure, the selection of steam pressure in power plant design must be tempered with consideration of technical and economic factors.

The use of reheating process is very common in steam power plants. The advantage is easily seen in the *T-s* diagram shown in Fig. 2-11. In the reheat cycle steam expands partially in the turbine and then returns to the boiler for reheating. The reheat is a constant pressure process and represented by the curve 3-4 in the *T-s* diagram. After reheating, steam continues its expansion in the turbine and eventually exhausts to the condenser. Reheating process may not substantially improve the cycle efficiency, but it does reduce the moisture content in the steam leaving the turbine. This may then improve the turbine internal efficiency and thus improve the cycle performance. The double reheat cycle is thermodynamically superior to the single reheat. But for various reasons the double reheat is seldom used in practice. In power plant design the reheat pressure (the pressure at which the reheat process takes place) is usually in the range of 20 to 28% of initial steam pressure, and the reheat temperature is frequently equal to the temperature of the steam leaving the boiler.

Finally, we consider the regenerative process in the Rankine cycle. To discuss this problem we must remember the average temperature for heat addition in the

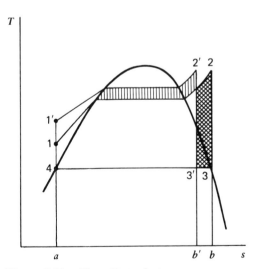

Figure 2-10. The effect of steam pressure on the Rankine cycle efficiency.

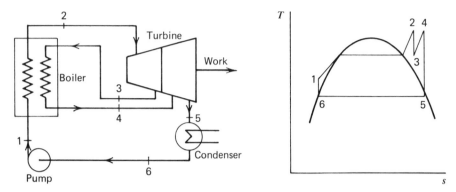

Figure 2-11. A reheat Rankine cycle.

Rankine cycle is usually lower than the maximum temperature. It is mainly due to the liquid heating in the boiler. If this liquid heating could be eliminated from the boiler, the average temperature for heat addition would be greatly increased and equal to the maximum cycle temperature in the limiting case. Figure 2-12 presents an ideal Rankine cycle with the water heating done outside the boiler. The water circulates around the turbine casing and flows in the direction opposite to that of the steam flow in the turbine. Because of the temperature difference, heat is transferred to the water from the steam. Let us consider that this is a reversible heat transfer process, that is, at each point the temperature of the steam is only infinitesimally higher than the temperature of the water. At the end of the heating process the water enters the boiler at the saturation temperature T_1. Since the decrease of entropy in the steam expansion line 2-3 is exactly equal to the increase of entropy in the water

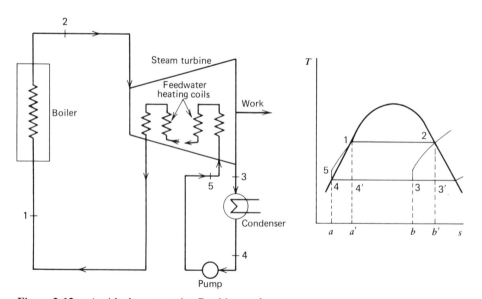

Figure 2-12. An ideal regenerative Rankine cycle.

heating process 5-1, the ideal regenerative Rankine cycle will have the same efficiency as the Carnot cycle, 1-2-3'-4'-1.

Obviously this ideal regenerative Rankine cycle is not practical. It is almost impossible to carry out the reversible heat transfer process just described. In addition, the moisture in the steam turbine will be too great for a safe and efficient operation. The practical regenerative cycle usually involves the use of feedwater heaters. In feedwater heaters water is heated by the steam that is extracted from the turbine. Figure 2-13 presents a Rankine cycle with one contact heater. Because of a reduction of water heating in the boiler, the cycle efficiency is expected to be improved.

EXAMPLE 2-2. Consider an ideal regenerative Rankine cycle as shown in Fig. 2-13. Saturated steam enters the turbine at the pressure 300 psia and exhausts at the pressure 1 psia to the condenser. Some of the steam at 60 psia is extracted for the purpose of the feedwater heating. Calculate the cycle efficiency and compare the result with that in Example 2-1.

Solution: We use the Mollier diagram to determine the steam properties at various locations shown in Fig. 2-13. The properties at locations 1, 3 were found in Example 2-1 to be

$$h_1 = 1204 \text{ Btu/lb} \qquad s_1 = 1.5105 \text{ Btu/lb-R}$$

$$h_3 = 844 \text{ Btu/lb} \qquad s_3 = 1.5105 \text{ Btu/lb-R}$$

To determine the extraction conditions, we locate the intersection of the turbine

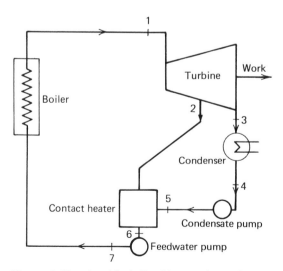

Figure 2-13. An ideal Rankine cycle with one contact heater.

process with the constant pressure curve (at 60 psia) in the Mollier diagram. The readings are

$$h_2 = 1077 \text{ Btu/lb}, \qquad s_2 = 1.5105 \text{ Btu/lb-R}$$

The water leaving the contact heater is expected to be in saturated conditions. Thus the properties at that location are

$$h_6 = 262.2 \text{ Btu/lb}, \qquad s_6 = 0.4273 \text{ Btu/lb-R}$$

Similarly, the properties at location 4 are

$$h_4 = 69.73 \text{ Btu/lb}, \qquad s_4 = 0.1326 \text{ Btu/lb-R}$$

To determine the enthalpy at locations 5 and 7, we use the identical procedure used in Example 2-1, that is,

$$h_5 = h_4 - \Delta w_{p1}$$

and

$$h_7 = h_6 - \Delta w_{p2}$$

where Δw_{p1} and Δw_{p2} are the pump work per unit mass. They are calculated by the equation

$$\Delta w_p = - \int_i^e v \, dp$$

Substituting the numerical values into the equation, we have these two enthalpies as

$$h_5 = 69.7 + 0.01614(60 - 1)\frac{144}{778}$$

$$= 69.88 \text{ Btu/lb}$$

and

$$h_7 = 262.2 + 0.01738(300 - 60)\frac{144}{778}$$

$$= 262.97 \text{ Btu/lb}$$

Next, we determine the steam extraction rate. This can be accomplished by applying the first law to the contact heater. The first-law equation is

$$\Delta m_2 h_2 + \Delta m_5 h_5 = \Delta m_6 h_6 \qquad \text{(a)}$$

Since $\Delta m_2 + \Delta m_5 = \Delta m_6$, Eq. (a) becomes

$$\frac{\Delta m_2}{\Delta m_6} h_2 + \left(1 - \frac{\Delta m_2}{\Delta m_6}\right) h_5 = h_6 \qquad \text{(b)}$$

Substituting the enthalpies into Eq. (b) gives us

$$\frac{\Delta m_2}{\Delta m_6} = 0.191$$

This ratio represents the fraction of the steam flow extracted for the purpose of heating the feedwater. To determine the cycle efficiency, we calculate the turbine work as

$$w_t = h_1 - 0.191 h_2 - (1 - 0.191) h_3$$

$$= 315.88 \text{ Btu/lb}$$

and the pump work as

$$w_p = (h_7 - h_6) + 0.809(h_5 - h_4)$$

$$= 0.89 \text{ Btu/lb}$$

Since the heat supplied to the cycle is calculated by

$$q_h = h_1 - h_7$$

$$q_h = 1204 - 262.97 = 941.03 \text{ Btu/lb}$$

the cycle efficiency is

$$\eta_{cy} = \frac{w_t - w_p}{q_h}$$

$$\eta_{cy} = \frac{315.88 - 0.89}{941.03} = 0.335 \qquad \text{or} \qquad 33.5\%$$

Compared with the results in Example 2-1, the cycle net work is reduced, but the cycle efficiency is increased.

Quite obviously the Rankine cycle efficiency will increase as the number of feedwater heaters is increased. In addition to the contact heater, surface heaters are frequently used for the purpose of feedwater heating. In a surface heater water flows inside heater tubes and absorbs the heat from the steam condensing outside. After condensing, steam leaves the heater either as saturated water or subcooled water; in

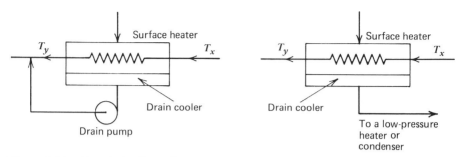

Figure 2-14. Two drain disposals.

either case, they are referred to as the heater drain, Figure 2-14 presents two heater drain disposals. In one arrangement the drain is pumped back to the feedwater circuit and in another arrangement, it is cascaded to a low-pressure heater or condenser. The terms, the heater terminal temperature difference (TTD) and drain cooler approach (DCA), are frequently used in the calculations. They are defined, respectively, as

$$\text{TTD} = T_{\text{sat}} - T_y \qquad (2\text{-}30)$$

and

$$\text{DCA} = T_{\text{drain}} - T_x \qquad (2\text{-}31)$$

where T_{sat} is the saturation temperature of the extracted steam, T_{drain} is the temperature of the drain leaving the heater, and T_x and T_y are, respectively, the feedwater inlet and outlet temperatures. In design the terminal temperature difference ranges from -3 to 10 F, and the drain cooler approach has a range of 10 to 20 F.

It should be pointed out that when the actual cycle efficiency is determined, various factors such as component efficiency and piping pressure drop must be taken into consideration. The procedure usually involves the calculation of the following items:

1. The thermodynamic properties at various cycle locations.
2. The extraction steam flow for feedwater heating.
3. The turbine output and the pump work.
4. The heat supplied to the steam and the cycle efficiency.

The example that follows will illustrate a Rankine cycle heat balance.

EXAMPLE 2-3. In a steam turbine plant steam enters the turbine at 1250 psia and 950 F and exhausts at 3.0 in. Hg abs. to the condenser. There are two extraction points at pressures 100 psia and 25 psia. The high-pressure steam is used in the contact heater while the low-pressure steam is directed to the surface heater. Other conditions are specified below:

Turbine internal efficiency	0.90
Pump efficiency	0.85
Terminal temperature difference	10 F
Drain cooler approach	16 F
Steam flow rate at the turbine inlet	1,300,000 lb/hr

For simplicity, the pressure drop in the turbine cycle is omitted from considera-tion. Calculate the turbine cycle net output and efficiency.

Solution: First, we determine the thermodynamic properties at various locations shown in Fig. 2-15. Using the Mollier diagram, we construct a turbine expansion curve as follows:

$$\eta_t = \frac{h_1 - h_4}{h_1 - h_{4s}}$$

With the numerical values, $h_1 = 1468.8$ Btu/lb, $h_{4s} = 915$ Btu/lb, and $\eta_t = 0.90$, the above equation gives

$$h_4 = 970.4 \text{ Btu/lb}$$

The turbine expansion curve is approximated by a straight line between states 1

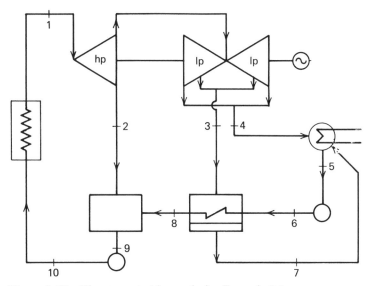

Figure 2-15.   The steam turbine cycle for Example 2-3.

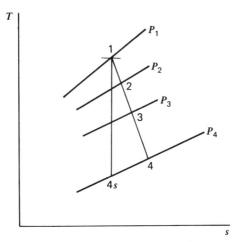

Figure 2-16. The turbine expansion curve
for Example 2-3.

and 4 as shown in Fig. 2-16. The interceptions between this line and the constant
pressure curves ($P_2 = 100$ psia and $P_3 = 25$ psia) will determine the states of two
extraction points. The readings from the Mollier diagram are

$$h_2 = 1226 \text{ Btu/lb}$$

$$h_3 = 1130 \text{ Btu/lb}$$

The water leaving the condenser is usually at the saturated temperature corresponding to the condenser pressure. In this case the water enthalpy and specific
volume at location 5 are

$$h_5 = 83.0 \text{ Btu/lb}$$

$$v_5 = 0.01618 \text{ ft}^3/\text{lb}$$

To determine the enthalpy and temperature at location 6, the following equation
is used:

$$h_6 = h_5 - w_p$$

or

$$h_6 = h_5 + \frac{v_5(P_6 - P_5)}{\eta_p} \tag{a}$$

Substituting the numerical values into Eq. (a) gives

$$h_6 = 83 + \frac{(0.01618)(100 - 1.474)(144)}{(0.85)(778)}$$

$$= 83.34 \text{ Btu/lb}$$

The water at location 6 is slightly compressed. From steam table, we find the temperature to be 115.4 F.

The conditions of heater drain are next determined. Utilizing the definition of drain cooler approach, we have

$$DCA = T_7 - T_6$$

or

$$T_7 = 115.4 + 16 = 131.4 \text{ F}$$

and from the steam table, $h_7 = 99.3$ Btu/lb. Using the definition of terminal temperature difference gives

$$TTD = T_{sat} - T_8$$

or

$$T_8 = 240 - 10 = 230 \text{ F}$$

The water leaving the surface heater is still slightly compressed and has the pressure higher than the saturation pressure at the temperature 230 F. In this example the pressure drop across the heater is neglected and, therefore, the pressure at 8 must be equal to the pressure at the condensate pump outlet or to the pressure of the contact heater. Similarly, the properties at states 9 and 10 are determined. For reference, the thermodynamic properties at various locations are summarized in the following table.

Location	Pressure (psia)	Temperature (F)	Enthalpy (Btu / lb)	Flow Rate (lb / hr)
1	1250	950	1468.8	1.3×10^6
2	100	397.4	1226.0	127,528
3	25	240	1130.0	130,478
4	1.474	115	970.4	1,041,994
5	1.474	115	83.0	1,172,472
6	100	115.4	83.3	1,172,472
7	25	131.4	99.3	130,478
8	100	230.0	198.0	1,172,472
9	100	327.8	298.8	1.3×10^6
10	1250	328.6	303.3	1.3×10^6

In the second step we calculate the extracted steam flows. Starting with the contact heater, we have the mass and energy balance equations as

$$m_2 + m_8 = m_9 \qquad\qquad (b)$$

$$m_2 h_2 + m_8 h_8 = m_9 h_9 \qquad\qquad (c)$$

Substituting the numerical values into these equations gives

$$m_2 + m_8 = 1.3 \times 10^6$$

$$1226m_2 + 198.1m_8 = 388.5 \times 10^6$$

and finally

$$m_2 = 127{,}528 \text{ lb/hr}$$

$$m_8 = 1{,}172{,}472 \text{ lb/hr}$$

Similarly, we have for the surface heater

$$m_3(h_3 - h_7) = m_8(h_8 - h_6)$$

or

$$m_3 = \frac{(1{,}172{,}472)(198.1 - 83.34)}{(1130 - 99.32)} = 130{,}478 \text{ lb/hr}$$

As the third step of the cycle heat balance, we calculate the turbine and pump work. Taking the steam turbine as a control volume, we have the first-law equation as

$$w_t = m_1 h_1 - m_2 h_2 - m_3 h_3 - m_4 h_4 \qquad \text{(d)}$$

where

$$m_4 = m_1 - m_2 - m_3 \qquad \text{(e)}$$

Substituting the numerical values into these equations gives us

$$w_t = (1.3 \times 10^6)(1468.8) - (127{,}528)(1226) - (130{,}478)(1130)$$

$$- (1{,}041{,}994)(970.4)$$

$$w_t = 594.5 \times 10^6 \text{ Btu/hr} \qquad \text{or} \qquad 174{,}187 \text{ kW}$$

Similarly, we calculate the pump work as

$$w_p = m_{10} \int_9^{10} v \, dp + m_6 \int_5^6 v \, dp$$

$$= \frac{(1.3 \times 10^6)(0.0177)(1250 - 100)(144)}{(0.85)(778)}$$

$$+ \frac{(1{,}172{,}472)(0.01618)(100 - 1.47)(144)}{(0.85)(778)}$$

$$= 6.17 \times 10^6 \text{ Btu/hr} \qquad \text{or} \qquad 1808 \text{ kW}$$

In the final step we calculate the heat supplied to the steam and the cycle efficiency. The amount of heat required is

$$Q_h = m_1(h_1 - h_{10})$$

$$= 1.3 \times 10^6(1468.8 - 303.3) = 1515 \times 10^6 \text{ Btu/hr}$$

and the cycle efficiency is

$$n_{cy} = \frac{w_t - w_p}{Q_h}$$

$$n_{cy} = \frac{(594.5 - 6.17)10^6}{1515 \times 10^6} = 0.388 \qquad \text{or} \qquad 38.8\%$$

It should be pointed out that the turbine cycle adopted in power plant design is much more complicated than that in Example 2-3. There will be many more feedwater heaters. A reheat process is frequently used to improve the cycle efficiency. For large generating units, feedwater pumps are driven by auxiliary steam turbines rather than by electric motors. Occasionally, steam is also extracted for the purpose of preheating the combustion air. The steam turbine cycle is discussed further in Chapter 7.

2.4 BRAYTON CYCLE

At the turn of the century work was initiated to develop the turbine system using air or gas as the working substance. Figure 2-17 shows a simple gas turbine system that consists of a compressor, combustor, and turbine. As seen in Fig. 2-17, air is first compressed and then mixed with fuel and burned in the combustor. After leaving the combustor, the products of combustion enter the turbine and produce useful work in an expansion process. Some of the turbine output is used to drive the compressor and the remainder is delivered outside to meet the power demand. Because of the combustion process, there is a change in the composition of the working substance. Also, the working substance does not go through a thermodynamic cycle like the water vapor in the steam turbine system. In order to analyze the gas turbine system in a convenient form, the assumptions listed below are frequently used.

1. The working substance is treated as the air of fixed composition. The air is an ideal gas and has a constant specific heat.

2. The combustion process is replaced by a heat transfer process from an external source. In other words, the mass flow rate remains unchanged throughout the system.

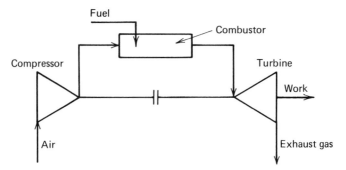

Figure 2-17. A simple gas turbine system.

3. The inlet and exhaust processes are replaced by a constant pressure process that will in turn complete the gas turbine cycle.

4. All processes are internally reversible.

The combination of these assumptions is called the air-standard cycle approach. The idealized gas turbine cycle is the air-standard Brayton cycle. Figure 2-18 shows the cycle *T-s* and *P-v* diagrams. To evaluate the cycle thermal efficiency, the best way is first to calculate the temperatures at the various cycle locations. For the isentropic processes 1-2 and 3-4, we have

$$\frac{T_2}{T_1} = \frac{T_3}{T_4} = \left(\frac{P_b}{P_a}\right)^{(k-1)/k} \tag{2-32}$$

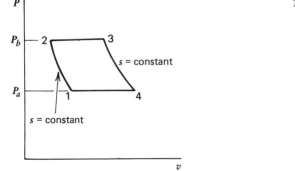

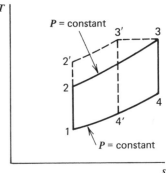

Figure 2-18. The air-standard Brayton cycle.

Let the pressure ratio r_p be defined by

$$r_p = \frac{P_b}{P_a}$$

and the isentropic temperature ratio by

$$\rho_p = \left(\frac{P_b}{P_a}\right)^{(k-1)/k}$$

Equation (2-32) becomes

$$\frac{T_2}{T_1} = \frac{T_3}{T_4} = r_p^{(k-1)/k} = \rho_p \tag{2-33}$$

The cycle thermal efficiency, based on a unit mass of air passing through the system, is

$$\eta_{cy} = \frac{w_t - w_c}{q_h} \tag{2-34}$$

Applying the first-law equation to the compressor, the combustor, and the turbine, we can change Eq. (3-34) to

$$\eta_{cy} = \frac{c_p(T_3 - T_4) - c_p(T_2 - T_1)}{c_p(T_3 - T_2)} = 1 - \frac{T_4 - T_1}{T_3 - T_2}$$

or

$$\eta_{cy} = 1 - \frac{T_1\left(\dfrac{T_4}{T_1} - 1\right)}{T_2\left(\dfrac{T_3}{T_2} - 1\right)} \tag{2-35}$$

Combining Eq. (2-33) and (2-35) yields

$$\eta_{cy} = 1 - \frac{1}{r_p^{(k-1)/k}} = 1 - \frac{1}{\rho_p} \tag{2-36}$$

Thus the thermal efficiency of the air-standard Brayton cycle is a function of the pressure ratio. Figure 2-19 shows the increase of cycle efficiency with the pressure ratio. This is also evident by examining the T-s diagram in Fig. 2-18. If the maximum temperature T_3 is kept constant, an increase in the pressure ratio will increase the cycle efficiency. However, there is a limit to the pressure ratio. The ratio

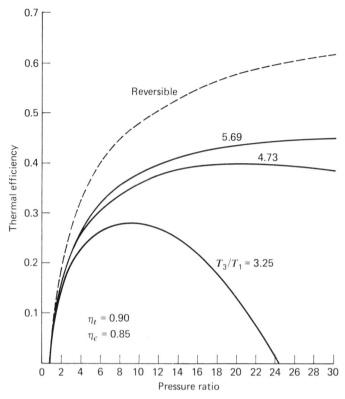

Figure 2-19. The thermal efficiency of a simple gas turbine cycle.

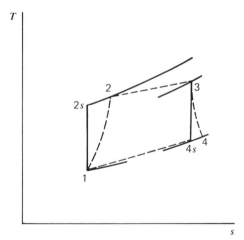

Figure 2-20. T-s diagram for an actual gas turbine system.

will reach the upper limit when the air temperature at the compressor outlet is equal to the design turbine inlet temperature. In this limiting case, the cycle net work tends toward zero, and the cycle efficiency approaches the Carnot efficiency.

The actual gas turbine system differs significantly from the idealized system. Figure 2-20 shows the actual gas turbine T-s diagram, taking into consideration the inefficiencies in the compressor and turbine, and the pressure drop in various locations. The following example illustrates the cycle efficiency calculation of the actual gas turbine.

EXAMPLE 2-4. Consider a simple gas turbine system with air entering the compressor at 14.7 psia, 60 F and exhausting at 150 psia. The maximum cycle temperature is 2000 F at the turbine inlet. Calculate the cycle efficiency and net work per pound of air, using

Compressor internal efficiency	0.85
Turbine internal efficiency	0.88
Average constant pressure specific heat	0.25 Btu/lb-R
Specific heat ratio	1.4

The pressure drops in the combustor, compressor inlet, and turbine outlet are assumed to be negligible.

Solution: The T-s diagram for this example is similar to that in Fig. 2-20 except for a negligible pressure drop. Equation (2-35) is still valid even though the compression and expansion processes are not reversible. To calculate the compressor exit temperature T_2, we make use of the isentropic relationship and compressor efficiency as follows:

$$\frac{T_{2s}}{T_1} = \left(\frac{P_2}{P_1}\right)^{(k-1)/k}$$

$$T_{2s} = 520\left(\frac{150}{14.7}\right)^{(1.4-1)/1.4} = 1009.8 \text{ R}$$

Since

$$\eta_c = \frac{T_{2s} - T_1}{T_2 - T_1}$$

we have the compressor exit temperature

$$T_2 = \frac{T_{2s} - T_1}{\eta_c} + T_1$$

$$= \frac{1009.8 - 520}{0.85} + 520 = 1096.2 \text{ R}$$

Similarly, we calculate the turbine exit temperature as below:

$$\frac{T_{4s}}{T_3} = \left(\frac{P_4}{P_3}\right)^{(k-1)/k}$$

$$T_{4s} = 2460\left(\frac{14.7}{150}\right)^{(1.4-1)/1.4} = 1266.8 \text{ R}$$

and

$$T_4 = T_3 - \eta_t(T_3 - T_{4s})$$

$$= 2460 - 0.88(2460 - 1266.8) = 1410 \text{ R}$$

After the temperatures at various locations are determined, the cycle efficiency is calculated by using Eq. (2-35):

$$\eta_{cy} = 1 - \frac{1410 - 520}{2460 - 1096.2} = 0.347 \quad \text{or} \quad 34.7\%$$

and the cycle net work is

$$w_{net} = w_t - w_c$$

$$= 0.25(2460 - 1410) - 0.25(1096.2 - 520)$$

$$= 118.4 \text{ Btu/lb}$$

It may be of interest to derive a general expression of the cycle network and efficiency. Starting with the compressor work, we have

$$w_c = c_p(T_{2s} - T_1)\frac{1}{\eta_c}$$

$$= \frac{c_p T_1}{\eta_c}\left(\frac{T_{2s}}{T_1} - 1\right)$$

If we use the definition of isentropic temperature ratio ρ_p, the above equation becomes

$$w_c = \frac{c_p T_1}{\eta_c}(\rho_p - 1) \qquad (2\text{-}37)$$

Similarly, the turbine work is given by

$$w_t = \eta_t c_p (T_3 - T_{4s})$$

$$= c_p \eta_t T_3 \left(1 - \frac{T_{4s}}{T_3} \right)$$

$$= c_p \eta_t T_3 \left(1 - \frac{1}{\rho_p} \right) \tag{2-38}$$

Combining Eqs. (2-37) and (2-38) yields

$$w_{net} = \frac{c_p T_1}{\eta_c} \left(1 - \frac{1}{\rho_p} \right) (\alpha - \rho_p) \tag{2-39}$$

where α is a dimensionless term defined by $\eta_c \eta_t k_1$, and k_1 is the cycle maximum temperature ratio T_3 / T_1. The heat supplied to the cycle on a unit mass of air is

$$q_h = c_p (T_3 - T_2)$$

$$= c_p T_1 \left(\frac{T_3}{T_1} - \frac{T_2}{T_1} \right) \tag{2-40}$$

Also, from Eq. (2-37)

$$\frac{T_2}{T_1} = \frac{\rho_p - 1}{\eta_c} + 1 \tag{2-41}$$

Combining Eq. (2-40) and Eq. (2-41) gives

$$q_h = \frac{c_p T_1}{\eta_c} \left[\eta_c (k_1 - 1) - \rho_p + 1 \right] \tag{2-42}$$

Finally, the cycle efficiency is expressed by the equation

$$\eta_{cy} = \frac{w_{net}}{q_h}$$

$$\eta_{cy} = \frac{\left(1 - \frac{1}{\rho_p} \right) (\alpha - \rho_p)}{\eta_c (k_1 - 1) - \rho_p + 1} \tag{2-43}$$

Equation (2-43) indicates that there are at least four major parameters affecting the cycle efficiency of a simple gas turbine. These include the maximum temperature ratio, pressure ratio (or isentropic temperature ratio), compressor and turbine efficiency. Some effects of these parameters on the cycle efficiency are shown in Fig. 2-19. In general, the cycle efficiency is relatively low, because of the high exhaust gas temperature, and because a significant portion of the turbine output is used for compressor operation. For a given turbine and compressor efficiency, the cycle performance is determined by the turbine inlet temperature and pressure ratio. The turbine inlet temperature is usually fixed by the metallurgical temperature limit of the first row of turbine blade. As this temperature increases, the cycle efficiency is greatly improved. In modern designs the gas turbine inlet temperatures range between 1800 and 2200 F.

The impact of the pressure ratio on the cycle efficiency is quite different. There is an optimal pressure ratio that produces the maximum cycle efficiency. To determine the optimal value for the pressure ratio, we could differentiate Eq. (2-43) with respect to r_p and equate the result to zero. As seen in Fig. 2-19, the optimal pressure ratio tends to increase as the turbine inlet temperature increases.

The cycle efficiency of the gas turbine system can be improved by introducing a regenerator. Figure 2-21 shows a simple gas turbine system with regenerator and the corresponding T-s diagram. In the regenerator heat is transferred from the turbine exhaust gas to the compressed air leaving the compressor. When the regenerator is an ideal heat exchanger, the temperature of the air leaving the regenerator T_x is equal to the temperature T_4, the temperature of the gas leaving the turbine. In this case the heat transfer process from an external source will take place between state x and state 3. Quite obviously, the amount of heat supplied to the cycle is reduced. Similarly, the heat rejection process is cut short from the curve 4-y-1 to the curve y-1.

The actual regenerator is quite different from the ideal. Because of the finite temperature difference in the heat exchanger, the temperature of compressed air leaving the regenerator is always less than the exhaust gas temperature. The regenerator efficiency is defined by

$$\eta_{reg} = \frac{h_x - h_2}{h_4 - h_2} \tag{2-44}$$

If the specific heat is assumed to be constant, as is true for both gas and air, Eq. (2-44) is further simplified to

$$\eta_{reg} = \frac{T_x - T_2}{T_4 - T_2} \tag{2-45}$$

The regenerator in the gas turbine system is relatively inefficient. The heat transfer surface area is usually large in order to keep the pressure drop across it as small as possible. The pressure drop represents a loss. This must be taken to account when a regenerative gas turbine is selected. The cycle efficiency of the regenerative gas

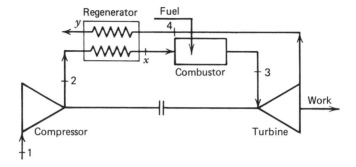

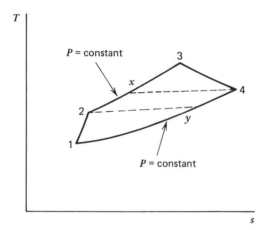

Figure 2-21. A regenerative gas turbine system.

turbine system is

$$\eta_{cy} = \frac{w_t - w_c}{q_h}$$

$$= \frac{c_p(T_3 - T_4) - c_p(T_2 - T_1)}{c_p(T_3 - T_x)}$$

$$= \frac{(T_3 - T_4) - (T_2 - T_1)}{(T_3 - T_2) - \eta_{reg}(T_4 - T_2)}$$

By further arrangement, it can be expressed as

$$= \frac{\dfrac{T_3}{T_1} - \dfrac{T_4}{T_3}\dfrac{T_3}{T_1} - \dfrac{T_2}{T_1} + 1}{\dfrac{T_3}{T_1} - \dfrac{T_2}{T_1} - \eta_{reg}\left(\dfrac{T_4}{T_3}\dfrac{T_3}{T_1} - \dfrac{T_2}{T_1}\right)} \qquad (2\text{-}46)$$

All these temperature ratios can be expressed in terms of the isentropic temperature ratio and maximum temperature ratio. From Eq. (2-37), we have the first temperature ratio T_2/T_1 in the form of

$$\frac{T_2}{T_1} = \frac{\rho_p - 1}{\eta_c} + 1 \tag{2-47}$$

A similar result was obtained for the second temperature ratio

$$\frac{T_4}{T_3} = 1 - \eta_t\left(1 - \frac{1}{\rho_p}\right) \tag{2-48}$$

The third temperature ratio T_3/T_1 is simply the maximum temperature ratio k_1. Thus the cycle efficiency of the regenerative system is determined by almost the same parameters as those for the simple gas turbine system with no regenerator.

In the case that the regenerative system is reversible (i.e., $\eta_c = \eta_t = \eta_{reg} = 1.0$), Eq. (2-46) can be further simplified and expressed as

$$\eta_{cy} = 1 - \frac{T_1}{T_3}\rho_p$$

or

$$\eta_{cy} = 1 - \frac{1}{k_1}\rho_p \tag{2-49}$$

Figure 2-22 presents a plot of both Eqs. (2-46) and (2-49) for two different maximum

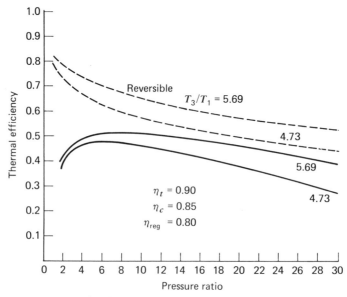

Figure 2-22. The thermal efficiency of a regenerative gas turbine cycle.

temperature ratios. It is seen that the irreversibility of the system significantly lowers the cycle efficiency. Compared with the simple gas turbine system (see Fig. 2-19), the optimal pressure ratio is smaller for the regenerative system. The small pressure ratio means a small cycle net output. Thus the cost associated with this output reduction must be weighted against the saving that can be effected by the cycle efficiency improvement.

EXAMPLE 2-5.   Consider the regenerative Brayton cycle as shown in Fig. 2-21. Calculate the cycle net output and thermal efficiency using the following conditions:

Compressor inlet conditions	14.7 psia
	60 F
Turbine inlet temperature	2000 F
Pressure ratio	4
Compressor efficiency	0.85
Turbine efficiency	0.88
Regenerator efficiency	0.70

Solution: Assuming that there is no pressure drop across the combustor and regenerator, we have

$$P_2 = P_x = P_3 \quad \text{and} \quad P_1 = P_4 = P_y$$

The temperatures at the various locations around the cycle are determined as follows. Using Eq. (2-47), we get

$$T_2 = T_1 \left(\frac{\rho_p - 1}{\eta_c} + 1 \right)$$

where

$$\rho_p = \left(\frac{P_2}{P_1} \right)^{(k-1)/k}$$

Substituting the numerical values into the above equation yields

$$T_2 = 520 \left(\frac{4^{(1.4-1)/1.4} - 1}{0.85} + 1 \right) = 817.3 \text{ R}$$

Using Eq. (2-48) gives us

$$T_4 = T_3 \left[1 - \eta_t \left(1 - \frac{1}{\rho_p} \right) \right]$$

Substituting T_3 and η_t into the equation, we obtain

$$T_4 = 2460\left[1 - 0.88\left(1 - \frac{1}{4^{0.286}}\right)\right]$$

$$= 1752 \text{ R}$$

To determine the temperature of the air leaving the regenerator, we make use of the regenerator efficiency as defined by Eq. (2-45):

$$T_x = T_2 + \eta_{\text{reg}}(T_4 - T_2)$$

$$= 817.3 + 0.70(1752 - 817.3)$$

$$= 1471.6 \text{ R}$$

Next, the compressor and turbine work are calculated on a unit mass of air passing through the system. The working substance is assumed to be an ideal gas, and the constant pressure specific heat is a constant. The compressor work is given by the equation

$$w_c = c_p(T_2 - T_1)$$

$$= 0.25(817.3 - 520)$$

$$= 74.3 \text{ Btu/lb}$$

and the turbine work is

$$w_t = c_p(T_3 - T_4)$$

$$= 0.25(2460 - 1752)$$

$$= 177 \text{ Btu/lb}$$

Thus the cycle net output is

$$w_n = w_t - w_c$$

$$= 177 - 74.3 = 102.7 \text{ Btu/lb}$$

Finally, the cycle efficiency is given by

$$\eta_{\text{cy}} = \frac{w_t - w_c}{q_h} = \frac{w_t - w_c}{c_p(T_3 - T_x)}$$

$$= \frac{177 - 74.3}{0.25(2460 - 1471.6)}$$

$$= 0.416 \quad \text{or} \quad 41.6\%$$

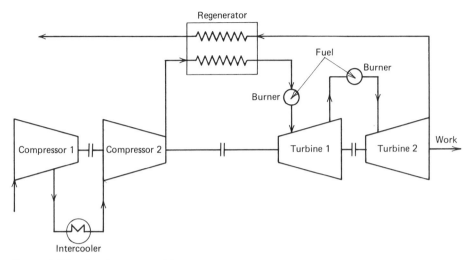

Figure 2-23. A reheat gas turbine system.

Thus the introduction of a regenerator increases the cycle efficiency. This happens mainly because of an increase in the average temperature of heat reception and at the same time a decrease in the average temperature of heat rejection. One practical way to continue to improve the cycle efficiency is to use intercooling between compressor stages and reheating between turbine cylinders. Figure 2-23 shows a cycle with one intercooler and one reheater. In most cases the temperature of compressed air leaving the intercooler is equal to the temperature of the air entering the first stage of compressor. The pressure ratio across these two compressors is usually the same for optimal operation. On the expansion side the temperature after reheating is equal to the temperature at the turbine inlet. The pressure at which the reheating takes place is approximately $\sqrt{P_i P_e}$ where P_i and P_e are, respectively, the turbine inlet and outlet pressure.

Recently underground compressed air storage is considered in connection with a gas turbine system. Figure 2-24 shows a typical arrangement for a gas turbine air storage plant. The ambient air is compressed by an axial flow compressor, inter-cooled, and boosted up to high pressure in a high speed centrifugal blower. The compressed air then flows through an aftercooler and enters an underground air storage facility. During the air compression operation the generator is used as a motor. At the time when the gas turbine system is needed to meet the power demand, compressed air will be led to the combustor through an expansion valve. After receiving heat from an external source, the working substance expands partially in one turbine cylinder, is reheated and completes the expansion process in another turbine cylinder. For economical reasons, the pressure ratio in the system is much higher than that for the simple gas turbine system and is usually in the range of 40 to 70.

The gas turbine air storage plant is attractive in terms of the system load management. During periods of low demand, excess generating capacity is used to drive large air compressors to pump air into an underground reservoir. During

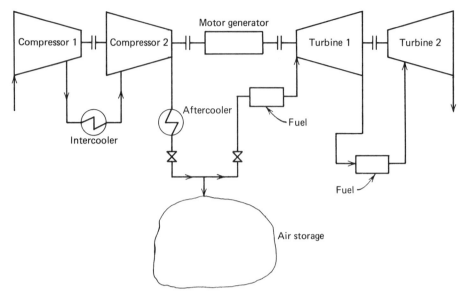

Figure 2-24. An air storage gas turbine plant.

periods of high demand, the compressed air is used for gas turbine operation. Since compressors are not operated, the net system output will be equal to the turbine output. In most cases, it means a two to threefold gain in the electrical output per unit of turbine fuel expended.

So far, our attention has been focused on the open-cycle gas turbine systems. Opposite to this are the so-called closed-cycle systems. In a closed-cycle system the working substance actually receives heat from an external source by a heat transfer

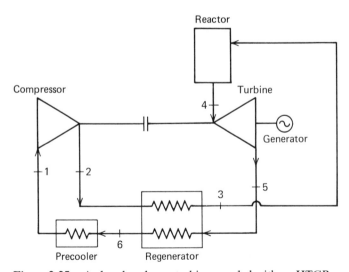

Figure 2-25. A closed-cycle gas turbine coupled with an HTGR.

process. Since there is no combustion involved, the working substance has a constant composition. There is also no intake or exhaust process. In place of it is the heat rejection process in which the working substance will return to the initial state at the beginning of the cycle.

Figure 2-25 shows the direct coupling of a closed-cycle gas turbine system with a high-temperature gas-cooled reactor (HTGR). In this system helium is the working substance. Helium is first compressed and heated in a regenerator before entering the nuclear reactor. Helium is also the coolant of the reactor. It receives heat in the reactor and enters the turbine for work production. As seen in Fig. 2-25, the working substance exhausts from the turbine and returns to the initial state through the regenerator and precooler. The cycle efficiency can be determined in the same manner as the open-cycle system.

EXAMPLE 2-6. Calculate the cycle efficiency for the closed-cycle gas turbine coupled with a HTGR. The system is shown in Fig. 2-25. The values listed below should be used for calculation.

State	Pressure (psia)	Temperature (F)
1	445	105
2	1000	
3	980	
4	960	1500
5	460	
6	450	
Turbine efficiency 0.90		
Regenerator efficiency 0.82		
Compressor efficiency 0.90		

Solution: In this example the pressure drops in the reactor, the regenerator and precooler have not been neglected. The working substance is helium, whose specific heat ratio k is approximately 1.667. The temperatures at the various locations around the cycle are determined as follows. By Eq. (2-47)

$$T_2 = T_1\left(\frac{\rho_p - 1}{\eta_c} + 1\right)$$

where ρ_p is the isentropic temperature ratio and previously defined as $(P_2/P_1)^{(k-1)/k}$. Using the numerical values, we have the temperature of the helium leaving the compressor as

$$T_2 = 565\left[\frac{\left(\dfrac{1000}{445}\right)^{\frac{1.667-1}{1.667}} - 1}{0.90} + 1\right] = 805.2 \text{ R}$$

The temperature of the helium exhausting from the turbine is by Eq. (2-48)

$$T_5 = T_4\left[1 - \eta_t\left(1 - \rho_p^{-1}\right)\right]$$

or

$$T_5 = 1960\left[1 - 0.90\left(1 - \left(\frac{960}{460}\right)^{\frac{1-1.667}{1.667}}\right)\right] = 1510.2 \text{ R}$$

The inlet temperature to the nuclear reactor is obtained by the regenerator efficiency as defined by Eq. (2-45):

$$T_3 = T_2 + \eta_{\text{reg}}(T_5 - T_2)$$

$$= 805.2 + 0.82(1510.2 - 805.2) = 1383.3 \text{ R}$$

and the inlet temperature to the precooler is simply

$$c_p(T_5 - T_6) = c_p(T_3 - T_2)$$

or

$$T_6 = T_5 + T_2 - T_3 = 1510.2 + 805.3 - 1383.3$$

$$T_6 = 932.1 \text{ R}$$

For reference, the temperatures at the various cycle locations are summarized below:

Locations	1	2	3	4	5	6
Temperatures (R)	565	805.2	1383.3	1960	1510.2	932.1

Finally, the cycle efficiency for this system is

$$\eta_{cy} = \frac{w_t - w_c}{q_h}$$

$$\eta_{cy} = \frac{c_p(T_4 - T_5) - c_p(T_2 - T_1)}{c_p(T_4 - T_3)}$$

$$= \frac{(1960 - 1510.2) - (805.2 - 565)}{1960 - 1383.3}$$

$$= 0.363 \quad \text{or} \quad 36.3\%$$

2.5 OTTO AND DIESEL CYCLES

This section is principally concerned with Otto and Diesel cycles. These two thermodynamic cycles are the idealized cycles, respectively, for the spark-ignition engine and the compression-ignition engine. These engines are frequently referred to as the internal combustion engines. Unlike the gas turbine and steam turbine systems, the entire thermodynamic cycle for the internal combustion engine occurs within one element (i.e., in a cylinder with a piston). Because internal combustion engines are compact and dependable, they usually serve as standby or emergency generating units in an electric power system.

Similar to the gas turbine system, internal combustion engines have replaced steam as the working substance with air. Since there is a combustion process, the working substance changes from air and fuel to products of combustion. Even though the engines operate in a mechanical cycle, the working substance does not go through a thermodynamic cycle because of intake and exhaust processes. To analyze the internal combustion engines, we again utilize the air-standard cycle. As indicated in the previous section, the air-standard cycle is mainly based on several assumptions. These include replacement of combustion process by a heat transfer process from an external source, and the intake and exhaust processes by a heat rejection to the environment. Also, the working substance is assumed to be air, which obeys the ideal-gas equation and has constant specific heat. Like the air-standard Brayton cycle, all processes are internally reversible. There is no question that the results thus obtained will be quite different from those of the actual engine. However, the approach of air-standard cycle is a convenient means to qualitatively analyze the performance of internal combustion engines.

Figure 2-26 shows the *P-v* and *T-s* diagrams for an air-standard Otto cycle. This cycle is the idealized cycle for a spark-ignition internal combustion engine. Process 1-2 is a reversible and adiabatic compression. It occurs just as the piston moves from crank-end dead center to head-end dead center. Process 2-3 is reversible and takes place at constant volume. In this process the working substance (air plus fuel) ignites

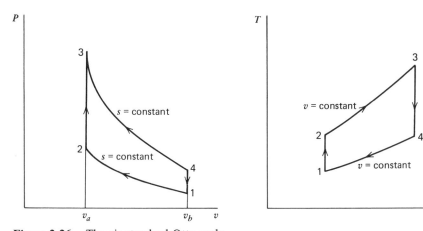

Figure 2-26. The air-standard Otto cycle.

and burns very rapidly. After these, the working substance expands isentropically in Process 3-4 and produces useful work. Process 4-1 is reversible and at constant volume. This process is used to replace the exhaust and intake processes of the actual engine.

As for the air-standard Brayton cycle, the thermal efficiency of the Otto cycle can be best evaluated by first calculating the temperatures at the various cycle locations. For the isentropic processes 1-2 and 3-4, we have

$$\frac{T_2}{T_1} = \frac{T_3}{T_4} = \left(\frac{V_b}{V_a}\right)^{k-1} \tag{2-50}$$

Let the compression ratio r_v be defined by

$$r_v = \frac{V_1}{V_2} = \frac{V_4}{V_3} = \frac{V_b}{V_a} \tag{2-51}$$

Equation (2-50) becomes

$$\frac{T_2}{T_1} = \frac{T_3}{T_4} = r_v^{k-1} \tag{2-52}$$

The thermal efficiency of the air-standard Otto cycle is then given by the equation

$$\eta_{cy} = \frac{q_h - q_l}{q_h} = 1 - \frac{q_l}{q_h}$$

$$= 1 - \frac{c_v(T_4 - T_1)}{c_v(T_3 - T_2)} = 1 - \frac{T_1}{T_2}\frac{\left(\dfrac{T_4}{T_1} - 1\right)}{\left(\dfrac{T_3}{T_2} - 1\right)} \tag{2-53}$$

Combining Eqs. (2-52) and (2-53), we have

$$\eta_{cy} = 1 - \frac{1}{r_v^{k-1}} \tag{2-54}$$

This expression indicates that the efficiency of the air-standard Otto cycle is a function of the compression ratio. As the compression ratio increases, the cycle efficiency increases. However, there is an upper limit beyond which undesirable detonation of the fuel will take place. In practice, the compression ratio for a spark-ignition engine is determined by the fuel antiknock characteristics and is usually between 6 and 12.

Figure 2-27 shows the P-v and T-s diagrams for an air-standard Diesel cycle. This cycle is the idealized cycle for a compression-ignition engine. The compression and

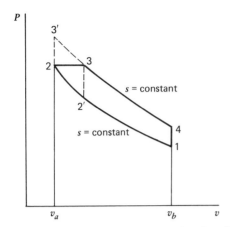
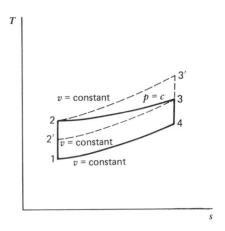

Figure 2-27. The air-standard diesel cycle.

expansion processes are assumed to be isentropic as they were in the Otto cycle. The combustion process is approximated by a reversible and constant pressure process. The fuel injection rate is maintained so that the pressure will remain unchanged even in the expansion process. Because there is only air being compressed, the compression ratio r_v is usually higher than that for an Otto cycle. The heat rejection is a reversible and constant volume process, which approximates the exhaust and intake processes of the actual Diesel engine.

The efficiency of the Diesel cycle is again best evaluated by first calculating the temperatures at the various cycle locations. We have

$$\frac{T_2}{T_1} = \left(\frac{V_1}{V_2}\right)^{k-1} = r_v^{k-1} \tag{2-55}$$

for the isentropic process 1-2 and

$$\frac{T_3}{T_2} = \frac{v_3}{v_2} = \alpha \tag{2-56}$$

for the constant pressure process 2-3. The term α is called the fuel cutoff ratio or load ratio. To determine the temperature ratio T_4/T_3, we obtain

$$\frac{T_4}{T_3} = \left(\frac{v_3}{v_4}\right)^{k-1}$$

or

$$= \left(\frac{v_3 v_2}{v_2 v_4}\right)^{k-1} \tag{2-57}$$

Making use of the definition of the compression ratio r_v and the load ratio α, we can express Eq. (2-57) as

$$\frac{T_4}{T_3} = \left(\frac{\alpha}{r_v}\right)^{k-1} \tag{2-58}$$

Finally, combining Eqs. (2-55), (2-56), and (2-58) will determine the temperature ratio T_4/T_1 as follows:

$$\frac{T_4}{T_1} = \left(\frac{T_4}{T_3}\right)\left(\frac{T_3}{T_2}\right)\left(\frac{T_2}{T_1}\right)$$

$$= \left(\frac{\alpha}{r_v}\right)^{k-1}(\alpha)(r_v)^{k-1}$$

$$= \alpha^k \tag{2-59}$$

The efficiency of the Diesel cycle is given by

$$\eta_{cy} = \frac{q_h - q_l}{q_h} = 1 - \frac{q_l}{q_h}$$

Since heat is added in a constant pressure process and rejected in a constant volume process, the above equation can be simplified and expressed as

$$\eta_{cy} = 1 - \frac{c_v(T_4 - T_1)}{c_p(T_3 - T_2)} = 1 - \frac{1}{k}\frac{T_1}{T_2}\left[\frac{\dfrac{T_4}{T_1} - 1}{\dfrac{T_3}{T_2} - 1}\right]$$

In terms of the compression ratio and the load ratio, the cycle efficiency of the Diesel cycle becomes

$$\eta_{cy} = 1 - \frac{1}{r_v^{k-1}}\left[\frac{\alpha^k - 1}{k(\alpha - 1)}\right] \tag{2-60}$$

The efficiency of the air-standard Diesel cycle does not depend only on the compression ratio (r_v), but also on the cutoff ratio (α). For the same compression ratio, the Diesel cycle efficiency is always less than the Otto cycle efficiency, because the term in the square bracket of Eq. (2-60) is always greater than unity. This is also evident from the T-s diagram in which the enclosed area 1-2-3-4-1 is smaller than the area 1-2-3'-4-1. However, it should not be concluded that the compression-ignition engines (also referred to as diesel engines) are less efficient than the spark-ignition engines. In fact, the reverse is true. Because only air is compressed in the compression process, diesel engines always have a much higher compression ratio than the Otto engines and therefore a higher efficiency. For a given maximum temperature as shown in Fig. 2-27, the Diesel cycle network represented by the area 1-2-3-4-1 is greater than the Otto cycle network represented by the area 1-2'-3-4-1.

In addition to the cycle thermal efficiency, the mean effective pressure is frequently used to measure the performance of the Otto and Diesel engines. The mean effective pressure is defined by

$$\text{MEP} = \frac{w_{\text{net}}}{v_b - v_a} \tag{2-61}$$

In other words the MEP is the engine output per unit swept volume, or the product of mean effective pressure and the piston displacement is equal to the cycle network. A high mean effective pressure indicates a compactness of the heat engine.

EXAMPLE 2-7. An air-standard Otto cycle has a compression ratio of 9. The pressure at the beginning of the compression stroke is 14.7 psia and the temperature is 60 F. The heat transfer to the air per cycle is 850 Btu/lb. Calculate the cycle efficiency, the mean effective pressure, and the maximum temperature in the cycle.

Solution: The P-v and T-s diagrams are identical to those in Fig. 2-26. We first calculate the maximum temperature in the cycle as follows:

$$v_1 = \frac{RT_1}{P_1}$$

$$v_1 = \frac{53.34 \times 520}{14.7 \times 144} = 13.1 \text{ ft}^3/\text{lb}$$

$$v_2 = \frac{v_1}{r_v} = \frac{13.1}{9} = 1.46 \text{ ft}^3/\text{lb}$$

$$T_2 = T_1\left(\frac{v_1}{v_2}\right)^{k-1} = 520(9)^{0.4} = 1252 \text{ R}$$

$$P_2 = P_1\left(\frac{v_1}{v_2}\right)^{k} = 14.7(9)^{1.4} = 318.6 \text{ psia}$$

$$q_h = c_v(T_3 - T_2) \qquad \text{or} \qquad T_3 = T_2 + \frac{q_h}{c_v}$$

$$T_3 = 1252 + \frac{850}{0.171}$$

$$T_3 = 6223 \text{ R}$$

This is the maximum temperature that occurs at the end of the heat addition process. To determine the cycle efficiency,we have by using Eq. (2-54):

$$\eta_{\text{cy}} = 1 - \frac{1}{r_v^{k-1}}$$

$$\eta_{\text{cy}} = 1 - \frac{1}{(9)^{0.4}} = 0.585 \qquad \text{or} \qquad 58.5\%$$

The mean effective pressure is given by

$$\text{MEP} = \frac{w_{net}}{v_b - v_a} = \frac{\eta_{cy} \times q_h}{v_1 - v_2}$$

$$= \frac{(0.585)(850)(778)}{(13.1 - 1.46)(144)} = 230.8 \text{ psia}$$

EXAMPLE 2-8.   An air-standard Diesel cycle has a compression ratio of 15 and a cutoff ratio of 3. At the beginning of the compression process the conditions are 14.7 psia and 60 F. Calculate the cycle efficiency, the mean effective pressure, and the cycle maximum temperature.

Solution: Designating the states as shown in Fig. 2-27, we first calculate the maximum temperature in the cycle as follows:

$$v_1 = \frac{RT_1}{P_1}$$

$$= \frac{53.34 \times 520}{14.7 \times 144} = 13.1 \text{ ft}^3/\text{lb}$$

$$v_2 = \frac{v_1}{r_v} = \frac{13.1}{15} = 0.873 \text{ ft}^3/\text{lb}$$

$$T_2 = T_1 \left(\frac{v_1}{v_2}\right)^{k-1} = 520(15)^{0.4} = 1536 \text{ R}$$

$$T_3 = \alpha T_2 = 3 \times 1536 = 4608 \text{ R}$$

The temperature T_3 is the maximum temperature in the cycle. To calculate the cycle efficiency, we simply use Eq. (2-60) and have

$$\eta_{cy} = 1 - \frac{1}{r_v^{k-1}} \left[\frac{\alpha^k - 1}{k(\alpha - 1)}\right]$$

$$= 1 - \frac{1}{15^{0.4}} \left[\frac{3^{1.4} - 1}{1.4(3 - 1)}\right]$$

$$= 0.558 \quad \text{or} \quad 55.8\%$$

The cycle network is the product of the cycle efficiency and the heat transfer to the air. That is,

$$w_{net} = \eta_{cy} q_h = \eta_{cy} c_p (T_3 - T_2)$$

$$= 0.558 \times 0.24 \times (4608 - 1536)$$

$$= 411.4 \text{ Btu/lb}$$

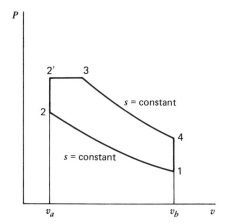

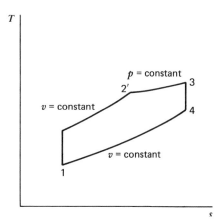

Figure 2-28.   The air-standard dual cycle.

Finally, the mean effective pressure of the cycle is

$$\text{MEP} = \frac{w_{\text{net}}}{v_1 - v_2}$$

$$= \frac{411.4 \times 778}{(13.1 - 0.873) \times 144} = 181.8 \text{ psia}$$

It should be emphasized that the thermal efficiency calculated by the air-standard cycle approach is always greater than the actual efficiency. This is simply because the assumptions in the air-standard cycle analysis are not compatible with reality and very difficult to implement in practice. In an actual engine, the combustion may not be complete, and the compression and expansion processes are not isentropic because of friction and heat loss. The engine operation involves an inlet and an exhaust process, and certain amount of work is usually required to overcome the friction in the processes. However, as previously mentioned, the main value of the air standard cycle is to enable engineers to identify the important parameters and qualitatively determine their influence on the engine performance.

It should be noticed that the main difference between the Otto engine and the Diesel engine is in the combustion. The Otto engine has a constant volume process for combustion, while the Diesel engine has a constant pressure process. Consequently, there is an intermediate class of engine whose performance may lie between these two extremes. In this kind of engine, combustion initially takes place at constant volume and finishes at constant pressure process. The corresponding air standard cycle is frequently referred to as the dual cycle. Fig. 2-28 shows the P-v and T-s diagrams for a dual cycle. The thermal efficiency can be determined in the same fashion as that for the Otto and Diesel cycles.

2-6 COMBINED AND BINARY CYCLES

Improving the cycle efficiency has been an important objective in any cycle analysis. One convenient approach is to combine two different cycles to form a new

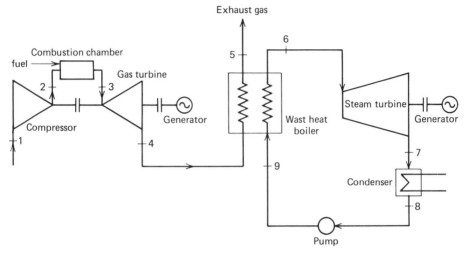

Figure 2-29. Schematic diagram for a combined-cycle system.

power-generating cycle. One of the most popular schemes is the combination of gas turbine cycle and steam turbine cycle as shown in Fig. 2-29. It is seen that the hot exhaust gas from the gas turbine is utilized to generate the steam that is in turn used to drive the steam turbine. In this system, combustion of the fuel is effected only at one point in the cycle, namely, in the combustion chamber of the gas turbine and the cycle work is produced at two different places. The overall thermal efficiency of the combined-cycle system is

$$\eta_{cy} = \frac{w_{gt} + w_{st}}{Q} \tag{2-62}$$

Let

$$\eta_{gt} = \text{gas turbine plant efficiency}$$

$$\eta_{st} = \text{steam turbine plant efficiency}$$

The gas turbine and steam turbine work, respectively, are

$$w_{gt} = \eta_{gt} \times Q \tag{2-63}$$

and

$$w_{st} = \left(1 - \eta_{gt}\right) \times Q \times \eta_{st} \tag{2-64}$$

Substituting these two terms into Eq. (2-62) gives us the combined-cycle efficiency in terms of the single-cycle efficiencies:

$$\eta_{cy} = \eta_{st} + \left(1 - \eta_{st}\right)\eta_{gt} \tag{2-65}$$

Equation (2-65) shows that the thermal efficiency of the combined cycle is greater than the steam turbine plant efficiency by an amount equal to $(1 - \eta_{st})\eta_{gt}$. With $\eta_{st} = 0.33$ and $\eta_{gt} = 0.26$ as typical values, the combined-cycle efficiency will be approximately 0.5. This cycle efficiency represents an optimistic estimate. When detailed design of a combined-cycle plant is made, it usually shows the plant efficiency in the range of 38 to 42%.

The waste heat steam generator is the component that couples the gas part of the system with the steam part. The turbine exhaust gas enters the bottom of the heat exchanger, moves upward, and releases its energy. At the end of the process, the turbine exhaust gas will leave the plant through a short stack. Fig. 2-30 shows the temperature variation for both hot gas and steam. Water enters the steam generator in the form of compressed liquid. As water receives heat from the hot exhaust gas, it becomes saturated, evaporated, and eventually superheated. The temperature difference between these two streams varies throughout the waste heat steam generator. The minimum value $(T_x - T_s)$ is frequently defined as the pinch point. The pinch point selection is important and can greatly affect the physical size of the heat exchanger. For economic reasons, the pinch point usually ranges between 40 to 80 F. To determine the amount of steam generated in the heat exchanger, we take the evaporator and superheater as a control volume and apply the first law to it. That is,

$$m_g c_p (T_{hi} - T_x) = m_s (h_{ce} - h_1)$$

or

$$\frac{m_s}{m_g} = \frac{c_p (T_{hi} - T_x)}{h_{ce} - h_1} \qquad (2\text{-}66)$$

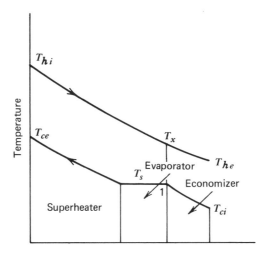

Figure 2-30. Temperature variation in a waste heat steam generator.

This equation gives the amount of steam generated by a unit mass of exhaust gas. In calculation, the pinch point is first selected; the temperature T_x is simply the pinch point plus the steam saturation temperature. The temperature of the gas entering the plant stack is also important and estimated by

$$m_g c_p (T_x - T_{he}) = m_s (h_1 - h_{ci})$$

or

$$T_{he} = T_x - \frac{1}{c_p} \frac{m_s}{m_g} (h_1 - h_{ci}) \tag{2-67}$$

To avoid corrosion from moisture formation in economizer and stack, the minimum gas temperature in the steam generator is always kept higher than the dew point temperature.

EXAMPLE 2-9. Consider the combined-cycle system as shown in Fig. 2-29. The gas side of the system is identical to that in Example 2-4. On the steam side steam enters the turbine at 1200 psia 800 F and exhausts to the condenser at a pressure of 4 in. Hg abs. The steam turbine and pump efficiency are, respectively, 88 and 85%. Calculate the combined-cycle efficiency and the temperature of the gas leaving the stack.

Assume that the boiler pinch point is 60 F.

Solution: Designating the states as shown in Fig. 2-29, we first calculate the amount of steam generated by a unit mass of the hot gas. Since the boiler pinch point is 60 F and the saturation temperature of steam at 1200 psia is 567.2 F, the intermediate gas temperature T_x must be

$$T_x = 60 + 567.2 = 627.2 \text{ F}$$

Also,

$$h_6 = 1379.7 \text{ Btu/lb}$$

$$h_1 = 571.9 \text{ Btu/lb (enthalpy of saturated water at 1200 psia)}$$

$$T_4 = 1410 \text{ R} \quad \text{or} \quad 950 \text{ F (from Example 2-4)}$$

Substituting these values into Eq. (2-66) gives us

$$\frac{m_s}{m_g} = \frac{0.25(950 - 627.2)}{1379.7 - 571.9} = 0.099 \text{ lb/lb}$$

Next, we calculate the steam turbine and pump work based on a unit mass of hot gas passing through the boiler. The turbine work is given by

$$w_{st} = \frac{m_s}{m_g} \eta_t (h_6 - h_{7s})$$

$$w_{st} = 0.099 \times 0.88(1379.7 - 894)$$

$$= 42.7 \text{ Btu/lb of gas}$$

and the pump work is

$$w_p = \frac{m_s}{m_g} \frac{1}{\eta_p} v(P_9 - P_8)$$

$$= \frac{0.099 \times 0.01623(1200 - 1.96) \times 144}{0.85 \times 778}$$

$$= 0.42 \text{ Btu/lb of gas}$$

It is seen that the pump work is negligibly small as compared with the steam turbine work. Therefore, it is omitted from the cycle analysis. In this system the combustion chamber of the gas turbine is only one place where the fuel is burned. The heat supplied is given by

$$g_h = c_p(T_3 - T_2)$$

$$= 0.25(2460 - 1096)$$

$$= 341 \text{ Btu/lb}$$

Then, the overall thermal efficiency of the combined cycle is

$$\eta_{cy} = \frac{(w_{gt} - w_c) + w_{st}}{q_h}$$

$$= \frac{118.4 + 42.7}{341} = 0.472 \qquad \text{or} \qquad 47.2\%$$

Finally, to determine the temperature of the gas entering the stack, we use Eq. (2-67) and have

$$T_{stack} = 627.2 - \frac{1}{0.25} \times 0.099 \times (571.9 - 93.7)$$

$$= 437.8 \text{ F}$$

The above calculations are based on the assumption of no pressure drop at the various cycle locations. When these pressure drops are taken into account, the efficiency of the combined cycle is greatly reduced.

In recent years, other arrangements of waste heat steam generator have been developed. In addition to the unfired boiler just described, the supplemental fired boiler and the exhaust-fired boiler are available. In the supplemental boiler, additional fuel is injected and burned in the furnace. Because of the additional firing, the temperature of the steam is expected to be somewhat higher than that in the unfired boiler and therefore to improve the performance of the steam side in the combined cycle. The exhaust-fired boiler is similar to the conventional steam generator equipped with a complete set of combustion equipment. Additional fuel is burned in the furnace, and the combustion air is supplied through the gas turbine compressor. In general, the combined system with exhaust fired boiler has higher efficiency but the initial investment also costs much more. Chapter 13 discusses the waste heat boiler selection.

There is another approach in combining the gas turbine and steam turbine system. Figure 2-31 shows the schematic diagram of this combined system. The air supply through the compressor is used to pressurize the combustion chamber of the steam generator. The flue gases from the boiler would act as the working substance and expand in the gas turbine. The steam generated in the boiler would go through the turbine cycle as it does in the conventional steam plant. In this system combustion of the fuel is effected only in the furnace of the boiler, and the useful work is produced by gas and steam turbine. The cycle efficiency can be calculated in the same manner as that for the previous combined cycle.

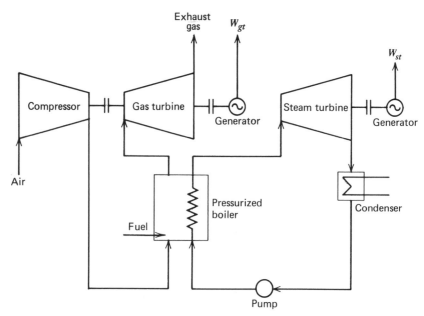

Figure 2-31. Schematic diagram for a combined-cycle system with pressurized boiler.

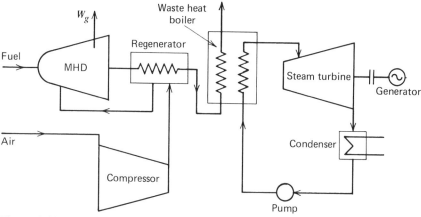

Figure 2-32.   Schematic diagram for a combined-cycle system with MHD generator.

In recent years work has been initiated to develop the combined gas-steam plant with magnetohydrodynamic (MHD) generator. Figure 2-32 shows the simplified flow diagram for this system. It is seen that the MHD generator replaces the gas turbine and produces useful work in the gas circuit. Then the gas passes through a regenerator on the way to the steam generator. The steam side of this combined cycle is similar to the steam sides just described.

The principle of MHD operation is based on the Faraday effect, and may be best illustrated in Fig. 2-33. The electrically conducting gas at high temperature enters the

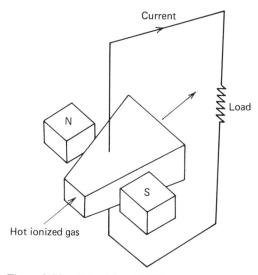

Figure 2-33.   Principle of MHD operation.

MHD generator and passes through the diverging channel. An intense magnetic field is created in the direction perpendicular to the direction of gas flow. Interaction of the conducting gas with the magnetic field will then induce an electric field in a direction normal to both the magnetic field and the gas flow. When electrodes are placed in the channel walls that are in contact with the gas stream, the current will flow through the gas, the electrodes, and the external load. In this fashion thermal energy is extracted from the gas stream and electric energy is produced.

The combined system with a MHD generator presents no new problem in a cycle analysis. When the calculation on the MHD generator is completed, the remainder of cycle analysis is similar to those we have discussed before.

It has been demonstrated that the thermal efficiency of the combined cycle is generally greater than the individual cycle efficiency. This is mainly because the combined cycle can take advantage of the best features of each individual cycle. For instance, the high-temperature feature of gas turbine is utilized in the heat addition process of the combined cycle. To avoid high temperature of heat rejection encountered in the gas turbine system, the combined cycle replaces it by a steam turbine system that is characterized by heat rejection in a low temperature. The concept of utilizing the best features from more than one cycle system is easy to understand. In fact, this concept is also utilized in binary cycles. In a binary cycle, two different working substances go through two separate cycles and produce useful work. There is one coupling device (or equipment) in which heat is transferred from one working substance to another. One of the most popular schemes for binary cycles is the mercury-steam cycle as shown in Fig. 2-34. It is seen that a Rankine cycle using dry saturated mercury is superposed on another Rankine cycle using superheated steam. The device coupling these two Rankine cycles is the heat exchanger in which mercury is condensed and water is changed from liquid to vapor phase. To increase the efficiency of the steam side, steam is usually superheated and the superheater is frequently located in the mercury boiler. Not shown in Fig. 2-34 is the economizer of the steam side. The economizer usually placed in the mercury boiler is used to raise the water temperature before the water enters the steam generator (or mercury condenser). The thermal efficiency of mercury-steam binary cycle is given by

$$\eta_{cy} = \frac{w_{Hg} + w_{st}}{Q} \tag{2-68}$$

where

$$w_{Hg} = \text{mercury turbine output}$$

$$w_{st} = \text{steam turbine output}$$

$$Q = \text{heat supplied to the binary cycle}$$

For simplicity, the pump works are omitted from consideration. The amount of heat supplied to the binary cycle is divided into two portions; one (x) is given to the

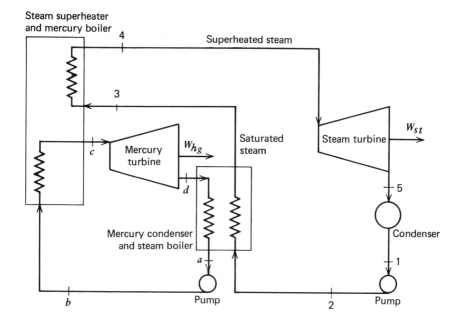

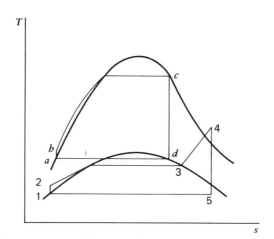

Figure 2-34. A mercury-steam binary cycle.

mercury, while another $(1 - x)$ is given directly to the steam in the superheater. Thus the mercury and steam turbine outputs are, respectively

$$w_{\text{Hg}} = xQ\eta_{\text{Hg}}$$

and

$$w_{\text{st}} = \left[(1 - \eta_{\text{Hg}})xQ + (1 - x)Q\right]\eta_{\text{st}}$$

where η_{Hg} is the thermal efficiency of the mercury side and η_{st} is the efficiency of the steam side. Substituting these two expressions into Eq. (2-68), we have

$$\eta_{cy} = x\eta_{Hg} + (1 - \eta_{Hg})x\eta_{st} + (1 - x)\eta_{st}$$

After rearranging, the thermal efficiency of the binary cycle becomes

$$\eta_{cy} = \eta_{st} + x\eta_{Hg}(1 - \eta_{st}) \qquad (2\text{-}69)$$

Equation (2-69) indicates that the binary cycle has greater efficiency than the steam cycle by the amount equal to $x\eta_{Hg}(1 - \eta_{st})$. When there is no superheater and economizer in the mercury boiler, the fraction of heat supplied to the mercury will become a unity (i.e., $x = 1$). Then Eq. (2-69) becomes

$$\eta_{cy} = \eta_{st} + \eta_{Hg}(1 - \eta_{st}) \qquad (2\text{-}70)$$

Mercury is one of the few working substances used in power plant cycles. For use in binary cycles mercury exhibits certain desirable characteristics that water may not have. These include a low specific heat of liquid mercury, large latent heat of vaporation, and low vapor pressure at high temperature. A low specific heat means a lesser need for feed heating in the mercury cycle. The low specific heat is also evident from the T-s diagram where the saturated liquid line for mercury has a very steep slope and is almost close to the vertical. A large latent heat means a heat addition process close to an isothermal. In other words, large latent heat will maximize the average temperature in which heat is added to the cycle and therefore improve the cycle efficiency. For a given power output the large latent heat also tends to reduce the equipment size. A low vapor pressure at high temperature is an important property for the working substance. It reduces not only the equipment cost, but also safety hazards generally associated with high-pressure operation.

EXAMPLE 2-10.   Consider the binary cycle as shown in Fig. 2-34. Dry saturated mercury vapor enters the mercury turbine at 225 psia and exhausts at the pressure 4 psia. In the steam side superheated steam enters the turbine at 680 psia and 900 F and exhausts to the condenser at 1 psia. Both turbine processes are assumed isentropic, and pump works are negligible. Calculate the thermal efficiency of this binary cycle using the following mercury properties:

P (psia)	T (F)	h_f (Btu / lb)	h_v (Btu / lb)	s_f (Btu / lb-R)	s_v (Btu / lb-R)
225	1138	32.20	156.32	0.03565	0.11852
4	557.8	17.16	143.44	0.02373	0.14787

Solution: Designating the states as shown in Fig. 2-34, we find the mercury properties at various cycle locations as follows:

$$h_c = 156.32 \text{ Btu/lb}$$

$$s_c = 0.11852 \text{ Btu/lb-R}$$

$$s_d = s_c = x_d s_{vd} + (1 - x_d) s_{fd}$$

$$0.11852 = 0.14787 x_d + 0.02373(1 - x_d)$$

$$x_d = 0.764$$

$$h_d = x_d h_{vd} + (1 - x_d) h_{fd}$$

$$= (0.764)(143.44) + (1 - 0.764)(17.16)$$

$$= 113.65 \text{ Btu/lb}$$

Neglecting the pump effects, we get

$$h_a = h_b = 17.16 \text{ Btu/lb}$$

The mercury turbine work is given by

$$w_{\text{Hg}} = h_c - h_d$$

$$= 156.32 - 113.65 = 42.67 \text{ Btu/lb}$$

and the heat supplied to the mercury is

$$q_1 = h_c - h_b$$

$$= 156.32 - 17.16 = 139.16 \text{ Btu/lb}$$

Next, we move to the steam side and find the steam properties as follows:

$$h_4 = 1460 \text{ Btu/lb}$$

$$s_4 = 1.6614 \text{ Btu/lb-R}$$

$$s_5 = s_4 = x_5 s_{v5} + (1 - x_5) s_{f5}$$

$$1.6614 = 1.9781 x_5 + 0.1326(1 - x_5)$$

$$x_5 = 0.828$$

$$h_5 = x_5 h_{v5} + (1 - x_5) h_{f5}$$

$$= (0.828)(1105.8) + (1 - 0.828)(69.73)$$

$$= 927.6 \text{ Btu/lb}$$

Again neglecting the pump work, we obtain

$$h_1 = h_2 = 69.73 \text{ Btu/lb}$$

The steam turbine work based on a unit of mass of steam passing through the turbine is

$$w_{st} = h_4 - h_5$$

$$= 1460 - 927.6 = 532.4 \text{ Btu/lb}$$

At this point, we calculate the amount of steam generated by one pound of mercury passing through the mercury-steam heat exchanger. The first law provides the following equation:

$$m_{Hg}(h_d - h_a) = m_s(h_3 - h_2)$$

Inserting the numerical values into the above equation gives

$$\frac{m_s}{m_{Hg}} = \frac{113.65 - 17.16}{1202.2 - 69.73}$$

$$= 0.0852 \text{ lb/lb}$$

The heat received by the steam in the superheater is

$$q_2 = \frac{m_s}{m_{Hg}}(h_4 - h_3)$$

$$= 0.0852(1460 - 1202.2)$$

$$= 21.96 \text{ Btu/lb (based on one pound of mercury)}$$

Finally, the binary cycle efficiency is given by

$$\eta_{cy} = \frac{w_{Hg} + \dfrac{m_s}{m_{Hg}} w_{st}}{q_1 + q_2}$$

$$= \frac{42.67 + 0.0852(532.4)}{139.16 + 21.96}$$

$$= 0.546 \quad \text{or} \quad 54.6\%$$

2-7 AVAILABILITY IN CYCLE ANALYSIS

This section presents an application of the second law to steady-state, steady-flow processes. It also briefly covers a use of the second law in power cycle analysis. The main value of the second-law approach is to enable engineers to discuss the energy in terms of its potential to work production. When applied in cycle analysis, the second law will also help engineers to identify the inefficiencies in the system and determine the potential areas for further improvement.

One of the important concepts in the second law application is the reversible work of the process. For a given change in state across a control volume, the reversible work is the maximum quantity of work that can be produced or the minimum quantity of work that must be supplied. This ideal situation arises only when all processes involved are completely reversible. Figure 2-35 shows a reversible process for a control volume called X that can exchange heat only with the environment of which the state conditions are P_0, T_0. To be completely reversible, not only all the internal processes through which the working substance goes must be reversible (called internal reversibility), but also there must be external reversibility. In other words, any heat exchange with the environment must take place through one or more reversible heat engines as shown in Fig. 2-35. The reversible work of the process is then the sum of the shaft work produced directly by the control volume X and the work produced by the reversible heat engines. Now, we evaluate the reversible work as follows.

First, we define the control volume Y, which includes the original control volume X and the reversible heat engine. The boundary of the control volume Y is shown in

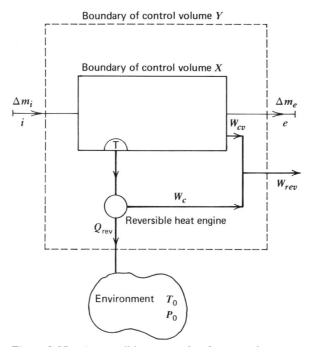

Figure 2-35. A reversible process for the control volume X.

Fig. 2-35. Applying the first law to this control volume gives

$$\Delta W_{rev} = \Delta m_i h_i - \Delta m_e h_e - \Delta Q_{rev} \tag{2-71}$$

Here, the kinetic and potential energy changes are omitted for simplicity. Since all processes in this control volume Y are reversible, entropy must be conserved. Applying the second-law equation (see Eq. 2-6) gives

$$\Delta m_i s_i - \Delta m_e s_e - \frac{\Delta Q_{rev}}{T_0} = 0 \tag{2-72}$$

When we combine Eqs. (2-71) and (2-72), we have the following expression for the reversible work of the process:

$$\Delta w_{rev} = (\Delta m_i h_i - \Delta m_e h_e) - T_0(\Delta m_i s_i - \Delta m_e s_e) \tag{2-73}$$

For a unit mass of the working substance it becomes

$$\Delta w_{rev} = (h_i - h_e) - T_0(s_i - s_e) \tag{2-74}$$

It should be pointed out that the reversible work is not a function of the process like the shaft work. It depends only on the end states of the process for a given condition in the environment. Therefore the reversible work is frequently referred to as the available energy of the process. It is the maximum amount of work that can be produced by a device between two given end states. In case the device is work-absorbing such as a compressor and pump, the available energy of the process is the minimum amount of work that must be supplied.

Another important concept in the second-law application is the thermodynamic availability (for simplicity, called availability in this text). The availability at a given state is defined as the reversible work produced when the working substance at that stage changes to the state that is in thermal and mechanical equilibrium with the environment. Thus the availability on the basis of unit mass is

$$a = \Delta w_{rev} = (h - h_0) - T_0(s - s_0) \tag{2-75}$$

The subscript zero denotes the conditions of the environment. It is seen from Eq. (2-75) that availability is a property that measures the potential of a working substance to do work under certain environmental conditions. The change of availability in a process means a change in the capacity of the substance to do work. It also represents the negative value of the reversible work in the process. In equation form it is expressed as

$$\Delta w_{rev}_{1 \to 2} = a_1 - a_2 \tag{2-76}$$

The reversible work is the maximum amount of work that could be produced in a process. It is generally greater than the useful work obtained in the real process with the same end states. The difference between the reversible and actual work is the lost work due to irreversibility (or simply the process irreversibility). If there is no work involved in the process, the reversible work is simply equal to the process irreversibility that is usually caused by the internal friction or the heat transfer through a finite temperature difference. In equation form the process irreversibility is given by

$$\Delta I = \Delta w_{rev} - \Delta w_{act} \tag{2-77}$$

To measure the quality of process design, or the process performance, the term process effectiveness is usually applied. The general definition of process effectiveness is

$$\varepsilon = \frac{\text{increase in the availability of the desired output}}{\text{decrease in the availability of the source}} \tag{2-78}$$

For a work-producing process, the effectiveness defined in Eq. (2-78) is simply the ratio of the actual useful work obtained to the maximum useful work obtainable. When the process approaches reversible, the process effectiveness will approach a unity. In other words, the term $1 - \varepsilon$ is a measure of potential for improvement. For a heat addition process, the effectiveness is the ratio of increase of availability in the working substance to the decrease of availability in the substance from which the heat is transferred. Because availability is not conservative in nature (similar to entropy), the heat exchanger effectiveness is always less than a unity.

Occasionally, the effectiveness is defined as an output to input ratio. That is,

$$\varepsilon' = \frac{\text{availability output}}{\text{availability input}} \tag{2-79}$$

This definition is particularly useful when the desired output of the process is difficult or impossible to identify. For instance, the effectiveness for a throttling process is simply the ratio of the availability leaving to the availability entering the valve.

EXAMPLE 2-11. A simple steam turbine system is shown in Fig. 2-36. It operates at steady-state, steady-flow conditions. Steam property data are given in the table below. Calculate the second-law efficiency for this generating system and show the availability losses in various system locations.

State	P(psia)	T(F)	h(Btu / lb)	s(Btu / lb-R)
1	320	(saturated vapor)	1204.3	1.5060
2	1		895.3	1.6031
3	1	(saturated liquid)	69.73	0.13266
4	320		70.92	0.13290

Assume that the coal availability is approximately equal to the coal heating value 13,200 Btu/lb and that the boiler efficiency is 82%.

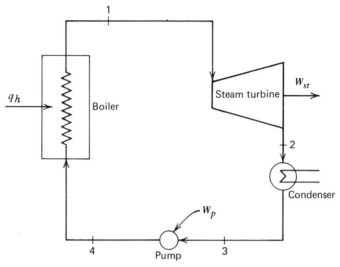

Figure 2-36. A simple steam turbine system for Example 2-11.

Solution: We determine the steam availability at various locations. The conditions of the environment are assumed to be $P_0 = 14.7$ psia and $T_0 = 77$ F.

$$a_1 = (h_1 - h_0) - T_0(s_1 - s_0)$$

$$= (1204.3 - 45.09) - 537(1.5060 - 0.0877)$$

$$= 397.6 \text{ Btu/lb}$$

Similarly, the availability at states 2, 3, 4 is calculated as

$$a_2 = (895.3 - 45.09) - 537(1.6031 - 0.0877)$$

$$= 36.5 \text{ Btu/lb}$$

$$a_3 = (69.73 - 45.09) - 537(0.13266 - 0.0877)$$

$$= 0.496 \text{ Btu/lb}$$

$$a_4 = (70.92 - 45.09) - 537(0.13290 - 0.0877)$$

$$= 1.56 \text{ Btu/lb}$$

Then, we calculate the turbine output as

$$\Delta w_{st} = (h_1 - h_2)$$

$$= 1204.3 - 895.3 = 309 \text{ Btu/lb}$$

and the fuel consumption per pound of steam generated as

$$m_f = \frac{h_1 - h_4}{\eta_b(HV)}$$

$$= \frac{1204.3 - 70.92}{0.82 \times 13200} = 0.105 \text{ lb/lb of steam}$$

The second-law efficiency is given by

$$\varepsilon = \frac{w_{st} - w_p}{m_f a_f}$$

$$= \frac{309 - 1.18}{0.105 \times 13200} = 0.222 \quad \text{or} \quad 22.2\%$$

To determine the availability loss in the boiler, we determine the reversible work that is the decrease of availability across the boiler. Since there is no work produced in the process, the reversible work must be equal to the availability loss or the process irreversibility. On the basis of a unit mass of steam passing through the boiler, the availability loss is

$$\Delta I_b = w_{rev,b}$$

$$= m_f a_f + a_4 - a_1$$

$$= 0.105 \times 13200 + 1.56 - 397.6$$

$$= 989.96 \text{ Btu/lb}$$

Similarly, the availability loss for the steam turbine, condenser, and pump are, respectively,

$$\Delta I_{st} = \Delta w_{rev,st} - \Delta w_{st}$$

$$= (a_1 - a_2) - (h_1 - h_2)$$

$$= (397.6 - 36.5) - (1204.3 - 895.3)$$

$$= 54.10 \text{ Btu/lb}$$

$$\Delta I_c = a_2 - a_3$$

$$= 36.5 - 0.496 = 36 \text{ Btu/lb}$$

and

$$\Delta I_p = \Delta w_{rev,p} - \Delta w_p$$

$$= (0.496 - 1.56) - (-1.18)$$

$$= 0.12 \text{ Btu/lb}$$

The table below presents the availability balance taking into account the availability loss (or destruction). Also included are the corresponding energy terms for comparison.

Increase Through Fuel Supply	Energy(Btu/lb) 1388.0	Availability(Btu/lb) 1388.0
Net Output	307.82	307.82
Losses		
Boiler	254.62	989.96
Turbine	—	54.10
Condenser	825.56	36.00
Pump	—	0.12
Total	1388.0	1388.0

It is seen that the availability analysis provides a different picture of performance from that of the energy analysis. The availability analysis identifies the boiler as the critical component in which the loss is largest. This is in contrast to the conclusion obtained in the energy analysis. Also, the availability analysis gives not only the availability loss but also availability destruction (for simplicity in this text availability destruction is counted as a part of the availability loss). The tabulated values indicate significant availability destruction for the turbine even though there is no energy loss from it at all.

EXAMPLE 2-12.   Consider a simple gas turbine system as shown in Fig. 2-37. The operating data are summarized in the table below. Estimate the effectiveness for the compressor, heat exchanger, gas turbine, and the entire system. The environmental conditions are 14.7 psia and 77 F.

State	Pressure (psia)	Temperature (R)
1	14.7	520
2	150.0	1096
3	150.0	2460
4	14.7	1410

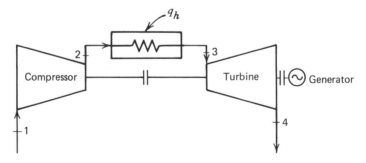

Figure 2-37.   A simple gas turbine system for Example 2-12.

Solution: The effectiveness for the compressor process is from Eq. (2-78)

$$\varepsilon_c = \frac{a_2 - a_1}{w_c} = \frac{(h_2 - h_1) - T_0(s_2 - s_1)}{h_2 - h_1}$$

Making use of the ideal gas conditions and constant specific heat, we have

$$\varepsilon_c = 1 - \frac{T_0\left(c_p \ln \dfrac{T_2}{T_1} - R \ln \dfrac{P_2}{P_1}\right)}{c_p(T_2 - T_1)}$$

Inserting the given data into the equation gives

$$\varepsilon_c = 1 - \frac{537\left(0.25 \ln \dfrac{1096}{520} - \dfrac{53.3}{778} \ln \dfrac{150}{14.7}\right)}{0.25(1096 - 520)}$$

$$\varepsilon_c = 0.898 \qquad \text{or} \qquad 89.8\%$$

Similarly, the turbine effectiveness is

$$\varepsilon_t = \frac{h_3 - h_4}{a_3 - a_4} = \frac{c_p(T_3 - T_4)}{c_p(T_3 - T_4) - T_0(s_3 - s_4)}$$

$$= \frac{0.25(2460 - 1410)}{0.25(2460 - 1410) - 537\left(0.25 \ln \dfrac{2460}{1410} - \dfrac{53.3}{778} \ln \dfrac{150}{14.7}\right)}$$

$$= 0.961 \qquad \text{or} \qquad 96.1\%$$

As to the effectiveness of the heat exchanger, it is the ratio of the availability increase in the working substance to the availability supplied to the heat exchanger. For this simple example the availability supplied is assumed to be the amount of energy transferred, that is, the quantity of $(h_3 - h_2)$. Thus the heat exchanger effectiveness is

$$\varepsilon_h = \frac{a_3 - a_2}{h_3 - h_2} = 1 - \frac{T_0(s_3 - s_2)}{h_3 - h_2}$$

$$= 1 - \frac{537\left(0.25 \ln \dfrac{2460}{1096} - \dfrac{53.3}{778} \ln \dfrac{150}{150}\right)}{0.25(2460 - 1096)}$$

$$= 0.682 \qquad \text{or} \qquad 68.2\%$$

Finally, the effectiveness of the entire gas turbine system is

$$\varepsilon_{sys} = \frac{w_t - w_c}{a_{fuel}} = \frac{c_p(T_3 - T_4) - c_p(T_2 - T_1)}{c_p(T_3 - T_2)}$$

$$= \frac{(2460 - 1410) - (1096 - 520)}{2460 - 1096.2}$$

$$= 0.347 \quad \text{or} \quad 34.7\%$$

Note that the first-law analysis of this system has been presented in Example 2-4. In both examples the combustion chamber has a 100% efficiency, that is, no energy loss from it. But the second-law analysis points out its effectiveness is only around 68.2%. In other words, about 68.2% of the fuel is utilized in the heat addition process of the gas turbine.

SELECTED REFERENCES

1. E. G. Obert, *Concepts of Thermodynamics*, McGraw Hill, 1960.

2. E. G. Cravalho and J. L. Smith, Jr., *Engineering Thermodynamics*, Pitman Publishing, 1981.

3. R. S. Benson, *Advanced Engineering Thermodynamics*, Pergamon Press, 1967.

4. R. W. Haywood, *Analysis of Engineering Cycles*, Pergamon Press, 1967.

5. M. J. Moran, *Availability Analysis: A Guide to Efficient Energy Use*, Prentice-Hall, 1982.

PROBLEMS

2-1. A heat engine operates in the ocean utilizing the temperature differential between the surface and some depth below. Assume that the water surface temperature is 55 F and the temperature 100 ft below is 40 F. What is the maximum thermal efficiency of such an engine?

2-2. A geothermal power plant utilizes an underground source of hot water. Assume that the source temperature is around 350 F. What is the maximum thermal efficiency for this plant?

  Assume that the temperature in the environment is 77 F.

2-3. Recent research and development has projected the gas turbine inlet temperature as high as 2500 F. What is the thermal efficiency of a Carnot engine using this source of energy and operating in a 60 F environment?

2-4. The simple steam turbine system shown in Fig. 2-36 has turbine inlet conditions of 1250 psia and 850 F and an exhaust pressure of 2 psia. The turbine and pump efficiency are, respectively, 85 and 72%. What is the cycle thermal efficiency? Also compare the result with the thermal efficiency of a Carnot cycle operating between the same two temperatures.

2-5. Repeat Problem 2-4 by adding a single reheat process to the turbine system. The temperature of steam leaving the reheater is the same as the turbine inlet temperature. Determine the reheat pressure at which the plant system has a maximum cycle efficiency.

Assume that there is no pressure drop in the reheat process.

2-6. A boiler feedwater pump receives 2800 gpm of saturated water at 130 psia and discharges at 1900 psia. The pump has an efficiency of 80%. Calculate the pump power and the temperature of the water leaving the pump.

2-7. The boiler feedwater pump in Problem 2-6 is driven by an auxiliary steam turbine. The turbine receives steam at 153 psia and 630 F and exhausts at 1.5 psia. Assuming the turbine isentropic efficiency is 82%, what is the required steam flow for this auxiliary turbine?

2-8. Consider the Rankine cycle with two feedwater heaters. One is a contact heater receiving the extracted steam at a pressure of 150 psia, while another heater is a surface-type heater without a drain cooler and uses extracted steam at 50 psia. Use the conditions below:

Turbine inlet pressure and temperature	1250 psia, 850 F
Condenser pressure	3.5 in. Hg abs.
Turbine efficiency	0.88
Pump efficiency	0.85
Heater terminal temperature difference	10 F

Calculate the cycle efficiency and the steam-flow rate for the plant net output of 200,000 kW. The cycle arrangement is similar to that in Fig. 2-15.

2-9. Consider a simple gas turbine cycle with the conditions described below:

Ambient air pressure and temperature	14 psia, 70 F
Compressor pressure ratio	11
Compressor efficiency	0.85
Turbine inlet temperature	2200 F
Turbine efficiency	0.90

Determine the thermal efficiency of this cycle and the required air flow for the cycle net output 50,000 kW.

2-10. Consider a regenerative gas turbine cycle. The regenerator has an effectiveness of 0.70 and the compressor has a pressure ratio of 4. Other conditions are identical to those in Problem 2-9. Calculate the thermal efficiency of this regenerative gas turbine cycle and estimate the cycle network per unit mass of the working substance.

2-11. A compressor receives air at the conditions of 14.7 psia and 60 F and exhausts it at 600 psia. Calculate the compressor work on the basis of a unit mass under three different conditions: (1) isentropic, (2) isothermal, and (3) isentropic with one optimum intercooling between two compressor stages.

2-12. Consider a gas turbine cycle with a reheater and regenerator. Use the conditions below:

Compressor inlet temperature and pressure	60 F, 14.7 psia
Compressor pressure ratio	7
Compressor efficiency	0.85
Turbine efficiency	0.90
Turbine inlet and reheat temperature	2200 F
Turbine reheat pressure	45 psia
Regenerator effectiveness	0.7

Calculate the thermal efficiency of this cycle and the air-flow rate for a 100,000 kW unit.

2-13. Consider the high temperature gas reactor (HTGR) system as shown in Fig. 2-25. If the working substance is air instead of helium and the temperature and pressure conditions are identical to those in Example 2-6, what is the thermal efficiency of this closed-cycle gas turbine system?

2-14. An air-standard Diesel engine intakes air at the conditions of 14.7 psia and 60 F. The compression ratio of the cycle is 13 and the cutoff ratio is 2.2. Calculate the cycle net work per pound of air, the mean effective pressure, and the thermal efficiency of this Diesel cycle.

2-15. The initial conditions for an air-standard Otto cycle are 14.7 psia and 60 F. The compression ratio is 8 to 1 and the heat transfer to the air per cycle is 820 Btu/lb. Calculate the mean effective pressure and the thermal efficiency of this Otto cycle.

2-16. Consider the air-standard dual cycle shown in Fig. 2-28. Derive an expression for the thermal cycle efficiency.

2-17. A combined-cycle system shown in Fig. 2-29 mainly consists of a gas turbine, a waste heat recovery boiler, and a steam turbine. The conditions are listed below:

Gas turbine	
Pressure ratio	9
Turbine inlet temperature	2100 F
Compressor efficiency	85%
Turbine efficiency	90%
Burner efficiency	95%
Boiler and steam turbine	
Steam pressure and temperature	600 psia, 750 F
Condenser pressure	3 in. Hg abs.
Boiler pinch point	60 F
Turbine efficiency	80%
Pump efficiency	78%

Calculate the thermal efficiency of this combined-cycle system.

2-18. Repeat Problem 2-17 with the boiler pinch point changed to 40 F and compare the results with that of Problem 2-17.

2-19. For the combined-cycle system described in Problem 2-17, calculate the temperature of the exhaust gas leaving the waste heat recovery boiler. Also discuss the method for further utilization of the waste heat if the stack gas is still too high.

2-20. A combined gas turbine and steam turbine plant is arranged in the scheme shown in Fig. 2-31. Air enters the compressor at 14.7 psia 70 F and exhausts at 100 psia. The gas enters the turbine at 1600 F and expands to the pressure close to the ambient air (14.7 psia). On the steam side, steam enters the turbine at 1250 psia and 850 F and exhausts to the condenser at the pressure 4 in. Hg abs. Other conditions are listed below.

Compressor efficiency	0.85
Gas turbine efficiency	0.90
Steam turbine efficiency	0.90
Pump efficiency	0.85
Boiler heat loss	2%
Boiler fuel	methane gas
Fuel heating value	21,502 Btu/lb
Boiler excess air	15%

Estimate the thermal efficiency of this combined cycle plant.

2-21. Steam at 1250 psia and 850 F is throttled to 900 psia. Calculate the steam availability before and after the throttling process. What is the process irreversibility?

2-22. For the steam turbine power plant described in Problem 2-8, calculate the process irreversibility for the turbine and the pump.

2-23. For the steam turbine power plant described in Problem 2-8, calculate the irreversibility and the effectiveness for the surface-type feedwater heater.

2-24. Air enters a compressor at 14.7 psia and 70 F and exhausts at 100 psia and 500 F. The compressor has a heat loss equal to 0.5% of the compressor work. Determine the reversible work, process irreversibility, and the effectiveness.

2-25. For the combined-cycle generating system described in Problem 2-17, calculate the second-law efficiency and the irreversibility at various plant locations.

2-26. For the regenerative and reheat gas turbine system described in Problem 2-12, calculate the second-law efficiency for the entire system and the irreversibilities at various cycle locations.

CHAPTER THREE

Economics

The basis of most design decisions is economic. Designing a system that functions properly is only a part of the engineer's task. The system must also be economical and show an adequate return on the investment. In the power plant design the engineer will search for the design that will have the minimum investment cost or will be able to generate electricity at the lowest overall cost.

3.1 COST OF ELECTRICITY

Many factors influence the cost of electricity. As is true of other commercial products, the cost of electricity is made up of both fixed and variable costs. The fixed cost generally remains constant regardless of the number of hours the facility is used. The variable cost is the cost related to the production level of the facility.

In the electric power business, the fixed cost is entirely dependent on the capital investment. The components of fixed cost are rate of return, depreciation rate, administrative and general expenses, insurance expenses, and taxes. These components are defined as follows.

1. *Rate of return.* It is the minimum acceptable percentage return on the invested capital. Sometimes it is referred to as the cost of capital, the discounted rate, or the interest rate.

2. *Depreciation rate.* There must be periodic depreciation charges to the income in order to recover the cost of equipment before its usefulness is exhausted. The annual depreciation charge in terms of the percentage of capital investment is

$$D_r = \frac{R}{(1 + R)^{n-1}} \tag{3-1}$$

where

D_r = depreciation rate expressed as a fraction
R = rate of interest expressed as a fraction
n = plant economic life in years

3. *Administrative and general expenses.* These expenses cover administrative and general salaries, miscellaneous materials and supplies, and any other expenses that are not accounted for in the other components. The administrative and general expenses will be expressed as a percentage of invested capital.

4. *Insurance expenses.* These cover insurance against accidents to equipment and personnel as a result of fire, windstorm, hail, flood, earthquake, and the like. Insurance expenses are expressed as a percentage of invested capital.

90

5. *Taxes other than income taxes (ad valorem taxes).* This expense deals with property-related taxes, payroll taxes, and other miscellaneous taxes other than income or franchise taxes. This tax component is expressed as a percentage of the invested capital.

6. *Income tax.* This component is determined by the rate of return. It is expressed as a uniform equivalent annual percentage of the invested capital.

The sum of these six components is frequently called the total fixed charge rate. Evidently, the total fixed charge rate is also expressed as a percentage of the invested capital. Table 3-1 shows the typical values of total fixed-charge rate and its components.

In an electric utility operation the variable cost mainly consists of two components: (1) fuel cost and (2) operation and maintenance cost.

1. Fuel Cost

The largest item of expense in the operation of thermal power plant is the original raw energy. This energy may be in the form of coal, nuclear, oil, natural gas, wood scrap, or other by-products. The fuel cost varies with the plant's efficiency, unit fuel cost, and the amount of electric energy produced. The fuel cost pattern is generally predicted over the economic life of the project after taking into account escalation in the cost of materials, labor, and transportation. From this information a levelized fuel price can be computed. The levelized concept is discussed later in this chapter.

2. Operation and Maintenance Cost

Operating and maintenance (O & M) expenses include operating labor, materials, and tools for plant maintenance on both a routine and emergency basis. These expenses are neither a function of plant capital cost nor plant generating capacity. They vary from year to year and generally become higher as the plant becomes older. These expenses also vary according to the size of plant, type of fuel used, loading schedule, and operating characteristics (peaking or base load). Accordingly, the operation and maintenance expenses are generally estimated in dollars per year, taking into consideration the previously mentioned factors, including estimated escalations. From such an estimate, a levelized operation and maintenance expense can be computed. In general, O & M expenses are approximately equal to one-fourth of the fuel expenses.

Table 3-1
Typical Values of Total Fixed Charge and Its Components

Rate of return	8–12%
Depreciation	0.25–0.40%
Administrative and general expenses	0.5–2.9%
Insurance	0.05–0.25%
All taxes	2.0–5.0%
Fixed-charge rate	10.80–20.55%

EXAMPLE 3-1. A coal-fired power plant of net output 582,600 kW was designed with the following economic factors:

Plant life	35 years
Fixed charge rate	11.638%
Fuel cost in the first year of commercial operation	$1.25/MBtu
Fuel cost escalation rate over the plant life	7.0%
Value of installed capacity in the first year of commercial operation	$866/kW
The rate of return	7.358%

This plant will have the following net plant heat rates:

Loading Number	Plant Net Output (kW)	Plant Net Heat Rate (Btu / kWh)
1	643,500	9,650
2	582,600	9,650
3	399,900	9,800
4	217,100	10,125
5	0	—

Operation hours are shown in the following table:

Plant Age (Years)	Loading Schedule (hr / year)				
	1	2	3	4	5 (Down Time)
1	438	517	1071	2102	4632
2–4	438	885	1927	2278	3232
5–7	438	1332	2628	2628	1734
8–11	438	4468	1314	876	1664
12–35	438	5081	876	438	1927

Calculate for the first year of operation (1) the fuel cost, (2) the operation and maintenance cost, (3) the fixed cost, and (4) the unit generation cost in terms of mills/kWh.

Solution: The following equation is used to calculate the fuel cost of each loading.

$$(\text{Fuel cost}) = (\text{plant net output})(\text{hours of operation})$$

$$\times (\text{plant net heat rate})(\text{unit fuel cost})$$

At loading 1

$$(FC)_1 = (643{,}500)(438)(9650)(1.25 \times 10^{-6})$$

$$= \$3.399 \times 10^6$$

At loading 2

$$(FC)_2 = (582{,}600)(517)(9650)(1.25 \times 10^{-6})$$

$$= \$3.633 \times 10^6$$

At loading 3

$$(FC)_3 = (399{,}900)(1071)(9800)(1.25 \times 10^{-6})$$

$$= \$5.246 \times 10^6$$

At loading 4

$$(FC)_4 = (217{,}100)(2102)(10{,}125)(1.25 \times 10^{-6})$$

$$= \$5.775 \times 10^6$$

The fuel cost for the first year is then

$$(FCOST) = (FC)_1 + (FC)_2 + (FC)_3 + (FC)_4$$

$$= (3.399 + 3.633 + 5.246 + 5.775) \times 10^6$$

$$= \$18.053 \times 10^6$$

This calculation is based on the assumption that the zero load input can be completely neglected. For the O & M cost it is approximated as 25% of the fuel cost. That is,

$$(OMCOST) = (\text{fuel cost})(0.25) = (18.035 \times 10^6)(0.25)$$

$$= \$4.514 \times 10^6$$

The fixed cost for the first year is

$$= (\text{unit capital cost})(\text{unit rating})(\text{fixed charge rate})$$

$$= (866)(582{,}600)(0.11638)$$

$$= \$58.717 \times 10^6$$

The unit generation cost for the first year is

$$(GCOST) = (\text{annual fuel cost} + \text{annual O \& M cost} + \text{fixed cost})$$

$$/(\text{kWh generated in the year})$$

$$= (18.053 + 4.514 + 58.717) \times 10^6$$

$$/(438 \times 643{,}500 + 517 \times 582{,}600 + 1071 \times 399{,}900$$

$$+ 2102 \times 217{,}000)$$

$$= \$0.05539/\text{kWh or } 55.39 \text{ mills/kWh}$$

3.2 PRESENT WORTH

The power plant economic life is generally 30 to 45 years. The value of annual operation expenses is related to the time these expenses occur. Because of this, the concept of present worth must be utilized. The present worth is the value of a sum of money at the present time that, with compound interest, will have a specified value at a certain time in the future. Let

$$S = \text{the sum of money at the } n\text{th year}$$

$$i = \text{annual interest rate}$$

$$n = \text{the year } n$$

Then, the present worth (P) of S dollars at the nth year is

$$P = \frac{1}{(1 + i)^n} S \tag{3-2}$$

The term $(1 + i)^{-n}$ is frequently referred to as the single payment present worth factor (PWF). Table 3-2 gives these factors for the compound interest of 6%.

On many occasions equal amounts of annual expenses are required. Then the present worth of a uniform annual series of payments is calculated by

$$P = \frac{1 - (1 + i)^{-n}}{i} A \tag{3-3}$$

where $A = $ annual payment.

The term $[1 - (1 + i)^{-n}]/i$ is often called the series present worth factor ($SPWF$). These factors for the compound interest of 6% are presented in Table 3-2.

Table 3-2
Values of (PWF) and (SPWF) at a Compound Interest of 6%

Year	Single Payment Present Worth Factor (PWF)	Uniform Annual Series Present Worth Factor (SPWF)	Year	Single Payment Present Worth Factor (PWF)	Uniform Annual Series Present Worth Factor (SPWF)
1	0.9434	0.943	16	0.3936	10.106
2	0.8900	1.833	17	0.3717	10.477
3	0.8396	2.673	18	0.3503	10.828
4	0.7921	3.465	19	0.3305	10.158
5	0.7473	4.212	20	0.3318	11.470
6	0.7050	4.917	21	0.2942	11.764
7	0.6651	5.582	22	0.2775	12.042
8	0.6274	6.210	23	0.2618	12.303
9	0.5919	6.802	24	0.2470	12.550
10	0.5584	7.360	25	0.2330	12.783
11	0.5268	7.887	26	0.2198	13.003
12	0.4970	8.384	27	0.2074	13.211
13	0.4688	8.853	28	0.1956	13.406
14	0.4423	9.925	29	0.1846	13.591
15	0.4173	9.712	30	0.1741	13.765

EXAMPLE 3-2. Calculate the present worth of lifetime fixed cost for the plant described in Example 3-1.

Solution: The annual fixed cost (A) is equal to

$$\$58.717 \times 10^6 \quad \text{(see Example 3-1)}$$

Then, the present worth of lifetime fixed cost (P) is expressed as

$$P = \frac{1 - (1 + i)^{-n}}{i} A$$

$$= \frac{1 - (1 + 0.07358)^{-35}}{0.07358} (58.717 \times 10^6)$$

$$= \$731.5 \times 10^6$$

EXAMPLE 3-3. Compute the present worth of lifetime fuel cost for the power plant described in Example 3-1.

Table 3-3
Present Worth Calculation for Example 3-3

Year	Energy Produced (10^6 kWh / yr)	Fuel Price ($ / MBtu)	Plant Net Heat Rate (Btu / kWh)	Single Payment Present Worth Factor	Present Worth Fuel Expense (Millions of Dollars)
1	281.85	1.25	9650	0.9315	3.1668
	301.20	1.25	9650	0.9315	3.3843
	428.29	1.25	9800	0.9315	4.8870
	456.34	1.25	10125	0.9315	5.3798
2	281.85	1.34	9650	0.8676	3.1563
	515.60	1.34	9650	0.8676	5.7739
	770.61	1.34	9800	0.8676	8.7636
	494.55	1.34	10125	0.8676	5.8108
3	281.85	1.43	9650	0.8082	3.1458
	515.60	1.43	9650	0.8082	5.7546
	770.61	1.43	9800	0.8082	8.7344
	494.55	1.43	10125	0.8082	5.7914
4	281.85	1.53	9650	0.7528	3.1353
	515.60	1.53	9650	0.7528	5.7354
	770.61	1.53	9800	0.7528	8.7053
	494.55	1.53	10125	0.7528	5.7721
5	281.85	1.64	9650	0.7012	3.1248
	776.02	1.64	9650	0.7012	8.6035
	1050.94	1.64	9800	0.7012	11.8325
	570.54	1.64	10125	0.7012	6.6367
6	281.85	1.75	9650	0.6531	3.1144
	776.02	1.75	9650	0.6531	8.5748
	1050.94	1.75	19800	0.6531	11.7931
	570.54	1.75	10125	0.6531	6.6146
7	281.85	1.88	9650	0.6084	3.1040
	776.02	1.88	9650	0.6084	8.5462
	1050.94	1.88	9800	0.6084	11.7537
	570.54	1.88	10125	0.6084	6.5925
8	281.85	2.01	9650	0.5667	3.0937
	2603.06	2.01	9650	0.5667	28.5715
	525.47	2.01	9800	0.5667	5.8573
	190.18	2.01	10125	0.5667	2.1902
9	281.85	2.15	9650	0.5278	3.0833
	2603.06	2.15	9650	0.5278	28.4763
	525.47	2.15	9800	0.5278	5.8377

Table 3-3—Continued

Year	Energy Produced (10^6 kWh / yr)	Fuel Price ($ / MBtu)	Plant Net Heat Rate (Btu / kWh)	Single Payment Present Worth Factor	Present Worth Fuel Expense (Millions of Dollars)
	190.18	2.15	10125	0.5278	2.1829
10	281.85	2.30	9650	0.4917	3.0731
	2603.06	2.30	9650	0.4917	28.3813
	525.47	2.30	9800	0.4917	5.8183
	190.18	2.30	10125	0.4917	2.1756
11	281.85	2.46	9650	0.4580	3.0628
	2603.06	2.46	9650	0.4580	28.2867
	525.47	2.46	9800	0.4580	5.7989
	190.18	2.46	10125	0.4580	2.1684
12	281.85	2.63	9650	0.4266	3.0526
	2960.19	2.63	9650	0.4266	32.0603
	350.31	2.63	9800	0.4266	3.8530
	95.09	2.63	10125	0.4266	1.0806
13[a]	3687.44	2.82	—	0.3973	39.9130
14	3687.44	3.01	—	0.3701	39.7800
15	3687.44	3.22	—	0.3447	39.6473
16	3687.44	3.45	—	0.3211	39.5190
17	3687.44	3.69	—	0.2991	39.3824
18	3687.44	3.95	—	0.2786	39.2520
19	3687.44	4.22	—	0.2595	39.1212
20	3687.44	4.52	—	0.2417	38.9906
21	3687.44	4.84	—	0.2252	38.8607
22	3687.44	5.18	—	0.2097	38.7311
23	3687.44	5.54	—	0.1954	38.6020
24	3687.44	5.94	—	0.1820	38.4733
25	3687.44	6.34	—	0.1695	38.3449
26	3687.44	6.78	—	0.1579	38.2172
27	3687.44	7.26	—	0.1471	38.0897
28	3687.44	7.77	—	0.1370	31.9627
29	3687.44	8.31	—	0.1276	37.8361
30	3687.44	8.89	—	0.1188	37.7099
31	3687.44	9.52	—	0.1107	37.5842
32	3587.44	10.18	—	0.1031	37.4590
33	3587.44	10.89	—	0.0960	37.3053
34	3587.44	11.66	—	0.0895	37.2096
35	3587.44	12.47	—	0.0833	37.0855
				Total	1221.58

[a] From the thirteenth to the thirty-fifth year the presentation is made on an annual basis.

Solution: The detailed calculations are presented in Table 3-3. They are self-explanatory. The present worth of lifetime fuel cost is the sum of all values in the last column and is found $1221.58 million.

3.3 LEVELIZING EQUATIONS

In power plant design many technical and economic factors vary from year to year. The fuel price and plant capacity factor are good examples. To compare various design alternatives, it is often desirable to have the equivalent, but the constant term that can represent the factor that changes throughout plant life. This term is frequently called the levelized value of this factor.

In this section a general levelizing equation will be developed that can be used to transform any factor that varies from year to year into an equivalent uniform factor. This factor is weighted by the time and the changing value of money in much the same way as the centroid is geometrically weighted. More specifically, the factor in year i, f_i, is multiplied by the single payment present worth factor $(PWF)_i$ and summed over the lifetime of the plant to give the lifetime present worth of the factor f. Then, the lifetime present worth of the factor f is divided by the sum of the present worth factors for year $i = 1, 2, \ldots, n$ to give the levelized value of the factor f. Mathematically the levelized value of the factor f is

$$f_{\text{levelized}} = \frac{\displaystyle\sum_{i=1}^{n} (f_i)(PWF)_i}{\displaystyle\sum_{i=1}^{n} (PWF)_i} \tag{3-4}$$

or

$$= \frac{\displaystyle\sum_{i=1}^{n} (f_i)(PWF)_i}{(SPWF)} \tag{3-5}$$

There are two special cases. If the factor f escalates at a constant rate, say k, Eq. (3-5) becomes

$$f_{\text{levelized}} = \frac{f \displaystyle\sum_{i=1}^{n} (1 + K)^{i-1}(PWF)_i}{(SPWF)} \tag{3-6}$$

and if f is constant throughout the plant life, Eq. (3-5) is expressed as

$$f_{\text{levelized}} = \frac{f \displaystyle\sum_{i=1}^{n} (PWF)_i}{(SPWF)} = f \tag{3-7}$$

Table 3-4
Preliminary Calculations for Example 3-4

Year	Single Payment Present Worth Factor	Fuel Price ($ / MBtu)	Present Worth Fuel Price ($ / MBtu)
1	0.9315	1.25	1.16
2	0.8676	1.34	1.16
3	0.8082	1.43	1.15
4	0.7528	1.53	1.15
5	0.7012	1.64	1.15
6	0.6531	1.75	1.15
7	0.6084	1.88	1.14
8	0.5667	2.01	1.14
9	0.5278	2.15	1.13
10	0.4917	2.30	1.13
11	0.4580	2.46	1.13
12	0.4266	2.63	1.12
13	0.3973	2.82	1.12
14	0.3701	3.01	1.11
15	0.3447	3.22	1.11
16	0.3211	3.45	1.11
17	0.2991	3.69	1.10
18	0.2786	3.95	1.10
19	0.2595	4.22	1.10
20	0.2417	4.52	1.09
21	0.2252	4.84	1.09
22	0.2097	5.18	1.09
23	0.1954	5.54	1.08
24	0.1820	5.93	1.08
25	0.1695	6.34	1.07
26	0.1579	6.78	1.07
27	0.1471	7.26	1.07
28	0.1370	7.77	1.06
29	0.1276	8.31	1.06
30	0.1188	8.89	1.06
31	0.1107	9.52	1.05
32	0.1031	10.18	1.05
33	0.0960	10.89	1.05
34	0.0895	11.66	1.04
35	0.0833	12.47	1.04
	Total 12.459		38.51

EXAMPLE 3-4. Compute the levelized fuel price for a project with a 35-year economic life. At the time the plant is commercially operated, the fuel price is $1.25 per million Btu and the escalation rate is expected to be 7% on the average. The rate of return for this project is 7.358%.

Solution: The preliminary calculations are presented in Table 3-4. With these results the levelized fuel price can be calculated as follows:

$$(\text{Fuel price})_{\text{levelized}} = \frac{\text{lifetime present worth of fuel price}}{SPWF}$$

$$= \frac{38.510}{12.459} = \$3.09 \text{ /MBtu}$$

The levelized fuel price can be also calculated directly from Eq. (3-6), which is expressed as

$$f_{\text{levelized}} = \frac{f\left(\dfrac{1 - P^n}{r - k}\right)}{(SPWF)} \tag{3-6a}$$

where

$$r = \text{rate of interest}$$

$$k = \text{escalation rate}$$

$$P = (1 + k)/(1 + r)$$

Substituting the numerical values into the equation gives

$$f_{\text{levelized}} = \frac{1.25\left(\dfrac{1 - 0.997^{35}}{0.07358 - 0.070}\right)}{12.459}$$

$$= \$3.09 \text{ /MBtu}$$

It should be noticed that the foregoing solution is valid only when the rate of fuel consumption is constant every year throughout the entire project life. In an electric power plant, this will never be the case. For fuel usage varying from year to year, the levelized fuel price should be calculated with the use of annual fuel consumption. The following examples illustrate this point.

EXAMPLE 3-5. Calculate the levelized annual fuel costs for the power plant described in Example 3-1.

Solution: The present worth of lifetime fuel costs is $1221.58 × 10^6 (see Example 3-3). The levelized annual expense is the present worth of lifetime fuel costs divided by the series payment present worth factor ($SPWF$). That is,

$$\text{Levelized annual fuel expense} = \frac{1221.58 \times 10^6}{12.459}$$

$$= \$98.05 \times 10^6/\text{yr}$$

EXAMPLE 3-6. The annual plant capacity factor is frequently defined as the ratio of actual plant output to the maximum possible plant output. (Note the maximum output is the product of plant rating and 8760 hours per year.) Using this definition, the plant described in Example 3-1 will have annual capacity factors as follows:

First year	28.70%
Next three years	40.40%
Next three years	52.50%
Next four years	70.55%
Next 24 years	72.25%

Calculate the levelized plant capacity factor and levelized annual generation.

Solution: The present worth factors for the foregoing period are

First year	= 0.9315
Next three years	= 2.4286
Next three years	= 1.9627
Next four years	= 2.0451
Next 24 years	= 5.0911

The levelized plant capacity factor by using Eq. (3-5) is

$$= \frac{1}{12.459}\left[(0.287)(0.9315) + (0.404)(2.4286) + (0.525)(1.9627)\right.$$

$$\left. + (0.7055)(2.0451) + (0.7225)(5.0911)\right]$$

$$= 0.594 \quad \text{or} \quad 59.4\%$$

The levelized annual generation is by definition

$$= (582,600)(8760)(0.594)$$

$$= 3031 \times 10^6 \text{ kWh/yr}$$

EXAMPLE 3-7. Calculate the levelized plant net heat rate for the power plant described in Example 3-1.

Solution: The levelized plant net heat rate is defined by

$$\frac{\text{levelized annual fuel consumption}}{\text{levelized annual generation}}$$

The levelized annual generation is 3031×10^6 kWh/yr as indicated in Example 3-6. The levelized annual fuel consumption is calculated according to the Eq. (3-4). That is,

$$\text{Levelized annual fuel consumption} = \frac{\sum_{1}^{35} F_i (PWF)_i}{\sum_{1}^{35} (PWF)_i}$$

$$= \frac{367.268 \times 10^6 \text{ MBtu}}{12.459}$$

$$= 29.480 \times 10^6 \text{ MBtu/yr}$$

The detailed calculations are presented in Table 3-5. Finally, the levelized plant net heat rate is

$$\frac{29.480 \times 10^6 \text{ MBtu/yr}}{3031 \times 10^6 \text{ kWh/yr}} = 9726 \text{ Btu/kWh}$$

EXAMPLE 3-8.   Calculate the levelized unit fuel cost and levelized fuel price for the power plant described in Example 3-1.

Solution: The levelized unit fuel cost

$$= \frac{\text{the levelized annual fuel expense}}{\text{the levelized annual generation}}$$

$$= \frac{\$98.05 \times 10^6/\text{yr}}{3031 \times 10^6 \text{ kWh/yr}}$$

$$= \$0.0323 /\text{kWh or } 32.3 \text{ mills/kWh}$$

The levelized fuel price is

$$= \frac{\text{the levelized unit fuel cost}}{\text{the levelized plant net heat rate}}$$

$$= \frac{\$0.0323 /\text{kWh}}{9726 \text{ Btu/kWh}}$$

$$= \$3.33 /\text{MBtu}$$

Table 3-5
Levelized Annual Fuel Consumption for Example 3-7

Yeat	Annual Fuel Consumption (10^9 Btu) F_i	Present Worth Factor $(PWF)_i$	Present Worth of Annual Fuel Consumption (10^9 Btu) $F_i(PWF)_i$
1	14444	.9315	13454
2	20254	.8676	17573
3	20254	.8082	16369
4	20254	.7528	15247
5	26284	.7012	18430
6	26284	.6531	17166
7	26284	.6084	15990
8	34914	.5667	19784
9	34914	.5278	18428
10	34914	.4917	17165
11	34914	.4580	15989
12	35681	.4266	15220
13	35681	.3973	14177
14	35681	.3701	13205
15	35681	.3447	12300
16	35681	.3211	11457
17	35681	.2991	10672
18	35681	.2786	9940
19	35681	.2595	9259
20	35681	.2417	8625
21	35681	.2252	8033
22	35681	.2097	7483
23	35681	.1954	6970
24	35681	.1820	6492
25	35681	.1695	6047
26	35681	.1579	5633
27	35681	.1471	5247
28	35681	.1370	4887
29	35681	.1276	4552
30	35681	.1188	4240
31	35681	.1107	3949
32	35681	.1031	3679
33	35681	.0960	3427
34	35681	.0895	3192
35	35681	.0833	2973
		12.459	367.268×10^6 MBtu

As shown in these examples, the plant capacity factor, net heat rate, and other parameters can be easily levelized. The levelized values are then used in economic evaluations. These will be shown in Section 3-4. In practice, however, other methods have been also used to describe the so-called levelized value or weighted value. For instance, the weighted plant net heat rate or capacity factor are frequently used. In the following an example is used to illustrate the procedure and calculations of these two terms.

EXAMPLE 3-9. Calculate the weighted plant capacity factor and the plant net heat rate for the power plant described in Example 3-1.

Solution: First we used the data in Example 3-1 and calculated the operation hours in terms of an annual percentage. These are summarized in Table 3-6. The plant annual capacity factor is the ratio of the actual plant output to the maximum possible output. Using this definition, we calculated the annual capacity factors and presented them in the last column of Table 3-6. The lifetime weighted plant capacity factor $(\overline{PCF})$ is simply the average of all factors. That is,

$$\overline{PCF} = [28.70 + (3 \times 40.40) + (3 \times 52.5) + (4 \times 70.55) + (24 \times 72.25)]/35$$

$$= 0.6639 \quad \text{or} \quad 66.39\%$$

The weighted plant net heat rate is the average heat rate with which the lifetime fuel consumption can be predicted. Using this definition, we have

$$(\overline{PNHR})(\text{unit rating})(8760)(\overline{PCF})(n)$$

$$= \sum_{i=1}^{n} \sum_{j=1}^{m} (PNHR)_j(\text{unit rating})(\text{loading index})_j(\text{hours})_{ij}$$

Table 3-6
Summary of Calculations for Example 3-9

	Plant Loadings					Annual
	1	2	3	4	5	Capacity
	643,500 kW	582,600 kW	399,900 kW	217,100 kW	0 kW	Factor
Year	110%	100%	69%	37%	0%	(%)
	Operation Hours (%)					
First year	5.0	5.9	12.2	24.0	52.9	28.70
Next 3 years	5.0	10.1	22.0	26.0	36.9	40.40
Next 3 years	5.0	15.2	30.0	30.0	19.8	52.50
Next 4 years	5.0	51.0	15.0	10.0	19.0	70.55
Next 24 years	5.0	58.0	10.0	5.0	22.0	72.25
k_j	5.0	47.94	13.38	10.06	23.63	

Rearranging it gives

$$\overline{PNHR} = \frac{1}{\overline{PCF}} \sum_{j=1}^{m} (PNHR)_j (\text{loading index})_j \sum_{i=1}^{n} \frac{(\text{hour})_{ij}}{(n)(8760)}$$

or

$$\overline{PNHR} = \frac{\sum_{j=1}^{m} (PNHR)_j (\text{loading index})_j K_j}{\overline{PCF}}$$

where

$$K_j = \frac{1}{n} \sum_{i=1}^{n} \frac{(\text{hour})_{ij}}{8760} = \frac{1}{n} \sum_{i=1}^{n} (\text{time})_{ij}$$

$(\text{time})_i$ = percentage of annual operation hours

$(\text{loading index})_j$ = loading index j

$(\overline{PNHR})_j$ = plant net heat rate at loading j

$i = i$th year

j = loading number

n = plant economic life

m = number of plant loadings

For this example the K_j terms have been determined and presented in Table 3-6. Finally the weighted plant net heat rate is

$$\overline{PNHR} = [(9650)(1.1)(5.0) + (9650)(1.0)(47.94) + (9800)$$

$$\times (0.69)(13.38) + (10125)(0.37)(10.06)] \frac{1}{66.39}$$

$$\overline{PNHR} = 9698 \text{ Btu/kWh}$$

3.4 ECONOMIC EVALUATION METHODS

Decision makings in the selection of design alternatives have been given the utmost importance in engineering practice. This is particularly true for power plant design where the equipment and fuel costs have been rapidly increasing in recent years.

Implementation of a poor choice from the design alternatives may result in serious financial consequences affecting the entire company operation. Although various methods for economic evaluation of design alternatives have been developed, the following three methods are commonly used to compare lifetime cost of alternative designs:

1. Annual cost method.
2. Present worth method.
3. Capitalized cost method.

Annual Cost Method

This method is used to compare the annual cost of alternative designs. The total annual cost generally consists of three terms: capital investment, fuel cost, and O & M cost. The relation is expressed by

$$C = C_c + C_f + C_{om} \tag{3-8}$$

or
$$C = I(AFCR) + C_f + C_{om} \tag{3-9}$$

where

$$C = \text{total annual cost}$$

$$C_f = \text{levelized annual fuel cost}$$

$$C_{om} = \text{levelized annual O \& M cost}$$

$$I = \text{capital investment}$$

$$(AFCR) = \text{annual fixed charge rate}$$

It must be pointed out that the terms C_f and C_{om} are the levelized values of the annual expenses. The method of calculation for these terms has been presented previously.

Figure 3-1 indicates a general relationship between the components of total annual cost and capital investments for various alternative designs. Generally, investment increases with efficiency, and the benefit of such increase will be realized in lower fuel costs. The O & M term frequently remains unchanged or at least insensitive to the change of capital investment.

The selection of alternative designs is made by comparing the total annual costs. The optimal design results in a minimum annual cost.

Present Worth Method

This method and the capitalized cost method described later may be regarded as the index methods of comparison. In the present worth method all annual costs (such as capital, fuel, and O & M costs) expended during the project lifetime are reduced into

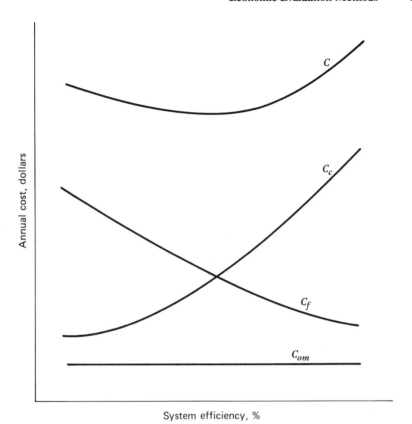

Figure 3-1.   General relationship between the annual cost and system efficiency.

present worths. The mathematical relation is expressed by

$$TPW = \sum_{i=1}^{n} (PWF)_i I(AFCR) + \sum_{i=1}^{n} (PWF)_i A_{f,i} + \sum_{i=1}^{n} (PWF)_i A_{om,i}$$

(3-10)

where

$$TPW = \text{total present worth of lifetime expenses}$$

$$(PWF)_i = \text{single payment present worth factor in } i\text{th year}$$

$$A_{f,i} = \text{annual fuel cost in the } i\text{th year}$$

$$A_{om,i} = \text{annual O \& M cost in the } i\text{th year}$$

$$n = \text{plant life in years}$$

When the levelized values of annual fuel cost and O & M cost are available, the total present worth of lifetime expenses can be calculated with the equation

$$TPW = \left[I(AFCR) + C_f + C_{om} \right](SPWF) \tag{3-11}$$

The selection of alternative designs is made by comparing the total present worth of lifetime expenses. The optimal design results in a minimum value of the total present worth.

Capitalized Cost Method

This method of comparison is similar to the present worth method. In the present worth method the assumption of a lump sum of money covers the actual schedule of total annual costs to be met during the project lifetime. In the capitalized cost method a hypothetical amount of capital could be added directly to the capital cost so that the lifetime fuel and O & M costs would be eliminated. Mathematically, it is

$$TEC = I + \frac{C_f}{AFCR} + \frac{C_{om}}{AFCR} \tag{3-12}$$

where TEC = total evaluated cost. For the selection of alternative designs engineers usually seek for the design having the minimum value of the total evaluated cost.

EXAMPLE 3-10. Two plant design alternatives have been proposed for a fossil-fuel power plant project. Use the present worth method to determine the economic choice. Here is the summary of these two design alternatives.

	Alternative 1	Alternative 2
Throttle steam pressure (psi)	2400	3500
Throttle steam temperature (F)	1000	1000
Reheat temperature (F)	1000	1000
Plant size (MW)	600	600
Net plant heat rate (Btu/kWh)		
100% load	9000	8860
80% load	9000	8860
60% load	9180	9040
35% load	10080	9920
Plant life	30	30
Construction cost ($/kW)	800	850
Levelized fuel cost ($/MBtu)	1.25	1.25
Interest rate (%)	6	6
Fixed charge rate (%)	11.5	11.5
Unit loading information		

		Years	
Load	1–10	11–20	21–30
100%	44%	23%	6%
80%	28	23	23
60%	13	34	34
35%	6	11	23
0%	9	9	14

Solution: The calculations are presented in four steps.

1. Annual fuel cost.
2. Present worth of lifetime fuel cost.
3. Present worth of lifetime investments.
4. Present worth of total evaluated cost.

Step 1. The annual fuel cost (*FCOST*) is obtained with the equation:

$$FCOST = (UCOST) \sum_{j=1}^{m} (LOAD)_j (HOUR)_j (PNHR)_j$$

where

$(UCOST)$ = unit fuel cost, \$/MBtu

$(LOAD)_j$ = plant net output in jth loading, kW

$(HOUR)_j$ = operation hours in jth loading, hr

$(PNHR)_j$ = plant net heat rate in jth loadings, Btu/kWh

The results are summarized in Tables A and B.

Step 2. The present worth of lifetime fuel cost is obtained using the equation

$$(PWFCOST) = \sum_{i=1}^{m} [(\text{fuel cost}) \times (\text{present worth factor for the period})]$$

The present worth factors can be found in Table 3-2 and m is the number of periods in the evaluation.

For Alternative 1:

$$(PWFCOST) = (45.356)(7.36) + (39.330)(11.47 - 7.36)$$

$$+ (32.060)(13.76 - 11.47)$$

$$(PWFCOST) = \$568.88 \times 10^6$$

Table A
Example 3-10 (Alternative 1)

$(LOAD)_j$ (kW)	$(HOUR)_j$ (hr)	$(PNHR)_j$ (Btu/kWh)	$(FCOST)_j$ ($\$10^6$/yr)	$FCOST$ ($\$10^6$/yr)
Years 1–10				
600,000	3854.4	9000	26.017	
480,000	2452.8	9000	13.245	
360,000	1138.8	9180	4.704	
210,000	525.6	10080	1.390	45.356
Years 11–20				
600,000	2014.8	9000	13.599	
480,000	2014.8	9000	10.879	
360,000	2978.4	9180	12.303	
210,000	963.6	10080	2.549	39.330
Years 21–30				
600,000	525.6	9000	3.547	
480,000	2014.8	9000	10.879	
360,000	2978.4	9180	12.303	
210,000	2014.8	10080	5.331	32.060

Table B
Example 3-10 (Alternative 2)

$(LOAD)_j$ (kW)	$(HOUR)_j$ (hr)	$(PNHR)_j$ (Btu/kWh)	$(FCOST)_j$ ($\$10^6$/yr)	$FCOST$ ($\$10^6$/yr)
Years 1–10				
600,000	3854.4	8860	25.612	
480,000	2452.8	8860	13.039	
360,000	1138.8	9040	4.632	
210,000	525.6	9920	1.368	44.651
Years 11–20				
600,000	2014.8	8860	13.388	
480,000	2014.8	8860	10.710	
360,000	2978.4	9040	12.116	
210,000	963.6	9920	2.509	38.723
Years 21–30				
600,000	525.6	8860	3.492	
480,000	2014.8	8860	10.710	
360,000	2978.4	9040	12.116	
210,000	2014.8	9920	5.246	31.564

For Alternative 2:

$$(PWFCOST) = (44.651)(7.36) + (38.723)(11.47 - 7.36)$$

$$+ (31.564)(13.76 - 11.47)$$

$$(PWFCOST) = \$560.06 \times 10^6$$

Step 3.   The present worth of lifetime capital investment is

$$(PWICOST) = (\text{unit rating})(\text{unit cost})(\text{fixed charge rate})(SPWF)$$

For Alternative 1:

$$(PWICOST) = (600{,}000)(800)(0.115)(13.76)$$

$$= \$759.55 \times 10^6$$

For Alternative 2:

$$(PWICOST) = (600{,}000)(850)(0.115)(13.76)$$

$$= \$807.02 \times 10^6$$

Step 4.   The present worth of total evaluated cost is

$$(PWTCOST) = \text{present worth of lifetime fuel cost}$$

$$+ \text{present worth of lifetime capital cost}$$

For Alternative 1:

$$(PWTCOST) = 568.88 \times 10^6 + 759.55 \times 10^6$$

$$= \$1328.43 \times 10^6$$

For Alternative 2:

$$(PWTCOST) = 560.06 \times 10^6 + 807.02 \times 10^6$$

$$= \$1367.08 \times 10^6$$

In terms of the present worth of lifetime total cost, the Alternative design 1 is superior to the Alternative 2. The difference is approximately \$38.65 million.

EXAMPLE 3-11.   Repeat Example 3-10 by using the capitalized method.

Solution: First, we calculate the average value of plant capacity factor as 64.93% and other parameters used for determination of the weighted plant net heat rate.

For convenience they are summarized below.
For Alternative 1:

j	LOADING INDEX	$PNHR$	k[a]
1	1.00	9000	24.33
2	0.80	9000	24.67
3	0.60	9180	27.00
4	0.35	10080	13.33

[a] Example 3-9 indicates the procedure for the calculation of k.

For Alternative 2:

j	LOADING INDEX	$PNHR$	k
1	1.00	8860	24.33
2	0.80	8860	24.67
3	0.60	9040	27.00
4	0.35	9920	13.33

Then the weighted plant net heat rate is for Alternative 1

$$\overline{PNHR} = [(9000 \times 1.0 \times 24.33) + (9000 \times 0.8 \times 24.67)$$

$$+ (9180 \times 0.6 \times 27.0) + (10080 \times 0.35 \times 13.33)]/64.93$$

$$= 9122.7 \text{ Btu/kWh}$$

and for Alternative 2

$$\overline{PNHR} = [(8860 \times 1.00 \times 24.33) + (8860 \times 0.80 \times 24.67)$$

$$+ (9040 \times 0.6 \times 27.00) + (9920 \times 0.35 \times 13.33)]/64.93$$

$$= 8981.3 \text{ Btu/kWh}$$

Second, we calculate the capitalized value of the fuel cost by

$$\frac{C_f}{AFCR} = \frac{\overline{PNHR} \times \overline{PCF} \times \text{unit rating} \times 8760 \times \text{fuel cost}}{\text{fixed charge rate}}$$

For Alternative 1, we have

$$\frac{C_f}{AFCR} = \frac{9122.7 \times 0.6493 \times 600{,}000 \times 8760 \times 1.25}{0.115 \times 10^6}$$

$$= \$338.4 \times 10^6$$

and for Alternative 2

$$\frac{C_f}{AFCR} = \frac{8981.3 \times 0.6493 \times 600,000 \times 8760 \times 1.25}{0.115 \times 10^6}$$

$$= \$333.2 \times 10^6$$

Finally, we determine the total evaluated cost (*TEC*) by using Eq. (3-12).We have for Alternative 1

$$TEC = 800 \times 600,000 + 338.4 \times 10^6$$

$$= \$818.4 \times 10^6$$

and for Alternative 2

$$TEC = 850 \times 600,000 + 333.2 \times 10^6$$

$$= \$843.2 \times 10^6$$

Based on the total evaluated cost, Alternative 1 is again the better economic choice. The conclusion is identical to that of Example 3-10.

SELECTED REFERENCES

1. E. Vennard, *The Electric Power Business*, McGraw-Hill, 1970.

2. K. A. Gulbrand, and P. Leung, "Power System Economics: A Sensitivity Analysis of Annual Fixed Charges," *Journal of Engineering for Power*, ASME Transaction Series A, Vol. 97, October 1975.

3. P. Leung and R. F. Durning, "Power System Economics: on Selection of Engineering Alternatives," *Journal of Engineering for Power*, ASME Transaction Series A, Vol. 100, April 1978.

4. P. Leung and L. E. Booth, "Power System Economics: On Evaluation of Availability," ASME Paper 78-JPGC-Pwr-3, 1978.

5. P. Leung and K. A. Gulbrand, "Power System Economics: An Evaluation of Plant Auxiliary System Incremental Kilowatt Consumption," *Journal of Engineering for Power*, ASME Transaction Series A, Vol. 99, July 1977.

6. B. Bornstein and P. Leung, "Nuclear Plant Turbine Cycles—A Sensitivity Analysis," ASME Paper 73-WA/Pwr-10, 1973.

7. E. L. Grant and W. G. Ireson, *Principle of Engineering Economy*, Fourth Edition, Ronald Press, 1960.

PROBLEMS

3-1. Find the present worth of $1 million paid annually for a period of 25 years if the interest rate is 15%.

3-2. What annual yearend payment will repay a present loan of one million dollars in 30 years if the interest rate is 15%.

3-3. Calculate the depreciation rate for the conditions when the economic life of equipment is 30 years and the interest rate is 15%.

3-4. Compute the levelized fuel price for a nuclear power plant. The base fuel price is $1.00 per million Btu and is expected to escalate at the constant rate of 10%. Assume that the economic life of nuclear power plant is 40 years and the interest rate is 15%.

3-5. Repeat Example 3-10 with the interest rate at 15%.

3-6. Repeat Example 3-10 by using the annual cost method.

3-7. Calculate the levelized plant capacity factor for a 50 MW gas turbine unit. Assume that the rate of return is 16%, and the load duration is expressed as below.

LOADINGS (MW)	PLANT NET HEAT RATE (BTU / KWH)	PLANT LIFE YEARS		
		1–10	11–20	21–30
		HOURS PER YEAR AT VARIOUS OUTPUT		
50.0	12,500	2000	1500	1000
37.5	13,500	200	200	100
0	—	6560	7060	7660

3-8. Calculate the weighted plant capacity factor for the gas turbine described in Problem 3-7.

3-9. Calculate the weighted plant net heat rate for the gas turbine described in Problem 3-7.

3-10. Calculate the levelized plant net heat rate for the gas turbine described in Problem 3-7. The gas turbine has an economic life 30 years. The turbine fuel cost is $3.00 per million Btu in the first year and expected to have an escalation rate 7.0%.

3-11. A 800 MW coal-fired unit has maximum initial and reheat temperature of 1000 F. It is proposed that a new alloy be used to increase these temperatures to 1100 F. With this improvement the plant net heat rate is expected to be reduced by 5% and the initial cost to be increased by $10 per kilowatt. Would you recommend use of this new alloy? The evaluation should be done on the basis of the following conditions:

The reference plant net heat rate (Btu/kWh)	9000
The referent plant construction cost ($/kW)	800
Levelized fuel cost ($/MBtu)	1.50
Levelized plant capacity factor	0.78
Annual fixed charge rate (%)	17
Economic life (yr)	35

CHAPTER FOUR

System Performance Characteristics and Selection

As indicated in Chapter 1, there are several generating systems available for electric power generation. Each of these systems has its own performance and operation characteristics. Some may have high construction cost but low operation expenses, while others may just have opposite characteristics. In this chapter these generating systems will be characterized in terms of construction cost, conversion efficiency, and other parameters. With this background information the general procedures used for system selection are briefly presented and discussed.

4.1 PERFORMANCE

The performance of generating plant can be expressed in terms of plant net heat rate, where

$$\text{Plant net heat rate } (PNHR) = \frac{\text{heat input } (I)}{\text{net kW output } (L)} \qquad (4\text{-}1)$$

expressed Btu/kWh. By definition, the plant net thermal efficiency is

$$\text{Plant net efficiency } (n_{th}) = \frac{\text{net output } (L)}{\text{heat input } (I)} \qquad (4\text{-}2)$$

and is related to the plant net heat rate by the equation

$$n_{th} = 3413/PNHR \qquad (4\text{-}3)$$

Table 4-1 indicates typical performance of various generating systems. It is seen that the plant efficiency varies in a wide range with the fossil fuel steam plant on the top and the simple gas turbine system at the bottom. The efficiencies in Table 4-1 are achieved only when the generating plants operate at a full load. In part-load operation the plants generally experience deteriorating performance. Figure 4-1 shows the variation of plant net heat rate with plant outputs. As the plant output decreases, the plant net heat rate increases. Also included in Figure 4-1 are the incremental heat rates (IR) defined as

$$IR = \frac{dI}{dL} \qquad (4\text{-}4)$$

115

Table 4-1
Typical Conversion Efficiency of Various Generating Systems

Generation Type	Unit Size (MW)	Thermal Efficiency (%)
Steam (oil)	200–800	32–40
Steam (coal)		30–38
Steam (BWR and PWR)	500–1100	31–34
Gas turbine (open cycle)	50–100	22–28
Combined gas/steam turbine	300–400	36–40
Diesel engine	10–30	27–30

With Eq. (4-1), the incremental heat rate is expressed as

$$IR = \frac{d(L \times PNHR)}{dL}$$

or

$$IR = PNHR + L\frac{d(PNHR)}{dL} \tag{4-5}$$

Equation (4-4) indicates that the incremental heat rate is the slope of the input–out-

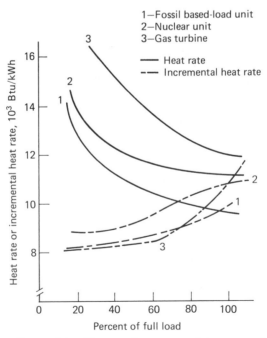

Figure 4-1. Plant net heat rate and incremental heat rate.

put curve at the given load. Physically the incremental heat rate represents the amount of energy needed to generate an additional unit of output at any given load. Unlike the plant net heat rate, the incremental heat rate will decrease as the plant net output decreases.

EXAMPLE 4-1. A 800 MW coal-fired power plant has an incremental heat rate curve defined by

Load, MW	240	400	560	680	800
IR, Btu/kWh	7650	7800	8300	8900	10,000

or by the equation

$$IR = 0.4818 \times 10^{-7}L^4 - 0.9089 \times 10^{-4}L^3 + 0.6842$$
$$\times 10^{-1}L^2 - 0.2106 \times 10^2 L + 9860$$

where IR is in Btu/kWh and L is in megawatts. Find the corresponding equation for the plant net heat rate.

Solution: Using Eq. (4-4), we have

$$dI = (IR) \, dL$$

Substituting (IR) and integrating from zero to an arbitrary load L would give

$$I = 9.636 \times 10^{-9}L^5 - 2.272 \times 10^{-5}L^4 + 2.281 \times 10^{-2}L^3$$
$$- 10.53L^2 + 9860L + I_0$$

where I is the heat input at the load L (10^3 Btu/hr), and I_0 is the heat input at zero load. Then, the plant net heat rate ($PNHR$) is

$$PNHR = \frac{I}{L}$$

$$= 9.636 \times 10^{-9}L^4 - 2.272 \times 10^{-5}L^3 + 2.281 \times 10^{-2}$$
$$\times L^2 - 10.53L + 9860 + I_0 L^{-1}$$

To determine the zero-load heat input I_0, it is best to make use of the fact that the plant net heat rate is approximately equal to the plant incremental heat rate at the full load. That is,

$$(IR)_{L=800} = (PNHR)_{L=800} = 10,000$$

This is substituted into the above equation. Then, it yields

$$I_0 = 1321 \times 10^3 \text{ Btu/hr}$$

EXAMPLE 4-2.   For the power plant described in Example 4-1 find the change in the plant input when the output increases from 600 to 700 MW.

Solution: Using Eq. (4-4), we have

$$dI = (IR)\, dL$$

Substituting (IR) and integrating from $L = 600$ to $L = 700$ MW, it gives

$$\Delta(I) = 9.636 \times 10^{-9}(700^5 - 600^5) - 2.272 \times 10^{-5}(700^4 - 600^4) + 2.281$$
$$\times 10^{-2}(700^3 - 600^3) - 10.53(700^2 - 600^2) + 9860(700 - 600)$$
$$\Delta(I) = 873{,}637 \times 10^3 \text{ Btu/hr}$$

Thus the plant input increase required is $837{,}637 \times 10^3$ Btu/hr. It is of interest to know that the calculation can be done by assuming the incremental heat rate curve as a straight line in the range of 600 to 700 MW. Then the average incremental heat rate in this range is at $L = 650$ MW. Substituting $L = 650$ into (IR) would lead to

$$(IR)_{ave} = 8718.2 \text{ Btu/kWh}$$

then, the change in the plant input is

$$\Delta(I) = \Delta(MW) \times (IR)_{ave}$$
$$= 100{,}000 \times 8718$$
$$= 871{,}800 \times 10^3 \text{ Btu/hr}$$

This value is 0.2% different from the value we get by using the integration method.

4.2 CONSTRUCTION COSTS

The construction cost of generating system mainly consists of the following expenses:

A. Planning and design.
B. Land and site preparation.
C. Building and machinery foundation.
D. Plant equipment, including all transportation fees.
E. Erection and startup.
F. Interest during the construction period.
G. Administrative work.

Figure 4-2 shows average approximate construction costs for 1985 start-up in U.S. dollars per kilowatt installed. It is obvious that the actual construction costs are likely to vary considerably with plant location, equipment origin, and system design.

In the last several years construction costs have risen considerably and will tend to rise further because of inflation and increasing demands for improving plant safety and protection of the environment.

In all types of power plants there is an economy of scale, that is, the construction costs per kilowatt decrease with unit size. Figure 4-2 indicates that the economy of scale is largest for nuclear plants. This is mainly because several structural and

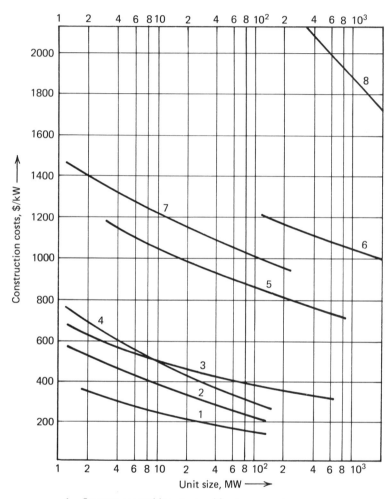

1—Jet type gas-turbine plants without recuperator.
2—Jet type gas-turbine plants with recuperator.
3—Industrial type gas turbine plants with recuperator.
4—Diesel plants for continuous service.
5—Conventional steam power plants, oil-fired.
6—Conventional steam power plants, bituminous coal-fired
  —2400 psia 1000 F/1000 F.
7—Low-pressure low-temperature nonreheat.
8—Nuclear LWR steam power plants.

Figure 4-2.   Average construction costs for 1985 startup.

Table 4-2
Components Cost Percentage for Various Generating Systems

Type of Plant	Diesel[a] Continuous Service		Steam— Conventional[b]			Nuclear— LWR[c]	
Unit Size MW	5	20	50	250	1000	600	1200
Total construction costs (1985 start-up) $/kW	600	430	950	1150	1050	2000	1840
Structures (%)	22	14	25	20	18	24	20
Steam supply system (%)	—	—	22	24	26	26	30
Turbine (diesel) and generator (%)	55	72	20	23	25	23	28
Auxiliaries and piping (%)			9	12	14	10	9
Electrical equipment (%)	14	8	12	11	10	7	6
Control equipment (%)			6	5	3	6	4
Miscellaneous (%)	9	6	6	5	4	4	3
Total (%)	100	100	100	100	100	100	100

[a] Combustion of heavy fuel oil.
[b] Coal firing, 50 MW unit for steam of 1250 psia 950 F, 250 MW with reheat, 1000 MW supercritical with double reheat.
[c] Average values of BWR and PWR plants.

equipment components such as containment, shielding, control and instrumentation cost nearly the same amount in large and small nuclear plants.

While construction cost varies from one project to another, the percentage share of the components does not change very much. Table 4-2 lists approximate percentages for three types of generating systems. All of these construction costs specified in Fig. 4-2 and Table 4-2 are based on the assumption that the once-through cooling is allowed. Use of wet cooling towers will increase these construction costs on the average of 3 to 4% and use of dry cooling towers by 12 to 15%.

The construction cost figures presented in this section are used to compare different generating systems. Before we can obtain real construction costs, we have to obtain commercial bids for equipment and calculate the costs of buildings, transport, and other expenses. Potential price escalation during construction must also be taken into consideration.

4.3 FUEL COST

Various fuels have been used for electric power generations. Table 4-3 lists typical fuel costs in 1983 dollars. It is obvious that the actual fuel costs vary considerably with plant location and with the time the fuel contract is signed. Like other commodities, fuel cost is subject to the rate of inflation.

Table 4-3
Typical Fuel Price in 1983 Dollars

Fuel Types	Fuel Price ($ / MBtu)
Nuclear	0.45
Coal	
High sulfur	1.00
Low sulfur	1.50
Natural gas	3.00
Crude oil	
High sulfur	6.00
Low sulfur	7.00
No. 2 distillate oil	7.00
No. 6 oil (Bunker C)	6.00

The fuel costs presented here are useful only for comparison. In general, the nuclear fuel cost is the lowest and does not vary much with the plant location. In contrast to nuclear fuel, the delivery cost of coal varies directly with the distance between the coal mines and plant site. In some cases the transportation cost constitutes a significant or even a major portion of delivered costs.

Of all fuels used for electric power generation, oil is the most expensive. Because of foreign sources, the oil price is subject to huge price fluctuations. The availability of fuel oil may also be questionable in the future.

4.4 OPERATION AND MAINTENANCE COSTS

Operation and maintenance (O & M) costs consist of expenses for salaries, wages, preventive maintenance, spare parts, lubricating oil, water, chemicals and services such as compressed air, fire fighting, and transport. For engineering analysis and study the O & M costs are broken down into two components. The first component is the variable component determined by the plant utilization. It can be expressed as

$$VAROMC = [A \times ENERGY + B \times SIZE \times (HROP + D \times HRSP)]/1000$$

$$(4-6)$$

where

$$VAROMC = \text{variable O \& M cost (dollars/yr)}$$
$$ENERGY = \text{energy generated (kWh/yr)}$$
$$SIZE = \text{unit rating (kW)}$$
$$HROP = \text{annual operation hours (hr)}$$
$$HRSP = \text{annual spinning hours (hr)}$$

In Eq. (4-6) the constants A, B, and D are to be determined by the existing data in the network system. Table 4-4 shows the typical values for the various generation types for the year 1980. However, these constants would vary from one company to another. They are useful only for comparison purposes.

Table 4-4
Typical Values For A, B, and D and Fixed O & M Cost

				Fixed O & M $/kW/yr (based on one unit)			
Generation Type	A (Mills / kWh)	B (Mills / kWh)	D	1 Unit Plant	2 Unit Plant	3 Unit Plant	4 Unit Plant
Steam (coal-fired)							
No scrubbers	0.640	0.602	1	15.5	10.2	7.8	6.2
With scrubbers	0.840	0.800	1	18.0	11.9	9.0	7.2
Steam (oil-fired)	0.400	0.400	1	12.0	7.9	6.0	4.8
Steam (nuclear)	0.858	0.807	1	23.5	14.1	11.8	9.4
Gas turbine	1.455	1.173	1	11.9	7.1	6.0	4.8
Combined cycle	0.640	0.602	1	15.5	9.3	7.8	6.2
Diesel engine	1.455	1.173	1	11.9	7.1	6.0	4.8

It should be noticed that the first term in Eq. (4-6) represents the O & M cost in terms of electrical energy production and the second term is related to the unit size and the number of operation hours per year.

The second component of the annual O & M costs is the fixed component. It is mainly determined by the type and size of the generating plant. In general, this fixed component constitutes 80 to 90% of the total O & M costs. It is customary to express it in terms of dollars per kW or as percentage of construction costs. Table 4-4 indicates the average values for the O & M fixed component.

4.5 AVAILABILITY AND FORCED OUTAGE RATES

In addition to the plant net heat rate, construction cost, fuel cost and O & M costs, other parameters indicating machine serviceability and dependability are needed in characterizing generating systems. These parameters include the output factor, capacity factor, service factor, availability rate, and forced outage rate. These terms are defined as follows:

$$\text{Output factor } (OF) = \frac{TG}{MDC \times SH} \times 100 \tag{4-7}$$

$$\text{Capacity factor } (CF) = \frac{TG}{MDC \times PH} \times 100 \tag{4-8}$$

$$\text{Service factor } (SF) = \frac{SH}{PH} \times 100 \tag{4-9}$$

$$\text{Availability rate } (AR) = \frac{PH - (POH + MOH + FOH)}{PH} \times 100 \tag{4-10}$$

$$\text{Forced outage rate } (FOR) = \frac{FOH}{FOH + SH} \times 100 \tag{4-11}$$

Table 4-5
Forced Outage Rates for Various Generation Types

Generation Type	Forced Outage Rate (%)
Steam (conventional)	10
Steam (nuclear)	13
Gas turbine (open cycle)	12
Combined cycle	11
Diesel engine	12

where

$$MDC = \text{maximum dependable capacity (MW)}$$

$$PH = \text{period hours (hr)}$$

$$SH = \text{service hour (hr)}$$

$$POH = \text{planned outage hours (hr)}$$

$$MOH = \text{maintenance outage hours (hr)}$$

$$FOH = \text{forced outage hours (hr)}$$

$$TG = \text{total generation (MWh)}$$

Of these parameters, the availability and forced outage rates are extremely important. They give general indication of serviceability, maintenance, and overhaul cost and reliability of the generation plant.

Table 4-5 shows the typical values of forced outage rate for various generation systems. These data are useful only for comparison purposes. To obtain an accurate value of one particular generating unit, one must collect the operating data for this type of machine and take into consideration the type of fuel, caliber of operating and maintenance personnel, and other local conditions.

EXAMPLE 4-3.   Consider a base-load unit of 800 MW in the network system in which the average energy replacement cost is 66 mills/kWh. Estimate the annual energy replacement cost if the unit has a forced outage rate of 10%. Assume that the unit has 6300 hours of service in a year.

Solution: Using Eq. (4-11), we calculate the forced outage hour as

$$0.1 = \frac{FOH}{FOH + 6300}$$

or

$$FOH = 700 \text{ hours}$$

Then, the annual energy replacement cost is

$$\text{Cost} = (\text{unit size})(FOH)(\text{energy replacement unit cost})$$

$$= (800{,}000)(700)(66/1000)$$

$$= \$36.96 \times 10^6/\text{year}$$

Evidently, the forced outage rate has a significant impact on the unit economic performance. If the unit in Example 4-3 improves its FOR by one percent, the annual energy replacement cost will be cut down by approximately $4 million. In general, an improvement of forced outage rate in the base-load coal-fired or nuclear units will reduce the oil and gas consumption used in peaking units and permit postponement of capital investment in new plants.

4.6 GENERATION MIX

Because of difference in load variation and system economics, each utility company's generating facilities will have a different composition. The composition of generating facilities is usually referred to as a generation mix, which is expressed in terms of output percentage for each type of generation. The company ABC located in New England area, for instance, may have 30% nuclear, 30% oil-fired, 20% coal-fired, and 20% gas turbine system. It must be pointed out that the system generation mix does not remain unchanged forever. In fact, it varies with the projected fuel cost, capital cost, and other parameters. This is also true for the generation mix in the country. The following table shows the composite generation mix for the United States over a period of recent years [10].

Year	Installed Capacity (Millions of kW)	Distribution (%)				
		Hydro	Steam (Conventional)	Gas Turbine	Steam (Nuclear)	Internal Combustion
1975	508	13.0	69.4	8.7	7.9	1.0
1976	531	12.8	69.3	8.9	8.8	1.0
1977	560	12.3	69.3	8.6	8.9	0.9
1978	579	12.3	69.1	8.5	9.3	1.0
1979	598	12.5	68.9	8.5	9.2	0.9
1980	614	12.4	69.2	8.3	9.1	1.0
1981	635	12.1	69.3	8.0	9.7	0.9

The generation mix of utility systems is usually dictated by the economic conditions under which it is operated. These can be best illustrated by the following example.

EXAMPLE 4-4. A power company is planning to construct a nominal 600 MW generating unit. The anticipated capacity factor is approximately 62%. The management has to make the decision on the type of generating systems, which include simple gas turbine, combined cycle, coal-fired and nuclear plant. Using the information tabulated below and assuming the annual fixed charge rate of 15%, determine the total annual cost (fuel and fixed charges) in terms of dollars per kilowatt and per year.

Type of Plant	Capital Cost ($ / kW)	Plant Net Heat Rate (Btu / kWh)	Levelized Fuel Cost ($ / MBtu)
Combined cycle	562	9,200	6.00
Simple gas turbine cycle	350	14,000	6.00
Coal-fired plant	1,100	9,600	1.10
Nuclear plant	2,000	10,000	0.45

Solution: The equation for this calculation is the total annual cost (TAC)

$$TAC = (\text{unit capital cost})(\text{annual fixed charge rate})$$

$$+ (8760)(\text{lifetime capacity factor})(\text{plant net heat rate})$$

$$\times (\text{levelized fuel cost}) \times 10^{-6}$$

For combined cycle, the total annual cost is

$$TAC = (562)(0.15) + (8760)(0.62)(9200)(6.00)(10^{-6})$$

$$= 384.1 \ \$/kW/yr$$

Similarly, for the simple gas turbine cycle

$$TAC = (350)(0.15) + (8760)(0.62)(14000)(6.00)(10^{-6})$$

$$= 508.7 \ \$/kW/yr$$

for coal-fired plant

$$TAC = (1100)(0.15) + (8760)(0.62)(9600)(1.10)(10^{-6})$$

$$= 222.3 \ \$/kW/yr$$

and for nuclear plant

$$TAC = (2000)(0.15) + (8760)(0.62)(10000)(0.45)(10^{-6})$$

$$= 324.4 \ \$/kW/yr$$

In summary, the economic evaluation indicates that the coal-fired plant leads all others under consideration.

It should be pointed out that in the real-world operation, other factors such as public acceptance and availability of financial resources must be taken into account when the type of generating plant is decided. It is evident that the plant lifetime capacity factor would play an important role in the plant selection. For a low capacity factor, the plant of less capital cost would have better chance to be selected. In Example 4-4 the gas turbine plant will have a lower annual cost than the coal-fired plant if the lifetime capacity factor decreases to 0.15.

The annual fixed charge rate is also an important factor in the plant-type selection. The annual fixed charge rate represents the annual cost of capital investment. For a high fixed charge rate, the plant of high capital cost would have less chance to be accepted. Using Example 4-4 again and only changing the annual fixed charge rate to 20%, we see that the nuclear plant will not be as favorable as the combined-cycle plant.

4.7 ECONOMIC SCHEDULING PRINCIPLE

The utility system always has more than one generating unit. Proper distribution of load among the generating units is a problem frequently encountered by engineers. If the load is not properly distributed, it will result in a decrease of the thermal efficiency as a whole. To select the generating unit properly for the utility network, one must consider not only the performance and operating characteristics, but also the economic impacts on the network. In other words, new units should be so selected that the total heat input to the network system will be minimized.

The economic scheduling principle presented in this section is established on the basis of an equal incremental heat rate. According to this principle, the load is so distributed that at any moment all generating units will have the same incremental heat rates. As the system load increases, the incremental heat rate at which the units operate will increase. In the following discussion this principle is illustrated by the example involving the two generating units in the network.

Let I_c be the combined input to units 1 and 2, and L_c be the combined output of units 1 and 2. Evidently, the combined input I_c is a function of either L_1 or L_2 for a given combined output L_c. When the combined input I_c is at a minimum, the following equation must hold.

$$\frac{dI_c}{dL_1} = 0 \qquad (4\text{-}12)$$

Since

$$I_c = I_1 + I_2 \qquad (4\text{-}13)$$

we have

$$\frac{dI_1}{dL_1} + \frac{dI_2}{dL_1} = 0 \qquad (4\text{-}14)$$

In the next step, we express the term dI_2/dL_1 as follows:

$$\frac{dI_2}{dL_1} = \frac{dI_2}{dL_2} \times \frac{dL_2}{dL_1} \qquad (4\text{-}15)$$

Since

$$L_c = L_2 + L_1 \qquad (4\text{-}16)$$

we have

$$\frac{dL_2}{dL_1} = -1 \qquad (4\text{-}17)$$

Substituting Eq. (4-17) into Eq. (4-15) gives

$$\frac{dI_2}{dL_1} = -\frac{dI_2}{dL_2} \qquad (4\text{-}18)$$

Combining Eq. (4-14) and Eq. (4-18), we finally have

$$\frac{dI_1}{dL_1} = \frac{dI_2}{dL_2} \qquad (4\text{-}19)$$

Thus, the combined input I_c is a minimum only if the incremental heat rate of unit 1 is equal to that of unit 2. This principle can be extended to the network of multiple units.

Figure 4-3 represents the input-output curves of two generating units that are to operate in parallel and to supply a common load. These curves are rather complicated and involve sudden changes of slope at the initiation of value openings. For most practical purposes it is sufficiently accurate to simplify the input-output curves by connecting the maximum valve-opening points by straight lines as shown in Fig. 4-3. The corresponding incremental heat rate curves are shown in Fig. 4-4. As expected, they are of step-function curves. The vertical sections of these curves have no physical meanings and merely connect two incremental heat rates of the same generating units.

In dividing the load between these units to achieve maximum fuel economy, we should utilize the principle of equal incremental heat rate. The loading procedures are as follows: starting at zero load, the turbine B picks up the load up to L_{B1} while the turbine A remains at zero load. When combined load is greater than L_{B1}, the turbine A starts to pick up load until it reaches L_{A1}. At the combined load $L_{A1} + L_{B1}$, the second valve of turbine B starts to open. As the load increases, the turbine B picks up the additional load while the turbine A remains at the load L_{A1}. For any increase beyond $L_{A1} + L_{B2}$, the turbine A starts to open the second valve and later the third valve. At the combined load $L_{A3} + L_{B2}$, the turbine A would be

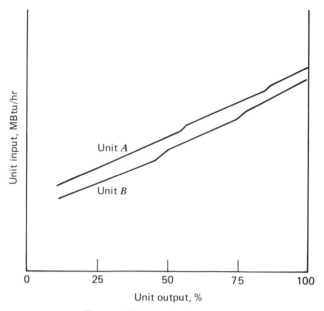

Figure 4-3.   Input–output curves.

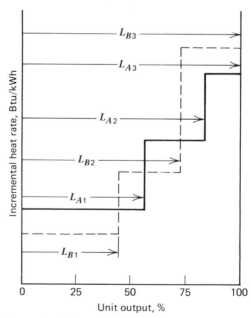

Figure 4-4.   Incremental heat rate curves for units *A* and *B*.

at a full load L_{A3} and the turbine B at the load L_{B2}. If the combined load increases beyond this level, the turbine B starts to open the last valve to meet the additional demand.

In practice, the principle of equal incremental heat rate is somewhat modified. When various fuels are used in the network system, the fuel costs must be taken into account. In other words, the load distribution should be made on the basis of equal incremental fuel cost. In addition, consideration must be also given to the system reserve capacity and operation characteristics of generating units.

EXAMPLE 4-5. Two coal-fired generating units A and B have the incremental heat rate defined by

$$(IR)_A = 0.4818 \times 10^{-7}L_A^4 - 0.9089 \times 10^{-4}L_A^3 + 0.6842$$

$$\times 10^{-1}L_A^2 - 0.2106 \times 10^2 L_A + 9860$$

for unit A and

$$(IR)_B = 0.9592 \times 10^{-7}L_B^4 - 0.7811 \times 10^{-4}L_B^3 + 0.2625$$

$$\times 10^{-1}L_B^2 - 0.2189 \times 10 L_B + 9003$$

for unit B; (IR) is in Btu/kWh and L is in megawatts. Determine the incremental heat rate at which the combined output of these two units is 1000 MW.

Solution: In addition to the two (IR) equations, the equation below must be also satisfied

$$L = L_A + L_B$$

where

$$L_A = \text{load of unit } A$$

$$L_B = \text{load of unit } B$$

$$L = \text{combined load of units } A \text{ and } B$$

To determine the incremental heat rate $IR = (IR)_A = (IR)_B$, we solve these three equations by the numerical method such as the Newton-Raphson algorithm. For simplicity, the results are simply presented as

$$L_A = 732.5 \text{ MW}$$

$$L_B = 267.5 \text{ MW}$$

$$(IR) = (IR)_A = (IR)_B = 9292 \text{ Btu/kWh}$$

The calculation details will be presented later in Chapter 10.

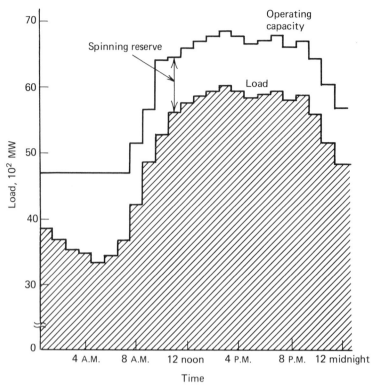

Figure 4-5. System load and spinning reserve.

Table 4-6
Generating Units Information

Unit No.	Unit Name	Fuel Type[a]	Rated Output (MW)	Minimum Load (MW)	Fuel Cost ($ / MBtu)	No Load Input (MBtu / hr)
1	Fbase1	F	800	240	1.76	800.0
2	Fbase2	F	750	225	1.76	735.0
3	Fbase3	F	750	225	1.76	712.5
4	Fbase4	F	800	240	1.76	704.0
5	Fcycle1	F	500	50	1.76	802.5
6	Fcycle2	F	450	50	1.76	675.0
7	Nuclear1	N	800	320	0.62	872.0
8	Nuclear2	N	800	320	0.62	792.0
9	Ccycle1	G	400	40	3.92	582.0
10	Ccycle2	G	350	35	3.92	456.7
11	Gas1	G	75	7.5	3.92	274.5
12	Gas2	G	75	7.5	3.92	274.5
13	Gas3	G	50	5	3.92	180.0
14	Gas4	G	75	7.5	3.92	265.5
15	Gas5	G	50	5	3.92	177.0
16	Gas6	G	50	5	3.92	175.5
17	Gas7	G	50	5	3.92	171.0

[a] F = fossil; N = nuclear; G = gas or oil.

130

Table 4-7
Summary of Incremental Heat Rates

Unit No.	Unit Name	Rated Load (MW)	Plant Incremental Heat Rate Information[a]				
1	Fbase1	800	30	50	70	85	100
			7650	7800	8300	8900	10000
2	Fbase2	750	30	50	70	85	100
			7450	7600	8100	8700	9800
3	Fbase 3	750	30	50	70	85	100
			7150	7300	7800	8400	9500
4	Fbase 4	800	30	50	70	85	100
			6450	6600	7100	7700	8800
5	Fcycle 1	500	10	30	50	75	100
			8950	9050	9250	9650	10700
6	Fcycle 2	450	10	30	50	75	100
			8250	8350	8550	8950	10000
7	Nuclear 1	800	40	55	70	90	100
			9000	9600	10200	10750	10900
8	Nuclear 2	800	40	55	70	90	100
			8000	8600	9200	9750	9900
9	Ccycle 1	400	10	30	50	75	100
			7350	7400	7500	8200	8700
10	Ccycle 2	350	10	30	50	75	100
			6350	6400	6500	7200	8700
11	Gas 1	75	5	30	55	80	100
			8550	8600	8850	10150	12200
12	Gas 2	75	5	30	55	80	100
			8550	8600	8850	10150	12200
13	Gas 3	50	5	30	55	80	100
			8350	8400	8650	9950	12000
			5	30	55	80	100

Table 4-7—Continued
Summary of Incremental Heat Rates

Unit No.	Unit Name	Rated Load (MW)	Plant Incremental Heat Rate Information[a]				
14	Gas 4	75					
			8150	8200	8450	9750	11800
			5	30	55	80	100
15	Gas 5	50					
			8150	8200	8450	9750	11800
			5	30	55	80	100
16	Gas 6	50					
			8050	8100	8350	9650	11700
			5	30	55	80	100
17	Gas 7	50					
			7750	7800	8050	9350	11400

[a] The numbers at the top of the table indicate the incremental heat rate (Btu/kWh) at the corresponding percentage of full load.

4.8 LOAD DISTRIBUTION

In a utility system with many generating units, decision is often needed to determine the order in which the generating units are dispatched into operation. In practice, these are usually done according to the unit relative efficiency. To maintain reasonable continuity of service, the generating capacity in operation must be greater than the system load. The difference between these two is defined as the spinning reserve. The minimum magnitude of spinning reserve varies from one utility to another, but is usually equal or greater than the capacity of the largest unit in operation. To minimize the total system fuel cost, the principle of equal incremental fuel cost as presented previously must be observed in distributing the system load among the units. In addition, certain operation constraints such as minimum load and zero-load heat input should be satisfied.

For example, a hypothetical utility system has a peaking demand 6000 MW and a typical daily load curve as shown in Fig. 4-5. The load factor and valley-to-peak ratio are approximately 0.829 and 0.56, respectively. The utility system has 17 generating units with a total capacity of 6825 MW. The composition of these 17 units is two nuclear, six fossil-fuel, two combined cycle, and seven gas turbines. Detailed unit information including rated output, fuel cost, minimum load, and zero-load heat input are presented in Table 4-6. Incremental heat rates are summarized in Table 4-7.

For these conditions, the load distribution is shown in Table 4-8. It is seen that not all generating units are always in operation. When the peaking demand occurs at

Table 4-8
Utility System Load Distribution in a 24-Hour Period

Hours	Load (MW)	Capacity in Operation (MW)	Spinning Reserve (MW)	1	2	3	4	5	6	7	8	9	10	11	12	13	14	15	16	17
										Load Distribution (MW)										
0100	3888	4700	812	483	513	574	719													
0200	3696	4700	1004	416	450	534	696													
0300	3552	4700	1148	341	418	508	685													
0400	3504	4700	1196	312	410	500	682													
0500	3360	4700	1340	240	382	472	665													
0600	3456	4700	1244	283	402	492	678													
0700	3696	4700	1004	416	450	534	696													
0800	4224	5150	926	566	568	624	749		117											
0900	4848	5650	802	687	664	695	800	63	339											
1000	5280	6400	1120	737	711	742	800	303	388	800	800	s	s							
1100	5616	6450	834	800	750	750	800	466	450	800	800	s	s							
1200	5760	6575	815	800	750	750	800	500	450	800	800	s	110								s
1300	5856	6675	819	800	750	750	800	500	450	800	800	s	206			s	s	s	s	s
1400	5904	6750	846	800	750	750	800	500	450	800	800	s	254			s	s	s	s	s
1500	6000	6825	825	800	750	750	800	500	450	800	800	77	273	s	s	s	s	s	s	s
1600	5904	6750	846	800	750	750	800	500	450	800	800	s	254	s		s	s	s	s	s
1700	5808	6675	817	800	750	750	800	500	450	800	800	s	158	s		s	s	s	s	s
1800	5856	6675	819	800	750	750	800	500	450	800	800	s	206			s	s	s	s	s
1900	5904	6750	846	800	750	750	800	500	450	800	800	s	254	s		s	s	s	s	s
2000	5760	6575	815	800	750	750	800	500	450	800	800	s	110				s	s	s	s
2100	5856	6675	819	800	750	750	800	500	450	800	800	s	206			s	s	s	s	s
2200	5568	6400	832	800	750	750	800	500	450	800	800	s	s			s	s	s	s	s
2300	5136	6000	864	716	692	722	800	418	376	800	800									
2400	4800	5650	850	680	658	689	800	239	323	800	800									

Note: "s" denotes the spinning reserve.

3:00 p.m., the total capacity in operation is 6825 MW. The variation of generating capacity in this 24-hour period is shown in Fig. 4-5.

SELECTED REFERENCES

1. B. G. A. Skrotzki and W. A. Vopat, *Power Station Engineering and Economy*, McGraw-Hill, 1960.

2. A. P. Fraas, *Engineering Evaluation of Energy Systems*, McGraw-Hill, 1981.

3. A. P. Priddy and J. J. Sullivan, "Engineering Considerations of Combined Cycle," Proceedings of the American Power Conference, 1972.

4. P. G. Hill, *Power Generation*, MIT Press, 1977.

5. L. K. Kirchmayer, *Economic Operation of Power Systems*, Wiley, 1967.

6. R. L. Sullivan, *Power System Planning*, McGraw-Hill, 1977.

7. K. W. Li and D. S. Lo, "Simulation of Energy Storage of Utility Networks," *Computer in Engineering 1982*, ASME publication, 1982.

8. U.S. Department of Energy, "Steam-Electric Plant Construction Cost and Annual Production Expenses," January 1978.

9. F. S. Aschner, *Planning Fundamentals of Thermal Power Plants*, Israel University Press, 1978.

10. U.S. Bureau of Census, "Statistical Abstract of the United States 1982–84," Washington, D.C., 1982.

PROBLEMS

4-1. The input-output equation of a coal-fired power plant is generally expressed by

$$I = a + b(L) + c(L)^2 + d(L)^3 + e(L)^4$$

where I is in MBtu/hr and L is in kilowatts. Derive the corresponding expressions for the plant net heat rate and incremental heat rate.

4-2. A 400 MW combined-cycle plant has an input-output curve expressed by

$$I = 1.9184 \times 10^{-8}L^5 - 1.9528 \times 10^{-5}L^4 + 8.7467 \times 10^{-3}L^3$$
$$- 1.0945L^2 + 9003L + 650$$

where I is in 10^3 Btu/hr and L is in megawatts. Obtain the corresponding expression for the plant incremental heat rate.

4-3. A 800 MW coal-fired power plant has the plant net heat rate curve defined by

$$PNHR = 0.5330 \times 10^{-7}L^4 - 0.1396 \times 10^{-3}L^3 + 0.1439L^2$$
$$- 0.7089 \times 10^2 L + 0.2427 \times 10^5$$

where $PNHR$ is in Btu/kWh and L is in megawatts. Derive the input-output and incremental heat rate equations.

4-4. For the power plant described in Problem 4-3, calculate the plant heat input at the loads, 600 and 700 MW. Compare the change of heat input with that obtained by integrating the incremental heat rate between 600 and 700 MW.

4-5. Two generating units have the incremental heat rates expressed by

$$IR = 0.2368 \times 10^{-7}L^4 - 0.6226 \times 10^{-4}L^3 + 0.5432$$

$$\times 10^{-1}L^2 - 0.1443 \times 10^2L + 9845 \qquad \text{if} \qquad 320 \le L \le 800 \text{ MW}$$

for unit A and

$$IR = -0.2196 \times 10^{-6}L^4 + 0.2528 \times 10^{-3}L^3 - 0.6672$$

$$\times 10^{-1}L^2 + 0.6604 \times 10L + 7177 \qquad \text{if} \qquad 40 \le L \le 400 \text{ MW}$$

for unit B, where IR is in Btu/kWh, and L is in megawatts. Using the principle of equal incremental heat rate, determine the incremental heat rate at which the combined output is 800 MW.

4-6. The two generating units of which the incremental heat rates are given in Problem 4-5, have fuel cost respectively \$0.8/MBtu and \$1.50/MBtu. Calculate the individual unit output at the time the combined output must be 800 MW. Compare the results with those obtained in Problem 4-5.

4-7. The third unit is added to those two described in Problem 4-5. The third unit has the equation below for its incremental heat rate.

$$IR = -0.1675 \times 10^{-3}L^4 + 0.4285 \times 10^{-1}L^3 - 0.2040$$

$$\times 10L^2 + 0.3260 \times 10^2L + 8454 \qquad \text{if} \qquad 40 \le L \le 75 \text{ MW}$$

where (IR) is in Btu/kWh and L is in megawatts. Again determine the incremental heat rate at which the combined output is 800 MW.

4-8. The information summarized below have been assembled for comparison of two proposed generating units (each 800 MW). These units have the incremental heat rate equations:

$$IR = 0.2368 \times 10^{-7}L^4 - 0.6226 \times 10^{-4}L^3 + 0.5432$$

$$\times 10^{-1}L^2 - 0.1443 \times 10^2L + 9845$$

for the nuclear unit and

$$IR = 0.4818 \times 10^{-7}L^4 - 0.9089 \times 10^{-4}L^3 + 0.6842 \times 10^{-1}L^2$$

$$-0.2106 \times 10^2L + 9860$$

for the coal-fired unit. IR is in Btu/kWh and L is in megawatts. The loading schedules for these two units are the same and expressed by

LOAD (MW)	HOURS
800	2500
600	2000
400	1500

The nuclear and coal-fired units have, respectively, an installation cost of $1000 and $750 per kW. The levelized fuel costs are $0.80 per million Btu for the nuclear unit and $1.80 per million Btu for the coal-fired unit. What is the economic choice if the annual fixed-charge rate is 10%?

4-9. Repeat Problem 4-8 with the annual fixed-charge rate changed to 18%.

4-10. A 75 MW gas turbine is connected with an utility network system. The daily load schedule for this unit is

TIME	LOAD (MW)
0100–1300	0
1300–1500	75
1500–1600	60
1600–2400	0

The unit has the incremental heat rate curve defined by:

Load (MW)	3.75	22.5	41.25	60	75
IR (Btu/kWh)	8550	8600	8850	10150	12200

Calculate the daily fuel consumption.

4-11. A 800 MW unit is needed for a base-load operation; a decision must be made whether a coal-fired or nuclear unit should be selected. Using the following parameters, estimate the generation cost in terms of mills per kWh.

PARAMETERS	COAL-FIRED UNIT	NUCLEAR UNIT
Total capital cost ($/kW)	800	1000
Plant capacity factor	0.7	0.8
Fuel cost ($/MBtu)	1.50	0.80
Rate of return	11	11
Economic life	35	40
Operation and maintenance cost (mills/kWh)	3	5
Plant net heat rate (Btu/kWh)	9300	11,500

4-12. Calculate the generation cost in terms of mills per kWh for the nuclear, coal-fired, and gas turbine units under the conditions listed below:

	NUCLEAR	COAL-FIRED	GAS TURBINE
Capital cost ($/kW)	1000	800	250
Fuel cost ($/MBtu)	0.80	1.50	3.20
Annual fixed charge rate	16	16	16
O & M cost ($/kW-yr)	13	13	12
Plant net heat rate (Btu/kWh)	11,500	9300	13,300

Compare these generating units using various plant capacity factors, say 0.1, 0.3, 0.5, and 0.8.

4-13. Repeat Problem 4-12 for the condition when all fuel prices increase by 20%.

4-14. Improving operating availability of a base-load unit will reduce the energy replacement costs and system reserve capacity. Consider a base-load unit of 1100 MW and estimate the present worth of the lifetime benefit derived from a 1.0% improvement of the unit availability. The following conditions should be used for evaluation:

Levelized energy replacement cost	26 mills/kWh
Unit economic life	45 year
Rate of return	12%
Levelized unit capacity factor	0.68
Levelized reserve capacity cost	$500/kW

CHAPTER FIVE

Steam Generation Systems

5.1 INTRODUCTION

The steam generating system, frequently called the boiler, is a system that transfers the heat from the products of combustion to water and produces hot water or steam. The combustion is accomplished in a furnace. Heat is transferred in the furnace mainly by radiation to water-cooled walls, which constitutes the evaporation section of the steam generation system. After leaving the furnace, the gases pass through a superheater in which steam receives heat from the gases and has its temperature raised above the saturation temperature. Since the temperature of the gases leaving the superheater section is still high, modern steam generators often employ additional heat transfer surfaces to utilize the thermal energy of the gases. These include the surfaces of reheaters, economizers, and air-preheaters.

Boilers may be classified into three categories according to their applications. These include industrial, marine, and central electric power station. Generally, the industrial boilers produce saturated steam or hot water with flow rates up to 50,000 lb/hr. The pressure condition is frequently 300 psia or lower. The marine boilers are much larger and usually produce superheated steam at the conditions around 900 psia and 1000 F. The boilers for electric power generation stations are quite different in terms of steam conditions and generation rates. These boilers can produce steam at the rate up to several million pounds per hour. The steam pressure may be either supercritical or subcritical and the temperature is frequently around 1000 F. In this chapter attention is only given to the boilers used for electric power generation.

Boilers may also be classified according to the relative positions of products of combustion. In one type boiler, called the fire-tube boiler, the products of combustion flow through tubes surrounded by water. This type of boiler is frequently used in most steam locomotives, in small factories, and sometimes in heating buildings. In another type of boiler, called the water-tube boiler, the products of combustion flow over water-filled tubes. Both ends of the water tubes are connected to the headers or the boiler drums. In the drum the steam is separated from the saturated water. Then, the saturated steam usually goes to the superheater in which the steam temperature is increased. All high-pressure and large boilers are of the water-tube type. The small tubes in the water-tube boiler can withstand high pressure better than the large drums of a fire-tube boiler.

Boilers are operated by firing various fuels. These fuels include bituminous coal, lignite, natural gas, and oil. Different fuels result in different boiler designs and operations. In the United States coal is the most prevalent fuel used in central electric power stations.

138

Use of steam for electric power generation in this country did not start until the year 1881. In that year the Brush Electric Light Company in Philadelphia started to generate steam from four 73 hp boilers. In 1903, the Commonwealth Edison Company became the first utility to run steam turbines exclusively for electric power generation. In the Commonwealth Edison plant, 96 boilers, each rated at 508 hp, were installed and used to supply the turbines with the steam at 170 psia and 434 F. In the last several decades progress has been made in steam generator development. Like turbine development, the progress is mainly in the areas of steam conditions and unit size. At the present most units generate steam at 2400 psia, 1000 F or 3500 psia, 1000 F. To attain high system efficiency, the steam generator usually consists of the evaporation section, superheaters, reheaters, economizers, and air preheaters. In power plant system design one steam generator is frequently used to match one turbine unit. Because of this, steam generator unit size increases as turbine unit size increases. For a 800 MW plant, a single steam generator produces almost 6 million pounds of steam per hour.

5.2 BOILER ARRANGEMENTS

All power station boilers are of the water-tube type. Water circulates within the tubes and partially becomes steam as it receives heat from the products of combustion. When water circulation within the boiler takes place due to its own density difference, it is called the natural-circulation boiler. In this type of boiler, water from the boiler drum first flows downward to the bottom of the heated evaporative tubes through several pipes (frequently called downcomers). Then, the water reverses its flow direction and returns to the drum as it receives the heat from the furnace. Since the evaporative tubes (frequently called risers) contain a mixture of steam and water, the average density in the riser is always lower than that in the downcomer. This density difference gives rise to a driving force that will overcome all friction in the water-steam circiut. Figure 5-1 shows a schematic diagram of water-tube boiler operating on the natural circulation principle. Natural circulation is a simple and efficient technique and is frequently employed in boiler designs.

As the boiler pressure becomes higher and higher, the difference in density of the fluid between the downcomers and the risers will become less and less. At a certain boiler pressure, the driving force, which is proportional to the density difference, is not sufficiently large to balance the frictional resistance. One alternative is to employ pumps to force the water through the evaporative tubes. The boiler using circulation pumps is called the forced circulation boiler. Figure 5-1 shows a schematic diagram of forced circulation water-tube boiler. It is seen that the circulation pumps take the water from the drum and supply it to the headers at the bottom of the boiler. From the headers water moves upward as it receives heat from the products of combustion. Because sufficient driving force is available, smaller diameter tubes can be used in the forced-circulation boiler. Furthermore, it is possible to apply an orifice to each tube so that more uniform flow and tube temperature can be achieved. These advantages frequently offset the cost of circulation pumps and their pumping power. Similar to the forced-circulation boiler is the once-through boiler shown in Fig. 5-1.

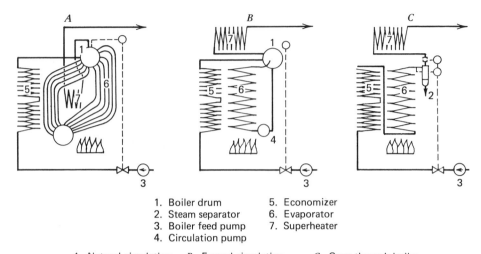

1. Boiler drum   5. Economizer
2. Steam separator   6. Evaporator
3. Boiler feed pump   7. Superheater
4. Circulation pump

A. Natural circulation   *B.* Forced circulation   *C.* Once-through boiler

Figure 5-1.   Boiler basic arrangements [7].

It is seen that there is no boiler drum. Water flows through the evaporation section without any recirculation. This arrangement is frequently employed when the steam pressure in the boiler is supercritical.

In the three boiler arrangements just introduced, each has its own economizer, evaporator section and superheater. Not shown in the diagrams is the reheater and air preheater, which are usually employed in modern boiler design. The economizer is a heat exchanger used to increase feedwater temperature. The evaporation section, which usually surrounds the boiler furnace, is to produce saturated steam and supply it to the superheater. In the superheater the steam is further heated and has its temperature raised to the level above the saturation temperature. Then, the super-heated steam flows to the turbine-generator throttle for power production. The reheater, when included in a steam generator, is usually installed in the location adjacent to the superheater. The reheater receives the steam from the high-pressure turbine after the steam partially expands. In the reheater, steam absorbs heat from the products of combustion and has its own temperature increased. Usually, the outlet temperature is identical to the temperature of the steam leaving the super-heater. To maintain high furnace temperature and boiler efficiency, an air preheater is frequently employed in boiler design. It is usually installed in the location just before the hot gases leave the steam generation system. More discussion on these components will be presented later.

The products of combustion are generated in the boiler furnace. The hot gases first transfer heat to the evaporation section by radiation and convection. Then, these gases exit the furnace and enter the superheater and the reheater zone. In these zones the gases further transfer heat away. The basic heat transfer mechanisms are still convection and radiation. Next along the gas path is the economizer. In the economizer heat is transferred to the feedwater from the gases. Because of the low temperature in the products of combustion, convective heat transfer is the prevalent mode. In the air preheater, the gas temperature is further reduced. The lower the gas temperature, the higher the boiler efficiency will be. However, the gas temperature should not be lower than the dew point of water vapor in the gases. Any water condensation will give rise to a formation of liquid acid, which results in a corrosion of the air heater surfaces.

Figure 5-2 illustrates the design of a typical water-tube boiler with natural circulation. For this type of boiler the capacity varies from 300,000 lb/hr to 7,000,000 lb/hr. The steam conditions are usually subcritical with throttle pressure 1800 to 2520 psia and the temperature around 1000 F. The boiler can use coal, lignite, oil and natural gas as the fuel. In case of burning coal or lignite, the boiler firing equipment is either a pulverizer-burner system or a cyclone furnace. Usually the boiler is completely automatic, including combustion, steam temperature, and feedwater flow.

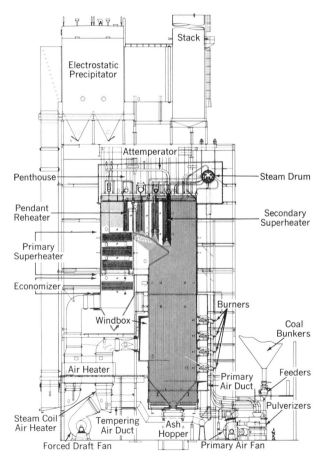

Figure 5-2. Radiant boiler for pulverized coal. (From *Steam/Its Generation and Use*, 1972)

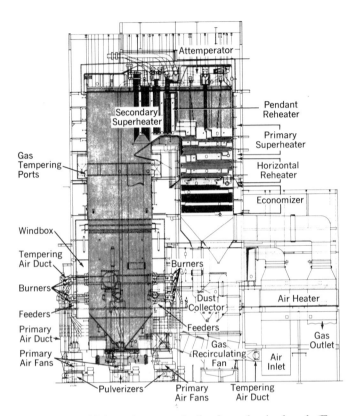

Figure 5-3. Universal-pressure boiler for pulverized coal. (From *Steam/Its Generation and Use*, 1972)

The design of a typical once-through boiler is illustrated in Fig. 5-3. This type of boiler is usually applied to a large turbine-generator unit size. The generating capacity can exceed 10,000,000 lb/hr. When the subcritical steam is generated, the conditions are usually 2400 or 2520 psia for throttle pressure and 1000 F for throttle temperature. For supercritical steam, the throttle pressure is 3500 psia or higher. Like the natural-circulation boiler, this type of boiler can burn coal, lignite, oil, and natural gas. In a once-through boiler the feedwater pump speed and turbine throttle are used to control the steam flow and steam pressure. Steam temperature is controlled by the fuel firing rate and the gas tempering. The temperature of steam leaving the reheater is also important. It is frequently controlled by gas-recirculation and/or attemperation. A further discussion of temperature control is presented later in this chapter.

Most central station boilers are equipped with air pollution control systems. These often include an efficient precipitator and sometimes an SO_2 removal system. In addition, sufficient stack height is frequently used to ensure an acceptable level of pollution concentration in the plant's surroundings.

5-3 BOILER FUEL CONSUMPTION AND EFFICIENCY

Energy is liberated within the boiler furnace by the chemical reaction of oxygen with the combustible elements of the fuel. For any fuel there is a minimum quantity of oxygen required for complete combustion. The amount of air that contains this minimum quantity of oxygen is termed theoretical air. The theoretical air varies according to the nature of the fuel. When the ultimate analysis of fuel is available, the theoretical air is calculated by

$$W_{ta} = 11.53\ C + 34.36\left(H - \frac{O}{8}\right) + 4.32\ S \qquad (5\text{-}1)$$

where the symbols C, H, O, S are weight fraction of the elements* and W_{ta} is the theoretical air in terms of pound per pound of fuel. Nitrogen and ash in the fuel are inert and thus do not enter into the combustion process. When the ultimate analysis of fuel is absent, the approximate relationship

$$W_{ta} = \frac{(a)\,(HHV)}{10,000} \qquad (5\text{-}2)$$

may be used for rough estimates. It is seen that the theoretical air is linearly proportional to the high heating value of the fuel. The constant a in this equation usually varies with the fuel. For instance, it is 7.6 for bituminous coal, 7.25 for lignite, and 7.45 for residual oil.

   Since it is impossible to achieve a complete combustion with theoretical air, additional amount of air must be supplied. The amount of excess air depends on many factors including the fuel characteristics, the type of burner, and the design of furnace. For a coal-fired steam generator the excess air varies in the range of 12 to 25%. The definition of excess air percentage is

$$\text{Excess air percentage} = \frac{W_a - W_{ta}}{W_{ta}} \times 100 \qquad (5\text{-}3)$$

where W_a = actual air (total air), pounds per pound of fuel. The total air for combustion is the sum of the theoretical air and the excess air and is expressed in pounds per pound of fuel. When the fuel consumption is determined, the total amount of air required for combustion can be easily calculated.

   The products of combustion are important in boiler design. They contain the components such as CO_2, CO, O_2, H_2O, and SO_2. The amount of SO_2 is relatively small and usually neglected in measurements. The gas analyzer determines the composition of gases on a dry basis. The readings include CO_2, CO, O_2, and N_2 in a volume percentage. The total gas weight is the sum of the total air and the fuel minus the ash content. The equation for the gas weight in pounds per pound of fuel

---

In this chapter C, H, O, N, S, A, W are used to represent the weight fraction of the elements as given in the fuel ultimate analysis.

burned is

$$W_g = W_a + (1 - A) \qquad (5\text{-}4)$$

where the symbol A is the weight fraction of the ash as given in the fuel ultimate analysis. Both total air (W_a) and gas weight (W_g) are needed to determine boiler fan capacities.

EXAMPLE 5-1. A bituminous coal has the following ultimate analysis:

$C = 71.6$	$N = 1.3$	$A = 9.1$
$H = 4.8$	$S = 3.4$	$W = 3.5$
$O = 6.3$		

Calculate the total air required for complete combustion per pound of coal. The excess air is assumed 20%.

Solution: First, the theoretical air is calculated by Eq. (5-1). It gives

$$W_{ta} = 11.53(0.716) + 34.36\left(0.048 - \frac{0.063}{8}\right) + 4.32(0.034)$$

$$W_{ta} = 9.78 \text{ lb/lb of coal}$$

Second, Eq. (5-3) is used for calculation of total air. That is,

$$0.20 = \frac{W_a - 9.78}{9.78}$$

$$W_a = 11.74 \text{ lb/lb of coal}$$

Boiler fuel comsumption is closely related to its output. In addition, it also depends on the heating value of the fuel and the boiler efficiency. The equation for fuel comsumption is

$$W_f = \frac{1}{HHV \times \eta_b}[m_s(h_2 - h_1) + m_r(h_4 - h_3) + m_b(h_s - h_1)] \qquad (5\text{-}5)$$

where

$W_f =$ lb of fuel/hr

$m_s =$ steam flow rate (lb/hr)

$m_r =$ reheat steam flow rate (lb/hr)

$m_b =$ boiler blowdown (lb/hr)

$\eta_b =$ boiler efficiency

HHV = high heating value (Btu/lb of fuel)

h_1 = enthalpy of feedwater at boiler inlet (Btu/lb)

h_2 = enthalpy of superheated steam at boiler outlet (Btu/lb)

h_3 = enthalpy of steam at reheater inlet (Btu/lb)

h_4 = enthalpy of steam at reheater outlet (Btu/lb)

h_s = enthalpy of saturated water at boiler pressure (Btu/lb)

The heating value of the fuel is the amount of heat released by one unit of fuel when it is completely burned and the products of combustion are cooled to the original fuel temperature. When the water from combustion is in a vapor form, the value is called the low heating value (LHV). If there is complete water condensation in the products of combustion, the heating value thus obtained is the high heating value (HHV). For coal, the higher heating value is frequently estimated by Dulong's equation

$$HHV = 14,600C + 62000\left(H - \frac{O}{8}\right) + 4050S \quad \text{Btu/lb} \tag{5-6}$$

where C, H, O, S are weight fraction of the elements as given in the ultimate analysis. This equation gives a reasonably good estimate for anthracite and bituminous coals. For other fuels, other approximations should be utilized. The low heating value can be obtained by substracting from the high heating value, the heat needed to vaporize the moisture of the coal and the moisture formed in combustion. That is,

$$LHV = HHV - 1040(W + 9H) \quad \text{Btu/lb} \tag{5-7}$$

where W is the moisture and H is the hydrogen as given in the fuel ultimate analysis.

Equation (5-5) can be treated as the definition of boiler efficiency. This boiler efficiency is based on the high heating value of the fuel. In some European countries such as Germany, the low heating value is frequently used. Therefore, it is important to define the boiler efficiency with a clear indication of whether HHV or LHV is used.

Boiler efficiency can be also expressed in terms of boiler heat losses. When these losses are in Btu per pound of fuel burned, the boiler efficiency is calculated by

$$\eta_b = \frac{HHV - \text{total losses}}{HHV} \times 100 \tag{5-8}$$

There are six major boiler losses including (1) the dry-gas loss (DGL), (2) the moisture loss (*ML*), (3) the moisture in combustion air loss (*MCAL*), (4) the incomplete combustion loss (*ICL*), (5) the unburned carbon loss (*UCL*), and (6) the radiation and unaccounted-for loss (*RUL*). To determine these losses the fuel and flue gas analysis must be available. The equations used for these calculations are summarized below.

In conjunction with the previous symbols, let:

t_a = temperature of air entering the boiler system (F)

t_g = temperature of flue gas leaving the boiler system (F)

t_f = temperature of fuel entering the boiler system (F)

h_s = enthalpy of superheated water vapor at t_g

h_w = enthalpy of water at fuel temperature at t_f

W_r = solid refuse (lb/lb of fuel burned)

UF = unburned fuel (lb/lb of fuel burned)

C_r = combustible in solid refuse (defined as UF/W_r)

W_{dg} = dry flue gas (lb/lb of fuel burned)

Dry-Gas Loss (DGL)

$$DGL = W_{dg}C_p(t_g - t_a)$$ (5-9)

The dry flue gas (W_{dg}) is different from the total gas weight (W_g) calculated by by Eq. (5-4). It can be obtained by substracting from the total gas weight, the unburned fuel (*UF*) and the moisture ($W + 9H$) in the gases. That is

$$W_{dg} = W_a + 1.0 - A - UF - (W + 9H)$$ (5-10)

The combination of ($A + UF$) is frequently referred to as the solid refuse (W_r), which is in a unit of pound per pound of fuel burned. Because of air infiltration to the boiler system, the actual air (W_a) in Eq. (5-10) cannot be accurately estimated. As an alternative, the weight of dry flue gas is expressed in terms of flue gas data (such as those generated by the Orsat analyzer). The derivation of this equation is in the following:

$$W_{dg} = \underbrace{\frac{\text{weight of dry gases}}{\text{lb of carbon burned}}}_{[1]} \times \underbrace{\frac{\text{lb of carbon burned}}{\text{lb of fuel}}}_{[2]}$$ (5-11)

The first term in the right-hand side is simply equal to

$$[1] = \frac{44CO_2 + 32O_2 + 28CO + 28N_2}{12CO_2 + 12CO} \tag{5-12}$$

The second term is the carbon in a pound of fuel minus the unburned fuel. That is,

$$[2] = C_f = C - UF$$

or

$$C_f = C - (W_r - A) \tag{5-13}$$

Substituting these equations into Eq. (10-11) gives

$$W_{dg} = C_f \times \frac{44CO_2 + 32O_2 + 28CO + 28N_2}{12CO_2 + 12CO} \tag{5-14}$$

The dry-gas loss (DGL) is the largest among the six losses. As indicated in Eq. (5-9), the loss is a function of the gas temperature t_g.

Moisture Loss (ML)

$$ML = (W + 9H)(h_s - h_w) \tag{5-15}$$

The term $(W + 9H)$ represents the amount of moisture formed during combustion. It is due to mechanical moisture and combustion of the hydrogen element in the fuel. The term $(h_s - h_w)$ is the enthalphy change of the moisture and is approximated as

$$h_s - h_w = 1066 + 0.5t_g - t_f \quad \text{if} \quad t_g > 575F \tag{5-16}$$

and

$$h_s - h_w = 1089 + 0.46t_g - t_f \quad \text{if} \quad t_g < 575F \tag{5-17}$$

Moisture in Combustion Air Loss (MCAL)

$$MCAL = W_a \omega C_{p,w}(t_g - t_a) \tag{5-18}$$

where ω is the humidity ratio of air entering the boiler system and has a unit of pounds of water vapor per pound of dry air. The humidity ratio is a function of the dry-bulb and wet-bulb temperatures, and can be easily determined by using a psychrometric chart.

The actual air (W_a) in Eq. (5-18) is not always available. As an alternative, Eq. (5-19) is used for an estimate.

$$W_a = W_{dg} + 8\left(H - \frac{O}{8}\right) - C_f - S - N \tag{5-19}$$

where C_f is defined by Eq. (5-13). Values of H, O, S, and N are obtained from the fuel ultimate analysis and should have a unit of pounds per pound of fuel. Equation (5-19) is based on the principle of mass conservation.

Incomplete-Combustion Loss (ICL)

$$ICL = W_{dg} \times \frac{28CO}{44CO_2 + 28CO + 28N_2 + 32O_2} \times 4380 \qquad (5\text{-}20)$$

Unburned Carbon Loss (UCL)

$$UCL = (UF)(14600) \qquad (5\text{-}21)$$

or

$$= (W_r)(C_r)(14600) \qquad (5\text{-}22)$$

Radiation and Unaccounted-for Loss (RUL)
This loss is mainly due to radiation and incomplete combustion resulting in hydrogen and hydrocarbons in the flue gas. It also includes those that are not previously taken into account. While the (RUL) loss is relatively small, it is difficult to be accurately determined. In practice, this loss ranges from 3 to 5%.

EXAMPLE 5-2.   An old boiler test provides the data as follows:
   Fuel ultimate analysis:

$$
\begin{array}{lll}
C = 57.7 & N = 1.0 & A = 16.5 \\
H = 3.7 & S = 3.3 & W = 12.0 \\
O = 5.8 & & \\
\end{array}
$$
$$HHV = 11{,}000 \text{ Btu/lb}$$

Flue gas analysis:

$$
\begin{array}{ll}
CO_2 = 13.0 & O_2 = 7.0 \\
CO = 1.0 & N_2 = 79.0 \\
\end{array}
$$

Refuse analysis:

$$C_r = 20\%$$

Flue gas temperature 360 F

Air temperatures   Dry-bulb   70 F
                    Wet-bulb   60 F

The radiation and unaccounted-for loss is assumed to be 3%. Calculate the boiler efficiency.

Solution: First, we estimate the solid refuse per pound of coal burned. It is equal to the sum of the unburned carbon and the ash contained in the coal. That is,

$$W_r = UF + A$$

$$W_r = C_r W_r + A$$

$$= 0.2W_r + 0.165$$

or

$$W_r = 0.21 \text{ lb/lb of coal}$$

Second, we estimate the dry flue gas by using Eq. (5-13) and Eq. (5-14). It gives

$$W_{dg} = 0.536 \times \frac{44(13) + 32(7.0) + 28(1.0) + 28(79)}{12(13) + 12(1.0)}$$

$$W_{dg} = 9.69 \text{ lb/lb of coal}$$

The actual air is determined using Eq. (5-19). Substituting the numerical values into the equation gives

$$W_a = 9.69 + 8\left(0.037 - \frac{0.058}{8}\right) - 0.536 - 0.033 - 0.01$$

$$W_a = 9.35 \text{ lb/lb of coal}$$

The humidity ratio for the boiler test is

$$\omega = 0.0088 \text{ lb of vapor/lb of dry air}$$

With these results, the boiler losses are calculated and summarized below:

$$DGL = (9.69)(0.24)(360 - 70)$$

$$= 674.4 \text{ Btu/lb of coal}$$

$$ML = (0.12 + 9 \times 0.037)(1089 + 0.46 \times 360 - 70)$$

$$= 536.6 \text{ Btu/lb of coal}$$

$$MCAL = (9.35)(0.0088)(0.47)(360 - 70)$$

$$= 11.21 \text{ Btu/lb of coal}$$

$$ICL = 9.69 \times \frac{28(1.0)}{44(13) + 28(1.0) + 28(79) + 32(7)} \times 4380$$

$$= 391.4 \text{ Btu/lb of coal}$$

$$UCL = (0.21)(0.2)(14600)$$

$$= 613.2 \text{ Btu/lb of coal}$$

$$RUL = (0.03)(11000)$$

$$= 330 \text{ Btu/lb of coal}$$

Then, the total losses are calculated as

$$\text{Total losses} = DGL + ML + MCAL + ICL + UCL + RUL$$

$$= 674.4 + 536.6 + 11.21 + 391.4 + 613.2 + 330$$

$$= 2556.8 \text{ Btu/lb of coal}$$

Finally, the boiler efficiency is

$$= \frac{HHV - \text{total loss}}{HHV} \times 100$$

$$= \frac{11000 - 2556.8}{11000} \times 100$$

$$= 77\%$$

Equation (5-8) is commonly used to evaluate the boiler performance. The boiler efficiency thus determined is called the overall gross efficiency. This efficiency should be distinguished from the net efficiency, which takes into account the auxiliary power needed for boiler operation. The complete detailed instructions for determination of boiler net efficiency can be found in the Power Test Code of the American Society of Mechanical Engineers (ASME).

The boiler efficiency depends on the boiler design parameters. In addition, it varies with boiler operating conditions such as its output. In modern boiler design, the boiler efficiency is generally around 87 to 90% for solid fuel firing. Natural gas fired boilers have a lower efficiency of 84 to 85% because of the higher water vapor in the products of combustion. Recently, the second-law boiler efficiency also appeared in some engineering analysis. The second-law boiler efficiency is formulated on the basis of fuel availability (i.e., fuel work potential). It indicates an utilization of fuel availability rather than fuel energy. In contrast to the conventional boiler efficiency, the second-law boiler efficiency is relatively low and frequently around 50% [8].

5-4 BOILER COMPONENTS

A steam generating system is large and complex. It consists of combustion equipment, furnace, and various heat transfer surfaces. In addition, the steam generating system has some auxiliary equipment needed for efficient operation. These auxiliaries include at least the boiler fans (forced-draft and induced-draft), stack, precipitator, and SO_2 removal system. This section presents materials related to the boiler major components, while the next section covers the boiler auxiliaries. The overall objectives in this area are to acquaint the reader with a steam generator system and to present some of the engineering calculations encountered in practice. A rigorous description of a boiler design is beyond the scope of this text.

5.4.1 Combustion Equipment

The selection of combustion equipment depends on the type of the fuel used. For solid fuels such as coal, three combustion systems (mechanical stoker, pulverizer-burner and cyclone-furnace) are generally suitable. Mechanical stokers were first developed in the history of the boiler. Almost any coal can be burned on some type of stoker. Other advantages of stokers include low power requirements and large operating range. They are usually used for the boiler capacity ranges from 75,000 to 400,000 lb of steam per hour. Because of the small capacity, they are seldom used for today's central electric power station.

The pulverizer-burner system was introduced in the third decade of this century. This system overcomes the size limitation of the mechanical stoker. Modern pulverizing systems are so well developed that they can burn almost any type of coal, particularly those in the higher grades and ranks. In addition, the system has improved response to the load change, higher combustion efficiency, and less manpower required in operation.

The cyclone furnace is the most recent advancement. This system is primarily designed to burn a variety of coals, particularly those in the lower grades and ranks. The cyclone furnace is applicable to coals having a higher ash content and an ash-softening temperature at 2600 F or lower. Compared with the pulverizer-burner system, the cyclone furnace will reduce the fly ash content in the flue gas and the size of the furnace. In addition, it costs less to prepare the fuel.

Below are descriptions of the pulverizer-burner and the cyclone furnace. Mechanical stokers are not discussed here because of their limited use in modern central power stations.

Figure 5-4 shows a typical firing system for pulverized coal. The function of this system is to pulverize the coal, deliver the coal powder to the burners, and accomplish complete combustion in the furnace. The system must operate in continuous process and can adjust itself to the load demand in a reasonable time. There are two major equipment components, pulverizer and burner, in the system. The pulverizer receives coal from the coal bunker through the coal feeder, and produces the coal powder according to the fitness requirement. At the same time the pulverizer receives the hot air from the primary-air fan for drying and transporting the coal powder to the burners. Each pulverizer is usually connected with several burners. In operation, the coal feed is proportioned to the load demand, and the primary air supply is adjusted to the rate of coal feed. The air-coal ratio is so determined that the air-coal mixture leaving the pulverizer should have a proper temperature and moisture. Generally, the temperature and moisture are, respectfully, 150 F and 1 to 2% for bituminous coals.

In addition to delivering a sufficient amount of air, the primary air fan is designed to maintain a high velocity of the air-coal mixture in pulverizer discharge lines. The velocity must be such that there is no settling and drifting of coal in the piping. At the burner the air-coal mixture is combined with secondary air and both injected into furnace. As indicated in Fig. 5-4, both primary and secondary air are from the boiler air preheater. When the moisture of coal is below the maximum level, or the boiler is in a low load condition, cold air is used to temper the primary air.

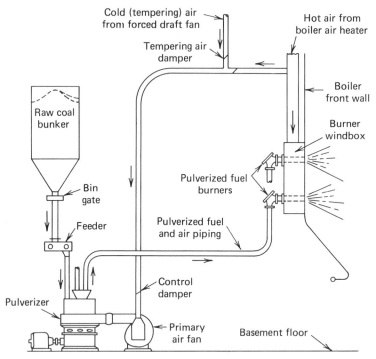

Figure 5-4.   Typical firing system for pulverized coal. (From *Steam/Its Generation and Use*, 1972)

Figure 5-5 shows a typical firing system for a cyclone furnace. It is different from the pulverized coal system, in that this system has a simple coal crusher that reduces coal size to $\frac{1}{4}$ in. or less. As shown in Fig. 5-6, hot primary air enters the cyclone and mixes the incoming coal. Secondary air is separately admitted at the roof of the cyclone main barrel. Both primary and secondary air in the cyclone furnace generate an intensive turbulent motion that results in a complete mixing between the coal and air. In the water-cooled horizontal furnace coal is burned at the heat release rate between 450,000 to 800,000 Btu per cubic feet and per hour. The velocity of the gases in the cyclone is about 300 ft/sec, and the temperature is in the order of 3000 F. The temperature is so high that the coal ash is melted into a liquid slag that continuously drains off the cyclone furnace floor into the water-filled slag tank. For this reason, boilers with a cyclone furnace generally have much smaller amounts of fly ash.

The cyclone furnace is operated in a high temperature environment. The cyclone furnace is protected by the water-filled tubes as indicated in Fig. 5-6. As the molten slag is formed, it also covers some furnace surface. The furnace surface is relatively small and the furnace heat absorption rate is in the range of 40,000 to 80,000 Btu per square feet per hour. This low heat absorption, combined with the high heat release rate, ensure a high furnace temperature, which is needed for good combustion. In cyclone furnaces, the excess air required is smaller than that for the pulverized-coal

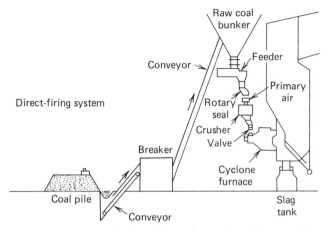

Figure 5-5. Typical firing system for cyclone furnace. (From *Steam/Its Generation and Use*, 1972)

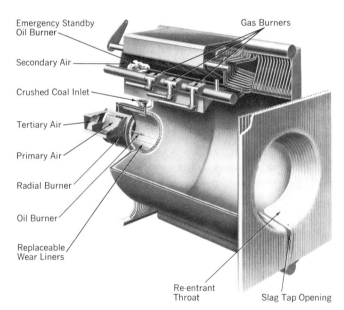

Figure 5-6. A cyclone furnace. (Courtesy Babcock and Wilcox)

system and generally around 10%. Also, the unburned combustible is relatively small and frequently amounts to less than a 0.1% loss in combustion efficiency.

The size and number of cyclone furnaces used for a given boiler depend mainly on the boiler size and the load flexibility. Today's cyclone furnace unit size varies in the range of 160 to 425 million Btu per hour. The firing arrangement is either one-wall firing or opposite-wall firing. For small boilers, sufficient firing capacity is

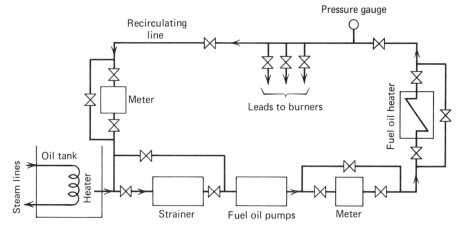

Figure 5-7.   Schematic diagram for fuel-oil system.

attained by locating all furnaces on one wall. For large boilers two opposite walls will be needed for installation of all furnaces. The present cyclone furnace is capable of burning a variety of coals. For the low-Btu coals such as lignite, the cyclone furnace is generally superior to the pulverized-coal system, because of its low auxiliary power requirement.

Combustion equipment for oil and natural gas is relatively simple. There is no need for a coal pulverizer, coal crusher, or other fuel preparation facilities. Figure 5-7 shows a schematic diagram for fuel oil system. Because of the high viscosity of the fuel oil, some types of heaters are usually needed in the oil storage tank to warm the oil and to facilitate pumping. Oil pumps receive the oil from the strainers and discharge it to the burners through heaters. To maintain a good combustion, the temperature of the fuel oil entering the burners should be around 150 F. As shown in Fig. 5-7, recirculation lines are provided in the fuel oil system. The recirculation lines are used to prevent stagnant oil from collecting in the piping system and cooling to the point of solidification. The burners for the fuel oil are similar to those for the pulverized coal. For more detailed information on combustion equipment, the readers should consult the publications by the manufacturers.

5.4.2 Boiler Furnace

A boiler furnace is construed to encompass the enclosure that surrounds the space needed for combustion and radiation heat transfer to the water-steam mixture. A boiler furnace also collects a portion of the coal ash at the furnace bottom and removes it. There are two types of furnace bottoms, depending whether the ash is in liquid or solid form. When the gas temperature in the furnace is higher than the ash fusion temperature, the ash is in a molten state, moving downward and eventually being trapped in the slag pool. In this arrangement, frequently called the wet-bottom furnace, about 50% of the total ash is trapped within the furnace. The remainder of the ash will leave the furnace. When the gas temperature in the furnace is lower than the ash fusion temperature, the ash remains in a solid state, falling into the bottom of the furnace. In this dry-bottom furnace, the ash is collected in a refractory-lined hopper. The hopper surface is cooled by placing water tubes on the refractory material. The dry-bottom furnace generally collects about 20% of the coal ash; thus comparatively more fly ash for the precipitators that are required for all boiler installations.

The furnace wall is protected by water-filled tubes, which are formed into a solid wall and are backed by refractory material. These tubes constitute almost the entire heat absorbing surfaces in the furnace. These surfaces receive heat from the products of combustion mainly by radiation. The amount of heat received depends on the quantity of energy released and the volume of the furnace. A large heat absorption means a large temperature reduction in the furnace, resulting in a low gas exit temperature. This furnace exit temperature plays an important role in boiler design. When the exit gas temperature exceeds the ash fusion, the ash becomes melted and the slag begins to build up on the boiler convective heat transfer elements such as the superheater and reheater. These slag deposits will adversely affect the performance of these elements and require periodic surface cleanings. On the other hand, low exit gas temperature will result in a large reduction of temperature differential for subsequent heat transfer elements and lead to an increase in the number of heating surfaces. In design, the temperature of the gas leaving the furnace should be a little lower than the fusion temperature of the coal fired. Regulation of this gas temperature is accomplished by controlling the amount of energy released and the size of the furnace. Experiences indicate that the heat release per unit area of heat absorbing surface is an important parameter. As the heat release increases, the gas exit temperature will increase. The general relationships are shown in Fig. 5-8.

Radiation heat transfer in the boiler furnace is complex, and accurate prediction of the gas temperature at the furnace exit is difficult. Many variables can affect the furnace radiation. These include at least the surface characteristics, and the composition of the combustion products. The surface characteristics is dependent not only on the tube material, but also the tube surface conditions such as the thickness of the slag and the ash deposits. The products of combustion contain various gases, water vapor, and solid particles. The exact composition depends on the type of fuel fired and the amount of excess air used. Some components including carbon dioxide, carbon monoxide, water vapor, and solid particles are the participating media, emitting and absorbing the radiation energy in the furnace. Other components such as oxygen and nitrogen are nonparticipating media, being transparent to thermal radiation. In addition, the furnace radiation heat transfer is affected by the furnace geometry complex and by variation in operating conditions. Evidently, a theoretical model for predicting the radiation and the furnace exit gas temperature is impossible at the present time. We now present an approximate method to solving this complex problem.

Before the furnace exit gas temperature can be estimated from Fig. 5-8, it is necessary to determine the energy release in the furnace. Based on the principle of energy conservation, the heat release for a coal-burning furnace is

$$Q_{re} = \left[HHV + 0.24W_a(t_f - t_a) - 1040(9H) - 14{,}600(UF) \right] W_f \quad (5\text{-}23)$$

where

$$Q_{re} = \text{furnace heat release (Btu/hr)}$$

$$W_a = \text{actual air (lb/lb of coal)}$$

$$UF = \text{unburned fuel (lb/lb of coal)}$$

$$W_f = \text{coal consumption (lb/hr)}$$

$$HHV = \text{high heating value of coal}$$

$$t_f = \text{temperature of air entering the furnace (F)}$$

$$t_a = \text{temperature of air entering the air preheater (F)}$$

Equation (5-23) indicates that the amount of heat released is basically the sum of fuel heating value and the energy of combustion air (represented by the first two terms) minus the latent heat for water vapor formation and the loss of the unburned

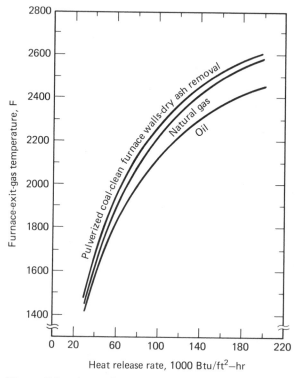

Figure 5-8. Approximate relationships of furnace exit gas temperature to heat release rate. (Courtesy Babcock and Wilcox)

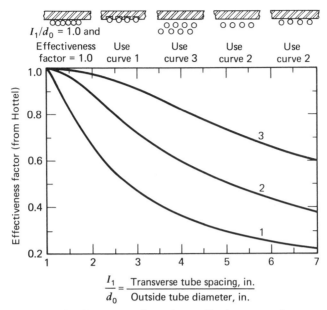

Figure 5-9.   Furnace-wall surface effectiveness. (Courtesy Babcock and Wilcox)

carbon. In case that the boiler radiation and unaccounted loss is taken into consideration, the term (HHV) should be replaced by $(HHV)(1 - RUL)$.

The furnace heat-absorbing surface is taken as the total enclosing surface, including the convective water-tube bank, minus the area of uncooled refractory. The calculation is tedious and also time-consuming. To utilize the curves in Fig. 5-8, it was recommended that the actual heat-absorbing surface is replaced by an equivalent surface area that can be obtained by multiplying the furnace enclosing surface by an effectiveness ratio. The surface effectiveness ratios for various furnace walls are presented in Fig. 5-9. When the wall is completely covered by water tubes, the effectiveness is 1.0. Other arrangements are less effective in heat absorption and, therefore, the effectiveness ratio is less than a unity.

The furnace heat absorption rate is the furnace available energy (Q_{re}) minus the quantity of heat leaving the furnace. The latter is equal to the product of the gas flow rate and its enthalpy. Since the products of combustion behave as an ideal gas, the enthalpy change is simply the specific heat times the temperature change (assuming a constant specific heat). Figure 5-10 indicates the flue gas enthalpy in relation to the temperature. For more accurate data, the readers should utilize the information given in Fig. 1 of Chapter 6 in the B & W publication [1].

EXAMPLE 5-3.   A steam generator has a coal-fired furnace of which the walls are completely covered by the water-filled tubes. The furnace dimensions are 27 ft × 24 ft × 100 ft. The furnace consumes coal at the rate of 125,000 lb/hr. Calculate the furnace exit gas temperature and heat absorption rate, using the following operating data:

> Coal high heating value (as fired) = 11,000 Btu/hr
> Coal H_2 content = 4.8%
> Coal ash content = 9.1%
> Solid refuse = 0.11 lb/lb of coal
> Actual air supplied = 12 lb/lb of coal
> Temperature of air entering the furnace = 300 F

Solution: The furnace heat release is first determined with Eq. (5-23). Using the given operating data and assuming the room temperature (t_a) 80 F, it gives

$$Q_{re} = [11,000 + 0.24 \times 12(300 - 80) - 1040(9 \times 0.048)$$

$$- 14,600\ (0.11 - 0.091)](125,000)$$

$$Q_{re} = 1363.4 \times 10^6\ \text{Btu/hr}$$

The furnace enclosing surface is assumed as the sum of the furnace walls and

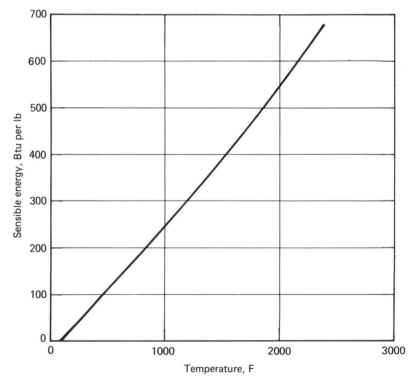

Figure 5-10. Approximate flue gas enthalpy in terms of temperature.

roof, neglecting the furnace bottom. That is,

$$A_f = (2)(24 + 27)(100) + (24 \times 27)$$

$$A_f = 10,848 \text{ ft}^2$$

The furnace walls are completely covered by water-filled tubes, and the surface effectiveness factor is therefore equal to 1.0 (see Fig. 5-9). Thus, the furnace heat release per unit surface area becomes

$$\frac{Q_{re}}{A_f} = \frac{1363.4 \times 10^6}{10,848}$$

$$= 125,682 \text{ Btu/ft}^2/\text{hr}$$

Figure 5-8 shows an approximate relationship between the furnace exit gas temperature and the heat release rate. For the heat release rate 125,682 Btu/ft²/hr, the exit gas temperature is approximately

$$T_{f,e} = 2380 \text{ F}$$

To calculate the furnace heat absorption rate, the enthalpy of the gases at the furnace exit must be first determined. Using Fig. 5-10, it gives

$$h_{f,e} = 670 \text{ Btu/lb}$$

The total energy leaving the furnace is the product of the enthalpy and the gas flow rate. The gas flow rate is approximated by

$$\dot{m}_g = \dot{m}_f(w_a + 1 - w_r) \tag{a}$$

Equation (a) is similar to Eq. (5-4) except the replacement of A by w_r when w_r is the solid refuse in the furnace. Using the given operating data, it gives

$$\dot{m}_g = (125,000)(12 + 1.0 - 0.11)$$

$$= 1.61 \times 10^6 \text{ lb/hr}$$

Thus, the quantity of energy leaving the furnace is

$$Q_{f,e} = \dot{m}_g \times h_{f,e}$$

$$= 1.61 \times 10^6 \times 670$$

$$= 1078.7 \times 10^6 \text{ Btu/hr}$$

Finally, the furnace heat absorption rate is

$$Q_{ab} = Q_{re} - Q_{f,e}$$

$$= (1363.4 - 1078.7)10^6$$

$$= 284.7 \times 10^6 \text{ Btu/hr}$$

The furnace heat-absorbing surface is an important parameter in the furnace design. As this heating surface area increases, the temperature of the flue gas leaving the furnace will decrease, therefore, resulting in an increase in the furnace heat absorption rate. Using Fig. 5-8 can only provide an approximate value for the furnace exit gas temperature. In boiler design a more complex procedure is required to do this kind of calculation. The furnace performance is not always in a steady state condition. As the boiler output increases, the furnace firing rate will increase. This will lead to an increase in the furnace heat release rate and thus to an increase in the exit gas temperature and the furnace heat absorption rate. Also, changes in excess air, combustion air temperature, and coal composition will have some effects on the furnace performance.

5.4.3 Superheater and Reheater

The superheater is a heat exchanger in which heat is transferred to the saturated steam to increase its temperature. Steam superheating is one of the design features accepted in central electric power stations. Superheating raises overall cycle efficiency. In addition, it reduces a moisture level in the last stages of the steam turbine and thus increases the turbine internal efficiency.

Superheaters are commonly classified as either radiant superheaters, convective superheaters, or combined superheaters, depending on how heat is transferred from the gases to steam. These superheaters have different performance characteristics. Figure 5-11 shows the outlet steam temperature change with the unit loads. The feature that the outlet steam temperature can stay essentially constant over a wide range of unit load is the most desirable. When the outlet steam temperature becomes excessive, it may cause failures from overheating parts of the superheater, reheater, or turbine. On the other hand, when the outlet steam temperature is below the design value, it will result in an erosion from excess moisture in the last turbine stages, and less efficient operation of the turbine.

The convective superheater is located in the furnace exit or in the zone where it can receive thermal energy from the high temperature products of combustion. The convective superheater is frequently screened from the furnace radiation by a bank of water-filled tubes. These tubes, when adequately spaced, can also intercept the slag particles and reduce slagging problems in superheaters. Convective superheaters in large steam generator systems are frequently split into two parts: the primary

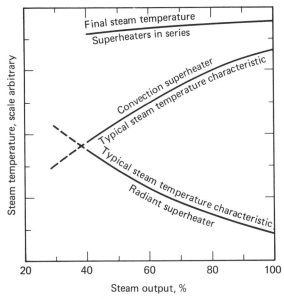

Figure 5-11. Outlet steam temperature for various types of superheater.

superheater and the secondary superheater. Saturated steam first enters the primary superheater and receives the initial superheating. The primary superheater is located in a zone of relatively low gas temperature. After the partial heating, steam moves to the secondary superheater and completes its superheating process. The main reasons for splitting the superheater are to provide space for the steam reheater and to achieve an effective heat transfer from the gases to steam.

The radiant superheater is not as commonly used as the convective superheater. When the radiant superheater is needed, it is usually placed on the furnace wall replacing a section of water-filled tubes. Another arrangement is to have the radiant superheater just behind the screen tubes. The radiant superheater is an integral part of the secondary superheater.

Central station boilers provide for steam reheating. The reheater is essentially a convective type and usually located in the space between the primary and secondary superheaters. After steam partially expands in the turbine, it returns to the boiler for reheating. The temperature of steam leaving the reheater is usually equal to the superheated steam temperature. Since the design and operation of reheaters are essentially the same as superheaters, the following discussion of superheaters will be equally applicable to reheaters.

A superheater is basically a counterflow heat exchanger. The heat transfer rate from the hot gas to steam is governed by

$$Q = UA(LMTD) \tag{5-24}$$

where

$$Q = \text{heat transfer rate (Btu/hr)}$$

$$U = \text{overall heat transfer coefficient (Btu/hr-ft}^2\text{-F)}$$

$$A = \text{heat transfer area (ft}^2)$$

$$LMTD = \text{Log-mean-temperature difference (F)}$$

The heat transfer rate Q is also equal to the heat gained by steam, or the heat lost by the hot gases. Both heat gain and heat loss can be expressed as the product of the mass flow rate and the change of enthalpy. They are

$$Q = \dot{m}_s \left(h_{s,e} - h_{s,i} \right) \tag{5-25}$$

and

$$Q = \dot{m}_g \left(h_{g,i} - h_{g,e} \right) \tag{5-26}$$

The subscripts (g and s) denote the gas and the steam, respectively. For the hot gases, the enthalpy change can be further expressed as the product of the specific heat and the temperature change.

In superheater thermal design, the steam temperature is first determined. This is generally accomplished in the plant system design, balancing the plant initial cost against the lifetime operating cost. In recent years the optimum steam temperature is approximately 1000 F for all large steam generation systems. In the second step, the amount of superheater surface required is approximated. This is usually done by using Eq. (5-24), Eq. (5-25), and Eq. (5-26). We must therefore first assume the overall heat transfer coefficient (U) in Eq. (5-24) and later compare it with the actual value.

After the amount of superheater surface is determined, the next consideration is to select the tube length, tube diameter, and the number of tubes. Evidently, the selection is an iterative process, generating a trial solution and checking to see whether all constraints are met. From several acceptable solutions, the optimum is found. The optimum superheater should have enough heat transfer surface necessary to give the design steam temperature. The tube parameters (length and diameter) are such that the steam pressure drop and tube metal temperature will not exceed the design values. The tube metal temperature is an important parameter and has a strong influence on the tube material selection. In addition, the optimum superheater should have its tubes so spaced that minimum ash and slag deposits will result.

Modern superheaters have many tube passes, and the tubes are arranged in-line rather than staggered. The tubes are usually cylindrical and have 2 or $2\frac{1}{2}$ in. outside diameter. There is no extended surface (i.e., fins) attached to the tubes. The material selection depends on the steam temperature and pressure. Carbon steel has an allowable temperature up to 800 F and is frequently used for low-temperature superheaters. Chrome-moly, stainless steel, or some similar heat resistant alloy can withstand the temperature up to 1200 F. Therefore they are selected for the superheater in a high-temperature zone.

In case that the superheater is split into the primary and secondary heaters, the procedures presented above are equally applicable. Each superheater should be individually designed, considering the constraints associated with each heater. The primary superheater may not have the same heat transfer rate as the secondary superheater. Because of different temperature encountered, these two heaters may have two different tube materials. The reheater is usually located in the zone between the primary and secondary superheaters. Design considerations given to reheaters are similar to those given to superheaters. For reheater design, particular attention must be given to the steam pressure drop. A large steam pressure drop will reduce the benefit due to the steam reheating.

Temperature regulation and control are important for both superheaters and reheaters. Steam temperature adjustments are frequently made at the time of the commissioning of a boiler. The principal methods are an addition or reduction of heat transfer surfaces. Steam temperature can be also adjusted by regulating the hot gas temperature and mass flow rate. These are generally accomplished by changing the excess air or the effectiveness of the evaporation section.

During a boiler operation, there are many factors affecting the temperature of steam leaving the superheater and reheater. These include a boiler load, excess air, feedwater temperature, and cleanliness of heating surfaces. Control of steam temperature during operation must be done without changing the arrangement of equipment. The most effective approaches are

Gas bypass
Burner control
Attemperation
Gas recirculation
Excess air
Divided furnace

A gas bypass is to control the gas flow rate to the superheater. The main disadvantages of this approach are the operating difficulties experienced by the movable dampers located in the high-temperature zone and the slow response to load change.

Burner control is used to control the flame location and combustion rate. Tilting burners can direct the flame toward or away from the superheater. These will result in a change of heat absorption in the furnace and change of gas temperature in the superheater. As the boiler load is reduced, burners are removed one by one from service. This will change the combustion rate and, thus, change the gas flow rate to the superheater.

Attemperation is one of the approaches frequently used. The attemperator is usually located at the point between the primary and the secondary superheaters.

There are two basic types of attemperator. The first is the tubular type in which some of superheated steam is passed through the tubes of a heat exchanger and has heat transferred to the boiler water (either boiler feedwater or water in the boiler drum). Subsequent to attemperation, the divided streams from the primary superheater will combine and enter the secondary superheater.

The second type of attemperator involves a spray of feedwater into the stream of superheated steam. The feedwater evaporates and reduces the steam temperature. Controlling the amount of feedwater will result in control of the steam temperature. Care must be exercised to ensure that the spray water has sufficient purity. The spray water should mix well with the superheated steam so that there are no water droplets in the inlet of the secondary superheater.

Gas recirculation is used to control the steam temperature by changing the heat absorption rates both in the furnace and in the superheater. When the steam temperature needs to be raised, some of the flue gas from the economizer outlet is recirculated back to the bottom of the furnace. Therefore, the furnace temperature will become lower, resulting in a lower heat absorption in the furnace and thereby a higher flue gas temperature in the furnace exit. This high gas temperature, combined with an increase in the gas flow rate, will increase the heat transfer rate in the superheater and thus increase the steam outlet temperature.

Temperature control can be affected by using different amounts of excess air. The more the excess air, the higher the steam outlet temperature would be. The reasons for this are similar to those for the gas recirculation method. It must be pointed out, however, that too much excess air will result in a reduction of boiler combustion efficiency. A divided-furnace boiler is usually arranged with a generation of saturated steam in one section and a superheating of steam in another section. The temperature of the superheated steam is regulated by controlling the firing rates in the two furnaces. This method is not economical and is seldom applied in a central electric power station.

EXAMPLE 5-4. A spray-type attemperator is furnished with 500 F water. The attemperator is connected in a steam line carrying 4,000,000 lb/hr at 750 psia. Calculate the amount of spray water that must be used to reduce the temperature from 800 to 760 F.

Solution: This example is solved by applying the principles of mass and energy conservation. The steam flow rate at the attemperator outlet is equal to the spray water plus the steam flow rate at the attemperator inlet. That is,

$$m_e = m_i + m_{\text{water}} \tag{a}$$

Based on the energy conservation principle, the steam energy at the attemperator outlet is equal to the sum of the energy of water and the energy of incoming steam. In terms of steam and water enthalpies, the equation is

$$m_e h_e = m_i h_i + m_{\text{water}} h_{\text{water}} \tag{b}$$

Assuming that there is no pressure drop in the attemperator, the enthalpies of steam and water are found as follows

$$h_i = 1400.6 \text{ Btu/lb}$$

$$h_e = 1377.3 \text{ Btu/lb}$$

$$h_{\text{water}} = 487.7 \text{ Btu/lb}$$

Combining Eqs. (a) and (b) and substituting the numerical values into the resulting equation gives us

$$(4{,}000{,}000 + m_{\text{water}})(1377.3) = (4{,}000{,}000)(1400.6) + m_{\text{water}}(487.7)$$

Thus the required spray water is

$$m_{\text{water}} = 104{,}766 \text{ lb/hr}$$

For this operation, the amount of spray water required is approximately 2.62% of the steam flow rate.

5.4.4 Economizer and Air Heater

Central station boilers are usually equipped with an economizer and air heater. Boiler efficiency rises about 1% for each 10 F increase produced by an economizer. Use of air heaters does not only improve boiler efficiency by lowering the stack temperature, but also improve the combustion conditions by raising the combustion air temperature.

The economizer is generally located ahead of the air heater in the gas stream. When a high combustion air temperature is desirable, it may be necessary to divide the air heater into two sections and place the economizer between them for an effective heat transfer. In some cases, especially those large high-pressure steam generators, an additional low-temperature economizer may be included and located in the gas stream after the air heater. This economizer, frequently called stack cooler, is used to replace one of the low-pressure feedwater heaters. The temperature of water leaving the stack cooler is in the range of 160 to 180 F.

The economizer is a tubular heat exchanger. The water from the last feedwater heater flows through the tubes and absorbs energy from the flue gas discharged from the superheater and the reheater. In recent designs steel tubes are usually used and have an outside diameter ranged from $1\frac{3}{4}$ to $2\frac{3}{4}$ in. All these tubes are continuous from inlet to outlet headers and have several horizontal sections in the gas stream. Feedwater enters the economizer through the bottom inlet header and moves upward, while the flue gas enters on the top of the economizer and flows downward outside the heat transfer tubes. The tubes in the economizer are either bare tubes or finned tubes. The finned tubes have some advantages over bare tubes. These include a low initial cost and a small space required for installation. However, attention must be given to the draft loss and potential fouling associated with the finned tubes.

An economizer is usually justified in central station boilers, because it can absorb some heat at less cost than other heat transfer surfaces in the boiler. The exact economizer size is determined by many variables such as the temperature of incoming feedwater and the boiler pressure. The feedwater temperature varies in the range of 400 to 560 F, depending on the number of feedwater heaters and the extraction steam conditions. The boiler pressure usually presets the upper limit of water temperature in the economizer. Steaming or nonsteaming also affects the economizer surface. In the steaming economizer, part of water flow is evaporated and the temperature of water leaving is naturally equal to the saturation temperature corresponding to the boiler pressure. The outlet water temperature in the nonsteaming economizer is usually lower than the saturated temperature.

In a boiler system design the economizer cannot be separated from the air heater. The distribution of the flue gas energy is frequently optimized, taking into account the equipment cost and fuel cost. For a given stack temperature there is an optimum gas temperature leaving the economizer. An increase in the gas temperature will reduce the size of the economizer and increase the size of the air heater. The selection of the stack temperature is important. From the viewpoint of energy utilization, the stack temperature should be as low as possible. But attention must be given to the corrosion of the heat transfer surfaces. Corrosion can be avoided by setting the flue gas temperature (i.e., stack temperature) above the dew point. Below the dew point water vapor in the flue gas will condense on the surface and combine with the sulfur dioxide in the flue gas to generate an acid.

Figure 5-12 shows several arrangements for tubular air heaters. Flue gas usually flows inside the tubes while air moves outside. These tubes can be placed vertically as well as horizontally. In both cases the tubes are either welded or expanded into tube sheets, one of which must float to allow for tube expansion. The tube diameter ranges from $1\frac{1}{2}$ to $2\frac{1}{2}$ in. and the tube material is frequently steel. Because of the low overall heat transfer coefficient, the tubular air heaters are relatively bulky and therefore occupy relatively large space.

In addition to the tubular air heater, plate-type air heaters are available for preheating the air for combustion. These heaters comprise parallel plates that provide alternative passages for flue gas and air. The spacing between the plates is approximately $\frac{1}{2}$ in. Like the tubular air heater, this heater also has a low overall heat transfer coefficient and therefore a large heat transfer surface. To provide high-temperature combustion air, steam-coil air heaters are sometimes used. In this case heat is transferred to the air from the steam that usually comes from the turbine extraction line. Steam-coil heaters are particularly suitable when the flue gas temperature is low for air preheating purposes.

Air heaters used in central station boilers may be classified as either the recuperative or regenerative type. The recuperative type are the air heaters we have just described. Generally, the flue gas is on one side of a plate or tube and air is on the other. Heat is transferred from the flue gas to air by convection and conduction. These are of static construction and there is only a nominal leakage from the air to the flue gas. In a regenerative air heater, flue gas flows through a closely packed matrix to raise its temperature and, then, air is passed through the same matrix to

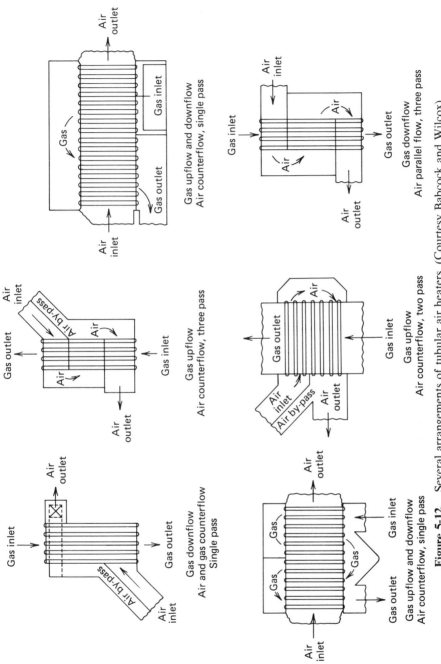

Figure 5-12. Several arrangements of tubular air heaters. (Courtesy Babcock and Wilcox)

absorb the heat. Either the matrix or the hood are rotated to achieve this in a continuous cycle. The regenerative air heater has several advantages over the recuperative type, including its compactness and subsequently its low initial cost. However, air leakage is more serious with the regenerative air heater.

Figure 5-13 shows a schematic diagram for regenerative-type air heater. A rotor, mounted with a box housing, turns slowly and moves the heat transfer matrix through the separated streams of flue gas and air. The heat transfer matrix receives heat in the flue gas stream and rejects heat in the air stream. The main disadvantage is the leakage of air into the gas. Leakage occurs at the radial seals and the annular space between the rotor and the housing. Another problem with the regenerative air heater is the entrainment of air and gas in the heat transfer matrix.

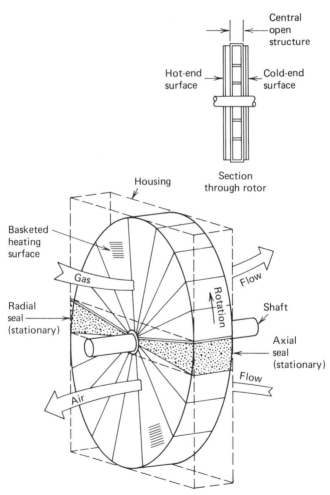

Figure 5-13. Schematic diagram for regenerative-type air heater. (Courtesy Babcock and Wilcox)

The heat transfer matrix used in regenerative air heater is usually a high performance type, producing a large heat transfer surface per unit volume and resulting in a compact structure. Because of the small space required for installation, the initial cost is usually lower than the tubular air heater.

5.5 BOILER AUXILIARIES

The purpose of this section is to discuss the auxiliaries required for a satisfactory boiler operation. It includes mainly boiler fans and air pollution control devices. The boiler fans are provided to induce a movement of combustion air and the flue gas inside the boiler. The pollution control devices are precipitators and scrubber systems. They are used to remove the particulates and various gaseous pollutants from the flue gas prior to its departure from the stack.

5.5.1 Fan Calculation

Figure 5-14 shows a typical air and flue gas path in a steam generator system. Air enters the system through the forced-draft fans (FD fans). After receiving thermal energy in the air heater, the combustion air is split into the primary and secondary air. As described in the previous section, the primary air is used to dry and transport the coal powder to the furnace. If a drier coal is used, the primary air is tempered by the cold air directly withdrawn from the forced-draft fans. Primary air fans are used to maintain a reasonably high velocity in the mills. If a wetter coal is used, the gas bypass around the economizer will be actuated. This leads to higher temperature gas to the air heater and thus to a high temperature of combustion air. The primary and secondary air eventually join together in the windbox and enter the furnace. After combustion, the flue gas starts its journey from the furnace. The flue gas will pass through superheater, reheater, economizer and air heater, transferring away its thermal energy along its path. Before the flue gas enters the induced-draft fans (ID fans), it usually flows through the precipitator for removing the particulates. In most cases the flue gas also flows through a scrubber system for removing various gaseous pollutants such as SO_2. The scrubber system is not shown in Fig. 5-14.

Many central station boilers employed both forced-draft and induced-draft fans. The FD fan overcomes the resistances in the air supply duct work and the coal pulverizer system, while the ID fan exhausts the products of combustion from the furnace. In this arrangement the furnace usually operates at a pressure slightly below the ambient. This makes it possible to have routine firebox inspection and maintenance even when the boiler is in a full operation. It must be pointed out that the air and gas flow are not constant. They are affected by the boiler output and excess air. In addition, the air leakage and infiltration from and to the boiler system must be taken into account.

Figure 5-15 shows the pressure variation in the boiler system. It is seen that the pressure is negative throughout the boiler except in the air supply ductwork and coal pulverizer system. The actual pressure drop for boiler components is supplied by the equipment manufacturers. This pressure drop is frequently given for a certain air (or gas) flow rate. For the flow rate different from this value, the parabolic law expressed

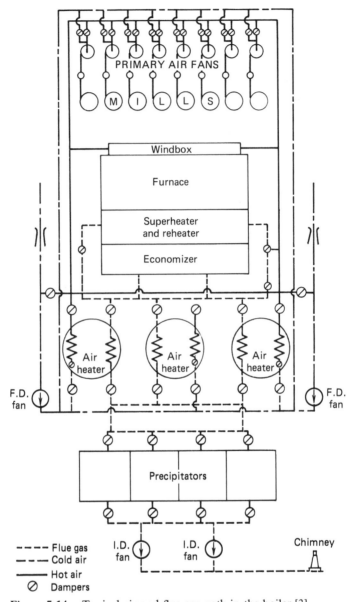

Figure 5-14.   Typical air and flue gas path in the boiler [3].

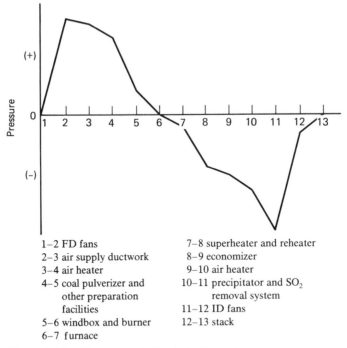

1-2 FD fans
2-3 air supply ductwork
3-4 air heater
4-5 coal pulverizer and
    other preparation
    facilities
5-6 windbox and burner
6-7 furnace

7-8 superheater and reheater
8-9 economizer
9-10 air heater
10-11 precipitator and SO_2
    removal system
11-12 ID fans
12-13 stack

Figure 5-15.   Pressure variation in boiler system.

below should be used:

$$D_o = D_r \left(\frac{W_o}{W_r} \right)^2 \tag{5-27}$$

where

$$D = \text{pressure drop}$$

$$W = \text{air or gas flow rate}$$

The subscripts o and r denote, respectively, the operating and reference condition. Since there is some uncertainty in predicting the pressure drop, it is customary to use a safety margin. In design, the safety margin may be up to 25% of the calculated value.

Air horsepower for a fan is a function of the air (or gas) flow rate times the developed pressure. The pressure developed by a fan is numerically equal to the pressure drop that the fan is designed to overcome. In terms of traditional units, the equation for fan horsepower is

$$\text{Air hp} = \frac{(Q)(\Delta p)}{6350} \tag{5-28}$$

where

$$Q = \text{air (gas) flow rate (ft}^3/\text{min)}$$

$$\Delta p = \text{developed pressure (inches of water)}$$

The air horsepower is the minimum power required for fan operation. The actual shaft power is obtained by using the fan efficiency (E). That is,

$$\text{Shaft hp} = \frac{\text{air hp}}{E} \tag{5-29}$$

Equations (5-28) and (5-29) are equally applicable regardless whether the developed pressure is static pressure or total pressure. When the developed pressure is static pressure, the air horsepower by Eq. (5-28) is called the static air horsepower. In this case the fan efficiency used must be the static efficiency. When the developed pressure is the total pressure (the total pressure is the sum of the static pressure and velocity pressure), the air horsepower by Eq. (5-28) will become the total air horsepower, sometimes called the power input. To determine the actual shaft horsepower, we must utilize the mechanical efficiency.

EXAMPLE 5-5.   A FD fan develops 25.0 in. static pressure and 1.0 in. velocity pressure when the air flow is 1,300,000 cfm. The fan shaft power is 6000 hp. Calculate the static and mechanical efficiencies.

Solution: By Eq. (5-28) the static air horsepower is

$$\text{Static air hp} = \frac{1,300,000 \times 25}{6350}$$

$$= 5118$$

Substituting the static air horsepower and the fan shaft horsepower into Eq. (5-29) gives the static efficiency as

$$E_s = \frac{5118}{6000} = 0.85$$

The total pressure is the sum of the static pressure and the velocity pressure. It gives

$$\Delta p_t = 25 + 1 = 26$$

Then the fan power by Eq. (5-28) is

$$\text{Power input} = \frac{1,300,000 \times 26}{6350}$$

$$= 5323 \text{ hp}$$

and the mechanical efficiency by Eq. (5-29) is

$$E_m = \frac{5323}{6000} = 0.89$$

$$= 0.89$$

It should be pointed out that the gas (or air) temperature or pressure have some effect on the fan horsepower. The developed pressure (Δp) in Eq. (5-28) varies directly with the gas density, which is a function of the gas temperature and pressure. Using the equation of state for ideal gas, it gives

$$\Delta p_f = \Delta p_i \frac{\rho_f}{\rho_i}$$

or

$$\Delta P_f = \Delta p_i \left(\frac{P_f T_i}{P_i T_f} \right) \tag{5-30}$$

The subscripts i and f denote the conditions before and after correction, respectively. Thus the fan power after correction will become

$$(\text{Air hp})_f = (\text{Air hp})_i \left(\frac{P_f T_i}{P_i T_f} \right) \tag{5-31}$$

The volumetric flow rate Q in Eq. (5-28) is the mass flow rate times the gas specific volume. This term is difficult to predict accurately, because of gas leakage or infiltration in the system. To make sure that the fans will not limit the boiler performance, a safety margin is frequently used to correct the calculated flow rate. The customary safety margin ranges from 15 to 20%.

5.5.2 Fan Characteristics and Control

Fans used in central station boiler are either the axial-flow or the centrifugal type. These fans are characterized by a very high flow rate but a low pressure rise. The centrifugal fan may have different curved blades such as forward-curved, backward-curved, and straight radial. The different blade will result in different fan performance. Table 5-1 summarizes the operating characteristics of centrifugal fans. The fan performance in terms of its capacity is shown in Fig. 5-16. It is seen that the fan design point is usually around the maximum static efficiency. In operation, the fan

Table 5-1
Operating Characteristics of Centrifugal Fans

Quantity	Forward-Curved Blade	Backward-Curved Blade	Straight Radial Blade
Size of impeller	Small	Medium	Medium
Mechanical efficiency	Average	High	Average
Stability	Poor	Good	Good
Fan speed	Low	High	Average
Impeller-tip speed	Low	High	Medium
Resistance to erosion	Poor	Medium	Good

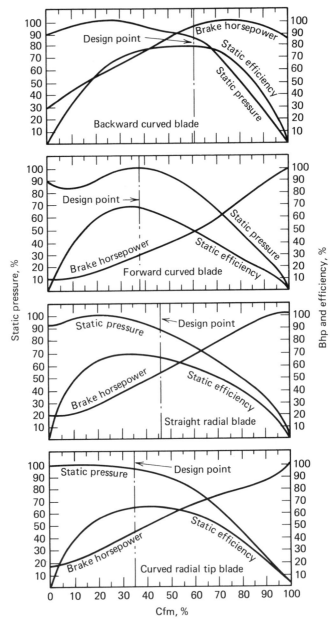

Figure 5-16.   Typical fan performance characteristics.

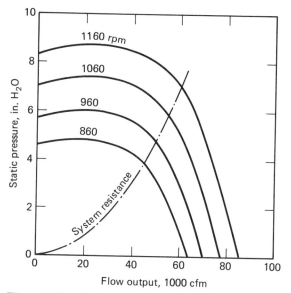

Figure 5-17. Typical system resistance and fan pressure head curves.

capacity will determine the fan static pressure head and horsepower. The static pressure head is expected to be equal to the total resistance in the system. If the system resistance is less than the pressure head developed, additional resistance such as a control damper must be added. Obviously, this approach is very wasteful.

The performance curves in Fig. 5-16 are based on a constant fan speed. As the fan speed varies, the fan characteristics will vary. However, the trends of performance curves remain unchanged. For instance, the pressure curves for various fan speeds shown in Fig. 5-17 are almost parallel to each other. Variable-speed fan calculations are carried out according to the so-called "fan laws." These laws are for constant efficiency

$$\frac{Q_1}{Q_2} = \frac{S_1}{S_2} \tag{5-32}$$

$$\frac{\Delta p_1}{\Delta p_2} = \left(\frac{S_1}{S_2}\right)^2 \tag{5-33}$$

$$\frac{\text{hp}_1}{\text{hp}_2} = \left(\frac{S_1}{S_2}\right)^3 \tag{5-34}$$

EXAMPLE 5-6. An ID fan is rated at 25 in. static pressure for 700,000 cfm of 80 F gas, 1200 rpm, and static efficiency of 84%. At what speed should this fan operate to develop 23 in. when the gas temperature is 500 F? Also estimate the fan capacity and horsepower.

Solution: From Eq. (5-30) it gives the pressure head at the temperature 500 F as

$$\Delta p_f = \Delta p_i \left(\frac{P_f T_i}{P_i T_f} \right)$$

$$\Delta p_f = 25 \left(\frac{540}{960} \right) = 14.06 \text{ in. of } H_2O$$

assuming that the gas pressure is identical to the rated value. While the gas temperature affects the developed pressure, it does not change the volumetric flow rate. To determine the fan speed required to develop 23 in. of H_2O, Eq. (5-33) is utilized. It gives

$$S_2 = S_1 \left(\frac{\Delta p_2}{\Delta p_1} \right)^{0.5}$$

$$S_2 = 1200 \left(\frac{23}{14.06} \right)^{0.5} = 1534 \text{ rpm}$$

At this new fan speed, the fan capacity is

$$Q_2 = Q_1 \left(\frac{S_2}{S_1} \right)$$

$$Q_2 = 700,000 \left(\frac{1534}{1200} \right) = 895,302 \text{ cfm}$$

and the fan horsepower is

$$\text{Shaft hp} = \frac{(895,302)(23)}{(0.84)(6350)} = 3860 \text{ hp}$$

assuming the static efficiency remains constant and equal to 84%.

Figure 5-17 also shows a general relationship between the system resistance and the gas flow. The system resistance increases with an increase in the gas flow. Varying the fan speed can effectively achieve the desirable capacity and have sufficient pressure head to match the system resistance. This speed control is the most efficient method of regulating the fan output. According to the fan laws, the fan output decreases by half when the fan speed is reduced by half. At that low speed, the fan pressure head and horsepower also drop, respectively, one fourth and one eighth. While this method of control is efficient, it has relatively high initial cost.

Variable-speed control can be realized by using a variable-speed motor or steam turbine. Some consulting engineers may prefer using constant-speed motors, with speed controlled by the use of magnetic or hydraulic coupling. In these cases, coupling losses must be taken into account.

Damper control is another method of regulating fan output. This method uses inlet louvers or discharge dampers to control flow in a constant speed fan. When the louvers or damper positions are changed, the system resistance is expected to change. The situation is similar to that in the pump operation shown in Section 10.8. Compared with the variable-speed control, this method will have high power consumption especially under low-flow conditions. This is mainly because of the energy dissipation caused by damper throttling. However, the advantages of the method are low initial cost, easy adaptation to automatic control, and convenience in maintenance.

Fan drive selection is one of the problems frequently encountered by consulting engineers. There are essentially two different fan drives: electric motor and steam turbine. Electric motor may be of either constant speed or variable speed and either ac or dc. Use of turbine-driven fans has been recently recognized. The advantage of the turbine drive is to increase plant electric output and improve the plant net heat rate. Fan-drive selection is generally based on an economic evaluation.

5.5.3 Pollution Control Equipment

One of the serious environmental concerns in boiler operation has been pollution of the air with sulfur dioxide, oxides of nitrogen, unburned hydrocarbons, and particulate matter. With pulverized coal firing, about 80% of the coal ash remains in the flue gas. This may be reduced to about 50% if the boiler furnace is equipped with a wet-bottom ash collector. Even in cyclone-furnace firing, the fly ash still amounts to 20 to 30% of the ash in the coal. To reduce the emission rate of the fly ash, central station boilers are frequently provided with electrostatic precipitators. To reduce the emission rate of SO_2, wet scrubbers are frequently required in addition to precipitators. These devices are usually located in the gas stream ahead of the induced-draft fans.

The principles upon which an electrostatic precipitator operates are based on the electrostatic forces created between the particulates of the flue gas and the collecting electrodes. The electrostatic forces then propel the charged particulates to the collecting electrodes. From these, the particulates are discharged by rapping into storage hoppers.

An electrostatic precipitator consists of a large chamber where the gas velocity is reduced and given enough resident time for the electric forces to remove the particulates from the gas stream. In the chamber there are suspended collecting electrodes that are frequently in the form of flat plates. Between these flat plates there are suspended discharge electrodes that produce a stream of ions and thus an electric charge on the particulates of the flue gas. Although most fly ash is deposited on the collecting electrodes, some fly ash is still on the discharge electrodes. Therefore, suitable rapping gear must be provided for both electrodes.

The rapping gear of modern precipitator is important. The precipitator efficiency depends on the effectiveness of rapping motions. In the past, when electric vibration or pneumatic devices were used they were found ineffective in most cases. Recently, manufacturers started to use some mechanical means for dislodging the dust. In any case, rapping is an action of dislocating the particulates from both electrodes. The electrode rapping is carried out periodically.

Hoppers are arranged below the main precipitator chamber and are made of either steel or concrete. Hoppers should have sufficient capacity for a 24 hour operation. Since an electrostatic precipitator is divided into several zones along the gas stream, each zone has its own hopper. The hopper associated for the first zone (the one closest to the precipitator inlet) is larger than those for the subsequent zones, because the first zone of the precipitator generally collects the largest portion of the particulates as compared with other zones.

In addition to the electrostatic precipitator, wet scrubbers are also used to remove the particulates from the flue gas. The operation of wet scrubbers is to mix a suitable liquid with the flue gas and transfer the suspended particulates from the flue gas to liquid. The wet scrubber can also remove some gaseous pollutants by gas absorption or chemical reaction. If a limestone slurry is used as the scrubbing fluid, the wet scrubber system can remove a minimum of 85% of the sulfur dioxide (SO_2) and about 99% of the particulates from the flue gas. However, attention must be given to the behavior of the wet plumes leaving the plant stack. Experience has shown that the wet plume from the plant might fall to the ground level under certain atmospheric conditions.

To remove sulfur dioxide from the flue gas, a dry scrubber system also can be used. The system generally consists of a fixed sorbent bed, which contains material to absorb the sulfur oxide (including the sulfur dioxide). The sorbent material may be char, activated carbon, or alumina impregnated with copper. The system is usually so designed that the sorbent material can be regenerated. Compared with the wet scrubber system, this system does not increase water content in the flue gas and, therefore, does not have the wet plume leaving the stack. However, the plant using a dry scrubber must be equipped with a precipitator for removal of the particulates.

The oxides of nitrogen are also some of the gaseous pollutants generated in fossil-fuel power plant. The formation of these gases generally occurs in the furnace where the temperature exceeds 3000 F. To reduce the amount of nitrogen oxides, the furnace temperature control seems most effective. However, low temperature in the furnace may result in an adverse effect on combustion.

The boiler chimney is an important element of air pollution control system. While precipitator and scrubber system are used to reduce the emission rate of various pollutants, the boiler chimney is frequently required to disperse gases and particulates adequately to avoid a high concentration at ground level. Both emission rate and ground-level concentration must be below those specified in various governmental regulations such as the Air Quality Act of 1967 in the United States. There are at least two main parameters in chimney design affecting the ground-level concentration. These include chimney height and flue gas exit velocity. The taller the chimney, the lower the ground-level concentration would be. Chimney height up to 1000 ft has been used for a 2 × 800 MW station. In design, the chimney height is determined by local topographical as well as meteorological conditions. The gas exit velocity is important because it affects the gas plume rise on the top of the chimney. The gas plume rise is mainly due to the gas momentum and buoyance. The plume rise plus the chimney height is called the effective chimney height, which is usually needed in the calculation of ground-level concentrations. As the effective chimney height increases, the ground-level concentrations are expected to decrease under normal atmospheric conditions. In recent central station design the chimney gas velocity up to 90 fps has been used. The draft loss resulting from the high velocity is partially offset by the greater natural draft induced by the tall chimney.

5.6 SYSTEM DESIGN CONSIDERATIONS

The boiler is a major component in the power plant. It is also a complex system by itself. There are certain predominant factors that must be regarded if a satisfactory design is to be evolved. First, the boiler must be capable of generating steam to meet the turbine need. In addition, steam conditions should be maintained constant over a wide range of the boiler output. Other factors to be considered in design are boiler efficiency, availability, and total cost. In the following are some explanations for these factors.

The boiler is usually designed to provide sufficient steam flow to the steam turbine. The steam flow requirements that are important to boiler design are peak flow, maximum continuous flow, and minimum flow. The peak flow is used to establish the top generating capacity of the boiler. Generally, the peak flow is obtained by assuming 5% overpressure at the turbine throttle and one or two feedwater heaters removed from operation. A 5% steam overpressure roughly means a steam flow that is 5% greater than the normal maximum flow and results approximately in an output that is 5% higher than the normal plant output. When one or two low-pressure feedwater heaters are removed from service, the steam normally used for feedwater heating would continue to pass through the turbine and, thus, to generate additional plant output. Because of higher steam flow and lower feedwater temperature at the boiler inlet, provision of additional firing must be provided in the boiler design. The boiler operates under peak-flow conditions only for a limited number of hours. The peak-flow operation is sometimes limited to short daily periods because of the boiler materials and other design parameters such as the potentiality for excessive slogging.

The boiler maximum flow is the steam flow a boiler is capable of generating in a continuous basis. When the maximum steam flow is provided to the steam turbine, the unit is expected to generate the rated output. The boiler minimum flow usually means the steam flow below which the combustion may become unstable. The boiler operation under this condition should be avoided.

The boiler may be designed for a base-load or cycling operation. A base-load boiler is expected to provide a normal maximum steam flow most of the time and experience little fluctuation in demand. A cycling boiler is entirely different. The boiler must be capable of full-load and part-load operation at a reasonable efficiency. In addition, it should have good response to the changes of the turbine demand.

The steam conditions at the turbine throttle are well defined for each generating system. When the steam temperature exceeds the design value, there will be adverse effects on the boiler and turbine materials. On the other hand, if the steam temperature is lower than the design point, a reduction in the plant efficiency will be experienced. Therefore, the boiler is expected to deliver steam at design conditions from both the superheater and the reheater in a wide range of boiler output. To compensate for the heat loss and the pressure drop in the steam lines, the steam temperature and pressure in the boiler outlet are higher than those in the turbine inlet. This difference is usually approximately 5 F, and the pressure drop varies between 8 and 15%. The consulting engineers usually determine the acceptable pressure drop by optimizing the main steam pipe diameters.

The reheater, as already seen in the previous section, is designed to raise the steam temperature back to the initial value and, thus, to improve the plant heat rate. The pressure drop in the reheater and associated pipings is important. A relatively large pressure drop can significantly offset the benefit due to the reheating. When large diameter piping is used, the pressure drop is reduced, but initial cost may increase proportionately. As in the main steam pipe, pressure drop in the reheater and its piping must be appropriately balanced against the initial cost.

Boiler efficiency is one of the factors to be considered in design. The firing equipment should be so chosen that there will be a minimum amount of unburned carbon in the ash and virtually complete combustion in the furnace. Air infiltration to the boiler should be minimum, because it results in additional loss to the chimney and, thus, to the steam generating system. To achieve a reasonable boiler efficiency, furnace size and shape are important variables. They should be so designed that the mixture of fuel and air will have sufficient residence time for complete combustion and the resultant products of combustion will have enough thermal energy at the furnace exit for subsequent heat transfer surfaces. Too much energy results in high chimney temperature and, thus, reduces boiler efficiency. On the other hand, if the furnace exit temperature is too low, there will be more heat transfer surface required, and some corrosion of the metal in the air heater. Therefore, the furnace exit temperature must be optimized in the boiler design.

The characteristics of fuel to be used here has important influences in boiler design. Fuel selection is usually determined on the basis of its availability and total cost. The fuel source should be dependable and have sufficient quantities of fuel for a plant lifetime supply. The fuel total cost is the sum of the material and transportation costs. Evidently, both are subject to inflation and other changes. In power plant design a lifetime levelized fuel cost is usually needed in economic evaluation. If the station fuel is coal, its utilization will require coal storage, coal handling, and preparation facilities before combustion. It also needs ash disposal and pollution control equipment after combustion. When the station fuel is oil or natural gas, its use generally involves a less complicated system. Compared with a coal-fired station, the station burning oil or natural gas usually has a lower initial, but a higher fuel cost.

There is a wide variation in the properties of coal and coal ash. Coals with high heating value and high fusion temperature are usually burned in the furnace with a pulverizer-burner system. Since the ash is still in solid form, the furnace bottom is usually equipped with dry-ash removal hoppers. When coals are of low heating value and low fusion temperature the boiler is usually equipped with cyclone furnace with slag-tap ash removal. In both cases, the temperature of the flue gas entering the convectional zone (i.e., superheater and reheater) should be lower than the ash fusion point. Otherwise, undesirable slag ash deposits may occur on the surfaces of superheater and reheater.

Boiler availability plays an important role in power plant operation. The boiler should have its availability as such that it would not lower the availability of the entire plant. In the past, poor availability has been often associated with the chokage of gas passes in tube banks by ash deposits. Recently, the failure of pressure parts such as superheater and furnace tubes is also one of the reasons for boiler outage. Accurate prediction of boiler availability is a complicated matter. In the absence of this information, consulting engineers should rely on their experiences with boiler systems. In comparative study or boiler proposal evaluation, credit must be given to the system that is expected to have the higher availability. The actual credit value varies in a wide range, depending mainly on the local conditions such as excess power in the network.

Like other major components in power plant system, the boiler should be efficient and reliable. In addition, the boiler must be compatible with other components in such a manner that it would result in a lowest production cost. The lowest total cost, not investment or operation cost, is one of the design objectives. When one boiler is compared with another boiler such as those encountered in equipment proposal evaluation, the more efficient boiler is not always automatically selected. Attention must be given to whether the lifetime saving in fuel cost can compensate for the additional equipment cost. If these two boilers have different performance in the rate of availability, it must be also taken into account.

In the boiler system design, there are many decisions to be made. Most cases can be eventually reduced to a point that the equipment efficiency is evaluated in terms of the equipment cost. For instance, a boiler fan with an efficiency of 85% costs little more than a fan with an 83% efficiency. In practice, the fan selection is not made on the basis of fan efficiency, but on the basis of total return, or the lowest total cost to the equipment user.

As already indicated in the previous section, the boiler fan drive is by either electric motor or auxiliary steam turbine. Use of motor drive will simplify the fan-drive arrangement and generally result in a relatively low initial cost. However, it does not mean that motor drive should be always selected. To make a decision on the fan drive, consulting engineers should determine the lifetime total cost for each fan drive. In this case, the total cost is the sum of initial and operating costs. Since the turbine drive usually results in an increase of plant net kilowatt and improved plant net heat rate, appropriate credits must be given to this arrangement in economic evaluation.

It should be pointed out that boiler operates as a component in power plant system, and its optimization cannot be separated from the entire plant optimization. In fact, boiler unit size and steam conditions are usually the results of the plant system optimization. The station fuel is determined at the time when the plant site is selected. These results are in turn used as the inputs to boiler system design. Other inputs required include feedwater temperature and the conditions of plant site and local climate.

SELECTED REFERENCES

1. Babcock and Wilcox Company, *Steam/Its Generation and Use*, 1972.

2. J. G. Singer (editor), *Combustion, Fossil Power Systems*, Combustion Engineering, Inc., 1981.

3. Central Electricity Generating Board, *Modern Power Station Practice*, Vol. 2, Pergamon Press, 1971.

4. P. J. Potter, *Power Plant Theory and Design*, Second Edition, Ronald Press, 1959.

5. B. G. A. Skrotzki and W. A. Vopat, *Power Station Engineering and Economy*, McGraw-Hill, 1960.

6. C. D. Swift, *Steam Power Plants*, McGraw-Hill, 1959.

7. F. S. Aschner, *Planning Fundamentals of Thermal Power Plants*, Israel University Press, 1978.

8. M. J. Moran, *Availability Analysis*, Prentice-Hall, 1982.

9. A. P. Fraas, *Engineering Evaluation of Energy Systems*, McGraw-Hill, 1982.

10. D. A. Berkowitz (editor), *Boiler Modeling*, MITRE Corporation, 1975.

11. E. MacNaughton, *Elementary Steam Power Engineering*, Third Edition, McGraw-Hill, 1948.

PROBLEMS

5-1. Coal ultimate analysis may be shown on three different bases: "as received," "dry-coal," and "combustible." In the following is the as-received ultimate analysis for a subbituminous coal:

$$C = 60.1 \qquad N = 1.1 \qquad A = 5.0$$
$$H = 4.2 \qquad S = 1.3 \qquad W = 12.2$$
$$O = 16.1$$

Calculate the analysis on the dry basis.

5-2. An anthracite coal has the following ultimate analysis:

$$C = 69.10 \qquad N = 1.50 \qquad A = 13.36$$
$$H = 3.71 \qquad S = 0.57 \qquad W = 3.0$$
$$O = 8.76$$

Estimate the coal high heating value on the dry basis and compare the result with that on the "as-received" basis.

5-3. A lignite coal has the following ultimate analysis on the dry basis:

$$C = 59.88 \qquad N = 1.04 \qquad A = 13.51$$
$$H = 4.21 \qquad S = 1.90$$
$$O = 19.46$$

Calculate the theoretical air and actual air per pound of lignite. The combustion is assumed to be complete and the excess air 18%.

5-4. For the lignite indicated in Problem 5-3, estimate the dry flue gas produced in combustion. The combustible in solid refuse is assumed to be 4%.

5-5. An Orsat analysis of the flue gas from the lignite specified in Problem 5-3 shows:

$$CO_2 = 10.1 \qquad\qquad CO = 0.1$$
$$O_2 = 8.3 \qquad\qquad N_2 = 81.5$$

The ashpit refuse is 0.15 lb/lb of coal. Estimate the dry flue gas by Eq. (5-14).

5-6. Estimate the coal consumption for the generating unit of which the turbine cycle arrangement is shown in Fig. 7-7. The coal heating value is 11,000 Btu/lb, and boiler efficiency is 90%.

5-7. Calculate the boiler heat balance on the dry basis for the data as follows:

Fuel ultimate analysis

$$C = 71.81 \qquad N = 1.20 \qquad A = 8.27$$
$$H = 5.23 \qquad S = 3.34$$
$$O = 10.15$$

High heating value = 13,082 Btu/lb
Flue gas analysis

$$CO_2 = 14.2 \qquad O_2 = 4.3$$
$$CO = 0.3 \qquad N_2 = 81.2$$

Refuse

$$w_r = 0.115 \text{ lb/lb of coal burned}$$

Flue gas temperature 463 F
Fuel and room temperature 85 F
Ambient relative humidity 50%

5-8. Repeat Problem 5-7 for the flue gas temperature 563 F.

5-9. A pulverized coal-fired furnace has a heat absorbing surface equivalent to 115,000 ft^2. The furnace receives the combustion air at 300 F and consumes the coal at the rate 690,000 lb/hr. The solid refuse is approximately 0.14 lb/lb of coal burned. The air and flue gas flow are, respectively, 11.8 and 12.7 lb/lb of coal. Estimate the furnace gas exit temperature and heat absorption rate.

The coal has the high heating value 12,300 Btu/lb and the ultimate analysis is

$$C = 68.36 \qquad N = 1.32 \qquad A = 13.75$$
$$H = 5.25 \qquad S = 0.82$$
$$O = 10.50$$

5-10. Repeat Problem 5-9 with the furnace heat absorbing surface increased by 10%. Would the furnace gas exit temperature decrease by 10%?

5-11. A pulverized coal furnace is assumed to be rectangular in plan and elevation views. The furnace has dimensions (width × depth × height)

$42 \times 21 \times 100$ ft and the wall completely covered by 3-in. tubes. The coal has the high heating value 14,100 Btu/lb and the ultimate analysis

$C = 79.71$	$N = 1.42$	$A = 5.19$
$H = 5.29$	$S = 1.26$	
$O = 7.13$		

Estimate the amount of steam produced in the furnace. The operating conditions are:

Excess air	20%
Radiation and unaccounted-for loss	2.0%
Combustion air temperature	300 F
Boiler room temperature	85 F
Boiler drum pressure	2300 psia
Coal firing rate	120,000 lb/hr

5-12. For the same conditions as Problem 5-11, estimate the temperature of the flue gas leaving the furnace.

5-13. A steam generator is designed to produce the superheated steam at the rate 1,000,000 lb/hr. The steam conditions are 1900 psia and 1000 F. The flue gas enters the superheater at 2000 F and has the flow rate 1,230,000 lb/hr. Estimate the superheater heat transfer surface, the number of tubes and the number of passes for the conditions below.

Tube diameter (OD)	$2\frac{1}{2}$ in.
Tube-wall thickness	0.25 in.
Tube center to center spacing	7 in.
Tube arrangement	in-line
Design steam mass flow	500,000 lb/hr-ft^2
Design gas mass flow	1,700 lb/hr-ft^2
Estimated overall heat transfer coefficient	8.8 Btu/hr-ft^2-F

Assume that the superheater is a pendant type and neglect the pressure drop for this calculation.

5-14. A spray-type attemperator is used to control the superheated steam temperature. Calculate the amount of spray water required for the conditions:

Steam pressure	2450 psia
Steam inlet temperature	920 F
Steam outlet temperature	905 F
Steam flow rate	5×10^6 lb/hr
Spray water temperature	200 F

5-15. Estimate the maximum metal temperature for Problem 5-13. The convective heat transfer coefficient on the steam side is approximately 110 Btu/hr-ft^2-F.

5-16. Repeat Problem 5-13 with the steam mass flow reduced to 450,000 lb/hr-ft^2. Assume that the overall heat transfer coefficient changes to 8.5 Btu/hr-ft^2-F.

5-17. A generating unit is operated on a reheat cycle. The steam conditions are identical to those in Fig. 7-7. What percentage of the total heat transfer to the steam occurs in the reheater?

5-18. A reheater is to be designed for the conditions as below:

Steam conditions at inlet	744 psia, 600 F
Steam conditions at outlet	690 psia, 1000 F
Steam flow rate	4×10^6 lb/hr
Flue gas temperature and flow rate at inlet	1600 F, 5.25×10^6 lb/hr
Overall heat transfer coefficient	8.5 Btu/hr-ft^2-F

Estimate the reheater surface area and the temperature of flue gas leaving the reheater.

5-19. In boiler design it is desirable to have the flue gas exit temperature above the dew point. Estimate the dew point temperature of the flue gas produced by a combustion of the coal with excess air of 20%. The coal ultimate analysis is:

$C = 74.0$	$N = 3.0$	$A = 8.0$
$H = 6.0$	$S = 1.0$	$W = 2.0$
$O = 6.0$		

Assume that air infiltration and leakage are negligible.

5-20. The Orsat analyzer provides the following readings before and after the air heater:

	BEFORE	AFTER
CO_2	14.9	13.5
CO	0.1	0.1
O_2	3.0	4.6
N_2	82.0	81.8

Estimate the air leakage in the air heater per pound of coal burned. The boiler coal has a carbon percentage of 74.0%.

5-21. An air heater is used to reduce the flue gas temperature from 600 F to 470 F. The flue gas flow rate is 1,230,000 lb/hr. Estimate the annual fuel saving for the conditions:

Coal heating value	12,000 Btu/lb
Boiler efficiency	90%
Operating hours per year	5000 hr
Unit cost	$1.20 per MBtu

5-22. An air heater is to be designed. The flue gas is available at the rate of 1,000,000 lb/hr and at 600 F. To avoid the metal corrosion and maintain a reasonable temperature in the pollution control equipment, we set the minimum flue gas temperature at 480 F. For the following conditions

Air-coal ratio	12.1 (by weight)
Room temperature	80 F
Overall heat transfer coefficient	4.4 Btu/hr-ft^2-F

estimate the heat transfer surface area and the temperature of combustion air leaving the heater. The air heater is designed for counterflow of the air and flue gas.

5-23. A counterflow economizer is arranged in 48 parallel loops. Each loop consists of a number of tubes in the gas stream. The tube length in the gas stream is 12 ft and the tube outside diameter is 2 in. Calculate the heat transfer surface area and the number of tubes in each loop for the following conditions:

Water temperature, inlet	300 F
outlet	360 F
Flue gas temperature, inlet	900 F
outlet	710 F
Flue gas flow	300,000 lb/hr
Overall heat transfer coefficient	12 Btu/hr-ft^2-F

Assume that the specific heat of flue gas is approximately 0.255 Btu/lb-F.

5-24. A boiler has two identical FD fans connected in parallel. Each fan is driven by variable-speed motor. The fan performance at 1200 rpm and 80 F air is:

FLOW RATE (CFM)	STATIC PRESSURE (IN.)	STATIC EFFICIENCY (%)
560,000	26.8	85.1
400,000	28.1	78.2
300,000	28.6	69.3
200,000	28.8	55.2
100,000	28.8	32.8
0	28.4	0

Determine and plot the characteristic curves at the fan speed 1300 rpm.

5-25. A boiler fan has a design point of 540,000 cfm, 80 F air, 27.1-in. static pressure, and 1200 rpm fan speed. The fan characteristic curves are identical to those in Problem 5-24, and the system resistance curve is a square curve. Determine the speed and the shaft power for the fan when the air flow rate is reduced by 25%.

5-26. A steam generator has a draft-balance furnace. Calculate the total fan power for the conditions below:

Air temperature and flow	85 F, 500,000 lb/hr
Gas temperature and flow	450 F, 581,000 lb/hr
Static pressure head	
FD fan	22 inches of water
ID fan	25 inches of water
Static fan efficiency (for both)	85%

5-27. A 600 MW coal-fired unit is operated under full-load conditions. The operation data are:

Unit output	600 MW
Unit net neat rate	9500 Btu/kWh

Coal ultimate analysis (as received)

$C = 65$ $N = 1.4$ $A = 12$
$H = 4.3$ $S = 1.2$ $W = 8$
$O = 8.1$ $HHV = 11,880$ Btu/lb

Orsat flue gas analysis

$$CO_2 = 13.8 \qquad CO = 0.7$$
$$O_2 = 4.8 \qquad N_2 = 80.7$$

Combustion air conditions

$DBT = 90$ F $WBT = 80$ F
Stack gas temperature   480 F
Percentage of combustible in the refuse   18%

Calculate:

a. The coal consumption per hour.

b. The combustion air volumetric flow rate.

c. The flue gas flow rate.

d. The dew point temperature of the flue gas.

e. The boiler losses assuming RUL $= 3\%$.

f. The boiler efficiency.

g. The horsepower for FD and ID fans, assuming there is a 25-in. water total pressure head and the mechanical efficiency is 86% for both fans.

5-28. A coal-fired boiler consumes the coal at the rate of 100,000 lb/hr. The coal has ash content 12.5% (by weight). The ashpit refuse is 0.1 lb/lb of coal. The combustibles in refuse and fly ash are, respectively, 25% and 40%. Estimate the total amount of unburned carbon per pound of coal burned and the fly ash to be removed by precipitator. Assume the precipitator efficiency is 99.6%.

5-29. A coal-fired boiler is equipped with a SO_2 removal system. The boiler consumes the coal at the rate of 100,000 lb per hour. The coal has the ultimate analysis on a dry basis as follows:

$C = 62.74$ $N = 1.04$ $A = 12.19$
$H = 5.25$ $S = 8.04$
$O = 10.74$

Calculate the amount of SO_2 to be removed in pounds per hour.

CHAPTER SIX

Nuclear Steam Supply Systems

6.1 INTRODUCTION

Nuclear energy is one of the important fuels for today's electric power generation. Nuclear energy results from changes in the nucleus of atoms. As a nucleus splits, it releases a tremendous amount of heat. This nucleus splitting process is frequently called the fission process. When one pound of pure uranium is completely fissioned, it will create as much heat as the burning of 1500 short tons of coal. In 1982, approximately 12% of electric power produced in the United States was generated from nuclear power plants.

The development of nuclear power progressed slowly in the nineteenth and early twentieth centuries. In 1911 the physicist Ernest Rutherford first discovered the existance of a subatomic particle, later referred to as the nucleus. Although Rutherford did not succeed in splitting a nucleus, he later showed the possibility of a fission process. In 1932, the physicist James Chadwick discovered the existance of a neutron. In 1938, two German chemists, Otto Hahn and Fritz Strassmann reported they had produced the element barium by bombarding uranium with neutrons. This experiment was later referred to as the first manmade fission reaction. This reaction had in fact split an uranium nucleus into two nearly equal fragments, one of which was a barium nucleus and another was a krypton nucleus. In this fission process two neutrons were also emitted. The mass of the two nuclei and two neutrons produced was somewhat less than that of an uranium nucleus and a neutron combined. The reaction had therefore produced a significant amount of energy.

In the same period Albert Einstein developed his famous relativity theory and related the matter to energy by the equation $E = mc^2$. The equation states that the energy (E) in a substance equals the mass (m) of that substance multiplied by the speed of light squared (c^2). The equation had been used by scientists to estimate the amount of energy released in a fission process.

The first manmade chain fission process was not produced until December 2, 1942, when the physicist Enrico Fermi and his associates constructed an atomic pile, using 50 short tons of natural uranium embedded in 500 short tons of graphite. Cadmium rods were used to control the chain reaction. With the successful development of atomic bombs in 1945 and the first nuclear-powered vessel, the submarine Nautilus in 1954, the first full-scale nuclear power plant began operations in 1956 at Calder Hall in northwestern England. Next year, the first large nuclear power station in the United States was completed in Shippingsport, Pennsylvania. The pressurized water reactor at this power station produced 60 MW of electricity.

By 1960, nuclear power generating systems in the range of 150 to 200 MW were in commercial operation. At that time, these reactors were still in the demonstration phase. By the middle 1960s nuclear reactor systems were being ordered by utility companies on the basis of favorable economic comparisons with fossil-fuel power plants. These nuclear reactor systems were in the range of 600 MW and were in commercial operation by 1970. By 1975, systems in the 1000 MW range were in operation. The most recent nuclear reactor systems produce 1300 MW, which is the current technical limit on the size of a single reactor system.

Present-day commercial nuclear reactors are of the "fission" type. Fission takes place when a fissionable nucleus (such as those of uranium) captures a free neutron. Capture upsets the internal force, which holds together the tiny particles called protons and neutrons in the nucleus. The nucleus splits into two fission fragments. Besides the heat energy produced, fission releases an average of two or three neutrons and such nuclear radiaton as gamma rays. The fission fragments give off beta rays. If one of the neutrons emitted is captured by another fissionable nucleus, a second fission takes place in the manner similar to the first. Another neutron may produce a third fission. When the fission becomes self-substaining, the process is called a chain reaction. The device in which this chain reaction is generated to produce nuclear energy is called a nuclear reactor.

6.2 NUCLEAR REACTORS AND THEIR CLASSIFICATIONS

Nuclear reactors used for electric power generation consist of four main parts. They are (1) the fuel core, (2) the moderator and coolant, (3) the control rods, and (4) the reactor vessel.

The fuel core contains the nuclear fuel and is the part of the reactor in which the fission takes place. The nuclear fuel may be either natural uranium or enriched uranium. The natural uranium contains 0.71% fissile U-235 and 99.28% fertile U-238 and fertile thorium Th-232. The enriched uranium is produced in a gaseous diffusion process and is expected to have a U-235 content up to 2 or 3%. This is three or four times the concentration in natural uranium.

In fission process the fertile materials are converted to fissile. For instance, the U-238 becomes Pu-239 and Pu-241 and the Th-232 becomes U-233. These converted fissile fuels are nuclear fuels themselves, which are then utilized in reactors.

The nuclear fuel is generally contained in cylindrical rods surrounded by cladding materials. The fuel-rod cladding materials must be able not only to maintain the fuel rod in shape, but also to hold under the reactor conditions. These materials include aluminum, magnesium, zirconium, stainless steel, and graphite.

The moderator is the substance used in nuclear reactor to reduce the energy of fast neutrons to thermal neutrons. Liquid and solid materials of small mass number and low neutron capture should be suitable. These include light water, heavy water, carbon, and beryllium.

The reactor coolant is used to remove heat from the reactor fuel core. The conditions for a good coolant include high specific heat, high thermal conductivity, and high boiling point at low pressure. The coolant should also have low power demand for pumping, low cost, and a high degree of stability in the reactor environment. For this purpose the coolants include light water, heavy water, air, carbon dioxide, helium, sodium potassium, and some organic liquids.

Control rods are long metal rods that contain such elements as boron, cadmium, or hafnium. These elements absorb fast neutrons and therefore help control a chain reaction. The control rods are attached to an elevator-like mechanism just outside the nuclear reactor. The mechanism inserts the rods into the fuel core or withdraws them to slow down or speed up a chain reaction. Three types of control rods are used. These include (1) shim rods, (2) regulating rods, and (3) safety rods. Shim rods are used for making occasional coarse adjustment in neutron density, while regulating rods are used for fine adjustment. Safety rods are designed for use in emergency. The safety rods are made of boron steel and are capable of coming into the reactor core very rapidly and stopping the chain reaction.

The reactor vessel is a tanklike structure that holds the reactor core and other internals. The walls of the vessel are designed for the high pressure and radiation environment. In most cases the vessel walls are lined with thick steel slabs to reduce the flow of radiation from the core. As indicated in the last section, nuclear fission generates large amounts of neutrons and gamma rays. Both of them are very harmful. Because of these, biological shielding is required around the reactor vessel. This shield consists of concrete blocks, which may be up to 6 ft thick.

Nuclear reactors can be classified according to their coolant and moderator. Table 6-1 presents seven different reactors. The two principal types are the pressurized water reactor (PWR) and the boiling water reactor (BWR). Both reactors use enriched uranium and light water as coolant as well as moderator. The difference between these two reactors is in the manner in which steam is being produced. The BWR generates steam within the reactor vessel. The steam is directly piped to a steam turbine and returned to the vessel after completing various processes. In the

Table 6-1
Nuclear Reactor Classifications

| | Moderator | | |
Coolant	Water	Heavy Water	Graphite
Carbon dioxide			GCR (Magnox)
			AGR
Helium			HTGR
	PWR		
Water		SGHWR	
	BWR		
Heavy water		CANDU	

PWR = pressurized water reactor
BWR = boiling water reactor
CANDU = canadian deuterium-uranium reactor
SGHWR = steam-generating heavy water reactor
HTGR = high-temperature gas-cooled reactor
AGR = advanced gas-cooled reactor
GCR (Magnox) = gas cooled reactor

Table 6-2
Typical Reactor Characteristics [6]†

	PWR	BWR	CANDU	SGHWR	Magnox	AGR	HTGR	LMFBR
Net electrical power output (MWe)	1100	1100	540	500	500	560	1200	350
Fuel								
Fuel material	UO_2	UO_2	UO_2	UO_2	U	UO_2	UO_2, ThC	PO_2, UO_2
Can material	Zr	Zr	Zr	Zr	Mg	SS[a]		SS[a]
Feed enrichment (percent U-235)	3.0	2.5	natural	2.0	natural	2.0–2.6		
Coolant								
Composition	H_2O	H_2O	D_2O	H_2O	CO_2	CO_2	He	Na
Pressure (psia/MPa)	2200/15.2	980/6.8	1350/9.3	1000/6.9	385/2.7	490/3.4	700/4.8	147/1.0
Inlet temperature (C)	283	275	250	274	247	300	340	380
Outlet temperature (C)	316	286	294	283	414	670	780	540
Core								
Diameter × height (m)	3.7 × 3.4	4.8 × 3.7	6.4 × 6	7 × 3.7	17.4 × 9.1	9.6 × 8.3	8.4 × 6	1.8 × 0.91
Average power density (MW thermal/m³)	95	51	9.2	12	0.9	2.6	8.4	400
Moderator	H_2O	H_2O	D_2O	D_2O	graphite	graphite	graphite	none
Pressure vessel	steel	steel	Zr tubes	Zr tubes	concrete	concrete	concrete	SS[a] tank
Inside diameter × height (m)	4.4 × 13	6.4 × 22	0.1 m dia.	0.13 m dia.	29 × 29	20 × 17.7	30.6 × 27.8	6.2 × 16
Wall thickness (mm)	215	160	5	5	3300	6400	4700	25
Power plant								
Steam pressure (psia/MPa)	860/5.9	970/6.7	570/3.9	900/6.3	665/4.7	2400/16.7	2400/16.7	2400/16.7
Temperature (C)	274	291	250	280	400	566	566	566
Station net thermal efficiency (percent)	33	33	31	33	31.4	43	43	43

[a] Stainless steel.
† Reprinted with permission for MIT Press.

PWR, however, only hot water is produced. The hot water is then transferred to a separate heat exchanger in which the thermal energy of hot water is utilized to generate steam. More description and discussion on these two systems will be presented later.

The prominent reactors using the heavy water as moderator are the Canadian deuterium-uranium reactor (CANDU) and the steam-generating heavy-water reactor (SGHWR). Both reactors use enriched uranium (UO_2) and have a separate heat exchanger for steam production. In the CANDU, the heavy water receives heat in the nuclear reactor and releases heat in the separate heat exchanger for steam production. Since the coolant heavy water flows through the tubes inside the reactor (i.e., pressurized tubes), there is no need for a pressure vessel like that in the PWR system. The SGHWR has a similar arrangement except that it used the light water as coolant in the reactor. While the heavy-water reactor does not have an efficiency as high as the PWR and BWR, it generally has considerably better overall utilization of the fission energy available in the nuclear fuel.

The gas-cooled reactors frequently use graphite as moderator, CO_2 as coolant and natural uranium as fuel. Since the fuel cladding is magnesium alloy (referred to as Magnox in England), the gas-cooled reactors (GCR) are often called Magnox reactors. The advanced gas-cooled reactor (AGR) is a result of continued development of the Magnox system. Again, the moderator is graphite and the coolant is CO_2. The AGR is designed to raise the system conditions to the level comparable to those in the fossil-fuel power plant.

Another kind of gas-cooled reactor is the high-temperature gas-cooled reactor (HTGR). Unlike the GCR and AGR, the HTGR uses helium as coolant which receives heat in the reactor and becomes a gas of high pressure and temperature. Next, the gas is passed through a steam generator in which the steam is generated for driving the turbines and producing mechanical work. After completing this process, the helium is returned to the reactor to start a new cycle. Because of high-temperature condition, the HTGR usually has an efficiency as high as the fossil-fuel system.

Table 6-2 summarizes the main characteristics for these reactors and for a typical fast breeder reactor. In addition to the fuel, coolant, and moderator, the table also includes the current maximum unit size, typical conditions of working substance and station thermal efficiency. For various reasons the light-water reactor (PWR and BWR) is the most popular one. Because almost all nuclear power systems are either PWR or BWR in the United States, the remainder of this chapter will be devoted to these two types of reactors.

6.3 PRESSURIZED WATER REACTOR SYSTEM

The steam supply system for a PWR plant consists of a pressurized water reactor, reactor coolant system, and associated auxiliary fluid systems. Figure 6-1 shows a typical steam supply system. The reactor coolant system is arranged as two or more closed coolant loops connected to the reactor vessel, each containing a reactor coolant pump and a steam generator. An electrically heated pressurizer is connected to one of the loops and is used to serve the whole system. The reactor vessel is

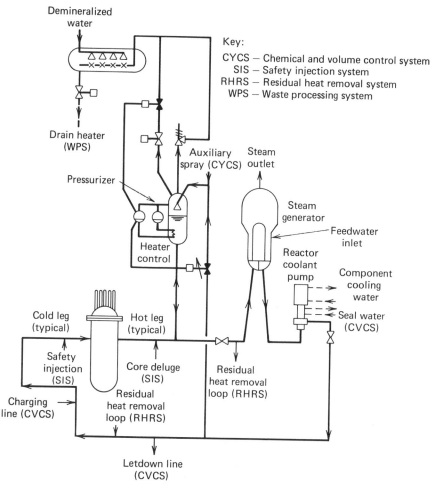

Figure 6-1. PWR steam supply system [2]. Reproduced with permission from the Keter Publishing House, Jerusalem, Israel.

cylindrical in shape with a hemispherical bottom and a flanged and gasketed removable upper head.

Figure 6-2 shows the cross section of a typical PWR reactor. The vessel contains the reactor core, and its support structures, coolant flow distribution devices, and the control rod assemblies. A typical reactor vessel for a 1100 MW reactor system is about 44 ft high, with an inner diameter of 15 ft and a vessel wall thickness of 9 in. The average coolant temperature in the vessel is about 600 F with a pressure of 2250 psia. This pressure provides a large amount of subcooling in the reactor core.

The reactor core is composed of uranium dioxide pellets enclosed in zircaloy tubes with welded end plugs. The tubes are supported in assemblies by a spring clip grid structure. The mechanical control rods consist of clusters of stainless steel clad absorber rod and zircaloy guide tubes located within the fuel assembly. Figure 6-3

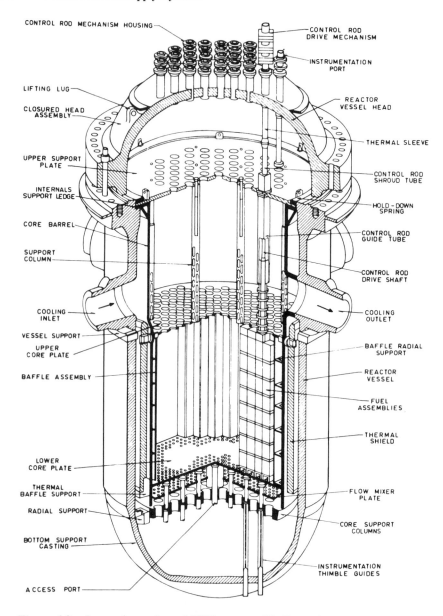

CONTROL ROD MECHANISM HOUSING
CONTROL ROD DRIVE MECHANISM
INSTRUMENTATION PORT
LIFTING LUG
REACTOR VESSEL HEAD
CLOSURED HEAD ASSEMBLY
THERMAL SLEEVE
UPPER SUPPORT PLATE
CONTROL ROD SHROUD TUBE
INTERNALS SUPPORT LEDGE
HOLD-DOWN SPRING
CORE BARREL
CONTROL ROD GUIDE TUBE
SUPPORT COLUMN
CONTROL ROD DRIVE SHAFT
COOLING INLET
COOLING OUTLET
VESSEL SUPPORT
BAFFLE RADIAL SUPPORT
UPPER CORE PLATE
REACTOR VESSEL
BAFFLE ASSEMBLY
FUEL ASSEMBLIES
THERMAL SHIELD
LOWER CORE PLATE
THERMAL BAFFLE SUPPORT
FLOW MIXER PLATE
RADIAL SUPPORT
CORE SUPPORT COLUMNS
BOTTOM SUPPORT CASTING
INSTRUMENTATION THIMBLE GUIDES
ACCESS PORT

Figure 6-2.   Isometric section of PWR reactor [2]. Reproduced with permission from the Keter Publishing House, Jerusalem, Israel.

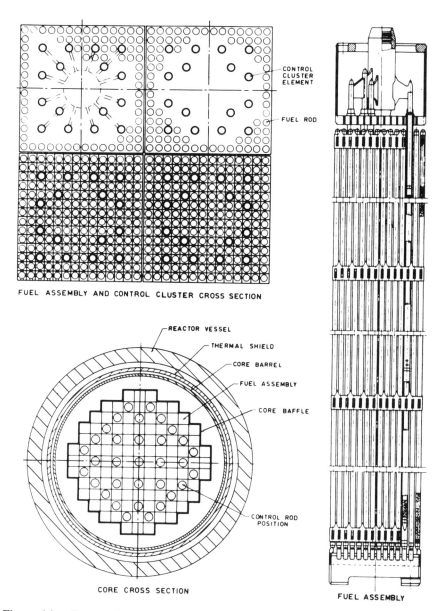

CONTROL CLUSTER ELEMENT

FUEL ROD

FUEL ASSEMBLY AND CONTROL CLUSTER CROSS SECTION

REACTOR VESSEL

THERMAL SHIELD

CORE BARREL

FUEL ASSEMBLY

CORE BAFFLE

CONTROL ROD POSITION

CORE CROSS SECTION

FUEL ASSEMBLY

Figure 6-3. Core section and fuel assembly of PWR reactor [2]. Reproduced with permission from the Keter Publishing House, Jerusalem, Israel.

presents a typical core section and fuel assembly. The fuel rods in a fuel assembly are generally arranged in a square array with 14 rod locations per side and a nominal centerline to centerline pitch of 0.556 in. between rods. Of the total possible 196 rod locations per assembly, 16 locations are occupied by control rods and one is for a central instrumentation sheath. The remaining positions are used for fuel rods.

The core fuel is located in three regions of varying enrichment. Figure 6-4 indicates that the outer assemblies have a 3.40% of the fuel, the intermediate 3.03%, and the central 2.27%. This variable enrichment is designed to generate a more even flux distribution. When the refueling is made, the fuel elements will move inward. The new fuel is introduced into the outer region, and the fuel in the central region is discharged to the spent fuel storage.

A one-piece thermal shield, concentric with the reactor core, is located between the core barrel and the reactor vessel. The shield is bolted and welded to the top of the core barrel. The shield, which is cooled by the coolant on its downward pass, protects the vessel by attenuating much of the gamma radiation and some of the fast neutrons that escape from the core. The shield also minimizes thermal stresses in the vessel, which result from the heat generated by the absorption of gamma energy.

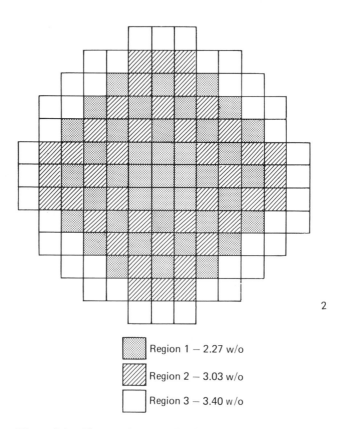

2

Region 1 — 2.27 w/o

Region 2 — 3.03 w/o

Region 3 — 3.40 w/o

Figure 6-4. Three region zone loading [13].

The reactor coolant enters the reactor vessel through inlet nozzles, which are located just above the core. The coolant first flows downward through the annular space between the vessel wall and the core barrel into a plenum at the bottom of the vessel. Then, the coolant reverses its direction and flows upward through the fuel core. The heated water (coolant) is collected at the upper plenum and exits the vessel through outlet nozzles, which are located at almost the same level as that of the inlet nozzles.

The pressurizer shown in Fig. 6-1 is used to maintain the coolant pressure during steady-state operation. It also limits the pressure changes caused by coolant thermal expansion and contraction during normal load transient and prevents the pressure from exceeding the design limit. The pressurizer contains replaceable direct immersion heaters, multiple safety and relief valves, a spray nozzle, and interconnecting piping and valves. The electric heaters located at the bottom of the vessel maintain the coolant pressure by keeping the water and steam in the pressurizer at the system saturation temperature. The pressurizer can also reduce the coolant pressure by spraying water into the steam volume.

Figure 6-5 presents a typical view of a pressurizer. A pressurizer is about 34 ft high with an inner diameter of 8 ft. The pressurizer contains saturated water and steam at the saturation temperature. It is usually constructed of low alloy steel with internal surfaces clad with austenitic stainless steel. The heaters are sheathed in austenitic stainless steel.

A reactor coolant system is usually designed with two or more coolant loops. Each loop contains its own steam generator and coolant pump. Steam generators are either a U-tube type or a once-through straight tube type. In both cases, the coolant passes through the tubes and the feedwater vaporizes outside. Figure 6-6 shows the design of a steam generator with U-tubes. The coolant enters the inlet side of the channel head at the bottom of the steam generator through the inlet nozzle, flows through the U-tubes to an outlet channel, and leaves the steam generator through another bottom nozzle. The inlet and outlet channels are separated by a partition. Manways are provided to permit access to the U-tubes and other internals.

Feedwater returning from turbine feedwater heater train enters the steam generator above the top of the U-tubes through a feedwater ring. The water first flows downward through the annulus between the tube wrapper and the generator shell and then, reverses its direction at the bottom and moves upward through the bundle. In the process, some of the water becomes steam. The steam-water mixture from the tube bundle passes through a steam swirl vane assembly in which steam is separated from the water. This moisture separator is usually located just above the tube bundle as shown in Fig. 6-6. The water from the moisture separator joins the incoming feedwater and again moves downward for another pass through the tube bundle. Meanwhile, the steam rises and passes through additional moisture separators. Under all load conditions the steam from the generator is expected to contain a moisture content not higher than 0.25%.

Figure 6-7 shows an once-through, straight-tube steam generator. The coolant enters at the top of the generator, passes through the straight-end bundles and leaves the generator at the bottom. Feedwater enters at the side of the shell through an inlet nozzle. The water first flows downward through an annulus where it is heated to the

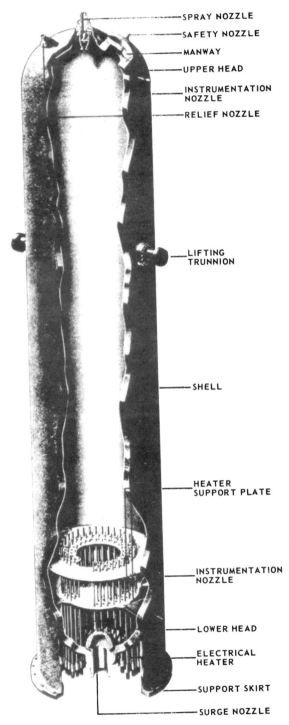

Figure 6-5. PWR pressurizer [13].

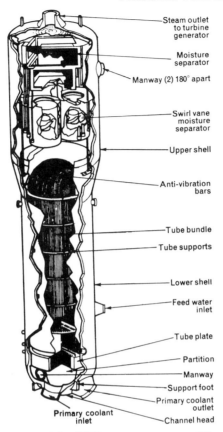

Steam outlet
to turbine
generator

Moisture
separator

Manway (2) 180° apart

Swirl vane
moisture
separator

Upper shell

Anti-vibration
bars

Tube bundle

Tube supports

Lower shell

Feed water
inlet

Tube plate

Partition

Manway

Support foot

Primary coolant
outlet

Primary coolant
inlet

Channel head

Figure 6-6. PWR steam generator [2]. Reproduced with permission from the Keter Publishing House, Jerusalem, Israel.

saturation temperature. Then, it reverses its direction at the bottom and moves upward through the tube bundle. In this process, feedwater is converted to steam and slightly superheated. As indicated in Fig. 6-7, steam leaves the tube bundle on the top of the generator, reverses its direction again, and exits the generator through an outlet nozzle.

The generator shell is usually constructed of low alloy steel. The heat transfer tubes are made of Inconel material. The surface in contact with the reactor coolant is clad with Inconel, while that in contact with feedwater is clad with austenitic stainless steel.

In addition to the steam generator, each coolant loop in the PWR has its own pump. Because of the PWR inherent characteristics, pumps are expected to circulate large amounts of coolant. Since the coolant is contaminated with radioactive materials, the pump must employ a well-designed seal system to restrict and control the leakage. Figure 6-8 shows a view of the controlled leakage pump frequently used in the PWR system. The coolant is pumped by the impeller attached to the bottom of the rotor shaft. The coolant is drawn up through the impellar, discharged through a discharge nozzle in the side of the casing. The rotor-impellar can be removed from

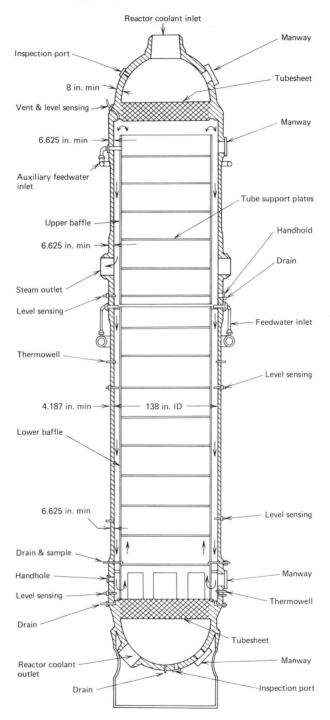

Figure 6-7. Section of once-through PWR steam generator [2]. Reproduced with permission from the Keter Publishing House, Jerusalem, Israel.

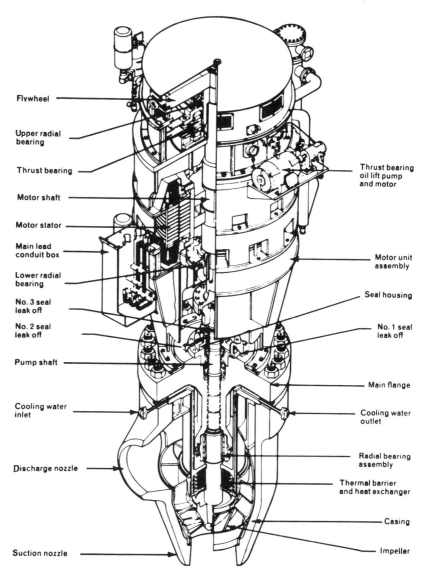

Figure 6-8. PWR coolant circulating pump with cascade shaft seals [2]. Reproduced with permission from the Keter Publishing House, Jerusalem, Israel.

the casing for maintenance or inspection without removing the casing from the piping. All parts of the pump in contact with the reactor coolant, except for the bearings and some special components, are austenitic stainless steel or equivalent corrosion resistant materials.

Several auxiliary fluid systems are needed in a nuclear steam supply system. These include those used to charge the reactor coolant system, to add makeup water, and to provide chemicals for corrosion inhibition and reactor control. Special cooling systems are also provided to cool the spent fuel storage pool and to remove residual heat when the reactor is shut down. Evidently, it is beyond the scope of this text to discuss such special topics. Those who are interested should consult the references [5,11].

6.4 BOILING WATER REACTOR SYSTEM

The nuclear steam supply system for a BWR plant mainly consists of reactor vessel and reactor coolant circuits. Figure 6-9 shows a schematic diagram of a steam supply system. Unlike the PWR, this system does not have the intermediate heat exchanger, or steam generator, between the coolant loop and the feedwater and steam system. Steam is generated within the nuclear reactor and transferred directly to the steam turbine. In other words, water acts as a coolant as well as a working substance in power plant cycle. Therefore, the pressure in the BWR vessel is generally much lower than that in the PWR and, thus, a smaller vessel wall thickness is used in the BWR.

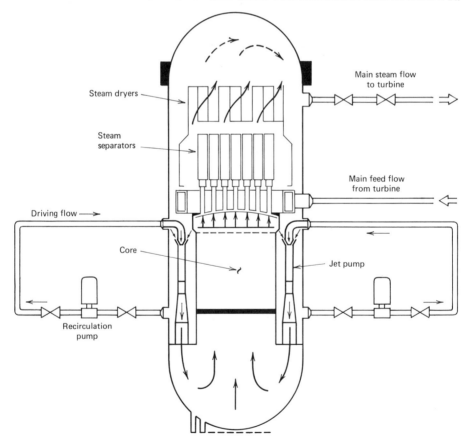

Figure 6-9.   Steam and recirculation water flow paths of BWR reactor [2]. Reproduced with permission from the Keter Publishing House, Jerusalem, Israel.

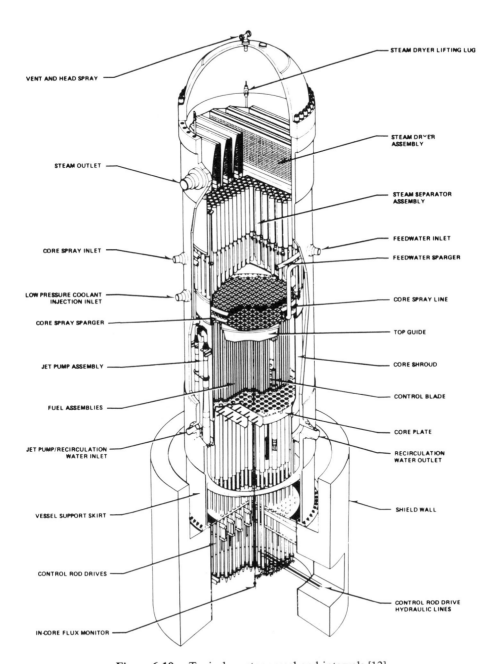

VENT AND HEAD SPRAY

STEAM OUTLET

CORE SPRAY INLET

LOW PRESSURE COOLANT
INJECTION INLET

CORE SPRAY SPARGER

JET PUMP ASSEMBLY

FUEL ASSEMBLIES

JET PUMP/RECIRCULATION
WATER INLET

VESSEL SUPPORT SKIRT

CONTROL ROD DRIVES

IN-CORE FLUX MONITOR

STEAM DRYER LIFTING LUG

STEAM DRYER
ASSEMBLY

STEAM SEPARATOR
ASSEMBLY

FEEDWATER INLET

FEEDWATER SPARGER

CORE SPRAY LINE

TOP GUIDE

CORE SHROUD

CONTROL BLADE

CORE PLATE

RECIRCULATION
WATER OUTLET

SHIELD WALL

CONTROL ROD DRIVE
HYDRAULIC LINES

Figure 6-10.   Typical reactor vessel and intervals [12].

The BWR pressure vessel contains the reactor core, its support structure, steam separator, and dryer assemblies. In addition, it also contains the control rod assemblies and the jet pumps that are used to circulate the coolant (water) internally. Figure 6-10 shows a view of a typical reactor vessel. For a 500 MW reactor system, the vessel is about 64 ft high, with an inner diameter of approximately 17 ft. The major connections to the pressure vessels include the four main steam lines, two feedwater lines, and pipings associated with jet pumps. The vessel is vertical and cylindrical in shape and is made of high-strength alloy carbon steel, designed for the conditions 1250 psia and 575 F. The vessel interior is clad with weld deposited E-308 electrode. The cladding thickness is approximately 0.125 in.

The reactor core is assembled in modules of four fuel assemblies set in the interstices of a cruciform control rod. The cross section of a fuel assembly is shown in Fig. 6-11. Each assembly has 49 fuel rods in square (7 × 7) array. Twenty-seven of these 49 rods have a reduced U-235 enrichment for an even power generation. The individual fuel rods are spaced and supported by the upper and lower tie plates.

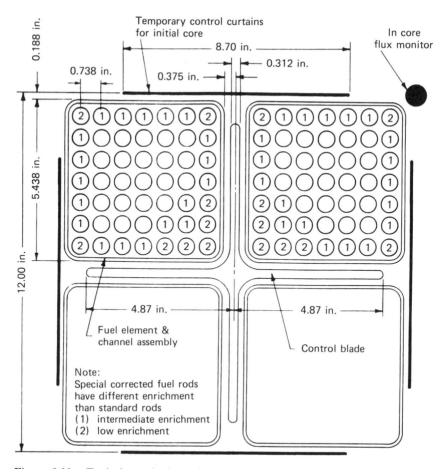

Figure 6-11. Typical core lattice unit [12].

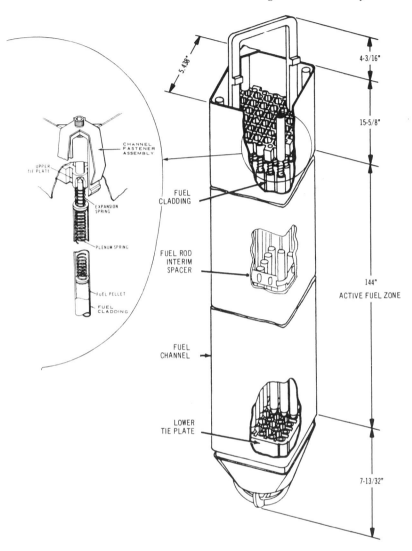

Figure 6-12.   BWR fuel assembly schematic [5].

Eight fuel rods are threaded into the bottom tie plate and attached to the upper tie plate by nuts. This arrangement facilitates a movement of fuel assembly in and out of the reactor core. The remaining rods have end plugs that are welded into the support plates. Figure 6-12 shows a typical fuel assembly.

The fuel assembly is contained in a canlike fuel channel that is made of zircaloy-4. The fuel channel is designed to provide rigidity and protection for the fuel rods during handling. In addition, it controls coolant flow paths and guides the control rods. For a 500 MW system, there are 484 fuel assemblies in the entire reactor core. The equivalent core diameter is approximately 12.5 ft.

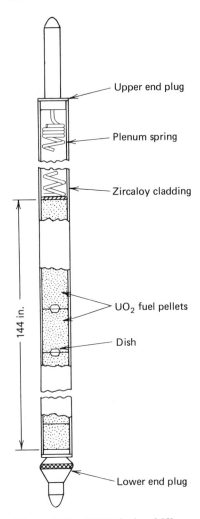

Figure 6-13.   BWR fuel rod [5].

As indicated in Figs. 6-11, 6-12, each fuel assembly contains 49 fuel rods. Each fuel rod contains uranium dioxide pellets approximately 0.5 inches in diameter. The pellets are enclosed in zircaloy-2 tubes. Figure 6-13 shows a schematic diagram for a fuel rod. In addition to fuel pellets, there is a plenum filled with helium and provided with a mechanical spring on top of the fuel rod. This arrangement is designed to accommodate axial expansion. To accomodate radial swelling, the pellet diameter is always somewhat less than the cladding inside diameter.

The zircaloy-2 tube is sealed by welding plugs in both ends. This cladding has a thickness of 0.032 in. and is capable of withstanding reactor pressure without collapsing against the fuel pellets. Dished pellets are used in the high-flux region near the center of the rod. The dish consists of a depression in the pellet and in the shape of a frustum of a cone. Undished pellets are used near the top and the bottom.

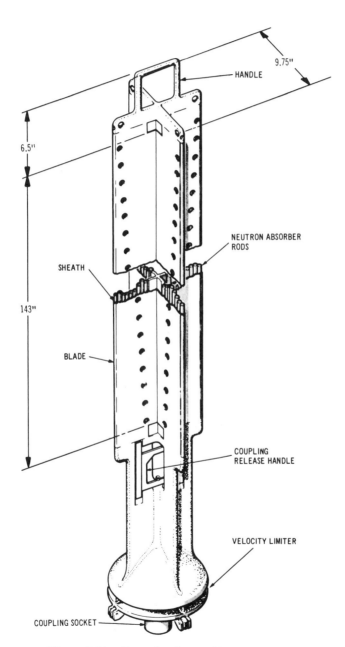

Figure 6-14.   Control rod assembly isometric [12].

Nuclear reaction is controlled by a combination of control rod movement and fixed control curtains. Both of these are shown in Fig. 6-11. The cruciform shaped control rods contain a number of vertical stainless steel tubes that are filled with boron carbide powder. Like the fuel rods just described, there is a plenum on the top of the control rod for helium and other gases generated during operation. These rods are held in cruciform array by a stainless steel sheath and have both ends well sealed as shown in Fig. 6-14. In the BWR, the control rods enter the reactor through the bottom. Their movement in the reactor core is controlled by individual, hydraulically operated driving mechanisms. Like the PWR system, provision is made to allow a high-speed insertion for rapid shutdown. In a 500 MW BWR system, there are 121 moveable cruciform shaped control rods. Each of these rods contains 84 tubes filled with control materials. These tubes have 0.188 in. outside diameter and 0.025 in. thickness.

The control curtains consist of boron-stainless steel sheets located in the narrow water gaps between fuel channels. Each curtain is approximately 141.25 in. long, 8.7 in. wide and 0.063 in. thickness. The curtains are supported by a hanger rod, which

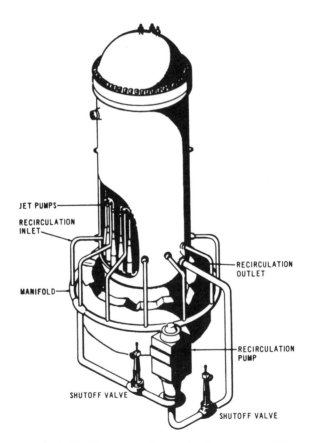

Figure 6-15.   Reactor coolant recirculation system [12].

attaches to the upper grid. The control curtains are used only at the time the initial core has excess reactivity.

Water (reactor coolant) enters the bottom of the reactor core and moves upward through the fuel assemblies where steam is generated. The steam is of low quality (approximately 9%) and is separated from the water by means of the steam separators and dryers in the top of the pressure vessel. The steam exits the vessel by passing through the main steam piping to the turbine. The unevaporated water mixes with the incoming feedwater and is returned to the core bottom inlet through jet pumps. These jet pumps are located in the gap between the core shroud and the pressure vessel. The driving force for the jet pumps is supplied by the water from the two coolant recirculation loops, which are located outside the pressure vessel.

The recirculation system is designed to force excess water through the reactor core and to improve the heat transfer performance. It consists of two recirculation pump loops and 20 jet pumps. Each loop has a motor-driven pump and several control valves. The loops are supported from the drywell structure. The two recirculation pumps discharge high-pressure water into manifolds from which connections are made to the jet pumps. Figure 6-15 shows an isometric view of this arrangement. As

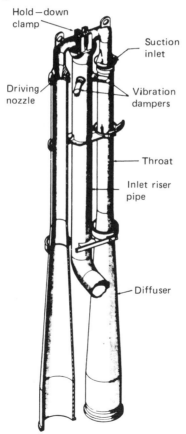

Figure 6-16. Jet pump [12].

indicated, these jet pumps are located inside the pressure vessel. There are 10 recirculation inlet pipes and only two recirculation outlet pipes. All these pipes penetrate the pressure vessel wall.

The jet pumps are designed to circulate the reactor coolant through the reactor core. The jet pumps have no moving parts and are operated on the principle of converting momentum to pressure. Figure 6-16 shows a pair of jet pumps supplied driving flow from a single riser pipe. These risers, as previously described, have individual vessel penetrations and receive flow from one of two recirculation inlet manifolds.

The jet pump material is stainless steel. The overall height from the top of the nozzle section to the discharge is approximately 20 ft and 10 in. In operation, more than 42% of the water flow goes through the recirculation loops and becomes the driving flow of the jet pumps. The remaining 58% is driven through the pumps by the driving water flow. Water flows out of the jet pumps at sufficient pressure to recirculate through the nuclear core.

6.5 CONTAINMENT SYSTEMS AND ENGINEERED SAFETY FEATURES

Radioactivity associated with nuclear power plant operations is generated in different ways. As previously indicated, nuclear fission produces radioactive fission fragments. While most of them stay within the nuclear fuel, small amounts leak through the cladding material and enter into the reactor coolant. In the nuclear fission process, large numbers of neutrons generated can also cause structural materials and coolant to become radioactive. Even though much of this induced radioactivity decays within the reactor, small amounts of radioactive liquids and gases are normally emitted from the power plants. The most potential radiation problem is, however, in the treatment of the spent fuel of which some elements such as plutonium could stay radioactive for many years. The disposal of the spent fuel is currently formulated at the national level.

Safety is of major concern to all people involved in the nuclear power generation business. The safety measures and regulations adopted for the design and operation of nuclear power plants have been far more restrictive than those in other fields. The actual safety record has also been good. In 1975, the Rasmussen report indicated that the risk of casualties in nuclear plant operations is much smaller than that in routine air and ground traffic [10]. The greatest potential danger with nuclear reactors has been, however, identified. It is the melting of the fuel core and subsequent release of radioactive materials into the atmosphere. The melting of the fuel core is generally caused by a loss of coolant in nuclear reactors. This event could result from a failure in the cooling system or a failure in the heat removal system following a shutdown of the reactor. The possibility of such a fuel core melting, therefore, must be taken into consideration in plant design. In fact, the plant containment systems and some engineered safety features are mainly designed for the event of a loss-of-coolant accident.

In the following sections are descriptions of the containment systems and some engineered safety features frequently employed in the PWR and BWR systems.

6.5.1 PWR System

The containment system consists of two separate structures: a reactor containment vessel and a shield building. The reactor containment vessel, sometimes called the primary containment, is a cylindrical steel pressure vessel with hemispherical dome and ellipsoidal bottom. It contains the reactor pressure vessel, the steam generators, the coolant pumps, the coolant loops, the reactor coolant pressurizer, and the accumulators of the safety injection system. The safety injection system is one of engineered safety features that will be discussed later. The shield building, sometimes called the secondary containment, has the shape of a right circular cylinder with a shallow dome roof. It houses the entire reactor containment vessel. An annular space is provided between these structures.

The containment systems are designed to provide protection for the plant operators and the public. They are designed according to the worst situation that results from a loss-of-coolant accident (LOCA). The reactor containment vessel is designed for a maximum internal pressure of 46 psia and a temperature of 268 F. For a 500 MW unit, the containment vessel has 105 ft inside diameter and 1.3 million ft^3 internal free space. In the normal plant operation the conditions inside the containment vessel are maintained by an air handling system. This air handling system removes the heat generated by the equipment inside the containment and maintains an environment of 120 F. In the event of loss of coolant, it can also remove the heat generated. The air handling system can also perform an internal cleanup function. It recirculates containment air through charcoal filters and provides fresh, tempered air to maintain the air quality. All air handling equipment, such as fans, are provided with an emergency power source.

The shield building is a reinforced concrete structure. The annular space between the reactor containment vessel and the shield building wall is used for construction operation and periodic inside inspection. For a 500 MW unit, this annular space may have 374,000 ft^3. The air conditions are maintained by a shield building ventilation system.

The shield building ventilation system consists of fans and ducts. The main function of this system is to collect the leakage from the containment vessel penetrations and discharge it through filters to the monitored containment system vent. The system is operated only after an abnormal event occurs. During this period the system is used to produce a slightly negative pressure within the annular space so that leakage from the shield building to the environment can be minimized. It also generates air movement or mixing in the annular space to reduce the average concentration level of radioactivity.

The normal air temperature within the shield building annulus is approximately the same as the air temperature inside the containment vessel. In the event of a loss-of-coolant accident, the air temperature will increase as the containment vessel wall temperature increases. This results in a slight pressurization of the shield building annulus. As soon as the ventilation system becomes operative, the annulus pressure will drop and stay at a pressure slightly lower than atmospheric pressure.

The engineered safety features are provided to minimize the consequences of the worst event (a loss-of-coolant accident). The PWR safety systems include:

1. Safety injection system.
2. Residual heat removal system.
3. Containment vessel spray system.

in addition to the containment air handling and the shield building ventilation systems. Basically, these systems are designed to move heat from the fuel in a loss-of-coolant accident, insert negative reactivity into the reactor, and reduce pressure of the containment vessel by removing thermal energy. They also remove the radioactive materials from the containment systems and minimize the radioactivity leakage from the nuclear power plant.

The safety injection system is used to automatically deliver cooling water to the reactor core in a loss-of-coolant event. This limits the fuel clad temperature and ensures that the fuel core does not melt. The injection process involves three different phases. First, the system injects borated water directly from the passive accumulators that are located inside the containment vessel. The second phase is the active injection in which the borated water is delivered by the safety injection pumps from the boric acid tank and from the refueling water tank. The final phase is the residual heat removal. The residual heat removal pumps move the spilled reactor coolant from the containment sump to the residual heat exchangers and then deliver some of the water to the nuclear core. The recirculation takes place in this closed water circuit. The balance of the water is used in the containment vessel spray system.

The heat generated in a loss-of-coolant accident must be removed from the containment vessel. This function is performed by two separate engineered safety feature systems: containment air handling system and containment vessel spray system. Either of these two systems alone will provide sufficient heat removal capability. These systems will serve as independent backups for each other. The air handling system including its cooling equipment has been described in connection with containment systems. The system is designed to cool the containment atmosphere in the event of a loss-of-coolant accident and to ensure that the conditions do not exceed its design values (46 psia and 268 F). The containment air handling system consists of four fan coil units, a duct distribution system, and its associated instrumentation and controls. The heat sink for the fan coils is provided by the cooling water system.

The containment vessel spray system is also designed to cool the containment atmosphere in the event of a loss-of-coolant accident. This system is to spray the borated water from the refueling water tank into the containment atmosphere. After the initial injection is completed, the borated water is recirculated by the containment pumps. To get rid of the heat received, a portion of the recirculation flow is diverted to the residual heat removal heat exchangers.

The containment vessel spray system consists of two spray pumps, spray nozzles and the associated piping and valves. The spray pumps are located in the auxiliary building and take suction directly from the refueling water storage tank. The containment vessel spray system also utilizes the two residual heat removal pumps and two residual heat exchangers of the safety injection system for long-term containment cooling.

In summary, the PWR system employs a safety injection system to limit the fission product from the nuclear core in the event of a loss-of-coolant accident. To minimize the radioactivity from the nuclear power plant, the PWR system employs two containment systems. In addition, it reduces the fission product concentration by spraying borated water in the containment system and by using the shield building ventilation system.

6.5.2 BWR System

Similar to the PWR system, the BWR system has two containments. The primary containment consists of a drywell, a pressure suppression chamber, and its associated pipings and control valves. The secondary containment is the reactor building that encloses the primary containment, the refueling facilities, and most components of the nuclear steam supply system.

The primary containment serves as a barrier to the release of fission product from the nuclear reactor in a normal operation or in the event of a loss-of-coolant accident. It can also reduce the pressure in the containment when the loss-of-coolant accident occurs. The drywell in the primary containment houses the reactor vessel and recirculation pumps. The drywell is a steel pressure vessel with a spherical lower portion and cylindrical upper portion. It is enclosed in reinforced concrete for shielding purposes. For a 500 MW unit, the drywell has a free space 134,000 ft^3 and an overall height of 106 ft. The wall thickness varies in the range of one and two inches. In addition, the primary containment also encloses a pressure suppression chamber and a connecting vent system between the drywell and the suppression chamber. The pressure suppression chamber stores a large volume of water. In the event of a loss-of-coolant accident within the drywell, reactor water and steam are released into the drywell air space. Then, the resulting increased pressure forces a mixture of air, steam, and water through the vents into the pool of stored water in the suppression chamber. There, the steam condenses very rapidly and the pressure in the drywell is therefore reduced.

The pressure suppression chamber is a steel pressure vessel in the shape of a torus below and encircling the drywell. It is supported on the reinforced concrete foundation slab of the reactor building. For a 500 MW unit, the suppression chamber contains up to 78,000 ft^3 of water.

The primary containment is designed for a pressure 56 psig and a temperature 281 F. The maximum pressure of the drywell and suppression chamber is 62 psig. The containment has its own cooling and ventilation system, which consists of four air coolers, ductwork, and fans. In a normal operation the drywell atmosphere is maintained at approximately 135 F.

The secondary containment houses the primary containment, fuel storage facilities, and other auxiliary systems. The main purpose is to minimize ground level release of airbourne radioactive materials to the environment, and to provide means for a controlled release of the building atmosphere if an accident should occur.

The air condition in the secondary containment is maintained by its own heating and ventilation system. The system includes fans, filters, cooling and heating coils. It has a capacity to remove the heat generated by the equipment and to maintain air temperatures for personnel comfort. The air pressure is at a slightly negative pressure to ensure a minimum leakage from the reactor building. Air flow in the containment is so directed that the air would move from areas of least potential for radioactive contamination to areas of greater potential contamination. The air is exhausted through a ventilation stack on the top of the reactor building. When radiation level in the containment is above a preset value, the normal ventilation system is shut down and exhausted air is directed to the standby gas treatment system.

The standby gas treatment system is designed to purge the primary and secondary containments. The system uses filters to remove radioactive particulates and charcoal absorbers to remove radioactive halogens. The flow from the standby gas treatment system is continuously monitored before it is released from the gas stack. The system has two separate full capacity units. In case that one unit fails to operate properly, another unit can be started automatically. Both units receive power from the emergency electrical supply.

The engineered safety features are provided to minimize the consequences of a loss-of-coolant accident. The BWR systems include:

1. Core spray cooling system.
2. Residual heat removal system.
3. High-pressure coolant injection system.
4. Automatic pressure relief system.

The main objectives of these systems are to prevent fuel cladding materials from melting and limit to a negligible amount of the metal water reaction in the event of a loss-of-coolant accident. Each safety system is designed to cover a specific range of accident conditions. All systems combined provide a redundancy of function to avoid undetected nuclear reactor failures.

The core spray cooling system is provided to deliver automatically cooling water to the reactor core in the event of a loss-of-coolant accident. The core spray starts when the reactor vessel pressure drops to 450 psig. The spray water limits the fuel clad temperature and ensures that the fuel core does not melt. The system has two independent spray cooling circuits. Each circuit is capable of performing the entire system cooling function. Cooling water is normally supplied from the suppression pool located inside the primary containment. In an accident, water is pumped to the circular sparger ring inside the reactor vessel inner shroud just over the core. Then, water is sprayed into the fuel bundles, receiving heat generated on the way down. The cooling water will cover the entire core in about eight minutes. Like other safety systems, the spray system is run by an emergency power source such as diesel generators.

The residual heat removal system removes heat generated from the nuclear reactor vessel in the event of a loss-of-coolant accident. This system also provides containment spray cooling and a heat removal from the suppression pool, which has been described in association with the primary containment. While the residual heat removal system is basically an engineered safety system, it is also used as a reactor normal shutdown cooling subsystem.

The entire system mainly consists of two heat exchangers, four main circulating pumps, and four service pumps. The equipment is connected by associated piping and valves. The circulating pumps are sized based on the flow required for the reactor cooling and the service pumps are sized based on the need of the heat exchangers. The system is so designed that one component failure would not affect the system performance.

Both the core spray cooling system and the residual heat removal system are independently designed to protect the reactor core in the event of a loss-of-coolant accident. These systems are automatically started when the reactor vessel pressure drops to a preset valve. However, there are some situations in which the line breaks are very small and the vessel pressure does not drop very rapidly. In these situations, the high-pressure coolant injection system would become operative. This system is a high-head, low-flow system, pumping water into the reactor vessel when the vessel pressure is still reasonably high. Water is either from the condensate storage tank or from the suppression pool. Water from either source is pumped into the reactor vessel through the feedwater line and the flow is distributed by the feedwater sparger. The cooling water is then mixed with the hot water in the reactor vessel. The vessel water eventually returns to the suppression pool for recycling. The pumps used in this system are turbine-driven, and the driving steam comes from the nuclear reactor.

When the high-pressure coolant injection system fails to deliver the required flow of cooling water to the reactor vessel, the automatic pressure relief system becomes operative. This system reduces the vessel pressure to such a level that the residual heat removal system starts to pump water into the reactor. Unlike the high-pressure coolant injection system, the system is a low-head and high-flow system. It is capable of delivering large amounts of water in a short time. When the residual heat removal system is in operation, the high-pressure injection terminates. The automatic pressure relief system is designed as a backup for the high-pressure coolant injection system. The pressure relief system is automatically actuated when the reactor simultaneously experiences high pressure and low water level. The system accomplishes reactor pressure reduction by releasing the steam to the suppression pool.

In summary, the BWR system employs several engineered safety features to minimize the impacts of a loss-of-coolant accident. The safety systems are so designed that they take various break sizes into consideration. In addition, the BWR system has a containment system to further control the release of radioactive materials from the plant to the environment.

6.6 SYSTEM DESIGN CONSIDERATIONS

The selection of the type of power plant that best meets the need of the utility company is a complex decision. Among the many factors that are considered are lifetime cost, reliability, availability of fuel, and the impact on the environment. This decision process is briefly described in Chapters 3 and 4. The nuclear reactor system, as the power plant's heat source, is one of the power plants that is evaluated. This section presents some of the system design considerations in the evaluation of reactor systems. These include fossil fuel versus nuclear fuel, and one reactor versus another reactor. Since saturated steam or slightly superheated steam is generated from nuclear reactors, the impacts on the balance of the plant must be considered.

Nuclear fuel is generally less expensive than fossil fuel. In the last 10 years (1973–1983), fossil fuel prices have increased drastically. Imported crude oil increased from $12 to $28 per barrel and domestic coal changed from $10 to $30 per ton. At the same period nuclear fuel has been relatively stable in price. It costs approximately $0.80 per million Btu for most utility companies. Unlike the cost of fossil-fuel power, nuclear power cost is relatively independent with the nuclear fuel price. Therefore, future change in fuel price will not so much affect the generation cost.

While a nuclear power plant has an advantage in fuel costs, it is generally more expensive in capital costs. In the last 10 years the cost of nuclear power plant construction has increased very rapidly. It changed approximately from $400 to $1100 per kilowatt. In addition to the usual inflation, the increase is mainly due to time-related items, which include the escalation and interest during the construction time. For the nuclear power plants recently completed, the escalation and interest costs during construction have been reported to be up to 50% of the total plant cost.

The decision process used to select nuclear fuel or fossil-fuel is very complex. Engineers must know the lifetime cost for each type. Also, they must know the impacts the new plants may have on the network system. These impacts include the system production cost, reliability, and operation. Obviously, the decision-making process is lengthy and requires the input information on the system load characteristics, present and future, the existing generating facilities, and possible future expansions. Other inputs may include the fuel cost and performance data for each individual generating unit. To accomplish this kind of generation expansion study including the selection of fossil or nuclear fuel, engineers frequently utilize the computer. They either develop their own computer programs or make use of computer programs available in the market. Among the most well-known programs are the Wien Automatic System Planning Package (WASP) and the Electri Generation Expansion Analysis System (EGEAS). The WASP was originally developed by R. T. Jenkins (Tennessee Valley Authority) and has been used in the last several years [7]. The EGEAS was recently completed and made available by the Electric Power Research Institute [8].

In addition to the above technical and economic considerations, engineers should take into account the environmental impacts of nuclear power plants. Unlike a fossil-fuel plant, there is no air pollution (such as SO_2 and particulates) associated with a nuclear power plant operation. The volume of gases remitted from a nuclear plant is much smaller than that for a fossil plant. While the amount of fuel used for a nuclear plant is relatively small, the amount of uranium ore handled at the mine is not necessarily small. The impacts of uranium mining must be counted as the part of the plant's environmental impact. Most nuclear plants operate at low thermal efficiency, 32 to 33% as compared with 39 to 41% for a fossil plant. Because of this, more energy is rejected in the cooling system of a nuclear plant than from a fossil-fuel plant. In other aspects (land use for power plants, and transmission line) the environmental problems are comparable for both fossil and nuclear plants.

In the recent years many nuclear construction projects have been delayed or cancelled. These decisions are definitely very costly. Among the many reasons for the delay and cancellation is a lack of public acceptance of nuclear plants. The public does not perceive the safety problems of nuclear plants in the same way as the people involved in nuclear power business. Before this social environment is changed, the public attitude toward the nuclear power plant must be taken into account.

Table 6-2 shows the typical reactor characteristics. To select a particular type of reactor for a given power plant project, engineers consider not only the plant efficiency, but also the actual cost. In other words, engineers should evaluate the design features of each reactor system and convert them to the fixed and variable costs for comparison. For example, the extra fixed cost for the PWR steam supply plus their periodic maintenance costs are evaluated against the fixed cost of extra shielding in the turbine building of a BWR plus the operation and maintenance cost resulting from radioactive steam in the turbine. The plant efficiency for reactor systems producing superheated steam may be little higher, but attention should be given to the cost associated with it. Unlike a fossil-fuel plant, the nuclear power plant fuel cost is related not only to the plant efficiency, but also to the fuel utilization in the reactor core. The reactor design determines the ratio of the fuel generated by conversion to the fuel burned. This ratio can be improved by minimizing the amount of neutron-absorbing material in the reactor core. This usually results in high capital cost.

In a PWR system evaluation, consulting engineers should study the reactor coolant pressure in the reactor. Higher pressure results in higher steam pressure and therefore higher plant efficiency. The temperature rise of the coolant through the reactor and other hydraulic design features (such as coolant flow and pressure drop) are also important. All these design features will eventually affect the plant performance and then the plant production cost. When the nuclear reactor is evaluated, its containment system must be included. The containment system design varies with the reactor. Different reactor requires different containment system. The containment cost will have some influence in the final decision.

Refueling of a nuclear reactor is one of the important considerations in the reactor evaluation. Reactor design will determine the duration and frequency of the periodic plant shutdown to replace the nuclear fuel. These shutdowns affect the variable cost of the network. The reactor requiring more time for refueling must be penalized and the reactor requiring less time should be given some appropriate credits. These penalty (credit) determinations are usually influenced by the actual situation of each network.

While the major design features should be optimized in terms of financial return, the reactor's safety features must be evaluated by using different criteria. These criteria include the protection of operating and maintenance staff against radiation, the protection of the environment, and consequently of the population. The nuclear reactor system is so designed that it must observe the accepted limits for the contamination of air and water. In the event of a loss-of-coolant accident, the nuclear system will not release radioactive material higher than that legally allowed.

The fact that today's nuclear reactor can only generate saturated steam or slightly superheated steam, has serious implications in the power plant system design. As steam expands inside the turbine, moisture forms along its path. The moisture must be removed for it will affect the turbine internal efficiency as well as its reliability. Generally, the moisture is limited to about 14%. For a nuclear plant the turbine stages are specially designed for effective moisture separation. The separated water is passed to the feedwater heating circuit after every three turbine stages. In addition, a large moisture separator is provided in the crossover connection between the high-pressure and low-pressure turbine cylinders. This separator is usually combined with a reheater in the turbine cycle system.

As indicated in Chapter 4, nuclear power plants are only economical if large units are installed. This large unit size plus relatively poor steam conditions makes it necessary to have a very large steam volume through the turbine cylinders. This is especially true at the low-pressure end. To avoid overstressed turbine blades, turbines in nuclear plants frequently have a speed of 1800 rpm rather than the 3600 rpm used for turbines in fossil plants.

Figure 6-17 shows a typical turbine cycle arrangement with a nuclear steam supply system. The cycle arrangement differs from the fossil one in two main areas. First, the moisture separator is needed and installed between the HP and the LP turbine cylinders. Second, the reheater is supplied with high-temperature steam as a heat source. For the cycle arrangement shown in Fig. 6-17, the reheating process takes place in two stages, first by the bleed steam and then by the throttle steam. Like the fossil-fuel power plant, the reheat temperature is equal to or less than the throttle steam temperature.

Every power plant has to discharge significant amounts of heat to its surroundings. This thermal discharge becomes much more serious when the nuclear reactor is used as a thermal energy source. This is because of the large unit size and low thermal cycle efficiency. In a fossil-fuel power plant a 10% of the thermal energy will leave through the tall chimney, and in a nuclear power plant the entire heat rejection must be borne by the condenser. Therefore, an adequate cooling system must be provided to handle the rejected heat and to minimize the impact on the plant environment. There is no doubt that additional cost is needed for nuclear plant cooling. This should be taken into account in the system design. Various cooling systems will be presented in Chapter 12.

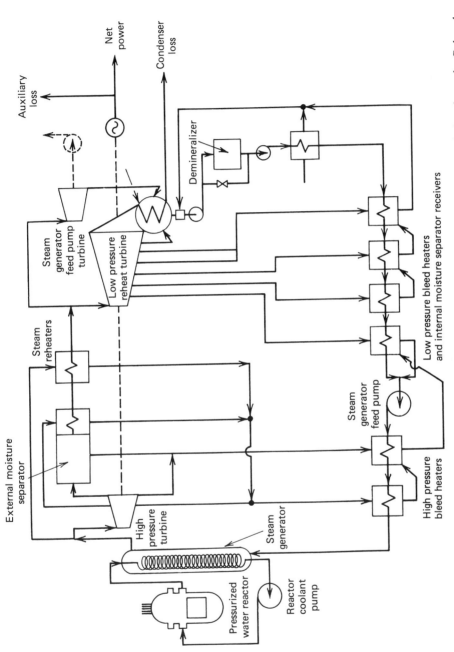

Figure 6-17. Typical turbine cycle arrangement for a PWR system [1]. Reproduced with permission from the Babcock and Wilcox Company.

These are just a few examples of many design system problems that are considered by engineers in the evaluation of nuclear reactor systems. In most cases it is important that the fixed and variable costs of the major design features be evaluated in order to determine the lifetime cost of each system.

SELECTED REFERENCES

1. Babcock and Wilcox Company, *Steam/Its Generation and Use*, 1973.

2. F. S. Aschner, *Planning Fundamentals of Thermal Power Plants*, Israel University Press, 1978.

3. Central Electricity Generating Board, *Modern Power Station Practice*, Vol. 8, Pergamon Press, 1971.

4. A. P. Fraas, *Engineering Evaluation of Energy Systems*, McGraw-Hill, 1982.

5. A. R. Foster and R. L. Wright Jr., *Basic Nuclear Engineering*, second edition, Allyn and Bacon, 1973.

6. P. G. Hill, *Power Generation*, MIT Press, 1977.

7. R. T. Jenkins and D. S. Joy, "Wien Automatic System Planning Package (WASP) —An Electric Utility Expansion Planning Code," Oak Ridge National Laboratory, July 1974.

8. Electric Power Research Institute, "Electric Generation Expansion Analysis System," an EPRI publication, 1982.

9. F. G. Baily, K. C. Cotton and R. C. Spencer, "Predicting the Performance of Large Steam Turbine-Generators Operating with Saturated and Low Superheat Steam Conditions," Proceedings of American Power Conference, 1967.

10. U.S. Nuclear Regulatory Commission, "Reactor Safety Study—An Assessment of Accident Risks in U.S. Commercial Nuclear Power Plant" (The Rasmussen Report), November 1975.

11. A. F. Henry, *Nuclear Reactor Analysis*, The MIT Press, 1975.

12. Northern States Power Company, "Final Safety Analysis Report for the Monticello Nuclear Generating Plant," 1971.

13. Northern States Power Company, "Final Safety Analysis Report for Prairie Island Nuclear Generating Plant," 1973.

PROBLEMS

6-1. It is known that the generating cost of nuclear plant is not sensitive to the fuel. Estimate the percentage change in power generation cost when the fuel cost is increased from 0.80 to $1.50/million Btu. The plant parameters are capital cost $1100/kW, capacity factor 62%, plant net heat rate 11,000 Btu/kWh, annual fixed charge rate 18%, and 0 & M cost 5 mills/kWh.

6-2. A helium-cooled high-temperature reactor (HTGR) is known to have thermal efficiency higher than the pressurized water reactor (PWR). However, the fuel for a HTGR is highly enriched and therefore is more expensive. Estimate the justifiable cost in competition with a PWR. The pressurized water reactor costs $1100/kW. The fuel costs are $0.80 and $1.30/million Btu, respectively, for PWR and HTGR.

Other financial and operating conditions are assumed to be identical.

6-3. The nuclear power plant is economical only if the plant has a high capacity factor. Estimate the power generation cost as a function of capacity factor, using the parameters specified for Problem 6-1.

6-4. It is frequent to compare a coal-fired plant with a nuclear plant for base-load operation. Estimate the generation cost for both types using the following information

	COAL	NUCLEAR
Capital cost ($/kW)	800	1100
Fuel cost ($/$10^6$ Btu)	3.00	0.80
Annual fixed charge rate (%)	18	18
Operation and maintenance cost, mills/kWh	6	4
Capacity factor (%)	62	62
Plant net heat rate (Btu/kWh)	9200	11,000

6-5. Compare the construction cost of a natural-draft cooling tower for coal and nuclear plants. The following parameters should be utilized.

	COAL	NUCLEAR
Unit size kW	800,000	800,000
Plant net heat rate (Btu/kWh)	9200	11,000
Tower unit cost ($/$10^3$ Btu)	5.2	5.2

Assume that the coal plant has 10% thermal discharge through the plant stack.

6-6. Consulting engineers receive two nuclear reactor proposals. In an average year, one reactor requires three days more than another reactor in refueling. Estimate the penalty cost for the reactor requiring more refueling time. The penalty cost should be capitalized. The data for this evaluation are:

System replacement cost (levelized value)	60 mill/kWh
Rate of return	13%
Annual fixed charge rate	18%
Unit size (same for both units)	800,000 kW
Plant economic life	35 yr

CHAPTER SEVEN

Steam Turbine Systems

7.1 STEAM TURBINE

A steam turbine operates by converting the thermal energy of steam into kinetic energy and then into mechanical energy. The history of steam turbine development is a long one. It started in about 150 B.C., when Hero of Alexandria built his first crude device, a rotating-reaction, nozzled-equipped sphere. This device was a pure reaction type and generated no useful work.

In 1831 Foster and Avery obtained a United States patent for a reaction wheel similar to Hero's. In 1882 Gustaf de Laval applied the turbine principle to a prime mover for his cream separator. A few years later he produced a series of small impulse turbines. Almost at the same time Sir Charles S. Parsons developed a reaction turbine and used it in a marine application. During the period of 1896, an American, C. G. Curtis, developed another kind of impulse turbine.

In the last 70 years, active development of steam turbine has made it the principal prime mover of generating stations. Figure 7-1 indicates steam turbine-generator size growth. At the present the average maximum unit size is approximately 600,000 kW for a single shaft fossil unit. It demonstrates the progress made from the 5000 to 30,000 kW turbines frequently built in the 1920s. There have also been some significant changes in steam conditions. In the 1920s most units used 200 psia and 550 F steam. But at the present most units are designed for steam of 2400 psia and 1000 F. Of 31 units completed in 1980–82 only three were designed for supercritical steam with an initial pressure of 3500 psia and initial and reheat temperatures of 1000 F.

Steam turbines are generally classified into two groups, impulse and reaction. In impulse turbines steam expands in stationary nozzle to attain a high velocity and then flows over the moving blades, converting some of its kinetic energy into mechanical work. In reaction turbines steam expands both in stationary nozzle and moving blades. The relative amount of expansion between them varies from one design to another. In practice, however, the turbines used in power generation always have both impulse as well as reaction sections.

Steam turbines have many stages, each of which generally consists of one row of stationary nozzles and one row of moving curved blades. Each stage is designed to convert a certain amount of thermal energy into mechanical work. Stage efficiency is defined as the ratio of mechanical work produced by the stage to the thermal energy available. Figure 7-2 indicates stage process in an h-s diagram. In term of steam enthalpy, the stage efficiency is

$$\eta_s = \frac{h_1 - h_2}{h_1 - h_{2s}} \qquad (7\text{-}1)$$

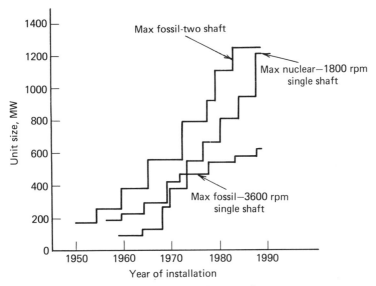

Figure 7-1. Steam turbine-generator size growth.

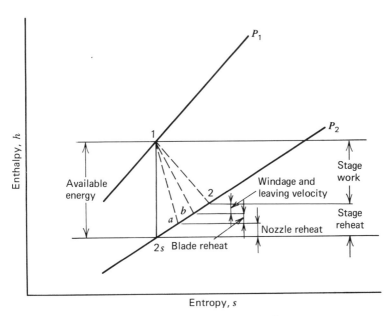

Figure 7-2. An *h-s* diagram for a turbine stage action.

Figure 7-2 also indicates friction effects on the enthalpy of steam. The frictional energy returns to steam in form of enthalpy. This phenomenon is frequently called reheat.

Similar to the stage efficiency, the diagram efficiency (η_d) frequently appears in literature. The diagram efficiency is defined as the product of nozzle efficiency (η_n) and blade efficiency (η_b). These efficiencies are in terms of the enthalpies in Fig. 7-2.

$$\eta_{diagram} = \eta_n \times \eta_b \qquad (7\text{-}2)$$

where

$$\eta_n = \frac{h_1 - h_a}{h_1 - h_{2s}} \qquad (7\text{-}3)$$

and

$$\eta_b = \frac{h_1 - h_b}{h_1 - h_a} \qquad (7\text{-}4)$$

The diagram efficiency is understood to include the effects of all losses due to friction in the nozzles and buckets. But the total stage loss is always larger than the sum of those losses in the nozzle and moving buckets. The additional losses are generally due to steam leakage through the diaphragm packing, windage, and nozzle-end losses for partial peripheral admission.

Figure 7-3 shows an *h-s* diagram for a multistage turbine. Because of the reheat in each stage and the diverging trend of constant pressure lines in the Mollier diagram, the sum of the available energy for all stages is always larger than the single isentropic expansion between the inlet and discharge pressure. The ratio of these two terms is frequently referred to as the reheat factor:

$$R = \frac{\sum (\Delta h)_s}{(\Delta H)_s} \qquad (7\text{-}5)$$

To measure the performance of a steam turbine as a unit, the internal efficiency η_i is frequently used. The term is defined as

$$\eta_i = \frac{\sum (\Delta W)}{(\Delta H)_s} \qquad (7\text{-}6)$$

where

$$\sum (\Delta W) = \text{sum of internal work generated in all turbine stages}$$
$$(\Delta H)_s = \text{isentropic enthalpy drop for the turbine}$$

If the stage efficiency remains unchanged throughout the steam turbine, the internal efficiency will simply become a product of the reheat factor and stage efficiency. That is,

$$\eta_i = R\eta_s \qquad (7\text{-}7)$$

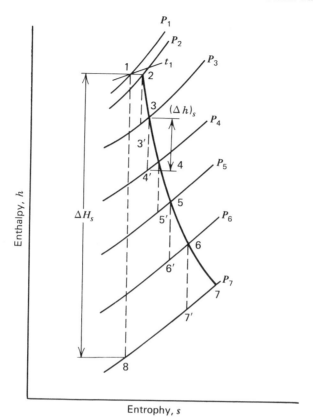

Figure 7-3.   An *h-s* diagram for a multistage turbine.

Table 7-1 indicates the range of values for these efficiencies. However, it is beyond the scope of this text to present the design variables affecting the internal and other efficiencies. There are several excellent books dealing with this subject. *Steam Turbine Theory and Practice* [2] is particularly recommended.

In practice, the internal efficiency of a steam turbine does not include the loss at the turbine exhaust end. The exhaust end loss occurs between the last stage of

Table 7-1
Estimates of Various Efficiencies

Stage efficiency	0.84–0.90
Diagram efficiency	0.86–0.92
Nozzle efficiency	0.92–0.95
Reheat factor	1.04–1.10
Internal efficiency	0.85–0.92
Blade efficiency	0.92–0.95

low-pressure turbine and the condenser inlet, and very much depends on the absolute steam velocity. The exhaust end loss generally includes (1) actual leaving loss, (2) gross hood loss, (3) annulus-restriction loss, and (4) turn-up loss. Figure 7-4 indicates a typical exhaust end loss curve showing the distribution of the component losses. Under full-load condition the exhaust end loss is generally around 3% of the turbine's available energy. One means of reducing the exhaust loss is to reduce the absolute steam velocity at the last stage by increasing the last stage blade length or the number of steam flows. As shown later in this text, the larger the exhaust end, the better the turbine performance will be.

It is obvious that because of the turbine exhaust end loss, the steam conditions at the condenser inlet are different from those at the exit of last turbine stage. The steam conditions at the exit of the last stage are referred to as those at the expansion-line-end point (ELEP). The steam conditions at the condenser inlet are those at the used-energy-end point (UEEP). By definition the enthalpy difference between these two points is approximately equal to the turbine exhaust end loss. This topic will be discussed further in Section 7.4.

In addition to the internal efficiency and exhaust end loss, other parameters affecting the turbine performance are the gland and valve stem leakage losses (packing losses), mechanical turbine efficiency (bearing losses), and generator efficiency. In general, they constitute an additional small loss. In modern design

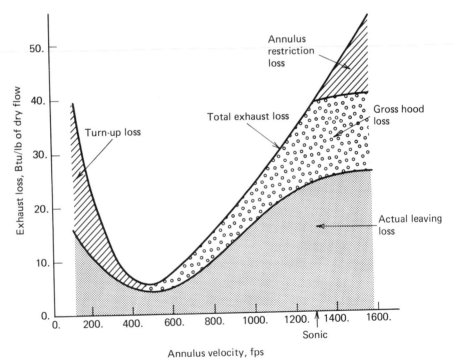

Figure 7-4. Typical exhaust loss curve showing distribution of component losses [4]. Reproduced with permission from McGraw Hill.

mechanical efficiency is somewhere between 99 and 99.5%, and generator efficiency is 98.5 to 99% for hydrogen-cooled generator.

Steam-turbine capacities range from a few kilowatts to over 1,000,000 kW. Inlet pressures range from a few pounds above atmosphere to the supercritical and temperatures from the saturated to over 1000 F. Speeds for generator drives are 3600 and 1800 rpm, while those for geared units can be 10,000 rpm or higher. To achieve a high capacity, a turbine may need more than one casing or one shaft. Figure 7-5 indicates typical arrangements of large condensing turbines. The arrangements *a* and *b* are often referred to, respectively, as the tandem-compound two flows and the tandem-compound four flows. The arrangements *c* and *d* are so called the cross-compound two flows and four flows. In general, the cross-compound arrangements are only considered for high-capacity machines.

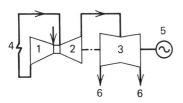

(*a*) Tandem-compound 2 flows from 150 to 400 Mw.

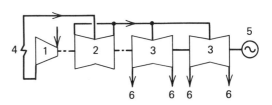

(*b*) Tandem-compound 4 flows from 300 to 800 Mw.

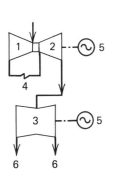

(*c*) Cross-compound 2 flows from 300 to 800 Mw.

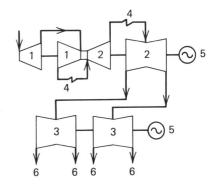

(*d*) Cross-compound 4 flows from 800 to 1200 Mw.

1. High pressure turbine	4. Reheater
2. Intermediate pressure turbine	5. Generator
3. Low pressure turbine	6. To condenser

Figure 7-5. Examples of casing and shaft arrangements of large condensing turbines [5]. Reproduced with permission from the Keter Publishing House, Jerusalem.

7.2 TURBINE-CYCLE HEAT BALANCE

The performance of a steam electric power plant is influenced not only by the steam turbines, but also by the choice of steam turbine cycle. The basic cycle used for electric power generation is the Rankine cycle, named after Professor William John Macquorn Rankine (1820–1872). Figure 7-6 shows a Rankine cycle with condensing turbine. It mainly consists of four components: steam generator, turbine, condenser and pump. Most thermodynamic textbooks show that the cycle's thermal efficiency would increase either by reducing the condenser pressure or by increasing the turbine inlet pressure and temperature.

The actual steam turbine cycle is much more complicated than that in Fig. 7-6. The principal cycle considerations are those of regenerative feedwater heating by turbine extraction steam and of interstage reheating. Figure 7-7 indicates a typical turbine cycle for a conventional fossil-fuel power plant. The plant has one stage of steam reheat and seven stages of regenerative feedwater heating. The steam conditions are supercritical, with an initial pressure of 3515 psia and initial and reheat temperatures of 1000 F. The condenser pressure in this cycle arrangement is set equal to 2.5 inches Hg abs. As indicated in Fig. 7-7, the feedwater pump is driven by an auxiliary steam turbine and the air preheater is supplied with the extracted steam from the main turbine. Omitting some minor cycle losses such as gland leakages, the turbine-generator system has a net output of 726,171 kW and net heat rate of 7823 Btu/kWh. The complete heat balance calculations are presented in Appendix A. Basically, the calculations are done on the basis of energy and mass conservation principles.

The general objective of heat and mass balances such as those in Fig. 7-7 is to determine the system performance that is frequently measured in terms of turbine net heat rate or gross heat rate. There is some confusion of these terms in literature. In this book the following set of definitions are used.

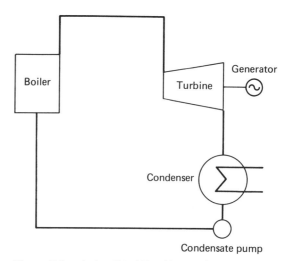

Figure 7-6.   A simplified Rankine cycle.

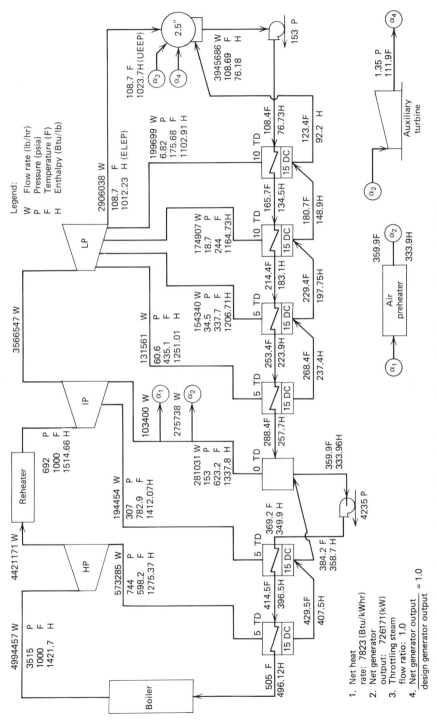

Figure 7-7. Heat balance diagram for the given design conditions (throttle steam flow ratio; 1.0).

Legend:

W	Flow rate (lb/hr)
P	Pressure (psia)
F	Temperature (F)
H	Enthalpy (Btu/lb)

1. Net heat rate: 7823 (Btu/kWhr)
2. Net generator output: 726171(kW)
3. Throttling steam flow ratio: 1.0
4. Net generator output = 1.0
   design generator output

229

The turbine gross heat rate (GHR) is

$$GHR = \frac{\text{heat input}}{\text{generator output}} \qquad (7\text{-}8)$$

for the cycle arrangement with motor-driven boiler feedwater pump,

$$GHR = \frac{\text{heat input}}{\begin{array}{c}\text{generator output} + \text{turbine shaft}\\\text{output for feedwater pumps}\end{array}} \qquad (7\text{-}9)$$

for the cycle arrangement with shaft-driven boiler feedwater pump, and

$$GHR = \frac{\text{heat input}}{\text{generator output} + \text{auxiliary turbine output}} \qquad (7\text{-}10)$$

for the cycle arrangement with turbine-driven boiler feedwater pumps.

In the power plant design the turbine net heat rate is frequently applied. This term is defined as

$$NHR = \frac{\text{heat input}}{\begin{array}{c}\text{generator output} - \text{electric power}\\\text{required for feedwater pumps}\end{array}} \qquad (7\text{-}11)$$

for the arrangement with motor-driven feedwater pumps, and

$$NHR = \frac{\text{heat input}}{\text{generator output}} \qquad (7\text{-}12)$$

for the cycle arrangement with shaft-driven or auxiliary turbine-driven feedwater pumps. It must be pointed out that the heat input in these definitions means the amount of heat received by the steam in the steam generator and should have a unit of Btu/hr. The generator output or power required for feedwater pump operation must be in unit of kilowatts.

To determine the performance of power plant as an integrated unit, one must also take into consideration the steam generator efficiency (sometimes called the boiler efficiency) and the plant auxiliary power requirement. The term used to measure the plant performance is the plant net heat rate ($PNHR$) defined as

$$PNHR = \frac{NHR}{\eta_{\text{boiler}}\left(1 - \dfrac{\%\ \text{auxiliary power}}{100}\right)} \qquad (7\text{-}13)$$

From Fig. 7-7, we can see many variables affecting the turbine net heat rate and net output. The following list shows the variables that frequently require consulting

Table 7-2
Effects of Cycle Parameters on Turbine Net Heat Rate and Net Output

A. *Condenser Pressure*						
CP, in. Hg abs.	2.0	2.5[a]	3.0	3.5	4.0	4.5
Relative net						
output (kW)	2,914	—	− 4,073	− 9,078	− 14,091	− 19,430
Relative net heat						
rate (Btu/kWh)	− 31.3	—	44.1	99.0	154.8	215.1
B. *Pressure Drop in Boiler*						
(including the piping						
to the turbine)						
PD (%)	15	17[a]	19	21		
Relative net						
output (kW)	346	—	− 345	− 732		
Relative net heat						
rate (Btu/kWh)	− 4.0	—	3.9	8.3		
C. *Pressure Drop in Reheater*						
PD (%)	5	7[a]	9	11		
Relative net						
output (kW)	1,403	—	− 1,445	− 2,935		
Relative net heat						
rate (Btu/kWh)	− 17.8	—	18.3	37.1		
D. *Pressure Drop in*						
Cross-Over Pipe						
(between IP and LP turbines)						
PD (%)	0[a]	3	5	7		
Relative net						
output (kW)	—	− 1,824	− 3,037	− 4,318		
Relative net heat						
rate (Btu/kWh)	—	19.7	33.2	46.8		
E. *Boiler Steam Temperature*						
(including superheated and						
reheat temperature)						
Steam temperature (F)	950	1,000[a]	1,050	1,100		
Relative net						
output (kW)	− 35,474	—	34,585	68,738		
Relative net heat						
rate (Btu/kWh)	118.9	—	− 106.0	− 202.9		
F. *Air Preheater*						
Use of air						
preheater	Yes[a]	No				
Relative net						
output (kW)	—	8,291				
Relative net heat						
rate (Btu/kWh)	—	− 88.3				

This figure denotes the value for the base cycle.

engineers' attention:

- Main steam pressure and temperature
- Reheat steam temperature and pressure
- Boiler and reheater pressure drops
- Condenser pressure
- Number of feedwater heaters
- Type of feedwater heaters
- Heater drain disposals
- Drain cooler approach
- Heater terminal temperature difference
- Steam extraction for auxiliary turbine
- Steam extraction for industrial usage
- Exhaust-end loss of low-pressure turbines

Effects of some of these parameters are presented in Table 7-2. These effects are expressed in terms of relative net output and net heat rate. The turbine-cycle arrangment shown in Fig. 7-7 is used as the base cycle. For instance, when the condenser pressure is increased from 2.5 to 4.5 in. Hg absolute, the turbine cycle system will experience a drop of 19,430 kW in net output and an increase of 215.1 Btu/kWh in net heat rate. This information coupled with the project economic factors will enable engineers to optimize the turbine exhaust pressure.

Turbine-cycle heat balance is time-consuming. Also, there are many possible cycle arrangements to be considered. Therefore, computer programs are frequently utilized for this particular purpose.

7.3 GENERALIZED HEAT BALANCE COMPUTER PROGRAM

There are at least two kinds of heat balance computer programs. One is designed and developed for a specific power plant while another is for a general study purpose. Evidently, the first kind will generate accurate information such as turbine net output and net heat rate. However, the program usefulness is limited to one system or one plant. The second kind of heat balance program is developed for general purpose study and with no particular power plant in mind. Therefore, it must take many different cycle arrangements into consideration. These may include type of heaters, drain disposals, and driving mechanism for feedwater pumps. To make the program manageable, minor losses such as gland leakage are generally omitted. The generalized heat balance program is frequently used to compare cycles that have been varied by changes in cycle parameters such as the number of feedwater heaters and terminal temperature differences. The differentials in the turbine net heat rate will reflect accurately the effects of changes in the cycle parameters.

The generalized heat balance program is expensive in its development and usually treated as proprietary material. One of the computer programs developed for this purpose is the Advanced Generalized Heat Balance Program (AHBP). The AHBP program consists of one main program and 23 subroutines. In addition, the subroutines prepared to cover the ASME steam table are included. The main program has 11 sections:

1. Nomenclature
2. Inputs and data file
3. Printing and checking of important input data
4. Calculation of turbine exhaust end conditions
5. Calculation of steam extraction conditions
6. Calculation of steam extraction rates
7. Calculation of pump work
8. Calculation of turbine work and generator output
9. Heat rate calculation
10. Partial-load calculation
11. Computer output printing

The simplified flowchart indicating these eleven sections is shown in Fig. 7-8. At the present the AHBP program can study the effects of variation of the following parameters:

- Main steam pressure and temperature
- Reheat steam pressure and temperature
- Boiler and reheater pressure drops
- Condenser pressure
- Number of feedwater heaters
- Type of feedwater heaters
- Method of heater drain disposal
- Drain cooler approach
- Feedwater heater terminal temperature difference
- Steam extraction for the auxiliary steam turbine
- Steam extraction for an industrial usage
- Turbine exhaust end loss
- Steam extraction pressures
- Different drive mechanism for the boiler feedwater pump
- Part-load performance calculations

The complete program document including the user's manual will be available upon request [6, 7].

For the turbine cycle arrangment shown in Fig. 7-7, the AHBP computer program can generate a complete heat and mass balance. The computer results are reproduced and presented in Table 7-3

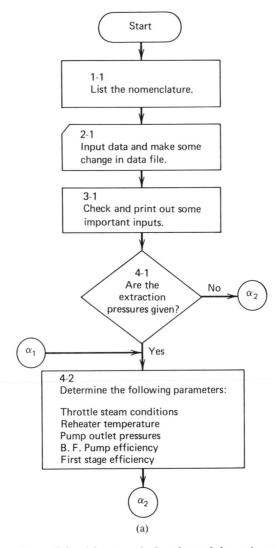

(a)

Figure 7-8. Macroscopic flowchart of the main program.

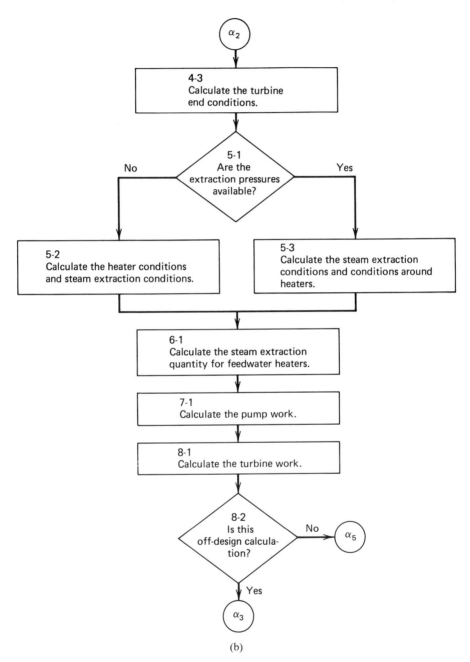

(b)

Figure 7-8.   (cont.)

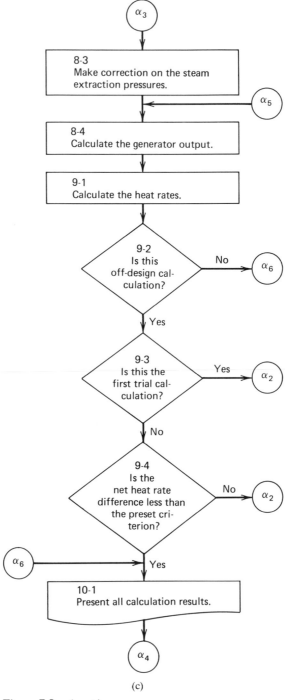

(c)

Figure 7-8. (cont.)

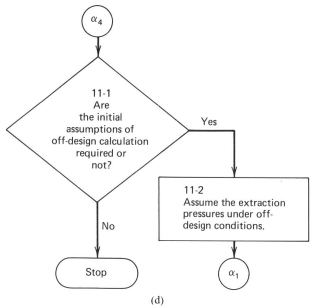

(d)

Figure 7-8.   (cont.)

7.4 PART-LOAD OPERATION

As indicated in Chapters 3 and 4, the demand to utility network is not constant and generating units do not always operate at full load. In fact, some units such as cycling coal-fired systems are expected to have capability to react continuously to the load change. The performance of these units will depend to a large extent on the particular method employed in load reduction. The principal methods are (1) throttle governing, (2) nozzle control governing, (3) bypass governing, and (4) some combinations of the above.

Figure 7-9 shows a schematic diagram of simple throttle governing system. A is the turbine stop valve; it is used to start and stop the turbine. When the machine is on load, the valve A is full open. Speed is actually controlled by the governor valve or throttle C. This is usually a double-seat valve having seats of equal or nearly equal area and is so designed that the forces of the valve due to static pressure and dynamic action are balanced. In general valve C is opened by an oil servomotor D which, in turn, is controlled by the centrifugal speed governor.

As indicated in Fig. 7-9, the steam pressure before the stop valve is P_{sv} and the steam pressure after the stop valve is P'_{sv}. In order to effect control, the pressure P_0 in the nozzle box must be lower than P'_{sv}. In general, the drop from P_{sv} to P_0 appears to vary from 2 to 5% of P_{sv} at full load.

The relationship between load and steam consumption for a turbine governed by throttling is given by the well-known Willans line shown in Fig. 7-10. Tests show that when turbine is governed by throttling, the Willans line is straight and expressed by

$$W = mk + W_0 \tag{7-14}$$

Table 7-3
Typical Output of Advanced Heat Balance Program

ADVANCED GENERALIZED HEAT BALANCE PROGRAM
(INPUT TO PROGRAM)

1. BOILER AND STEAM CONDITIONS
   TEMP (F)   1000.00
   PRESSURE(PSIA)   3515.00
   FLOW RATE(LB / HR)   4994457.00
   BOILER PRESSURE DROP   0.17
          (DIMENSIONLESS)

2. REHEATER
   OUTLET TEMP.(F)   1000.00
   PRESSURE DROP   0.07
          (DIMENSIONLESS)

3. STEAM TURBINES

	INLET P.(PSIA)	OUTLET P.(PSIA)	TURBINE EFF.
H.P.TURBINE	3515.00	744.00	0.830
I.P.TURBINE		153.00	0.884
L.P.TURBINE		2.50(IN. HG)	0.890

   NO. OF EXTRACTION FOR H.P.TURBINE   1
   NO. OF EXTRACTION FOR I.P.TURBINE   2
   NO. OF EXTRACTION FOR L.P.TURBINE   4
   H.P.TURBINE 1ST STAGE PRESSURE(PSIA)   2530.80
   PRESSURE DROP BETWEEN IP AND LP TURBINE   0.0
          (DIMENSIONLESS)

4. FEEDWATER HEATERS

	1	2	3	4	5	6	7
HEATERS NUMBER(SEE NOTE(E))							
TYPE OF HEATERS(SEE NOTE(A))	3.00	3.00	3.00	3.00	1.00	3.00	3.00
DRAIN DISPOSALS(SEE NOTE(B))	3.00	2.00	2.00	2.00	0.0	2.00	2.00
TERMINAL TEMP. DIFFERENCE(F)	10.00	10.00	5.00	5.00	0.0	5.00	5.00
DRAIN COOLER APPROACH(F)	15.00	15.00	15.00	15.00	15.00	15.00	15.00
EXTRACTION STEAM PRESSURE DROP (DIMENSIONLESS)	0.0	0.0	0.0	0.0	0.0	0.0	0.0

238

5. PUMPS

	INLET P.(PSIA)	OUTLET P.(PSIA)	PUMP EFF.	DRIVE
CONDENSATE PUMP	2.50(IN. HG)		0.82	MOTOR-DRIVEN
DRAIN PUMP			0.80	MOTOR-DRIVEN
FEEDWATER PUMP			0.85	AUX. TURBINE-DRIVEN

6. OTHER INPUTS (SEE NOTE(C))

MOTOR EFF.	0.950
MECH. COUPLING EFF.	1.000
ELECTRICAL LOSS	0.050
BOILER EFF.	0.900
AUXILIARY TURBINE EFF.	0.800
GENERATOR EFF.	0.985
PLANT AUX.POWER CONSUMPTION	0.070
AUX. TURBINE EXHAUST PRESSURE DROP (IN.HG.ABS)	0.250
AUX. TURBINE INLET PRESSSURE DROP (DIMENSIONLESS)	0.0

FEEDWATER PUMP IS LOCATED BEFORE HEATER NO. 6.0
STEAM EXTRACTION FOR AUXILIARY TURBINE IS AT EXTRACTION POINT (5)

7. INDUSTRIAL STEAM USAGE

USAGE NO.	INDEX	EXTRACTION LOCATION	RETURN LOCATION (NOTE(D))	RETURN CONDITION (SAT. WATER (F))	FLOW RATE (LB/HR)
NO.2	1.00	5.00	0.0	359.89	103400.00

(USE AS AIR PREHEATER)

8. INPUTS FOR OFF-DESIGN CALCULATION

THROTTLING STEAM FLOW RATIO 0.75

EXTRACTION NUMBER	1	2	3	4	5	6	7
DESIGN EXTRACTION PRESSURE (PSIA)	6.82	18.70	34.50	60.60	153.00	307.00	744.00

9. EXHAUST LOSS CALCULATION INPUTS

STEAM TURBINE ANNULUS EXHAUST AREA (SQ.FT) 55.60
NUMBER OF THE FLOW AT EXHAUST END 4.00
EXHAUST LOSS CURVE(SEE NOTE(G))

	1	2	3	4	5	6	7	8	9	10
EXHAUST VELOCITY (FT/SEC)	207.0	310.0	413.0	570.0	723.0	828.0	1035.0	1345.0	1655.0	2172.0
EXHAUST LOSS (BTU/LB)	21.0	12.0	8.0	5.5	8.0	11.0	19.5	33.0	44.0	58.5

Table 7-3—Continued

NOTE(A): 1.0 = CONTACT HEATER
         2.0 = SURFACE HEATER
         3.0 = SURFACE HEATER WITH DRAIN COOLER

NOTE(B): 0.0 = CONTACT HEATER
         1.0 = CHARGE UP-STREAM
         2.0 = CHARGE TO NEXT HEATER
         3.0 = CHARGE TO CONDENSER

NOTE(C): PRESSURE DROP, ELECTRICAL LOSS AND PLANT AUX. POWER CONSUMPTION ARE EXPRESSED IN A DIMENSIONLESS TERM

NOTE(D): 0.0 MEANS CONDENSER TO WHICH THE WATER IS RETURNED
         NON-ZERO NUMBER MEANS THE NUMBER OF HEATER TO WHICH THE WATER IS RETURNED

NOTE(E): THE NUMBER OF FEEDWATER HEATERS IS COUNTED FROM THE CONDENSER SIDE

NOTE(F): DEFINITIONS OF HEAT RATE ARE EXPRESSED AS FOLLOWS
         1. GROSS HEAT RATE = (TOTAL HEAT INPUT INTO SYSTEM) / (GROSS GENERATOR OUTPUT + AUX. TURBINE OUTPUT
            FOR B.F.PUMP)
         2. NET HEAT RATE = (TOTAL HEAT INPUT INTO SYSTEM) / (GROSS GENERATOR OUTPUT)
         3. PLANT HEAT RATE = (NET HEAT RATE) / (BOILER EFFICIENCY*(1.0-PLANT AUX. POWER CONSUMPTION))

NOTE(G): WHEN THE EXHAUST VELOCITY AND EXHAUST LOSS ARE EQUAL TO ZERO. IT MEANS THAT THE EXHAUST LOSS CURVE
         ORIGINALLY PREPARED FOR THE PROGRAM HAS BEEN USED IN THE COMPUTER PROGRAM

ADVANCED GENERALIZED HEAT BALANCE PROGRAM
PROGRAM OUTPUT (PART A)
** (DESIGN CONDITION) **

SUMMARY

STEAM TEMP.(F)	1000.00
STEAM PRESSURE (PSIA)	3515.00
STEAM FLOW RATE (LB/HR)	4994457.00
FIRST STAGE EFF. OF H.P.TURBINE	0.83
B.F.PUMP EFFICIENCY	0.85
PLANT AUX. POWER CONSUMPTION	0.07
BOILER EFF.	0.90
FEEDWATER PUMP POWER (KW)	23302.16
GROSS GENERATOR OUTPUT (KW)	726171.13
NET GENERATOR OUTPUT (KW)	726171.13
PLANT OUTPUT (KW)	675339.13
GROSS HEAT RATE (BTU/KW-HR)	7579.58
NET HEAT RATE (BTU/KW-HR)	7822.80

	TEMP. (F)	PRESSURE (PSIA)	ENTHALPY (BUT/LB)	FLOW RATE (LB/HR)
HIGH PRESSURE TURBINE				
THROTTLE CONDITIONS	1000.00	3515.00	1421.70	4994457.000
EXHAUST CONDITIONS	598.15	744.00	1275.37	4421171.000
(INLET OF REHEATER)				
EXTRACTION CONDITION(7)	598.15	744.00	1275.37	573285.313
INTERMEDIATE PRESSURE TURBINE				
INLET CONDITIONS	1000.00	691.92	1514.66	4421171.000
OUTLET CONDITIONS	623.22	153.00	1337.77	3566547.000
(INLET OF L.P.TURBINE)				
EXTRACTION CONDITION(6)	782.91	307.00	1412.07	194453.500
EXTRACTION CONDITION(5)	623.22	153.00	1337.77	660169.563
LOW PRESSURE TURBINE				
INLET CONDITIONS	623.22	153.00	1337.77	3566547.000
OUTLET CONDITIONS(ELEP)	108.69	1.23	1012.23	2906038.000
EXTRACTION CONDITION(4)	435.13	60.60	1251.01	131561.313
EXTRACTION CONDITION(3)	337.74	34.50	1206.71	154339.563
EXTRACTION CONDITION(2)	244.22	18.70	1164.73	174907.438
EXTRACTION CONDITION(1)	175.68	6.82	1102.91	199698.750
CONDENSER				
INLET CONDITIONS(UEEP)	108.69	1.23	1023.65	2906038.000
OUTLET CONDITIONS	108.69	1.23	76.18	3945686.000

ADVANCED GENERALIZED HEAT BALANCE PROGRAM
PROGRAM OUTPUT (PART B)
** (DESIGN CONDITION) **

	TEMP. (F)	PRESSURE (PSIA)	ENTHALPY (BTU/LB)	FLOW RATE (LB/HR)

FEEDWATER HEATER(1)
(DRAIN COOLER TYPE TTD(1)=10.0, DCA(1)=15.0)

Table 7-3—Continued

CONDENSATE INLET	108.42	153.00	76.73	3945686.00
CONDENSATE OUTLET	165.68	153.00	134.50	3945686.00
EXTRACTION STEAM INLET	175.68	6.82	1102.91	199698.75
DRAIN CONDITIONS	123.42	6.82	92.24	660507.06

FEEDWATER HEATER (2)
(DRAIN COOLER TYPE TTD(2) =10.0, DCA(2) =15.0)

CONDENSATE INLET	165.68	153.00	134.50	3945686.00
CONDENSATE OUTLET	214.41	153.00	183.07	3945686.00
EXTRACTION STEAM INLET	244.22	18.70	1164.73	174907.44
DRAIN CONDITIONS	180.68	18.70	148.91	460808.31

FEEDWATER HEATER(3)
(DRAIN COOLER TYPE TTD(3) = 5.0, DCA(3) =15.0)

CONDENSATE INLET	214.41	153.00	183.07	3945686.00
CONDENSATE OUTLET	253.44	153.00	223.86	3945686.00
EXTRACTION STEAM INLET	337.74	34.50	1206.71	154339.56
DRAIN CONDITIONS	229.41	34.50	197.75	285900.88

FEEDWATER HEATER(4)
(DRAIN COOLER TYPE TTD(4) = 5.0, DCA(4) =15.0)

CONDENSATE INLET	253.44	153.00	223.86	3945686.00
CONDENSATE OUTLET	288.36	153.00	257.66	3945686.00
EXTRACTION STEAM INLET	435.13	60.60	1251.01	131561.31
DRAIN CONDITIONS	268.44	60.60	237.40	131561.31

FEEDWATER HEATER(5)
(CONTACT TYPE   TTD(5) = 0.0)

CONDENSATE INLET	288.36	153.00	257.66	3945686.00
CONDENSATE OUTLET	359.89	153.00	333.96	4994457.00
EXTRACTION STEAM INLET	623.22	153.00	1337.77	281031.25
DRAIN CONDITIONS	359.89	153.00	333.96	4994457.00

Table 7-3—Continued

FEEDWATER HEATER(6)
(DRAIN COOLER TYPE TTD(6) = 5.0, DCA(6) = 15.0)

CONDENSATE INLET	369.19	4234.94	349.89	4994457.00
CONDENSATE OUTLET	414.47	4234.94	396.50	4994457.00
EXTRACTION STEAM INLET	782.91	307.00	1412.07	194453.50
DRAIN CONDITIONS	384.19	307.00	358.69	767738.81

FEEDWATER HEATER(7)
(DRAIN COOLER TYPE TTD(7) = 5.0, DCA(7) = 15.0)

CONDENSATE INLET	414.47	4234.94	396.50	4994457.00
CONDENSATE OUTLET	504.93	4234.94	496.12	4994457.00
EXTRACTION STEAM INLET	598.15	744.00	1275.37	573285.31
DRAIN CONDITIONS	429.47	744.00	407.45	573285.31

ADVANCED GENERALIZED HEAT BALANCE PROGRAM
PROGRAM OUTPUT (PART C)
** (DESIGN CONDITION) **

	INLET PRESSURE (PSIA)	OUTLET PRESSURE (PSIA)	FLOW RATE (LB / HR)
CONDENSATE PUMP	1.23	153.00	3945686.00
FEEDWATER PUMP	153.00	4234.94	4994457.00

EXTRACTED STEAM USAGE

	STEAM TEMP. (F)	STEAM PRESSURE (PSIA)	STEAM ENTHALPY (BTU / LB)
AUXILIARY TURBINE FOR FEEDWATER PUMP	623.22	153.00	1337.77
INDUSTRIAL USAGE (2) (USE AS AIR PREHEATER)	623.22	153.00	1337.77

	DRIVE	POWER (KW)
CONDENSATE PUMP	MOTOR	708.87
FEEDWATER PUMP	TURBINE	23302.16

	FLOW RATE (LB / HR)	RETURN (F)	CONDITIONS (PSIA)
INDUSTRIAL USAGE (2)	275738.31	111.99	1.35
	103400.00	359.99	(SAT. LIQ.)

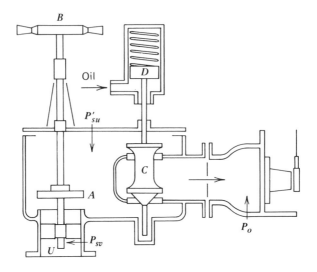

A = Turbine stop valve C = Governor valve
B = Hand control wheel D = Oil servo motor

Figure 7-9. Diagrammatic arrangement of simple throttle valve. Reproduced with permission from Pitman Publishing Ltd., London.

where

$$W_0 = \text{steam flow rate at no load (lb/hr)}$$

$$W = \text{steam flow rate at load } K \text{ (lb/hr)}$$

$$K = \text{load (kW)}$$

$$m = \text{constant}$$

The no-load steam consumption W_0 varies from one machine to another and is generally in the range of 10 to 14% of full-load value. The constant m is the slope of the Willans line and therefore is the change of steam flow rate per unit change of turbine load. In Fig. 7-10 the steam rate curve is also presented. The steam rate is defined as the ratio of the steam flow to the turbine output at the given load. Making use of Eq. (7-14), we can express the steam rate as

$$\frac{W}{K} = m + \frac{W_0}{K}$$

It is seen that the steam rate is not constant. As the load increases, the steam rate will decrease. The steam rate will asymptotically approach the value equal to m.

As a steam turbine system varies its operation to satisfy the demand, steam pressures at various turbine locations change accordingly. Figure 7-11 indicates typical variation of extraction stage-shell pressures. It is seen that the stage pressure is roughly proportional to the mass flow to the following stage, the throttle flow minus leakages, and all extractions from the preceding stages and stage in question, plus any steam returned to the turbine ahead of this stage. Mathematically, the

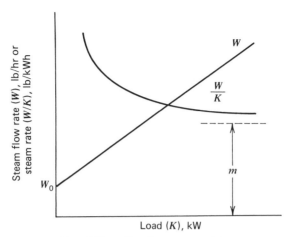

Figure 7-10. Willans line for steam turbine.

relationship is

$$P_i = P_{i,d}\frac{W_i}{W_{i,d}}$$ (7-15)

where

P_i = steam pressure at the nozzle of stage i

$P_{i,d}$ = design steam pressure at the nozzle of stage i

W_i = steam flow rate to the stage i

$W_{i,d}$ = design steam flow rate to the stage i

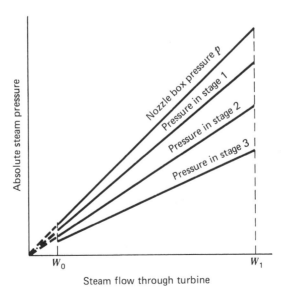

Figure 7-11. Variation of stage-shell pressure.

In part-load heat balance steam extraction pressures are first assumed (say, the design value multiplied by the throttle steam flow rate ratio) and the steam flows for feedwater heating are then determined by using the principle of energy conservation. With these, one can use Eq. (7-15) and check the assumption of steam extraction pressures. If the calculated values are not within a desirable range of the assumed, the new values for extraction pressures must be assumed and the new heat balance repeated. In general, it takes three or four trials before the extraction pressures are correctly estimated.

EXAMPLE 7-1. Referring to the part-load heat balance in Fig. 7-15, calculate and check the steam pressure at extraction point no. 4 (counted from the condenser side). The design conditions for this turbine system are shown in Fig. 7-7.

Solution: From Fig. 7-7 we find the design conditions at extraction no. 4 as

$$P_d = 60.6 \text{ psia}$$
$$W_d = 3,566,547 - 131,561$$
$$= 3,434,986 \text{ lb/hr}$$

The term W_d means the steam flow rate to the next stage. Under the part-load conditions this flow rate becomes

$$W = 2,701,136 - 105,869$$
$$W = 2,595,267 \text{ lb/hr}$$

Using Eq. (7-15), we calculate the extraction pressure as

$$P = P_d \frac{W}{W_d}$$
$$P = \frac{(60.6)(2,595,267)}{3,434,986}$$
$$P = 45.8 \text{ psia}$$

The calculated value is indeed very close to that in Fig. 7-15.

It is of interest to notice the effects of throttling control on the turbine expansion line. Figure 7-12 indicates that the turbine will experience different expansion lines as the load is decreased. But the part-load expansion lines are generally parallel to the full-load expansion line. In other words, the internal efficiency under part-load conditions is very close to that under full-load conditions.

Figure 7-12 also indicates that as the turbine load decreases, the first stage efficiency deteriorates very rapidly. The exact relationship is related to the turbine stage design. In practice, the first-stage performance is frequently approximated by using the thermal data provided by manufacturing companies.

It should be pointed out that the internal efficiency has been used as a measure of turbine internal losses. For a low-pressure steam turbine, the exhaust losses as

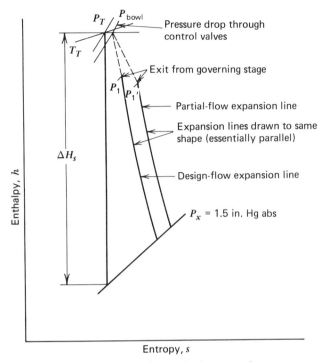

Figure 7-12. Expansion lines of the nonreheat condensing turbine.

described previously must be taken into account. As the load is decreased, the annulus velocities are expected to change and so are the exhaust losses.

Figure 7-13 shows a typical exhaust loss for General Electric machines. For a given turbine exhaust end design, the exhaust loss is determined by the annulus velocity, which is defined by

$$V_{an} = \frac{Q_a v (1 - 0.01Y)}{3600 A_{an}} \tag{7-16}$$

where

V_{an} = annulus velocity (ft/sec)

Q_a = condenser flow rate (lb/hr)

v = saturated dry specific volume (ft^3/lb)

A_{an} = annulus area (ft^2)

Y = percentage of moisture at the expansion line end point

It should be noticed that the exhaust loss from Fig. 7-13 is in Btu/lb of dry flow. To

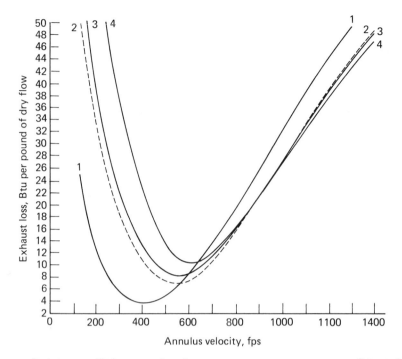

Curve Number	Bucket Length (inches)	Pitch Diameter (inches)	Last Stage Annulus Area Single Flow (ft²)	Exhaust Pressure (inches Hg absolute)	Saturated Dry Specific Volume (ft³/lb)
1	14.3	52.4	16.3	0.5	1256.4
1	16.5	57.5	20.7	1.0	652.3
1	17	52	19.3	1.5	444.9
1	20	60	26.2	2.0	339.2
2	23	65.5	32.9	2.5	274.9
3	26	72	41.1	3.0	231.6
4	30	85	55.6	3.5	200.0

1. Read the exhaust loss at the annulus velocity obtained from the following expression:

$$V_{an} = \frac{Q_a v (1 - 0.01Y)}{3600 A_{an}}.$$

2. The enthalpy of steam entering the condenser is the quantity obtained from the following expression:

$$UEEP = ELEP + (\text{exhaust loss})(0.87)(1 - 0.01Y)(1 - 0.0065Y).$$

3. This exhaust loss curve includes the loss in internal efficiency which occurs at light flows as obtained in tests.
4. For annulus velocity greater than 1400, refer to Reference [3].

Legend:

V_{an} = Annulus velocity in feet per second.

Q_a = Condensor flow in pounds per hour.

v = Saturated dry specific volume in cubic feet per pound corresponding to the actual exhaust pressure (when end point is in superheat region v = actual specific volume at end point).

A_{an} = Annulus area in square feet.

Y = Percent moisture of expansion line end point.

$ELEP$ = Expansion line end point at actual exhaust pressure in Btu per lb.

$UEEP$ = Used energy end point in Btu per lb.

Figure 7-13a. 3600-rpm condensing section, exhaust loss [3].

248

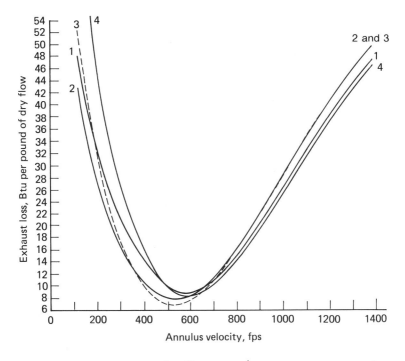

Curve Number	Bucket Length (inches)	Pitch Diameter (inches)	Last Stage Annulus Area Single Flow (ft^2)	Exhaust Pressure (inches Hg absolute)	Saturated Dry Specific volume (ft^3/lb)
1	35	110	84.0	0.5	1256.4
2	38	115	95.3	1.0	652.3
3	.43	132	123.0	1.5	444.9
4	52	152	172.4	2.0	339.2
				2.5	274.9
				3.0	231.6
				3.5	200.0

Figure 7-13b. 1800-rpm condensing section with downward exhaust, exhaust loss [3].

determine the turbine exhaust end loss per pound of steam flow, the following equation has been recommended [3].

$$\text{Exhaust end loss per pound of steam flow} = (\text{exhaust loss})(0.87)(1 - 0.01Y)(1 - 0.0065Y) \quad \cdot (7\text{-}17)$$

In heat balance calculation the enthalpy of steam entering the condenser is always desirable. This value, as previously referred to as the used energy end point (UEEP), is calculated by

$$\text{UEEP} = \text{ELEP} + (\text{exhaust end loss per unit steam flow}) \quad (7\text{-}18)$$

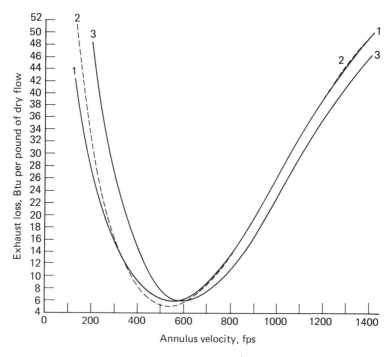

Curve Number	Bucket Length (inches)	Pitch Diameter (inches)	Last Stage Annulus Area Single Flow (ft²)	Exhaust Pressure (inches Hg absolute)	Saturated Dry Specific Volume (ft³/lb)
1	38	115	95.3	0.5	1256.4
2	43	132	123.6	1.0	652.3
3	52	152	172.4	1.5	444.9
				2.0	339.2
				2.5	274.9
				3.0	231.6
				3.5	200.0

Figure 7-13c. 1800-rpm condensing section with axial exhaust, exhaust loss [3].

EXAMPLE 7-2. Calculate the low-pressure turbine output and exhaust loss under the part-load conditions shown in Fig. 7-14. The turbine is a tandem-compound 4 flows and has last stage blade length of 30 inches. It is assumed that the exhaust loss can be approximated by curve no. 4 in Fig. 7-13a.

Solution: We calculate the low-pressure turbine output by the first-law equation

$$W_{LP} = \sum m_i h_i - \sum m_e h_e$$

Substituting the numerical values into the foregoing equation, we have

$$W_{LP} = (1{,}795{,}334)(1325) - (61{,}537)(1240.6)$$
$$- (68{,}025)(1197.8) - (80{,}208)(1157.4)$$
$$- (55{,}481)(1099.4) - (1{,}530{,}083)(1042.3)$$
$$W_{LP} = 472.36 \times 10^6 \text{ Btu/hr} \quad \text{or} \quad 138{,}401 \text{ kW}$$

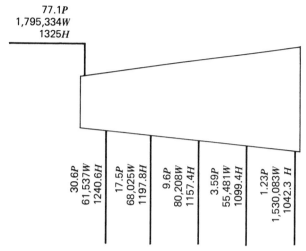

77.1P
1,795,334W
1325H

30.6P
61,537W
1240.6H

17.5P
68,025W
1197.8H

9.6P
80,208W
1157.4H

3.59P
55,481W
1099.4H

1.23P
1,530,083W
1042.3 H

Figure 7-14. Schematic diagram for a *LP* turbine for Example 7-2.

To calculate the turbine exhaust end loss, we first calculate the annulus velocity by Eq. (7-16). Since

$$Q_a = \text{total condenser flow rate/number of condenser flows}$$

$$= 1{,}530{,}083/4 = 382{,}520 \text{ lb/hr}$$

$$v = 274.9 \text{ ft}^3/\text{lb, based on the pressure 2.5 in. Hg abs.}$$

$$A_{an} = 55.6 \text{ ft}^2 \text{ (see Fig. 7-13a)}$$

$$Y = 6.4\%$$

we have the annulus velocity as

$$V_{an} = \frac{(382{,}520)(274.9)(1 - 0.01 \times 6.4)}{(3600)(55.6)}$$

$$V_{an} = 491.7 \text{ ft/sec}$$

Reading the exhaust loss at this annulus velocity from the curve (no. 4) in Fig. 7-13a, we obtain the value approximately 14 Btu/lb. Then, the turbine exhaust end loss per pound of steam flow is

$$= (\text{exhaust loss})(0.87)(1 - 0.01Y)(1 - 0.0065Y)$$

$$= (14)(0.87)(1 - 0.01 \times 6.4)(1 - 0.0065 \times 6.4)$$

$$= 10.93 \text{ But/lb}$$

The total turbine exhaust loss is the product of the unit loss and the steam flow. Subtracting this turbine loss from the *LP* turbine output will yield the actual

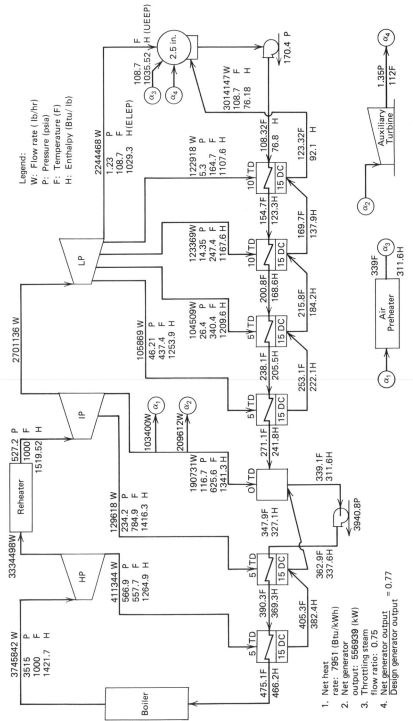

Figure 7-15. Heat balance diagram for the off-design conditions (throttle steam flow ratio, 0.75).

252

output as

$$W_{Lp} = 472.36 \times 10^6 - 10.93 \times 1,530,083$$
$$= 472.36 \times 10^6 - 16.72 \times 10^6$$
$$= 455.64 \times 10^6 \text{ Btu/hr} \quad \text{or} \quad 133,500 \text{ kW}$$

In addition to the throttle governing, the nozzle control governing is another method for regulating the turbine output. The steam consumption rate is much smaller for the nozzle control than for the throttling control. In the nozzle control arrangement the nozzles are divided into groups under the control of valves. The number of nozzle groups may vary from three to five, or more. At full load, all nozzles will be delivering steam at full pressure and the turbine will operate at maximum efficiency. At some part-load conditions one group of nozzles may be shut off while other nozzles are fully operated. Because of these arrangements, the throttling effects on steam will be either eliminated or minimized. Details of nozzle control and other governing methods can be found in Kearton's book [2].

For example, the turbine cycle shown in Fig. 7-7 is reexamined under the part-load conditions. When the throttle steam rate is reduced to 75% of the design value, approximately at 3.75×10^6 lb/hr, the turbine system will have its net heat rate changed to 7951 Btu/kWh and its output to 556,939 kW. Figure 7-15 indicates this part-load heat balance. Compared with the full-load heat balance in Fig. 7-7, the extraction steam pressures are seen decreasing according to Eq. (7-15). As the throttle steam rate decreases, the turbine net heat rate is expected to deteriorate.

7-5 SECOND-LAW ANALYSIS OF STEAM TURBINE CYCLES

In recent years the second-law analysis of electric-power generating system has increasingly attracted the engineer's attention [8, 9]. While the first-law analysis provides an indication of system performance, the second-law analysis would pinpoint the losses as well as the efficiency of each component in the system. The second-law analysis is essentially based on the concept of thermodynamic availability, which is sometimes referred to in the literature as available energy, useful energy, or exergy.

The thermodynamic availability of working substance including steam is defined by

$$a = (h - h_0) - T_0(s - s_0) \tag{7-19}$$

where h and s are, respectively, the enthalpy and entropy of substance at given state. The subscript 0 represents the dead state or reference state to which the availability calculations are referred. A brief review of thermodynamic availability was presented in Chapter 2.

Another concept frequently used in the second-law analysis is the system effectiveness (or process effectiveness), which is defined by

$$\varepsilon = \frac{\text{increase in the availability of desired output}}{\text{decrease in the availability}} \tag{7-20}$$

Equation (7-20) can also be expressed in a form similar to that of the first-law efficiency, which is the ratio of the desired energy output to the energy supplied. In that case, Eq. (7-20) becomes

$$\varepsilon = \frac{A_{\text{output}}}{A_{\text{input}}} \tag{7-21}$$

where

A_{output} = availability produced by a system

A_{input} = availability supplied to a system

To illustrate the calculations of thermodynamic availability and process effectiveness, let us examine the conventional steam turbine cycle shown in Fig. 7-7. Using Eq. (7-19), we can calculate the thermodynamic availabilities under various steam conditions. The results are shown in Fig. 7-16. It is seen that the steam availability is a function of steam temperature and pressure. The steam with higher availability is more valuable than that with lower availability.

To calculate the process effectiveness, the thermodynamic availability shown in Fig. 7-16 must be utilized. For instance, the availability input ($\mathring{A}_{\text{input}}$) for the high-pressure turbine process is

$$\mathring{A}_{\text{input}} = \sum W_i a_i - \sum W_e a_e \tag{7-22}$$

where

$$a = \text{unit availability, Btu/lb}$$

$$W = \text{steam flow rate, lb/hr}$$

The subscripts i and e, denote turbine inlet and exit, respectively. Using the numerical values in Fig. 7-16, we have

$$\mathring{A}_{\text{input}} = 4{,}994{,}457 \times 633.92$$

$$- 573{,}285 \times 473.04 - 4{,}421{,}171 \times 473.04$$

$$= 803.51 \times 10^6 \text{ Btu/hr}$$

The availability output $\mathring{A}_{\text{output}}$ for the high-pressure turbine is simply the turbine output. That is,

$$\mathring{A}_{\text{output}} = \sum W_i h_i - \sum W_e h_e \tag{7-23}$$

or

$$\mathring{A}_{\text{output}} = 4{,}994{,}457 \times 1421.7$$

$$- 573{,}285 \times 1275.4 - 4{,}421{,}171 \times 1275.4$$

$$= 730.84 \times 10^6 \text{ Btu/hr}$$

The process availability loss $\mathring{A}_{\text{loss}}$ is defined as the difference between $\mathring{A}_{\text{input}}$ and $\mathring{A}_{\text{output}}$. For the high-pressure turbine, the availability loss is

$$\mathring{A}_{\text{loss}} = \mathring{A}_{\text{input}} - \mathring{A}_{\text{output}}$$

$$= 803.51 \times 10^6 - 730.84 \times 10^6$$

$$= 72.67 \times 10^6 \text{ Btu/hr} \tag{7-24}$$

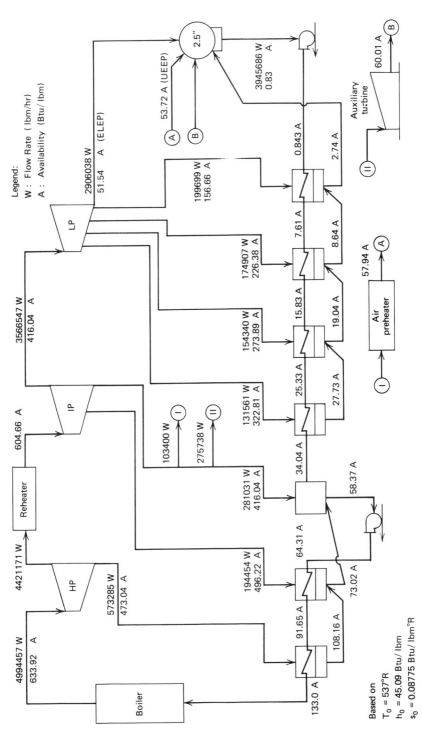

Figure 7-16. Steam availability of a conventional power plant.

Legend:
W : Flow Rate (lbm/hr)
A : Availability (Btu/lbm)

Based on
$T_0 = 537°R$
$h_0 = 45.09$ Btu/lbm
$s_0 = 0.08775$ Btu/lbm°R

255

Finally, using Eq. (7-21), we have the effectiveness for the high pressure turbine

$$\varepsilon = \frac{730.84 \times 10^6}{803.51 \times 10^6}$$

$$= 0.91 \quad \text{or} \quad 91\%$$

The same procedures can be used to calculate $\mathring{A}_{input}$, $\mathring{A}_{output}$, and $\mathring{A}_{loss}$ for various processes. Table 7-4 summarizes the calculated results for the entire turbine cycle specified in Fig. 7-7. Detailed calculations are presented in Appendix B.

It is interesting to note from Table 7-4 that most availability loss takes place in the steam generator (3217 MBtu/hr) and the boiler effectiveness is only in the

Table 7-4
Second-Law Analysis of a Modern Steam Turbine Cycle

	Item Description of Component	$\mathring{A}_{Input}$ (10^6 Btu / hr)	$\mathring{A}_{output}$ (10^6 Btu / hr)	$\mathring{A}_{loss}$ (10^6 Btu / hr)	Device Effectiveness (%)
1.	Steam generator	6300.0	3084.0	3217.0	48.95
2.	Steam turbine system	3085.8	2556.9	528.9	82.9
3.	HP turbine	803.5	730.8	72.7	90.95
4.	IP turbine	817.8	767.5	50.30	93.85
	a. Section 1	479.4	453.6	25.80	94.62
	b. Section 2	338.4	313.9	24.50	92.76
5.	LP turbine	1172.1	1021.6	150.50	87.16
	a. Section 1	332.5	309.5	23.00	93.08
	b. Section 2	168.0	152.2	15.80	90.60
	c. Section 3	155.9	137.7	18.20	88.33
	d. Section 4	216.5	192.0	24.50	88.68
	e. Section 5	299.15	230.2	68.95	76.95
6.	Condenser	177.19		177.19	
7.	Air preheater	37.03			
8.	Auxiliary turbine	98.14	79.52	18.62	81.03
9.	Feedwater train	770.58	719.88	50.70	93.40
	a. Heater 1	31.30	24.53	6.77	78.40
	b. Heater 2	41.06	32.45	8.61	79.03
	c. Heater 3	40.48	37.49	2.99	92.61
	d. Heater 4	38.84	34.36	4.48	88.47
	e. Heater 5	303.40	291.50	11.90	96.08
	f. Heater 6	106.30	93.05	13.25	87.54
	g. Heater 7	209.20	206.50	2.70	98.71
10.	Feedwater pump	79.61	73.17	6.44	91.91
11.	Condensate pump	2.17	0.05	2.12	20.10

Note: 1. The calculations were made with the information $a_f = 13186$ Btu/lbm of coal, and $w_f = 475,0$ lbm/hr.

2. The exhaust end loss is taken into consideration at LP turbine section 5.

neighborhood of 50%, which is much lower than the first-law efficiency. This happens because the second-law analysis takes into account not only the availability loss, but also the availability destruction in the process. The condenser is also relatively ineffective and has approximately 177.2 MBtu/hr of availability loss. Compared with typical results in the first-law analysis, the loss calculated above is much smaller. In summary, the second-law analysis is very much needed when the real inefficient components are to be identified in the electric-power generating system.

7-6 STEAM TURBINE SYSTEM PERFORMANCE

The steam turbine systems consist of turbine cylinders, electric generators, feedwater heaters, pumps and its connecting pipes. The performance of the turbine system is generally expressed in terms of turbine net heat rate, which has been previously defined. There are many factors affecting the turbine net heat rate. While some of them have major influences, others have only minor effects. In practice, the factors listed below always demand the consulting engineer's attention:

- Condenser pressure
- Steam inlet conditions
- Low-pressure turbine exhaust end

It is well known from thermodynamics that the turbine net heat rate will increase as the exhaust pressure (approximately equal to condenser pressure) increases. Figure 7-17 indicates general trends for various condenser pressures. Generally, turbines operate in the range of 1.0 to 5.0 in. Hg abs. When turbines operate in connection with dry-type cooling tower (or air-cooled condenser), the exhaust pressure is expected to be much higher for economical reasons. Further discussion of this unusual high pressure can be found later in this text.

As indicated in Fig. 7-17, the turbine net heat rate is not only a function of condenser pressure, but also a function of turbine output. At a given throttle steam flow rate, an increase in turbine net heat rate is always associated with a decrease in turbine output. Using the definition of turbine net heat rate, we can express the relationship between these two terms as

$$\Delta(NKW) = \frac{-\Delta(NHR)}{1 + \dfrac{\Delta(NHR)}{100}} \tag{7-25}$$

where

$\Delta(NKW)$ = change of turbine system kW output in percentage

$\Delta(NHR)$ = change of turbine net heat rate in percentage

The following example illustrates a use of this equation.

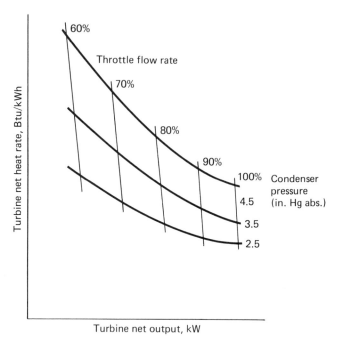

Figure 7-17. Effects of condenser pressure on the turbine net heat rate.

EXAMPLE 7-3. A 700 MW turbine system has a net heat rate of 7826 Btu/kWh at the condenser pressure of 2.5 in. Hg abs. For some reasons consulting engineers decide to select this turbine system and operate it at the condenser pressure of 4.5 in. Hg abs. What would be the turbine output change if the turbine net heat rate changes to 7980 Btu/kWh at the new condenser pressure?

Solution: The percent change of turbine net heat rate in this case is

$$\Delta(NHR) = \frac{7980 - 7826}{7826} \times 100$$

$$= 1.97\%$$

Applying Eq. (7-25), we have

$$\Delta(NKW) = \frac{-1.97}{1 - \dfrac{1.97}{100}}$$

$$= -2.01\%$$

That is, the turbine net output is expected to decrease by 2.01% if the throttle steam flow rate remains unchanged.

In the United States the manufacturing companies usually indicate the guaranteed and expected performance of steam turbines. In the guaranteed performance the steam turbine is specified to produce a certain number of kilowatts while operating at rated steam conditions, 3.5 inches of mercury absolute exhaust pressure, 0% cycle makeup, and other cycle feedwater heating conditions. To assure that the steam turbine will pass the guaranteed throttle flow, the turbine is frequently designed for a steam flow rate larger than the guaranteed value. This new value is sometimes called the expected steam flow and is usually around 105% of the guaranteed value. For this reason, the actual output of the turbine is expected to be larger than the guaranteed value.

The turbine is guaranteed to be safe for continuous operation with valves wide open. Furthermore, the turbine is also capable of operating continuously with valves wide open and at the same time at 105% of rated initial pressure. Under these conditions the expected steam flow would become maximum (approximately 110% of the guaranteed value) and thus the expected turbine output.

Steam inlet conditions strongly influence the turbine performance. At a given pressure the turbine performance can be improved by increasing the initial steam

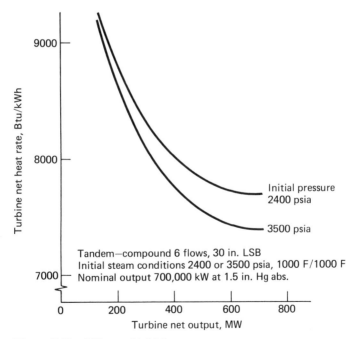

Figure 7-18. Effects of initial steam pressure.

temperature. From the thermodynamic viewpoint, the higher the steam temperature, the better the turbine performance will be. However, there is a temperature limit beyond which turbine and steam generator will become less reliable. The modern design will have the initial temperature between 1000 and 1200 F.

An increase in steam pressure at turbine inlet will also increase the plant thermal efficiency. Figure 7-18 shows the effects of initial steam pressures using a 700 MW unit as an example. These trends are also true for units of similiar size.

The low-pressure turbine exhaust end is one of the important factors affecting the turbine performance. The size of the exhaust end is determined by the number of exhaust flows and the length of the last-stage blades. In general, the larger the exhaust end, the lower the full-load net heat rate. These are clearly shown in Fig. 7-19. Under the part-load conditions, however, turbines with a large exhaust end will deteriorate more rapidly in performance.

In the United States manufacturing companies usually produce a steam turbine with 2, 4, 6, and 8 exhaust flows. The most common lengths of last-stage blades for one U.S. manufacturer are 23, 26, 30, and 33.5 inches for tandem-compound 3000 and 3600 rpm and 38, 43, and 52 inches for cross-compound 1500 and 1800 rpm arrangements. Another U.S. manufacturer designates the 3000 and 3600 rpm blade length as 21, 25, 28, and 31 inches, but has the same exhaust annulus areas as the former. In selecting steam turbine systems, attention must be paid to the full-load as well as part-load performance. The selection procedures are to be discussed in the next section.

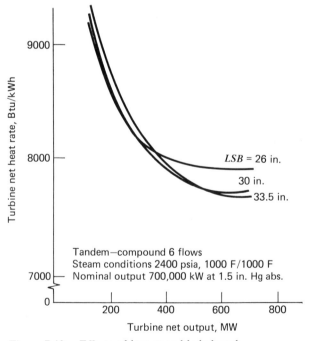

Figure 7-19.   Effects of last-stage blade length.

7.7 TURBINE SYSTEM SELECTION

Chapter 3 describes three methods commonly used to evaluate various design alternatives. In this section the present worth method is used to select the steam turbine system. The basis for this particular evaluation is restricted to the initial investment and the O & M costs associated with each alternative selection.

Initial investment of each turbine system is usually provided by the manufacturer. The present worth of initial cost can be calculated once the manufacturer's equipment installation cost is known and the utility specifies a rate of return, turbine economic life, and a fixed charge rate. Mathematically, the present worth of lifetime initial cost ($ICOST$) is

$$(ICOST) = \text{(initial cost) (annual fixed charge rate)}$$

$$(SPWF) \tag{7-26}$$

While the initial cost is usually provided by the manufacturer, the determination of O & M is the responsibility of the consulting engineer. In most cases, the significant O & M cost differential between two alternatives is in the cost of fuel. In this section only fuel cost is considered.

The present worth of lifetime fuel cost is the sum of the present worth of fuel cost. Mathematically, the present worth of lifetime fuel cost ($FCOST$) is calculated with the equation

$$FCOST = \sum_{i=1}^{n} P_i F_i \tag{7-27}$$

where

$$F_i = \text{annual fuel consumption in } i\text{th year (Btu/yr)}$$

$$P_i = \text{present worth of unit fuel cost in } i\text{th year (\$/Btu)}$$

$$n = \text{turbine system economic life (years)}$$

The annual fuel consumption F_i is simply the sum of fuel consumptions from the various turbine loadings. That is,

$$F_i = \sum_{j=1}^{k} (LOAD)_j (PNHR)_j (HOUR)_j \tag{7-28}$$

where

$$(LOAD)_j = \text{generating unit output at } j\text{th loading (kW)}$$

$$(PNHR)_j = \text{plant net heat rate at } j\text{th loading (Btu/kWh)}$$

$$(HOUR)_j = \text{number of hours at } j\text{th loading (hr)}$$

$$k = \text{number of loadings in a year}$$

The present worth of unit fuel cost is generally determined by the first-year unit fuel cost, escalation rate, and present worth factor. When the escalation rate of fuel cost is constant, the present worth of unit fuel cost in ith year is

$$P_i = (FC)(1 + ES)^{i-1}(PWF)_i \tag{7-29}$$

where

$$FC = \text{unit fuel cost in the first year, \$/Btu}$$

$$ES = \text{average escalation rate of fuel cost, dimensionless}$$

Then, the present worth of lifetime evaluated cost $(TOCOST)$ is

$$(TOCOST) = (ICOST) + (FCOST) \tag{7-30}$$

It should be pointed out that the above fuel cost calculation assumes no effect on the economic dispatch because of different turbine selections. This means that the number of operating hours for each loading will not be changed in the evaluation process.

The turbine system selection is usually complicated by the fact that the plant may have different generating capabilities at the maximum loading conditions such as those of valve wide open or valve wide open and 5% overpressure at the turbine inlet. Because of this, credit (or penalty) must be given to the plant that can generate more electricity (or less electricity). To make our evaluation more amenable to computer programming, we calculate the relative penalty cost based on the generation loss. Using the reference value, the loss of generating capability ΔKW is

$$\Delta KW = RKW - MKW \tag{7-31}$$

where

$$RKW = \text{reference value (kW)}$$

$$MKW = \text{maximum output of turbine system (kW)}$$

In a sense the term ΔKW represents a relative deficiency in power generation. To make up this loss of generation, the network either purchases energy or constructs additional generating facility. The associated cost is defined as the penalty that must be taken into account in the turbine system selection. In case of purchase, the present worth of lifetime penalty $(PCOST)$ is given by

$$PCOST = \sum_{i=1}^{n} (\Delta KW)(UCP)(1 + ESI)^{i-1}(PWF)_i$$

$$+ \sum_{i=1}^{n} (\Delta KW)(HOUR)_i(UCE)(1 + ES2)^{i-1}(PWF)_i \tag{7-32}$$

where

$$UCP = \text{first year demand cost (\$/kW)}$$

$$UCE = \text{first year energy cost (\$/kWh)}$$

$$ES1 = \text{escalation rate for demand cost (dimensionless)}$$

$$ES2 = \text{escalation rate for energy cost (dimensionless)}$$

Evidently, the first term in Eq. (7-32) is related to the power demand cost while the second is the energy cost. If additional facility is used to replace the loss of generation, the penalty must involve the initial cost as well as fuel cost. These costs may be entirely different from those of the turbine system under evaluation. The present worth of lifetime penalty cost for the replacement can be obtained by the same procedures as those for the (*TOCOST*).

EXAMPLE 7-4. Two steam turbine systems are considered for a power plant project. Both systems have the steam inlet conditions, 2400 psia, 1000/1000 F, and seven feedwater heaters with final feedwater temperature around 480 F. The rating of these two turbine systems is the same and equal to 300,000 kW at the condenser pressure pressure 1.5 in Hg abs. However, the main difference between these two is in the last stage blade length (23 vs. 30 in.). This means that they have different turbine net heat rates as indicated below.

Loading Number	1	2	3	4	5
T/G output (kW)	312,000[a]	300,000	240,000	180,000	120,000
Turbine NHR LSB = 23 in.	7912	7899	7901	8019	8349
(Btu/KWh)					
LSB = 30 in.	7715	7716	7809	8069	8574

[a] Expected turbine output.

The annual operation hours are the same for both systems and summerized in the following table.

Plant Age (Year)	Loading Number				
	1	2	3	4	5
1–2	438	517	1071	2120	4652
3–5	438	885	1927	2278	3232
6–8	438	1332	2628	2628	1734
9–12	438	4486	1314	876	1664
13–35	438	5081	876	438	1927

Calculate the present worth of lifetime fuel cost for these two systems and estimate the breakeven capital cost differential. The economic factors listed below should be used for this evaluation.

Plant economic life	35 years
Rate of return	10%
Annual fixed charge rate	15%
First year fuel cost	$1.31/MBtu
Fuel cost escalation rate	5%

Solution: The first step is to estimate the plant net output and net heat rate. Assuming boiler efficiency 90% and constant auxiliary power 8%, we have these results.

Loading Number	1	2	3	4	5
T/G output (kW)	312,000	300,000	240,000	180,000	120,000
Plant net output (kW)	287,100	276,000	220,800	165,600	110,400
$PNHR$, (Btu/kWh)					
LSB = 23 in.	9555.6	9539.9	9542.3	9771.1	10083.3
LSB = 30 in.	9317.6	9318.8	9431.2	9745.2	10355.1

The second step is to utilize Eq. (7-28) to calculate the annual fuel consumption. For instance, the first year fuel consumption for alternative A is

$$F_1 = \sum_{j=1}^{k} (LOAD)_j (PNHR)_j (HOUR)_j$$

$$F_1 = 287,100 \times 9555.6 \times 438 + 276,000 \times 9539.9 \times 517$$

$$+ 220,800 \times 9542.3 \times 1071 + 165,600 \times 9771.1 \times 2102$$

$$+ 110,400 \times 10083.3 \times 4652$$

$$F_1 = 13.4 \times 10^{12} \text{ Btu/yr or } 13.4 \times 10^6 \text{ MBtu/yr}$$

and the first year fuel consumption for alternative B is

$$F_1 = 287,100 \times 9317.6 \times 438 + 276,000 \times 9318.8 \times 517$$

$$+ 220,800 \times 9431.2 \times 1071 + 165,600 \times 9745.2 \times 2102$$

$$+ 110,400 \times 10355.1 \times 4652.$$

$$F_1 = 13.44 \times 10^{12} \text{ Btu/yr or } 13.44 \times 10^6 \text{ MBtu/yr}$$

Similarly, the fuel consumption for other years are calculated. The results are summarized in Table 7-5. Also included are the present worth of unit fuel cost

Table 7-5
Summary of Calculations for Example 7-4

Year	P_i ($ / MBtu)	$F_i (10^6$ MBtu / yr) Alternative A	Alternative B	$P_i F_i (10^6 \$$ / yr) Alternative A	Alternative B
1	1.1932	13.40	13.44	15.99	16.04
2	1.1389	13.40	13.44	15.26	15.31
3	1.0872	14.88	14.83	16.18	16.12
4	1.0378	14.88	14.83	15.44	15.39
5	0.9906	14.88	14.83	14.74	14.69
6	0.9456	16.43	16.29	15.54	15.40
7	0.9026	16.43	16.29	14.83	14.70
8	0.8616	16.43	16.29	14.16	14.03
9	0.8224	19.05	18.76	15.67	15.43
10	0.7850	19.05	18.76	14.95	14.73
11	0.7493	19.05	18.76	14.27	14.06
12	0.7153	19.05	18.76	13.63	13.42
13	0.6828	19.28	18.97	13.16	12.95
14	0.6517	19.28	18.97	12.57	12.36
15	0.6221	19.28	18.97	11.99	11.80
16	0.5938	19.28	18.97	11.45	11.26
17	0.5668	19.28	18.97	10.93	10.75
18	0.5411	19.28	18.97	10.43	10.26
19	0.5165	19.28	18.97	9.96	9.80
20	0.4930	19.28	18.97	9.51	9.35
21	0.4706	19.28	18.97	9.07	8.93
22	0.4492	19.28	18.97	8.66	8.52
23	0.4288	19.28	18.97	8.27	8.13
24	0.4093	19.28	18.97	7.89	7.76
25	0.3907	19.28	18.97	7.53	7.41
26	0.3729	19.28	18.97	7.19	7.07
27	0.3560	19.28	18.97	6.86	6.75
28	0.3398	19.28	18.97	6.55	6.45
29	0.3243	19.28	18.97	6.25	6.15
30	0.3096	19.28	18.97	5.97	5.87
31	0.2955	19.28	18.87	5.70	5.61
32	0.2821	19.28	18.97	5.44	5.35
33	0.2693	19.28	18.97	5.19	5.11
34	0.2570	19.28	18.97	4.96	4.88
35	0.2454	19.28	18.97	4.73	4.65
				$FCOST =$ 370.92	366.49

and the annual fuel cost. It is seen that the present worth of lifetime fuel cost is $370.92 million for alternative A and $366.49 million for alternative B. The difference is $4.33 million in favor of the steam turbine with last stage blade length 30 inches. The breakeven cost differential is calculated by

$$BECD = \frac{\text{present worth of lifetime fuel saving}}{(AFCR)(SPWF)}$$

$$= \frac{3.44}{(0.15)(9.64)}$$

$$= \$2.38 \text{ million}$$

EXAMPLE 7-5. Two steam turbine systems are considered for a power plant project. Both systems have the same steam conditions and the same number of feedwater heaters. In fact, they are almost identical to each other except for different maximum expected outputs. Alternative A has a maximum expected output 25 MW larger than alternative B. To have a fair comparison between these two, a purchase option is used to compensate for the generation deficiency of alternative B.

Calculate the present worth of lifetime purchase cost for alternative B. The input information listed below should be used for this evaluation.

First-year demand cost	20 $/kW
First-year energy cost	20 mills/kWh
Escalation rate for demand cost	3%
Escalation rate for energy cost	5%
Rate of return	10%
Plant economic life	35 year
Average operation hours	500 hr

Solution: Using Eq. (7-32), we express the present worth of lifetime purchase cost as

$$COST = (\Delta KW)(UCP) \sum_{i=1}^{35} (1 + ES1)^{i-1}(PWF)_i$$

$$+ (\Delta KW)(HOUR)(UCE) \sum_{i=1}^{35} (1 + ES2)^{i-1}(PWF)_i$$

Since $ES1 = 0.03$, $ES2 = 0.05$, and the rate of return $R = 0.1$, we have

$$\sum_{i=1}^{35} (1 + ES1)^{i-1}(PWF)_i = 12.85$$

and

$$\sum_{i=1}^{35} (1 + ES2)^{i-1}(PWF)_i = 16.07$$

Then, the present worth of lifetime purchase cost is

$$COST = (25,000)(20)(12.85)$$

$$+ (25,000)(500)(0.02)(16.07)$$

$$= \$10.45 \times 10^6$$

It should be pointed out that the purchase cost in the above example is not exactly equal to the previously defined penalty cost. Because of additional generating capability, more fuel will be needed for alternative A. This additional fuel cost must be substracted from the purchase cost of alternative B before the penalty cost can be determined.

In both examples condenser pressure has been assumed constant throughout the year. No consideration is given to the effects of the seasonal weather change. It is well known that the condenser pressure will decrease as the cooling water temperature decreases (due to the change of weather conditions). To better approximate the turbine system operation, variation of condenser pressure should be taken into account. For the evaluation purposes such as those in these examples, the monthly or seasonal average is frequently used. The following example will illustrate this aspect.

EXAMPLE 7-6. Consulting engineers compare two similar turbine-cycle systems for a nominal 650 MW unit. Both systems have the same configuration except different driving arrangements for boiler feedwater pumps. Alternative A is auxiliary turbine driven and alternative B is motor-driven. They have the same loading schedule presented below.

Seasons	Plant Loads (kW)	Operation Years			
		1–2	3–10	11–20	21–30
Winter	651,000	920.00	985.50	526.25	194.50
	488,250	591.25	635.00	526.25	591.25
	325,500	306.25	307.00	809.00	832.25
Spring and fall	651,000	1,840.00	1,971.00	1,052.50	389.00
	488,250	1,182.50	1,270.00	1,052.50	1,182.50
	325,500	612.50	614.00	1,618.00	1,664.50
Summer	651,000	920.00	985.50	526.25	194.50
	488,250	591.25	635.00	526.25	591.25
	325,500	306.25	307.00	809.00	832.25

The previous study of the cooling system indicates that the average condenser pressure is 2.5 in. Hg abs. for winter, 3.5 in. Hg abs. for spring and fall, and 4.5 in. Hg abs. for summer. The plant net heat rates obtained in heat balance calculation are given below.

Alternative A (Auxiliary Turbine-Driven Pumps)

Condenser Pressure (in. Hg abs.)	Plant Net Output (kW)	Plant Net Heat Rate (Btu / kWh)
2.5	651,000	9,356.9
	488,250	9,554.7
	325,500	9,971.6
3.5	651,000	9,476.7
	488,250	9,728.8
	325,500	10,210.1
4.5	651,000	9,609.9
	488,250	9,891.9
	325,500	10,419.9

Alternative B (Motor-Driven Pumps)

Condenser Pressure (in. Hg abs.)	Plant Net Output (kW)	Plant Net Heat Rate (Btu / kWh)
2.5	651,000	9,402.2
	488,250	9,574.5
	325,500	9,968.1
3.5	651,000	9,494.1
	488,250	9,727.6
	325,500	10,204.5
4.5	651,000	9,614.7
	488,250	9,892.5
	325,500	10,415.3

Calculate the present worth of lifetime fuel cost for these two design alternatives and estimate the acceptable capital cost differential. The economic factors used for this evaluation are

Unit economic life	30 years
First-year fuel cost	$1.50/MBtu
Fuel cost escalation	6% per year
Rate of return	13%
Annual fixed charge rate	18.24%

Solution: First, we calculate the annual fuel consumption by Eq. (7-28):

$$F_i = \sum_{j=1}^{9} (LOAD)_j (PNHR)_j (HOUR)_j$$

The subscript j denotes different sets of conditions under which the unit is operated. In this case the number of combinations is nine with three different loadings and each at three different condenser pressures. For instance, the first-year fuel consumption for the design alternative A is

$$F_1 = (651,000)(9,356.9)(920.00)$$
$$+ (488,250)(9,554.7)(591.25)$$
$$+ (325,500)(9,971.6)(306.25)$$
$$+ (651,000)(9,476.7)(1,840.0)$$
$$+ (488,250)(9,728.8)(1,182.5)$$
$$+ (325,500)(10,210.1)(612.50)$$
$$+ (651,000)(9,609.5)(920.00)$$
$$+ (488,250)(9,891.5)(591.25)$$
$$+ (325,500)(10,419.9)(306.25)$$
$$F_1 = 38.0099 \times 10^{12} \text{ Btu}$$

Using Eq. (7-29), we can determine the present worth of unit fuel cost. For the first year, it is

$$P_1 = (1.50 \times 10^{-6})(1)(0.885)$$
$$P_1 = \$1.33 \times 10^{-6}/\text{Btu}$$

Then, the first-year fuel cost is the product of P_1 and F_1. That is,

$$(FCOST)_1 = \$50,455,592$$

Similarly, the annual fuel cost for other years can be calculated. The results for alternatives A and B are summarized in Table 7-6. It is seen that the design alternative A is superior to the alternative B in terms of lifetime fuel cost. On the basis of the fuel saving, the breakeven capital cost diff▢ ▢ntial must be

$$\Delta(\text{Fixed cost}) = \frac{\text{fuel saving}}{(AFCR)(SPWF)}$$
$$= \frac{768,458}{(0.1824)(7.8889)}$$
$$= \$534,000$$

Table 7-6
Summary of Calculations for Example 7-6

Year	Reference Case Present Worth ($)	Motor-Driven Feedwater Pump Present Worth ($)
1	50,455,592	50,527,905
2	47,330,024	47,397,858
3	47,268,617	47,337,405
4	44,340,473	44,405,000
5	41,593,718	41,654,247
6	39,017,116	39,073,896
7	36,600,126	36,653,389
8	34,332,862	34,382,825
9	32,206,047	32,252,916
10	30,210,983	30,254,947
11	23,624,050	23,644,172
12	22,160,643	22,179,489
13	20,787,832	20,805,538
14	19,500,090	19,516,700
15	18,292,120	18,307,700
16	17,158,980	17,173,595
17	16,096,034	16,109,744
18	15,098,934	15,111,795
19	14,163,602	14,175,666
20	13,286,211	13,297,528
21	10,007,790	10,011,749
22	9,387,838	9,391,553
23	8,806,291	8,809,775
24	8,260,768	8,264,036
25	7,749,039	7,752,105
26	7,269,010	7,271,886
27	6,818,718	6,821,415
28	6,396,319	6,398,850
29	6,000,087	6,002,461
30	5,628,400	5,630,627
Total	659,848,314	660,616,772

Present worth of lifetime fuel saving = $768,458.

It is well known that the auxiliary steam turbines (including the associated pipings and valves) for feedwater pumps are more expensive than the electric motors used for the same purpose. The breakeven cost differential calculated above indicates an acceptable increase of capital expenditure if the auxiliary steam turbines are selected. While there are many factors affecting the decision of selecting the driving mechanism for feedwater pumps, the fuel cost and unit size are obviously the important factors. As the fuel cost increases, the auxiliary turbine will generally become more economical for driving the feedwater pumps. Recent experiences indicate that the auxiliary turbine is frequently selected if unit size is greater than 300 to 400 MW.

It must be pointed out that the break-even cost differential in the above example is conservative. If the loss of generation capability is taken into consideration, this cost differential is expected to change significantly. More on the loss of generation capability will be discussed in the next section.

7-8 CONSIDERING DIFFERENTIAL GENERATION CAPABILITY AND ENERGY LOSS

In turbine-cycle configuration design, situations frequently arise that the system performance could be further improved by increasing the equipment expenditure. To make an intelligent decision, engineers must determine the benefit and cost for each design alternative. When the benefit-to-cost ratio is larger than unity, the design alternative is considered for adoption.

To determine the benefit of the design alternative, credit must be assigned to the unit differential of generation capability. What is the value for one additional kilowatt generation capability? Evidentally, there is no general answer to this question. The value varies from network to network, and is dependent with many factors such as system reserve and fuel costs. If this value is denoted by (UC) and expressed in terms of dollars per year and per kilowatt, the annual benefit will simply become

$$B1 = (\Delta PNKW)(UC) \qquad (7\text{-}33)$$

where

$$B1 = \text{annual benefit due to differential capability, dollar/yr}$$

$$\Delta PNKW = \text{differential generation capability, kW}$$

$$UC = \text{unit credit for differential generation capability, dollar/kW-yr}$$

In practice, the term (UC) is sometimes referred to as the combined capital charge for generation capability and energy loss. More discussion about this term will be made later.

A change in turbine-cycle configuration affects not only the station output, but also the station net heat rate. Reduction in station net heat rate means a reduction in fuel cost, which represents additional benefit. The annual benefit due to the change of station net heat rate is calculated by

$$B2 = (8760)(CF)(KW)(\Delta PNHR)(UFCOST)(10^{-6}) \qquad (7\text{-}34)$$

where

$B2$ = annual benefit due to the change of station net heat rate, dollar/yr

CF = plant capacity factor

KW = reference station output, kW

$\Delta PNHR$ = change of station net heat rate, Btu/kWh

$UFCOST$ = unit fuel cost, dollar/MBtu

Then the benefit to cost ratio is

$$\text{Ratio} = \frac{B1 + B2}{(COST)(AFCR)}$$

where

$COST$ = additional equipment expenditure for design alternative, dollars

EXAMPLE 7-7. A base-load 700 MW unit is being designed. Two pipe sizes are proposed for main steam service. The pressure drops between the boiler super-heater outlets to the turbine stop valve are 150 psi and 100 psi, respectively, for plan A and plan B. Because of this, plan B has additional 5000 kW generation capability over plan A and advantage of 6 Btu/kWh in station net heat rate. Estimate the maximum allowable increase in capital expenditure for plan B. The economic factors used for this evaluation are

Plant capacity factor	70%
Annual fixed charge rate	16%
Unit credit for differential generation capability	$300/kW-yr
Unit fuel cost	$1.50/MBtu

Solution: First, we determine the annual benefit due to the increase of genera-tion capability. By Eq. (7-33), we have

$$B1 = (\Delta PNKW)(UC)$$

$$= 5000 \times 300 = \$1.5 \times 10^6/\text{yr}$$

Second, we calculate the annual saving in fuel cost. Using Eq. (7-34) leads to

$$B2 = (8760)(0.70)(700{,}000)(6)(1.50)(10^{-6})$$

$$= \$38{,}631/\text{yr}$$

In this case, the benefit due to the fuel saving is relatively small as compared with that due to the differential generating capability. The sum of these two (i.e., $1.538 million/yr) is the total annual benefit received if the plan B is adopted. The maximum allowable increase of capital expenditure is the value when the benefit-to-cost ratio is set equal to unity. Thus, the final result is

$$COST = (B1 + B2)/(AFCR)$$

$$= 1.538 \times 10^6/0.16$$

$$= \$9.61 \times 10^6$$

It should be emphasized that the foregoing approach is valid only if the generating unit under consideration is of base-load type. The unit is expected to operate full-load whenever it is available. In this example it is seen that the annual fixed charge rate is an important factor. For a high fixed charge rate, the plan with a higher capital cost would have less chance to be accepted. Using Example 7-7 again and only changing the annual fixed charge rate to 20%, the maximum allowable increase of capital expenditure will become $7.69 million.

In conducting a turbine-cycle configuration design, consulting engineers frequently select a base cycle. This base cycle does not have to be an optimal cycle and is used only as a starting point when the effect of cycle parameters is studied. Through the cost-benefit analysis the more optimal cycle is gradually evolved. In practice, consulting engineers often have to determine the penalty cost for the system configuration that suffers a loss of generation capability. For instance, one system has a 10,000 kW generation capability less than another system. To have a fair comparison, some penalty cost must be added to the evaluated cost of the first system. The unit penalty cost generally consists of two components. The first is the demand charge that represents the capital expenditure needed to make up one kilowatt generation capability loss. This demand charge varies in a wide range, depending on the actual network conditions. It can be treated either as unit plant cost or as incremental unit plant cost. Sometimes, the demand charge may be simply assumed as the unit cost of peaking generation plant or the unit price charged by the neighboring companies. The unit loading pattern definitely has an impact on the demand change. Throughout this text the unit of the demand charge is dollar per kilowatt and per year.

The second component of penalty cost is the capital charge due to the replacement energy cost. Because of generation capability loss, additional electric energy must be purchased from outside the system or generated in less efficient units within the system. This additional energy cost is called the replacement energy cost. When the annual replacement cost is capitalized, it becomes the capital charge of replacement energy cost. Like the demand charge, this charge also has a unit of dollar per kilowatt and per year.

The unit penalty cost is simply the sum of the demand charge and the above-defined capital charge. It is used to produce the annual penalty cost for the system with loss of generation capability. In that sense, the unit penalty cost should be equal to the unit credit cost defined at the beginning of this section.

EXAMPLE 7-8. The network system has economic factors identical to those in Example 7-7. Calculate the capital charge of replacement energy cost in terms of dollar per kilowatt and per year. Assume that the replacement energy cost is 66 mills/kWh.

Solution: We first determine the annual replacement energy cost for one kilowatt generation capability loss by

$$\text{Annual cost} = (8760)(\text{unit capacity factor})(1 \text{ kW})(\text{unit energy cost})$$

Substituting the numerical values, we have

$$\text{Annual cost} = (8760)(0.70)(1)\left(\frac{66}{1000}\right)$$

$$= \$404.7/\text{kW-yr}$$

To capitalize the annual replacement energy cost, we must divide it by the annual fixed charge rate 16%. That is,

$$\text{Capital charge} = 404.7/0.16$$

$$= \$2529.4/\text{kW-yr}$$

EXAMPLE 7-9. A base turbine cycle configuration has been specified showing the station net output of 720,000 kW and net heat rate of 9500 Btu/kWh at the condenser pressure 2.5 in. Hg abs. An investigation is needed to determine the penalty cost for a loss of generation capability when the condenser pressure is increased to 3.5 in. Hg abs. The heat balance at the new condenser pressure indicates that the station will decrease its output by 4.76% and increase its heat rate by 5%. Calculate the present worth of lifetime penalty cost if the higher condenser pressure is chosen. The economic factors listed below should be used.

Plant capacity factor	68%
Plant economic life	35 years
Rate of return	12%
Levelized fuel cost	$1.60/MBtu
Combined capital charge for capability and energy loss	$400/kW-yr

Solution: The annual penalty cost consists of two components: the first is due to the fuel cost increase resulting from the station heat rate increase and the second is the cost for a loss of generation capability. The annual fuel cost increase is calculated by

$$P1 = (8760)(\text{plant capacity factor})\,(\text{base kW output})$$

$$(\text{station heat rate change})(\text{levelized fuel cost})$$

In this case the station heat rate increases by 5%. That is, the net heat rate change is 475 Btu/kWh. Substituting the numerical values into the above expression, we have the first penalty

$$P1 = (8760)(0.68)(686{,}000)(475)(1.60)/10^6$$

$$= \$3.26 \text{ millions per year}$$

The annual penalty cost from the loss of generation capability is simply calculated by

$$P2 = (\text{station kW output change})(\text{unit penalty cost})$$

The unit penalty cost is identical to the combined capital charge for generation capability and energy loss. In this case this value is given \$400/kW-yr. The station output decreases by 34,000 kW. Using these values, we have the second penalty cost as

$$P2 = (34{,}000)(400)$$

$$= \$13.6 \text{ million per year}$$

Thus, the sum of $P1$ and $P2$ represents the annual penalty cost when the condenser is increased from 2.5 to 3.5 in. Hg abs. When this penalty cost is multiplied by the series present worth factor (SPWF), we have the present worth of lifetime penalty cost as

$$= \frac{1 - (1 + i)^{-n}}{i} A$$

$$= \frac{1 - (1 + 0.12)^{-35}}{0.12} (3.26 + 13.6)$$

$$= \$137.8 \text{ million}$$

It is well known that the plant cooling system design (including condenser, cooling tower and other associated equipment) is closely related to the selection of condenser pressure. As the condenser pressure increases, the cooling system will be reduced in size and, therefore, the capital expenditure will decrease. However, it does not mean that a higher condenser pressure is always preferred. The above penalty cost must be taken into consideration. The cooling system design is discussed in Chapter 12.

SELECTED REFERENCES

1. J. K. Salisbury, *Steam Turbines and Their Cycles*, Wiley, 1950.

2. W. J. Kearton, *Steam Turbine Theory and Practice*, Pitman, 1973.

3. R. C. Spencer, K. C. Cotton, and C. M. Cannon, "A Method for Predicting the Performance of Steam Turbine Generator 16,500 kW and Larger," ASME Transaction, Series A, *Journal of Engineering for Power*, October 1963.

4. R. L. Bartlett, *Steam Turbine Performance and Economics*, McGraw-Hill, 1958.

5. F. S. Aschner, *Planning Fundamentals of Thermal Power Plants*, Israel University Press, 1978.

6. K. W. Li, "Computer-Aided Design of Turbine Cycle Configuration," *Computer in Engineering 1983*, ASME publication, 1983.

7. K. W. Li and P. P. Yang, "User's Manual for Generalized Turbine Cycle Heat Balance Program," an internal report, Mechanical Engineering Department, North Dakota State University, 1980.

8. M. J. Moran, *Availability Analysis: A Guide to Efficient Energy Use*, Prentice-Hall, 1982.

9. J. E. Ahern, *The Exergy Method of Energy Systems Analysis*, Wiley, 1980.

PROBLEMS

7-1. Referring to the heat balance diagram in Fig. 7-7, calculate and check the turbine net output. The generator efficiency is assumed to be 98.5%.

7-2. Referring to the heat balance diagram in Fig. 7-7, estimate the heat input to the turbine system and the fuel consumption. The boiler efficiency is assumed to be 90% and the coal heating value is 13,186 Btu/lb.

7-3. Estimate the condenser cooling water requirement in terms of gallons per minute for the system described in Fig. 7-7. The cooling water enters the condenser at 65 F and leaves at 85 F.

7-4. Calculate the turbine net heat rate and output for the cycle arrangement shown in Fig. 7-20. Assume that all pressure drops can be omitted from consideration and the exhaust end loss can be approximated as 10 Btu/lb. For this heat balance calculation the data listed below should be utilized.

HP turbine efficiency	90%
IP turbine efficiency	92%
LP turbine efficiency	89%
Generator efficiency	98%
Pump efficiency	90%
Throttle steam flow rate	4×10^6 lb/hr
Feedwater pump drive	motor-driven

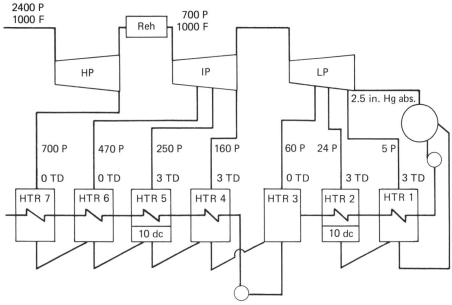

Figure 7-20. A turbine cycle for Problem 7-4.

7-5. Repeat Problem 7-4 by using turbine-driven feedwater pumps. The steam is withdrawn from the crossover piping between the intermediate-pressure turbine and low-pressure turbine. It is assumed that the exhaust pressure of the auxiliary turbine is higher than the main condenser by 0.25 inch of mercury.

7-6. A 700 MW steam turbine system has steam inlet conditions, 3500 psia, and 1000 F/1000 F. The turbine is a tandem compound six flow and has 30 inches last-stage blade length. At a condenser pressure of 2.5 in. Hg abs. the turbine net heat rate is given by

$$NHR = 87{,}000\left(\frac{NKW}{1000}\right)^4 - 178{,}600\left(\frac{NKW}{1000}\right)^3$$

$$+ 138{,}600\left(\frac{NKW}{1000}\right)^2 - 49{,}400\left(\frac{NKW}{1000}\right) + 14{,}650$$

where NHR = turbine net heat rate (Btu/kWh) and NKW = turbine net output (MW). Calculate the present worth of lifetime fuel cost for this turbine system. Assume that the boiler efficiency is 92% and plant auxiliary power 7%. The average turbine loading pattern during the entire plant economic life is approximated by

Loading Number	1	2	3	4
Percentage of turbine-generator full load	100	80	60	40
Operation hour	2000	1500	1500	1000

The economic factors used for this calculation are identical to those in Example 7-4.

7-7. Repeat Problem 7-6 for the same-size turbine system with inlet steam conditions 2400 psia and 1000 F/1000 F. The turbine is also tandem compound six flow and has 30 inches last-stage blade length. Because of different steam conditions it has the turbine net heat rate given by

$$NHR = 66,030\left(\frac{NKW}{1000}\right)^4 - 138,100\left(\frac{NKW}{1000}\right)^3 + 110,200\left(\frac{NKW}{1000}\right)^2$$

$$- 40,780\left(\frac{NKW}{1000}\right) + 13,810$$

at a condenser pressure of 2.5 in. Hg abs. Other data are identical to those in Problem 7-6.

7-8. An engineering project requires a 700 MW steam turbine system with inlet conditions of 2400 psia and 1000 F/1000 F. Previous studies have identified two choices: a cross-compound four flow with 52 inches LSB and a tandem-compound four flow with 33.5 inches LSB. The turbine net heat rates provided by a manufacturing company are

$$NHR = 67,430\left(\frac{NKW}{1000}\right)^4 - 139,700\left(\frac{NKW}{1000}\right)^3 + 110,200\left(\frac{NKW}{1000}\right)^2$$

$$- 40,960\left(\frac{NKW}{1000}\right) + 14,030$$

for the cross-compound turbine and

$$NHR = 82,360\left(\frac{NKW}{1000}\right)^4 - 168,200\left(\frac{NKW}{1000}\right)^3 + 129,500\left(\frac{NKW}{1000}\right)^2$$

$$- 45,460\left(\frac{NKW}{1000}\right) + 14,120$$

for the tandem-compound turbine. NHR is in unit of Btu/kWh and NKW is in MW. The loading schedule below are the same for these two systems.

Loadings	1	2	3
Percentage of T/G full load	100	75	50
Operation hours			
First 10 years	2000	2000	2500
Second 10 years	1500	2500	1500
Third 10 years	1000	3000	1000

Use a boiler efficiency of 92% and a plant auxiliary power of 8.5% to estimate the plant net heat rate and net output. The economic factors listed below are provided for this study.

Plant economic life	30 years
Annual fixed charge rate	17%
Rate of return	13%
First-year fuel cost	$2.00 per MBtu
Fuel cost escalation rate	6%
Initial cost	$700/kW for TC4F system
	$750/kW for CC4F system

Estimate the present worth of the lifetime total cost for these systems. Which system would you recommend?

7-9. The two turbine systems described in Problem 7-8 have different maximum outputs under valve wide-open conditions. The system with the cross-compound turbine has 11 MW more in generation capacity. Because of this, purchase option is used to compensate for the deficiency of the system with a tandem compound turbine. Estimate the present worth of lifetime penalty cost using the following input information.

First-year demand cost	$40 per kW
First-year energy cost	30 mills per kWh
Escalation rate for demand and energy cost	4%
Average operation hours per year	400 hr

Other conditions are identical to those in Problem 7-8. With these new calculations would the recommendation made in Problem 7-8 be still correct?

7-10. Consulting engineers proposed two similar turbine system configurations for a 800 MW power plant project. In fact both turbine cycle systems have the same arrangement in every aspect except the size of turbine exhaust end. One system has a four flow to the condenser while another has a six flow to the condenser. The plant net heat rates are summarized below. For the four-flow turbine plant:

	Plant Net Output (MW)		
Condenser Pressure (in. Hg abs.)	800	640	480
2.5	9491.0	9518.5	9665.5
3.5	9538.8	9627.2	9896.1
4.5	9636.8	9792.1	10145.8

For the six-flow turbine plant:

	Plant Net Output (MW)		
Condenser Pressure (in. Hg abs.)	800	640	480
2.5	9297.5	9413.4	9702.5
3.5	9495.8	9675.4	10068.1
4.5	9709.7	9951.0	10347.7

Estimate the present worth of lifetime fuel cost for these two designs. The following data should be used for evaluation:

Economic life	30 years
Fuel cost	$1.50 per MBtu
Fuel cost escalation	6%/yr
Rate of return	13%
Annual fixed charge rate	17%
Operation patterns	

Seasons	Plant Loads (MW)	Year			
		1–2	3–10	11–20	21–30
Winter	800	920.00	985.50	526.25	194.50
	640	591.25	635.00	526.25	591.25
	480	306.25	307.00	809.00	832.25
Spring and	800	1,840.00	1,971.00	1,052.50	389.00
Fall	640	1,182.50	1,270.00	1,052.50	1,182.50
	480	612.50	614.00	1,618.00	1,664.50
Summer	800	920.00	985.50	526.25	194.50
	640	591.25	635.00	526.25	591.25
	480	306.25	307.00	809.00	832.25

The average condenser pressure is assumed 2.5 in. Hg abs. for winter, 3.5 in. Hg abs. for spring and fall, and 4.5 in. Hg abs. for summer.

7-11. Based on the calculations done in Problem 7-10, what is your recommendation? Specify the conditions under which the recommendation is valid.

7-12. Repeat Problem 7-10 using the annual average condenser pressure 3.5 in. Hg abs. Discuss and compare the results with that in Problem 7-10.

7-13. Estimate the capital charge for replacement energy cost in terms of dollars per kilowatt and per year for a base-load unit. The economic factors in the network are 25 mills/kWh for replacement energy cost and 16% for annual fixed charge rate. Assume that the base-load unit has plant capacity factor 68%.

7-14. Reduction of capital expenditure for a fossil-fuel unit results in a loss of generation capability 25,000 kW. Estimate the present worth of lifetime penalty cost. The economic conditions are

Demand charge	$200/kW-yr
Replacement energy cost	25 mills/kWh
Annual fixed charge rate	16%
Unit capacity factor	70%
Unit economic life	35 years
Rate of return	10%

CHAPTER EIGHT

Evaporative Cooling Tower

As indicated in Chapter 7, the amount of heat rejected from the plant condensers is significant. Waste heat will eventually find its way to the earth's atmosphere. There are two basic paths that heat can follow. In once-through cooling, which most power plants have used in the past, heat is first removed from condensers and transferred to water bodies such as oceans, rivers, and natural lakes. The added heat is then transferred to the atmosphere by evaporation, convection, or radiation. The body of water may be seriously affected in this waste-heat transfer process. This may, therefore, be a matter of concern. In recent years, the federal and local governments have established very stringent regulations controlling the use of bodies of water for this purpose.

In contrast to once-through cooling, cooling towers are being used to transfer the waste heat directly to the earth's atmosphere. In this approach heat removed from condensers is carried by water to the cooling tower from which the heat is rejected to the atmosphere. Because of this direct path to the atmosphere, there is no adverse effect on water bodies. Cooling towers have been used for generating stations for many years in locations for which once-through coolings have not been available. It is estimated that more than 50% of new generating stations will require cooling towers or similar systems in the future.

8.1 DESCRIPTION OF EVAPORATION COOLING TOWERS

Evaporative cooling towers are devices that cool water by bringing water into contact with air. The water and air flow are directed in such a way as to provide maximum heat transfer. One of the configurations used is the counterflow arrangement. In this arrangement, the water falls downward while the air is drawn upward by a fan. Figure 8-1 is a schematic diagram of the counterflow. The second type of configuration is the crossflow arrangement. Here water again falls downward, but the air moves perpendicular to the water flow, as indicated in Fig. 8-2.

The purpose of the evaporative cooling tower is to cool the circulating water to an acceptable temperature level. An induced-draft cooling tower is usually composed of the following components:

- Inlet water distributors
- Tower fills (tower packing)
- Air moving equipment such as fans
- Air inlet louvers
- Drift eliminators
- Water storage basins

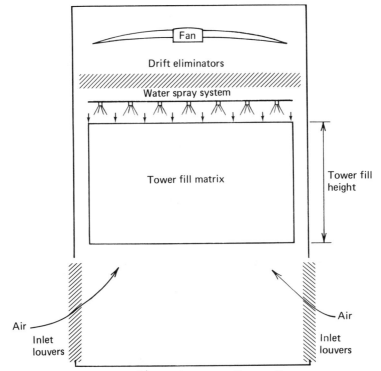

Figure 8-1. Schematic diagram for counterflow cooling tower.

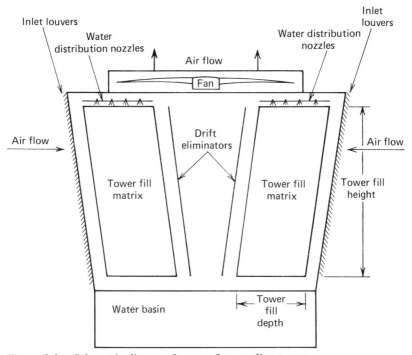

Figure 8-2. Schematic diagram for crossflow cooling tower.

283

Figure 8-3. Construction photo of induced draft-counterflow cooling tower. The cooling tower has 13 cells each 42 by 42 ft and cools 145,000 gpm of water from 115.5 to 92 F when the wet-bulb temperature is 79 F. In the left foreground are two vertical motor-driven circulating water pumps. (Courtesy Chas. T. Main, Inc., Boston)

Figures 8.1 and 8.2 indicate the relative positions for these components. The physical appearance of typical towers is shown in Figs. 8-3 to 8-5. One of the most important tower components is the tower fill. The purpose of the tower fill is to increase the air-water contact area and the water resident time in the tower. Various fills achieve this purpose in different ways. Details are presented later in this chapter. The mechanical-draft fan is used to ensure that the proper amount of air flow is maintained for cooling. The fan is placed at the top of the tower, inducting the air through the tower. Louvers are placed at the tower air entrance. Their function is to direct air into the tower and to produce an even air distribution through the fill. The drift eliminators are placed at the tower fill exit to remove the suspended water particles from the air stream. The natural-draft cooling tower has a similar arrangement except it has no mechanical-draft fans. In the natural-draft cooling tower air movement is induced by the bouyant force (i.e., chimney effects).

The definition of several terms is important in the tower fill description. For counterflow tower, the important fill dimensions are the fill matrix height and cross-sectional area or volume. The fill matrix height refers to the fill dimension in the vertical direction (i.e., air and water path length). The tower fill matrix cross-sectional area is the air and water frontal area; it refers to the area over which the

Figure 8-4. Photo of the fan deck of the cooling tower shown in Fig. 8-3. (Courtesy Chas. T. Main, Inc., Boston)

Figure 8-5. Construction photo of concrete hyperbolic shell for natural draft cooling tower to serve the condenser of an 850 MW steam turbine. The tower is designed to cool 310,000 gpm of water from 123 to 95 F when the wet-bulb temperature is 78 F and the relative humidity is 50%. (Courtesy Chas. T. Main, Inc., Boston)

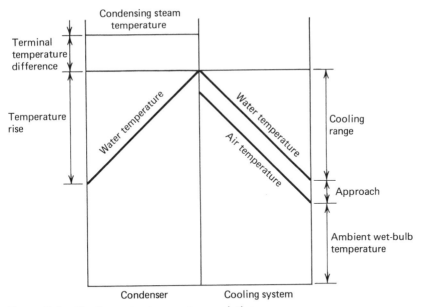

Figure 8-6. Cooling water temperature variation.

water and air flows are distributed. The tower fill matrix volume corresponds to the volume of space occupied by the fill.

The important dimensions for the crossflow tower fill include the fill matrix height, depth, length, and volume. The fill matrix height is the dimension in the vertical direction, or the water path length. The fill matrix depth is the length of the air path in the horizontal direction. The fill matrix length refers to the distance along the inlet louvers. And again, the matrix volume is the space occupied by the tower fill. In addition to these dimensions, the configuration parameter (CP) is frequently used to describe the crossflow fill. The configuration parameter is defined as the ratio of the fill depth to fill height.

In addition to these dimensions necessary to describe the cooling towers, the terms cooling range and tower approach are frequently used. The cooling range is the water temperature drop inside the tower. The tower approach is defined as the temperature difference between the cold water leaving and the ambient air temperature (wet-bulb). In the evaporative cooling tower the lowest water temperature attainable is the ambient wet-bulb temperature. Figure 8-6 indicates the cooling water temperature variation in terms of cooling range, tower approach, and ambient air temperature in combination with the condenser.

8.2 COUNTERFLOW TOWER EQUATIONS

Figure 8-7 indicates a differential control volume of the counterflow cooling tower. The control volume has one square foot of plan area and cooling volume dV. There are L pounds of water and G pounds of dry air per hour flowing through the tower

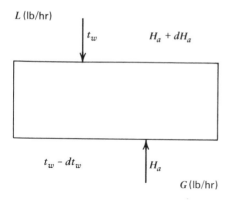

L (lb/hr)

t_w $H_a + dH_a$

$t_w - dt_w$ H_a

G (lb/hr)

Figure 8-7. A differential control volume of cooling tower.

under steady-state and steady-flow conditions. Assuming that the evaporation rate in the control volume is negligible, the product of water temperature drop dt_w and water flow rate L must be equal to the product of the air enthalpy rise dH_a and flow rate G. That is,

$$LC_{pw}\, dt_w = G\, dH_a \tag{8-1}$$

Equation (8-1) also represents the amount of heat transferred from the water droplets to the surrounding air. Applying the principles of heat transfer and mass transfer to the interface between the water and air, we can write

$$LC_{pw}\, dt_w = ha\, dV\,(t_w - t_a) + Ka\, dV\,(W_w - W_a)h_{fg} \tag{8-2}$$

where

h = heat transfer coefficient at the interface

K = mass transfer coefficient at the interface

a = contact surface area per unit volume of tower packing

C_{pw} = specific heat for water

h_{fg} = latent heat of water

W_w = humidity ratio of saturated air at the water temperature

W_a = humidity ratio of moist air at the air stream temperature

The first term on the right represents the heat transferred to air by convection and the second term the heat transferred by evaporation. At the interface the following Lewis approximation generally holds:

$$K = h/C_p \tag{8-3}$$

C_p is the specific heat of moist air at the interface. Combining Eqs. (8-2) and (8-3), we obtain

$$LC_{pw}\, dt_w = Ka\, dV \left(C_p t_w - C_p t_a + h_{fg}W_w - h_{fg}W_a \right) \qquad (8\text{-}4)$$

or

$$LC_{pw}\, dt_w = Ka\, dV \left(H_w - H_a \right) \qquad (8\text{-}5)$$

if

$$H_w = C_p t_w + W_w h_{fg} \quad \text{and} \quad H_a = C_p t_a + W_a h_{fg} \qquad (8\text{-}6)$$

H_w is the enthalpy of saturated air at the local water temperature, and H_a is the average enthalpy of the air stream. To determine the overall performance of the cooling tower, an integration over the entire packing volume is required. The final result, which is the Merkel's equation, can be expressed as

$$\int_0^v \frac{Ka\, dV}{L} = \int_{t_e}^{t_i} \frac{C_{pw}\, dt_w}{H_w - H_a} \qquad (8\text{-}7)$$

or

$$\frac{KaV}{L} = \int_{t_e}^{t_i} \frac{C_{pw}\, dt_w}{H_w - H_a} \qquad (8\text{-}8)$$

It must be pointed out that the derivation of Eq. (8-8) is based on several assumptions. These assumptions include the omission of evaporation loss, neglecting the resistance at the air-water interface, constant latent heat, and the validity of the Lewis relationship. In the literature, the term KaV/L is frequently referred to as the tower characteristics. The meaning of this term is similar to that of NTU (number of transfer unit) in heat exchanger design. In other words, the value of KaV/L (or NTU) specifies the size of the equipment necessary to achieve the maximum possible effectiveness. Cooling towers with a value of KaV/L greater than necessary are too large and therefore too expensive.

Equation (8-8) has no closed-form solution. This is mainly because the enthalpy of saturated air does not vary linearly with the water temperature. This is best illustrated in Fig. 8-8, which shows the enthalpy of saturated air (H_w) as a function of the local water temperature and the enthalpy of moist air (H_a) as a straight line. The straight-line relationship is suggested by Eq. (8-1). In a finite form Eq. (8-1) becomes

$$H_{a,e} - H_{a,i} = \frac{L}{G}(CR) \qquad (8\text{-}9)$$

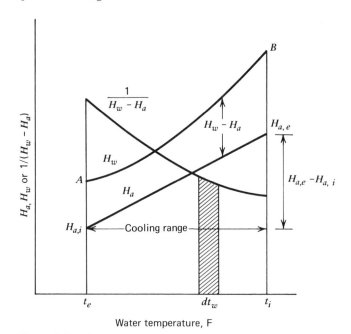

Figure 8-8.   Cooling tower enthalpy diagram.

where

$$H_{a,e} = \text{air enthalpy at the tower exit, Btu/lb}$$

$$H_{a,i} = \text{air enthalpy at the tower entrance, Btu/lb}$$

$$CR = \text{cooling range, F}$$

In Fig. 8-8 the vertical distance between the curves H_w and H_a represents the local enthalpy difference. The inverse of this term is also a function of the local water temperature, as indicated in Fig. 8-8. To determine the tower characteristics KaV/L, as defined in Eq. (8-8), it is best to evaluate the area under the curve of $1/(H_w - H_a)$. The following example demonstrates this numerical procedure.

EXAMPLE 8-1.   Calculate the tower characteristics KaV/L for the conditions: hot water temperature 110 F, cold water temperature 84 F, ambient wet-bulb temperature 69 F, and flow ratio (L/G) 1.3.

Solution: Values of H_w, H_a and $1/(H_w - H_a)$ are plotted in Fig. 8-9 as functions of the local water temperature from 110 F to 84 F. In drawing these curves the enthalpy of saturated air H_w was obtained from Table C-3 according to the local water temperature. The enthalpy of moist air H_a is determined by Eq. (8-9). With these values, the curve $1/(H_w - H_a)$ is then prepared. The details are summarized in Table 8-1.

The area under the curve $1/(H_w - H_a)$ in Fig. 8-9 represents the value of the tower characteristics KaV/L. Table 8-1 contains the numerical calculation of this term. The final result (tower characteristics) is approximately 1.51.

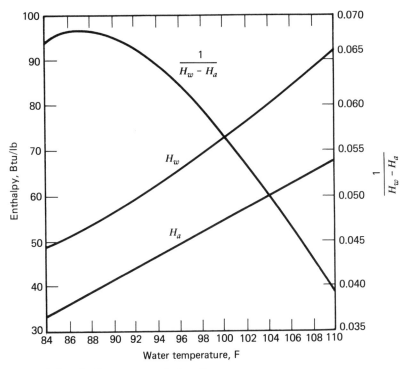

Figure 8-9. Cooling tower enthalpy diagram.

Table 8-1
Summary of Calculations for Example 8-1

T_w(F)	H_w(Btu / lb)	H_a(Btu / lb)	$1 / (H_w - H_a)$(lb / Btu)	Average Value of $1 / (H_w - H_a)$(lb / Btu)
84	48.22	33.75	0.0668	0.0672
86	50.66	35.85	0.0675	0.0676
88	53.23	38.45	0.0677	0.0674
90	55.93	41.05	0.0672	0.0667
92	58.78	43.65	0.0661	0.0653
94	61.77	46.25	0.0644	0.0633
96	64.92	48.85	0.0622	0.0609
98	68.23	51.45	0.0596	0.0581
100	71.73	54.05	0.0566	0.0550
102	75.42	56.65	0.0533	0.0516
104	79.32	59.25	0.0498	0.0481
106	83.42	61.85	0.0464	0.0447
108	87.76	64.45	0.0492	0.0412
110	92.34	67.05	0.0395	

Sum = 0.757

$$\frac{KaV}{L} = \int \frac{C_{pw}dt_w}{H_w - H_a} = 2 \times 0.757 = 1.51$$

Evaluation of the tower characteristics is time-consuming, as indicated by the above example. In practice, this can be avoided by using the charts published by the Cooling Tower Institute in Houston [4]. In these charts the tower characteristics is expressed in terms of the cooling range, tower approach, ambient wet-bulb temperature, and flow ratio (L/G). Appendix E includes some typical charts available in that reference. The computer programs for calculation of counterflow tower characteristics are also available and will be discussed later in this chapter.

EXAMPLE 8-2. Repeat Example 8-1 by using the charts published by the Cooling Tower Institute.

Solution: First we calculate the cooling range

$$CR = HWT - CWT = 110 - 84 = 26 \text{ F}$$

and the tower approach

$$AP = CWT - WBT = 84 - 69 = 15 \text{ F}$$

Using $CR = 26$ F, $AP = 15$ F, $WBT = 69$ F and $L/G = 1.3$, we find from Fig. E-1 (see Appendix E) that

$$\frac{KaV}{L} = 1.51$$

8.3 CROSSFLOW TOWER EQUATIONS

Figure 8-10 indicates a typical cross section of a crossflow tower. The water enters the top and flows uniformly downward through the tower fill. The air flow is induced by a fan and moves horizontally. A heat balance on the differential volume as shown

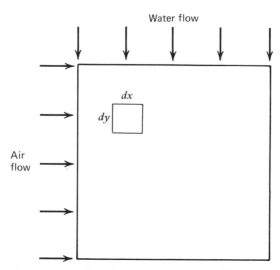

Figure 8-10. Crossflow tower section with incremental volume.

by Fig. 8-10 yields

$$-(dx)(1)(L)(C_{pw})(dT_w) = (dy)(1)(G)(dH_a)$$

Rearranging it yields

$$-LC_{pw}\left(\frac{\partial T_w}{\partial y}\right) = G\left(\frac{\partial H_a}{\partial x}\right) \tag{8-10}$$

where

$$T_w = \text{water temperature, F}$$

$$H_a = \text{air enthalpy, Btu/lb}$$

$$L = \text{water loading, lb/hr-ft}^2$$

$$G = \text{air loading, lb/hr-ft}^2$$

$$C_{pw} = \text{water specific heat, Btu/lb-F}$$

Equation (8-10) is the partial differential equation relating the heat lost by water to the heat gained by air. Examining the differential volume again and equating the heat lost by water to the heat transfer rate at the water-air interface, we have

$$-(dx)(1)(L)(C_{pw})(dT_w) = (Ka)(dy)(dx)(1)(H_w - H_a)$$

or

$$-LC_{pw}\left(\frac{\partial T_w}{\partial y}\right) = Ka(H_w - H_a) \tag{8-11a}$$

where

$$K = \text{mass transfer coefficient, lb/hr-ft}^2$$

$$a = \text{contact surface area per unit tower fill volume, ft}^2/\text{ft}^3$$

$$H_w = \text{enthalpy of saturated air at the water temperature, Btu/lb}$$

Combining Eq. (8-11a) with Eq. (8-10) gives

$$G\left(\frac{\partial H_a}{\partial x}\right) = Ka(H_w - H_a) \tag{8-11b}$$

Equations (8-11a) and (8-11b) relate the heat transfer rate at the interface, respectively, to the heat lost by water and the heat gained by air. As indicated in the last section, the drive force in the heat transfer process is the enthalpy differential $(H_w - H_a)$. Completing the mathematical model that defines the cooling tower process requires two boundary conditions and property relationships. These are

$$H_a(0, y) = C_1 \tag{8-12}$$

$$T_w(x, 0) = C_2 \tag{8-13}$$

and

$$H_w = \left(0.4233 \times 10^{-10}T_w^6 - 0.6294 \times 10^{-7}T_w^5 + 0.1849\right.$$

$$\times 10^{-4}T_w^4 + 0.1457 \times 10^{-2}T_w^3 - 0.1075T_w^2 + 62.209$$

$$\left.T_w + 168.9737\right)/\left(212 - T_w\right)$$

$$\text{for } 35 \text{ F} \le T_w \le 200 \text{ F} \tag{8-14}$$

It must be pointed out that the derivation of these equations is based on assumptions similar to those used in Merkel's equation. These include the omission of evaporation loss, negligible resistance at the water-air interface, constant latent heat, and specific heat. In addition, the empirically observed inverse relationship between the hot water temperature and mass transfer coefficient is not taken into account.

The mathematical model for a crossflow tower is nonlinear and therefore has no exact solution. The numerical method frequently used is the finite difference. The algebraic equations for water temperature and air enthalpy calculations are derived as follows.

Consider a two-dimensional body that is to be divided into equal increments in both the x and y direction, as shown in Fig. 8-11. The nodal points are designated as shown, the m locations indicating the x increment and the n locations indicating the y increment. We wish to establish the water temperature and air enthalpy at any of these nodal points within the body. First, the conditions on the top of the body are determined. Applying Eq. (8-11b) to the point a, we have

$$G\left[H_a(1,0) - H_a(0,0)\right] = Ka\,\Delta x\left[H_w - H_a\right]_a$$

The term $\left[H_w - H_a\right]_a$ is the enthalpy differential at the point a. Making use of the approximation

$$\left[H_w - H_a\right]_a = \frac{H_w(0,0) - H_a(0,0)}{2} + \frac{H_w(1,0) - H_a(1,0)}{2}$$

we have

$$H_a(1,0) = \frac{H_a(0,0) + \dfrac{M_x}{2}\left[2H_w(0,0) - H_a(0,0)\right]}{1 + \dfrac{M_x}{2}}$$

where

$$M_x = \frac{Ka\,\Delta x}{G}$$

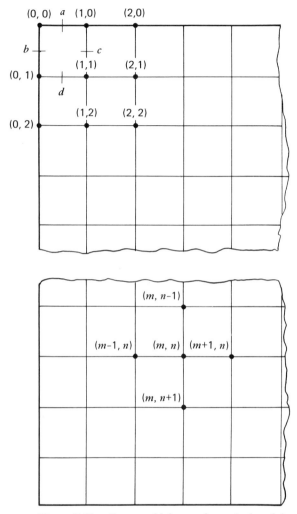

Figure 8-11. Tower grid for mathematical nodal.

In a generalized form the above equation becomes

$$H_a(m,0) = \frac{H_a(m-1,0) + \dfrac{M_x}{2}\left[2H_w(m-1,0) - H_a(m-1,0)\right]}{1 + \dfrac{M_x}{2}} \quad (8\text{-}15)$$

Equation (8-15) is used to estimate the air enthalpy on the top of the tower fill. Since the air enthalpy at nodal point $(0,0)$ is always available, the enthalpy at the subsequent nodal points can be easily determined. To estimate the water temperature at the air entrance, we apply (Eq. 8-11a) to the point b and have

$$-LC_{pw}[T_w(0,1) - T_w(0,0)] = Ka\,\Delta y\,(H_w - H_a)_b$$

Again, making use of the approximation

$$(H_w - H_a)_b = \frac{H_w(0,0) - H_a(0,0)}{2} + \frac{H_w(0,1) - H_a(0,1)}{2}$$

we obtain

$$T_w(0,1) = T_w(0,0) - \frac{M_y}{2}[H_w(0,0) + H_w(0,1) - 2H_a(0,0)]$$

where

$$M_y = \frac{Ka\,\Delta y}{L} \quad \text{and} \quad C_{pw} = 1.0$$

In a generalized form, the above equation becomes

$$T_w(0,n) = T_w(0,n-1) - \frac{M_y}{2}[H_w(0,n-1) + H_w(0,n) - 2H_a(0,n-1)]$$

$$(8-16)$$

Equation (8-16) is used to calculate the water temperature at the air entrance. Since there are two unknowns $[T_w(0,n), H_w(0,n)]$ in this equation, Eq. (8-14) relating the enthalpy of saturated air to the water temperature must be utilized in this particular step. To determine the conditions in the interior nodal points, we apply Eq. (8-11a) to the point c and obtain by using the same procedure

$$T_w(m,n) = T(m,n-1) - \frac{M_y}{2}[H_w(m,n-1) - H_a(m,n-1)$$

$$+ H_w(m,n) - H_a(m,n)] \qquad (8-17)$$

Similarly, applying Eq. (8-11b) to the point d yields

$$H_a(m,n) = H_a(m-1,n) + \frac{M_x}{2}[H_w(m,n) - H_a(m,n)$$

$$+ H_w(m-1,n) - H_a(m-1,n)] \qquad (8-18)$$

Solving Eq. (8-17) and Eq. (8-18) simultaneously will give the water temperature and air enthalpy at the interior nodal point. Because of the nonlinear nature in the relationship between the enthalpy of saturated air and the temperature, an iterative technique is recommended.

In calculation, the tower design conditions first establish the water temperature across the top and the air enthalpy along the air inlet side of the tower. These conditions set the water temperature and the air enthalpy at the node $(0,0)$. This is the only point at which the air enthalpy and water temperature are both known. Equation (8-15) is used to calculate the air enthalpy at the node $(1,0)$ and Eq. (8-16) determines the water temperature at node $(0,1)$. The air enthalpy and water temperature information are used with Eqs. (8-17) and (8-18) to determine the water and air conditions at the node $(1,1)$. Thus, the air and water conditions at the nodes $(0,0)$, $(0,1)$, $(1,0)$ and $(1,1)$ are determined. The average water temperature at $(0,1)$ and $(1,1)$ is then calculated and compared with the desired exit water temperature. If the average water temperature is greater than the desired exit water temperature, the procedure continues by adding another column and row of nodes and using Eqs. (8-15), (8-16), (8-17), and (8-18). The procedure is repeated until the average water temperature leaving the last row of the grid is sufficiently close to the desired exit water temperature. The crossflow tower characteristics (or tower NTU) is then the sum of the M_y mesh values needed to obtain the tower grid size that gives the desired exit water temperature.

For accuracy the M_y value or mesh size should be initially set at not less than 0.1. Since the M_x and M_y are not independent with each other, the value of M_x must be calculated with the equation

$$M_x = M_y \frac{L'}{G'} \tag{8-19}$$

where

$$L' = \text{water flow rate, lb/hr}$$

$$G' = \text{air flow rate, lb/hr}$$

As an illustrative example, we calculate the value of KaY/L for a crossflow tower for the conditions: ambient wet-bulb temperature 75 F, hot water temperature 120 F, cold water temperature 102 F, and total flow rate ratio $(L'/G')1.0$. In this calculation the mesh size $M_y = M_x$ is assumed to be 0.1. The calculated temperature at various nodal points are shown in Fig. 8-12. It is seen that the number of nodal points in the horizontal direction is equal to that in the vertical direction. Also indicated in Fig. 8-12 is the order at which the nodal temperatures are determined. When the number of nodal points in one direction reaches 5, the average of water temperatures leaving the bottom row is 101.93 F. If this is sufficiently close to the desired value (102 F), the calculation procedure will terminate. Then, the tower characteristics KaY/L is obtained by calculating the sum of the tower meshes that are required to construct the tower grid. In this case the value is equal to 0.5.

Similar to the publication by the Cooling Tower Institute, Kelly's handbook contains many charts for calculating crossflow towers [3]. Again, the tower character-

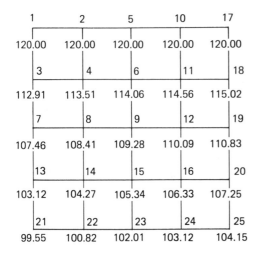

Figure 8-12. Illustrative example for crossflow KaY/L calculation.

istics is expressed in terms of cooling range, tower approach, ambient wet-bulb temperature, and the ratio of overall fluid flow. Typical charts are presented in Appendix E.

EXAMPLE 8-3. Use the Kelly's performance curves and calculate the crossflow tower characteristics for the conditions: $WBT = 75$ F, $HWT = 120$ F, $CWT = 102$ F, and $L'/G' = 1.0$.

Solution: First, we calculate the cooling range

$$CR = HWT - CWT = 120 - 102 = 18 \text{ F}$$

and the tower approach

$$AP = CWT - WBT = 102 - 75 = 27 \text{ F}$$

With $CR = 18$ F, $AP = 27$ F, $WBT = 75$ F, and $L'/G' = 1.0$, we can find the tower characteristics 0.475 from Fig. E-5 (see Appendix E). The discrepancy between this value and the previous is mainly due to the oversimplification of the numerical calculation and assumptions as previously mentioned in this text.

8.4 TOWER FILL DESCRIPTION

As stated previously, one of important cooling tower components is the tower fill or packing. Basically, the tower fill can be classified into at least two types: film type or splash type. In either case, the main function of the tower fill is to increase the contact area between the water and air and to maximize the resident time for the water in the tower. Although the function of these two fills is the same, the method by which they achieve the water-air contact is different. The method used by the film-type packing is to run a thin layer of water along a sheet of material, usually a type of plastic, and then have air directed past the water. Generally, several of these sheets are placed together at fixed angles to maximize the air-water contact area and water resident time in the fill.

The splash-type packings increase the air-water contact area and resident time in the tower by splashing. The water enters the tower, and is allowed to fall on splash boards called the tower packing. The splash boards serve two functions. The first is to break the larger water droplets into smaller ones, thus increasing the air-water contact area. The second function is to slow the fall of the water droplets, and increase the water resident time in the tower.

In general, the film-type packings have a denser configuration than do the splash-type fills. Thus, less volume of film-type packing is required to remove a given heat load. However, associated with the denser packing arrangements are higher fan power requirements. Therefore, when considering film- or splash-type fills, it must be decided whether it is the fill volume or fan power that is to be kept at a minimum.

The tower fill performance is affected not only by the fill arrangement, but also by the water and air loadings. Typical water and air loadings for tower fills can be found in Table 8-2. A low water loading results in poor water distribution, while a high water loading causes the tower to flood, producing excessive air pressure losses. In both cases, these conditions cause a deteriorating fill performance.

Another factor to consider in the selection of the tower fill is the actual physical shape. The tower dimensions should be kept in mind for the crossflow splash-type fill. For this fill the configuration parameter, as defined previously, should be in the range of 0.4 to 1. If the configuration parameter is outside this range, there will be water and air distribution problems within the fill.

The tower fill is one of the important components of the cooling tower system. The fill comprises not only the major physical part, but also the major financial portion of a cooling tower. Therefore, accurate tower fill performance information is necessary for intelligent design decisions. Unfortunately, the fill performance information is scattered throughout various references, and the method of presentation is not always uniform. We shall now briefly describe several of the tower fills frequently used in designs.

The splash-type fills presented here came from two sources. These are *Kelly's Handbook of Crossflow Cooling Tower Performance* [3] and the reference by Kelly

Table 8-2
Typical Tower Water and Air Loadings

Type of Packing	Water Loading (lb / hr / ft²)	Air Loading (lb / hr / ft²)
Counterflow and crossflow film packing	2000–10000	1600–3000
Counterflow splash type packing	1500–3000	1600–3000
Crossflow splash type packing	2000–12000	1600–3000

and Swenson [6]. When the fills from these sources are referred to, they will be called, respectively, Kelly's fill and Kelly and Swenson's fill.

The information on the Kelly's crossflow splash-type fill configurations is presented in Fig. 8-13. As can be seen, there are various configurations used to increase the water-air contact area and to increase the water resident time in the tower. The fill in these configurations, unless otherwise noted, is made of redwood. The redwood fills are horizontally spaced, and nailed to a redwood stringer to comprise one deck. The decks are then vertically spaced according to the tower configuration desired. For this crossflow configuration, the water is falling downward with the air moving perpendicular across the water flow.

The details on the counterflow splash-type fill presented by Kelly and Swenson are presented in Fig. 8-14. Again, there are various fill configurations to accomplish the desired amount of the air-water interaction. The fill material for the Kelly and

Figure 8-13. Kelly's fill information [3].

Fill Desig-nation	Code	Center to Center, Installed Vertical	Center to Center Installed Horizontal	Fill Configuration
A	W	$1\frac{3}{8}$ in.	4 in.	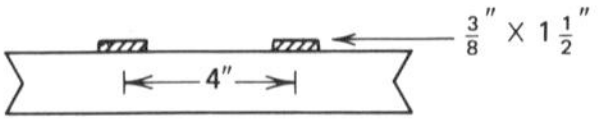
B	W	$1\frac{3}{8}$ in.	8 in.	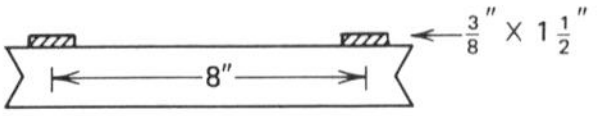
C	W	2 in.	4 in.	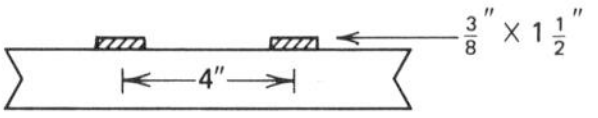
D[a]	W	2 in.	8 in.	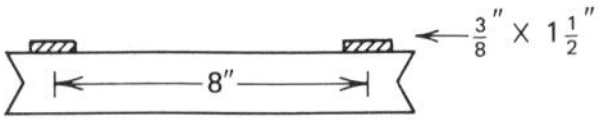
E[a]	W	4 in.	4 in.	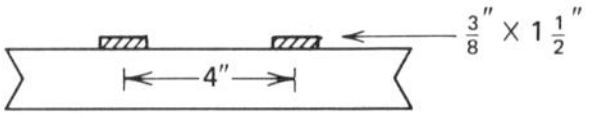
F[a]	W	4 in.	8 in.	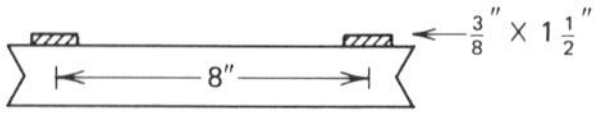
G[a]	P1[b]	8 in.	8 in.	
H	P2[b]	8 in.	8 in.	

[a] Indicates fills for which the mass transfer coefficient and the pressure drop correlations were made.
[b] These shapes are covered by U.S. patents.
Code: W Standard wood lath $\frac{3}{8}$ in. thick $\times 1\frac{1}{2}$ in. wide, mounted flat.
    P1 $2\frac{1}{2}$ in. $\times 2\frac{1}{2}$ in. plastic, 90° angle, $\frac{3}{8}$ in. rhomboid perforations both legs, mounted vertex up.
    P2 $1\frac{1}{4}$ in. high $\times$ 4 in. wide plastic airfoil (wedge), $\frac{3}{8}$ in. diameter perforations, mounted with vertex toward entering air.
Note: All fill members mounted horizontally, with member length perpendicular to air flow.

Figure 8-14. Kelly and Swenson's fill information [6].

Fill Designation	Fill Material	Fill Configuration
Deck A[a]	Redwood	Vertical spacing A = 9 in., B = 12 in.
Deck B[a]	Redwood	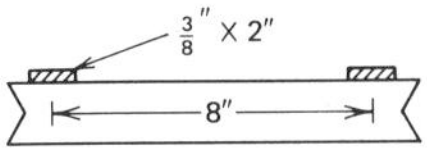
Deck C[a]	Redwood	Vertical spacing C = 15 in., D = 24 in.
Deck D[a]	Redwood	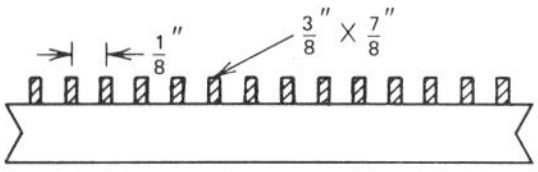
Deck E[a]	Redwood	Vertical spacing 24 in. 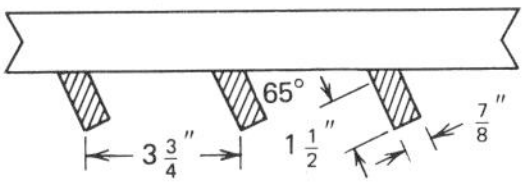
Deck F	Redwood	Vertical spacing 24 in. 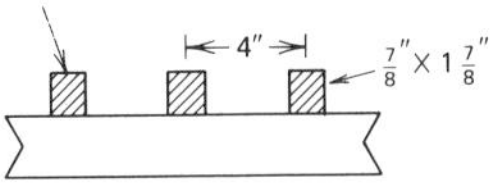
Deck G	Redwood	Vertical spacing 24 in. 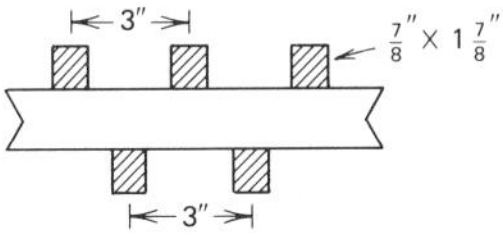
Deck H	Redwood	Vertical spacing 24 in. 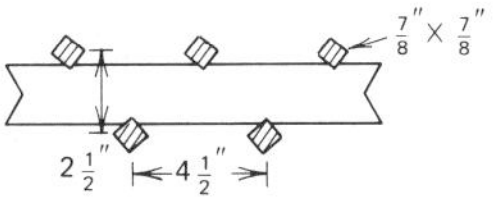
Deck I	Redwood	Vertical spacing 24 in. 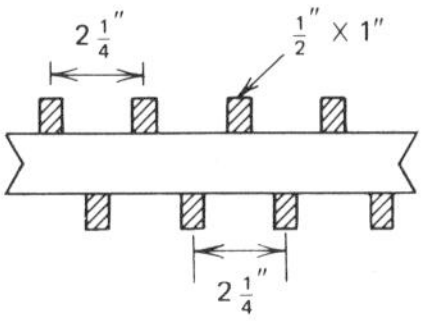
Deck J	Redwood	Vertical spacing 24 in. 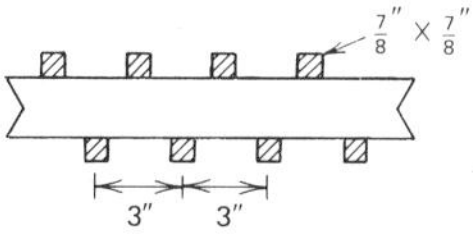

[a] Indicates fills for which the mass transfer coefficient and the pressure drop correlations were made.

Swenson fills is redwood. The redwood fill is spaced horizontally, and nailed to 1 by 2 inch redwood stringers to make up one deck. These decks are vertically spaced according to the fill configuration desired. The relative air-water flow has the water falling downward, and the air moving upward.

In design, the fill performance characteristics are needed. The performance characteristics for typical Kelly's crossflow fills are included in Appendix E. The fill characteristics are shown for various fill depth X (air travel distance) and fill height Y. These fill characteristic curves are straight-line logarithmic plots of KaY/L vs. L'/G'. In other words, the fill characteristics can be expressed in the form of

$$\frac{KaY}{L} = \alpha\left(\frac{L'}{G'}\right)^{\beta} \tag{8-20}$$

if the fill height and fill depth are specified. Making use of the definition of configuration parameter, we can change Eq. (8-20) to

$$Ka = C(L)^{m}(G)^{n} \tag{8-21}$$

where Ka = mass transfer coefficient per unit fill volume (lb/hr/ft^3), and the constants, C, m, n are related to α and β by the equations

$$C = \frac{\alpha}{Y}(CP)^{\beta} \tag{8-22}$$

$$m = \beta + 1 \tag{8-23}$$

$$n = -\beta \tag{8-24}$$

Table 8-3 presents these correlation constants for Kelly's fills. Similarly, the pressure drop for these fills can be expressed in the form of

$$DP = A(L)^{b}(G)^{d} \tag{8-25}$$

where

DP = static air pressure, inches of water per foot of air travel distance

A, b, d = empirical constants

For Kelly's Type D, E, F, and G, these constants can be found in Table 8-3. In general, these types of tower fills will have the mass transfer coefficient in the range of 200 to 500 lb/hr/ft^3 and the pressure drop in 0.005 to 0.02 in. of water per foot of fill depth.

The counterflow tower fill characteristics are similar to those for the crossflow fill. In the counterflow tower fill, however, the fill height is only one important parame-

Table 8-3
Splash-Type Tower Fill Performance Correlations

Source	Fill Type	Fill Designation	Constants for Ka Equation Eq. 8-21			Constants for DP Equation Eq. 8-25		
			C	m	n	A	b	d
Kelly	Splash type Crossflow fill	Type D	0.103	0.49	0.51	5.90 E-12	0.09	1.86
		Type E	0.088	0.42	0.58	9.86 E-12	0.82	1.84
		Type F	0.081	0.48	0.52	387.10 E-12	0.94	1.16
		Type G	0.090	0.47	0.53	20.05 E-12	0.58	1.44
Kelly and Swenson	Splash type Counterflow fill	Type A	0.083	0.38	0.61	2.12 E-9	0.34	1.83
		Type B	0.074	0.40	0.60	74.50 E-9	0.44	1.22
		Type C	0.078	0.41	0.59	6.42 E-9	0.52	1.48
		Type D	0.070	0.46	0.54	4.53 E-9	0.58	1.44
		Type E	0.059	0.56	0.44	12.40 E-9	0.46	1.44

ter. The air travel distance (X) is exactly equal to the fill height (Y). That is, the fill performance equations similar to those of crossflow types can be determined if the fill height is given. For instance, the Kelly and Swenson counterflow fill (Deck D) at the height 20 ft has

$$Ka = 0.07(L)^{0.46}(G)^{0.54} \qquad (8\text{-}26)$$

and

$$DP = 4.529 \times 10^{-9}(L)^{0.58}(G)^{1.44} \qquad (8\text{-}27)$$

To estimate the mass transfer coefficient and pressure drop for other counterflow tower fills, we again use Eq. (8-21) and Eq. (8-25). Table 8-3 summarizes the correlation constants for Kelly and Swenson counterflow fills (Types A, B, C, D, and E). In their paper, however, Kelly and Swenson suggested the correlation equation as

$$\frac{KaV}{L} = 0.07 + AN\left(\frac{L}{G}\right)^{-n} \qquad (8\text{-}28)$$

where N is the number of decks and A and n are constants for any given matrix. Table 8-4 summarizes the values of these constants to relate the data for the geometries in Fig. 8-14. It must be pointed out that this correlation is based on the assumption of negligible influence of hot water temperature. For an estimate of pressure drop, Kelly and Swenson suggested the equation [6]

$$\frac{\Delta P}{N} = BG^2\left(\frac{0.0675}{\rho_a}\right) + CL\sqrt{S_f}G_{eq}^2\left(\frac{0.0675}{\rho_a}\right) \qquad (8\text{-}29)$$

where ΔP is the air pressure drop in inches of water, and ρ_a is the density of the dry

Table 8-4
Values of A and n in Equation (8-28) [6]

Deck	A	n
A	0.060	0.62
B	0.070	0.62
C	0.092	0.60
D	0.119	0.58
E	0.110	0.46
F	0.100	0.51
G	0.104	0.57
H	0.127	0.47
I	0.135	0.57
J	0.103	0.54

Table 8-5
Values of _B_ and _C_ in Equation 8-29 [6]

Deck	Vertical Deck Spacing S ft	Plan Solidity Fraction	Vertical Free Mean Fall S_f ft	$B \times 10^8$	$C \times 10^{12}$
A	0.75	0.250	3.00	0.34	0.11
B	1.00	0.250	4.00	0.34	0.11
C	1.25	0.333	3.75	0.40	0.14
D	2.00	0.333	6.00	0.40	0.14
E	2.00	0.404	4.95	0.60	0.15
F	2.00	0.219	9.13	0.26	0.07
G	2.00	0.292	6.83	0.40	0.10
H	2.00	0.550	3.64	0.75	0.26
I	2.00	0.444	4.50	0.52	0.16
J	2.00	0.292	6.80	0.40	0.10

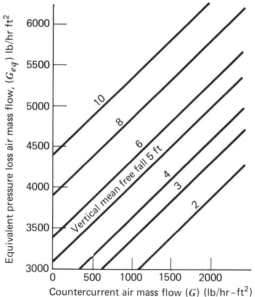

Figure 8-15. Effects of the countercurrent air mass flow and the vertical mean free fall of the water droplets on the equivalent pressure loss air mass flow [6].

Table 8-6
Film-Type Tower Fill Performance Correlations

Source	Fill Type	Fill Size (Height × Depth in Feet)	Constants for Ka Equation Eq. (8-21)			Constants for DP Equation Eq. (8-25)		
			C[a]	m[a]	n[a]	A	b	d
Munters	Crossflow XF12560/15	5 × 2	0.61	0.23	0.77	8.16 E-09	0.433	1.665
		5 × 3	0.60	0.20	0.80	14.10 E-09	0.210	1.849
		7.5 × 2	0.61	0.20	0.80	7.31 E-09	0.428	1.705
		7.5 × 3	0.54	0.22	0.78	7.09 E-09	0.371	1.757
		7.5 × 4	0.51	0.23	0.77	17.10 E-09	0.358	1.665
	Crossflow XF19060	7.5 × 3	0.19	0.54	0.46	0.20 E-09	0.739	1.682
		7.5 × 4	0.23	0.51	0.49	0.54 E-09	0.622	1.701
		(Height)						
Munters	Counterflow CF12060	1	1.08	0.25	0.75	44.10 E-12	0.305	2.545
		2	0.93	0.14	0.86	4.15 E-12	0.175	2.944
		3	0.80	0.12	0.88	1.32 E-12	0.148	3.103
		4	0.71	0.13	0.87	2.29 E-12	0.148	3.019
	Counterflow CF19060	2	0.50	0.16	0.84	1.01 E-09	0.272	2.065
		3	0.50	0.09	0.91	0.67 E-09	0.209	2.180
		4	0.49	0.04	0.96	0.76 E-09	0.257	2.120
		5	0.45	0.08	0.92	1.28 E-09	0.240	2.070

Note: The correlation constants are based on data published by the Munters Corporation [9].
[a]All Ka equation constants determined were based on an air velocity of 600 ft/min.

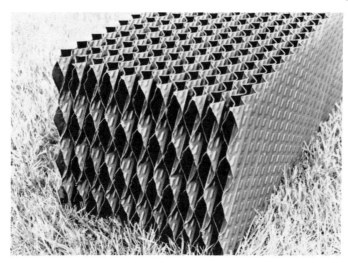

Figure 8-16. Typical plastic packing (Courtesy the Munters Corporation, Fort Myers, Florida).

air in pounds per cubic foot of air-vapor mixture. The constants B, C, S_f are related to the counterflow fill matrix and are presented in Table 8-5. The quantity G_{eq} is an equivalent air mass flow rate corresponding to the velocity of the air relative to the falling water droplets. The values of this term are presented in Fig. 8-15.

Figure 8-16 indicates typical film type packings. These tower packings are usually made of polystyrene, polypropylene, high-density polythene or p.v.c. These materials are very durable and also fire resistant. The Munters Corporation produced several film-type tower packings [9]. For crossflow towers fills XF 12560/15 and XF19060 are used and CF12060 and CF19060 for the counterflow towers. The performance of these tower fills can be approximated by Eq. (8-21) and Eq. (8-25). The constants needed for these equations are summarized in Table 8-6. The film-type packings usually have the mass transfer coefficient (Ka) in the range of 500 to 1000 lb/hr/ft^3 and the pressure drop 0.02 to 0.08 inch of water per foot of air traveling.

Tower fill information is scattered in literature and the method of presentation is not always uniform. Caution must be exercised in their use.

8.5 TOWER THERMAL DESIGN

In previous sections both counterflow and crossflow tower characteristics have been shown in terms of cooling range, tower approach, ambient wet-bulb temperature, and the ratio of fluid flow. These tower characteristics are the fill characteristics required for a specific job and, therefore, are sometimes called the required fill characteristics. In design, the required fill characteristics must be equal to the fill performance characteristics which, as shown previously, is a function of fluid flow ratio for a given fill matrix. This may be best explained in graphical form such as given in Fig. 8-17. The required tower characteristics corresponding to a set of

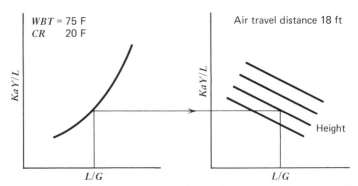

Figure 8-17. Graphical relationship between the tower and fill characteristics.

conditions is determined in the left-hand curves. Drawing a horizontal line through this value as shown in Fig. 8-17 would identify the fill height at which the required fill characteristics is equal to the performance characteristics. With this calculated fill height and heat load, one can easily estimate the volume of tower fills required.

Another approach to this problem is to simultaneously solve the appropriate fill characteristic equation and tower characteristic equation or equations (such as Eq. (8-8) for counterflow and Eqs. (8-10) through (8-14) for crossflow). Obviously, the method is iterative and can be easily adapted to computer use.

EXAMPLE 8-4. Estimate the air flow rate and the volume of crossflow tower fill for the project of cooling 1000 gpm from 105 F to 87 F. The ambient conditions are 75 F wet-bulb and 50% relative humidity. The water loading is 3000 $lb/hr/ft^2$, and the overall fluid flow ratio is 1.0.

Solution: We first calculate the cooling range

$$CR = HWT - CWT$$

$$= 105 - 87 = 18 \text{ F}$$

and tower approach

$$AP = CWT - WBT$$

$$= 87 - 75 = 12 \text{ F}$$

With the cooling range, tower approach, and ambient wet-bulb temperature, we can identify the tower characteristic curve from Fig. E-5. From this curve the tower KaY/L at $L'/G' = 1.0$ is found as

$$\left(\frac{KaY}{L}\right) = 1.25$$

The required tower characteristics must be equal to the fill characteristics. Let us

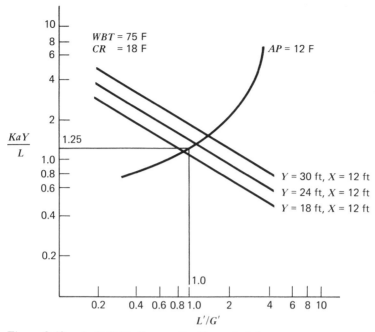

Figure 8-18. A KaY/L diagram for Example 8-4.

select Kelly's crossflow fill-type F with a configuration of $X = 12$ ft. The fill characteristic curves from Fig. E-7 are reproduced and presented in Fig. 8-18. Transporting the tower curve to the same figure, we can see the intersection at the point $KaY/L = 1.25$ and $L'/G' = 1.0$, indicating that the matrix height should be approximately 20 ft. To determine the cross-sectional area of fill matrix, we divide the water flow rate by the water loading. That is,

$$\text{Area} = \frac{1000 \times 500}{3000} = 166.7 \text{ ft}^2$$

Then, the volume of fill matrix is

$$\text{Volume} = \text{area} \times \text{height}$$

$$= 166.7 \times 20 = 3333 \text{ ft}^3$$

Since $L'/G' = 1.0$, the air flow rate must be equal to the water flow rate, that is, $G' = 500,000$ lb/hr. The air loading density, if desired, can be calculated by

$$G = G'/\text{frontal area to air flow}$$

$$G = 500,000/3333/12$$

$$G = 1799.6 \text{ lb/hr/ft}^2$$

In this example the water-to-air ratio is assumed available, and then the matrix height is calculated. In design, occasions may arise that the fill matrix is specified in the dimensions. If this is the case, the fill characteristic equation must be coupled with the tower equation and then solved simultaneously for KaY/L and L'/G'. Graphically, it means the fill characteristic curve will intersect with the tower curve. The intersection of these two curves indicate the desired KaY/L and L'/G'. The following example illustrates this aspect.

EXAMPLE 8-5.   The fill-type F from Kelly's handbook is selected as the tower fill. The packed height Y and depth X are, respectively, 36 and 18 ft. Estimate the air flow rate and fill requirement for the project of cooling 1000 gpm from 105 F to 87 F. The ambient wet-bulb temperature is 75 F. The water loading is assumed to be 3000 lb/hr/ft^2.

Solution: Let us use the graphical approach. Since the tower fill is specified, the fill characteristic curve can be found in Fig. E-8. Next, we calculate the cooling range as

$$CR = HWT - CWT$$

$$= 105 - 87 = 18 \text{ F}$$

and the tower approach as

$$AP = CWT - WBT$$

$$= 87 - 75 = 12 \text{ F}$$

Using these calculated values and ambient wet-bulb temperature 75 F, the tower characteristic curve can be constructed by using either the materials from the handbook or the computer program prepared for this purpose. In this example, the tower curve from Kelly's handbook is reproduced and presented in Fig. 8-19. Imposing the fill curve on the same figure, we can determine the intersection of these curves. The results are

$$KaY/L = 1.70 \qquad L'/G' = 1.36$$

The tower fill requirement is

$$\text{Volume} = (\text{cross-sectional area})(\text{height})$$

$$= \left(\frac{\text{water flow rate}}{\text{water loading density}} \right)(\text{height})$$

Using the numerical values provided in the problem statement, we have

$$\text{Volume} = \left(\frac{1000 \times 500}{3000} \right)(36)$$

$$= 6000 \text{ ft}^3$$

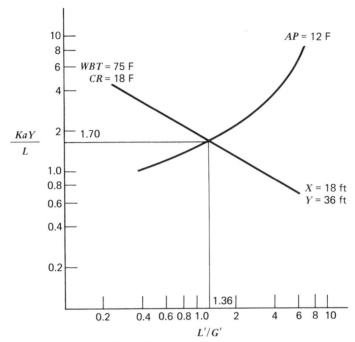

Figure 8-19. A KaY/L diagram for Example 8-5.

The air flow rate is

$$G' = (\text{water flow rate})/(L'/G') = 500{,}000/1.36$$

$$= 367{,}467 \text{ lb/hr}$$

The air loading density is then

$$G = G'/(\text{frontal area to air flow})$$

$$= 367{,}467/6000/18$$

$$= 1102 \text{ lb/hr/ft}^2$$

The air loading in this example is outside the normal range 1600 to 3000 lb/hr/ft^2 and therefore should not be acceptable. The tower fill requirement is 6000 ft^3 and is almost twice the value calculated in Example 8-4. All these indicate that the choice of matrix configuration ($x = 18$ ft, $y = 36$ ft) may not be appropriate. In normal conditions the tower designer must reselect the matrix configuration and repeat the above calculation.

When the tower matrix configuration is specified, the volume of tower fill is simply determined by selecting the water loading. As indicated in Example 8-5, the fill volume is inversely proportional to the water loading. Also, the calculation of air loading is important, because it will reveal whether or not the tower fill is effectively utilized. In algebra, the air loading (G) is simply the product of air density and velocity. That is,

$$G = 60\rho V$$

where

$$\rho = \text{air density lb/ft}^3$$

$$V = \text{air velocity ft/min}$$

At the standard ambient conditions ($P_0 = 14.7$ psia, $T_0 = 77$ F) the air loading 2000 lb/hr/ft^2 will correspond to the air velocity 451 ft/min.

In preliminary design the tower fill may not be specified. If this is the case, the mass transfer coefficient should be first estimated and then, the fill requirement is calculated. The following example will demonstrate the calculational procedure.

EXAMPLE 8-6.   Repeat Example 8-4 for the crossflow tower fill of which the mass transfer coefficient per unit volume (Ka) is 300 lb/hr/ft^3. The fill configuration parameter is 0.5.

Solution: The required fill characteristics is 1.25 as calculated in Example 8-4. The fill height can be determined as

$$\frac{KaY}{L} = 1.25$$

$$Y = 1.25 \times 3000/300$$

$$Y = 12.5 \text{ ft}$$

Since the configuration parameter is 0.5, the fill depth X (i.e., air travel distance) is $0.5 \times 12.5 = 6.25$ ft. As in Example 8-4, the frontal area to water flow is 166.7 ft^2. The volume of fill matrix is then

$$\text{Volume} = 12.5 \times 166.7 = 2083.7 \text{ ft}^3$$

The air loading density is calculated with the equation

$$\left(\frac{L}{G}\right)\left(\frac{X}{Y}\right) = \left(\frac{L'}{G'}\right)$$

or

$$G = (L)\left(\frac{X}{Y}\right)\Big/\left(\frac{L'}{G'}\right)$$

Substituting the numerical values into the above equation, we have the air loading density as

$$G = (3000)(0.5)/(1.0) = 1500 \text{ lb/hr/ft}^2$$

In cooling tower design the fan horsepower is an important item. It affects not only the initial cost, but also the operating cost. To estimate the fan horsepower required, one must take into account all pressure drops along the air path. The pressure drops across the fill matrix have been presented in Section 8-4. Those across inlet louvers and drift eliminators are included in Appendix E (see Figs. E-10 and E-11). Evidently, these values are approximate in nature. They should be used only in the absence of manufacturers' information.

In addition to this static pressure drop, the velocity pressure, which may be as much as 2/3 of total static pressure, must be considered. Velocity pressure is generally dependent on fan as well as recovery stack design. Details of these calculations are beyond the scope of this text.

Using the information just presented, the fan brake horsepower is calculated from the following equation:

$$BHP = \frac{(ACFM)(SP + VP)}{6356(EFF)} \qquad (8\text{-}30)$$

where

$$BHP = \text{fan brake horsepower}$$

$$ACFM = \text{actual air volume flow rate, cfm (cubic feet per minute)}$$

$$SP = \text{static pressure drop, inches of water}$$

$$VP = \text{effective velocity pressure, inches of water}$$

$$EFF = \text{fan mechanical efficiency, dimensionless}$$

The fan mechanical efficiency is generally provided by equipment vendors. In absence of vendor's information, the fan efficiency can be approximated as 75%. In Eq. (8-30) the actual air flow rate is determined at the fan inlet. This value is quite different from that at the tower air inlet because of an increase in air temperature and water vapor content.

EXAMPLE 8-7.   Calculate the fan shaft horsepower in the cooling tower described in Example 8-4. Assume that the fan efficiency is 0.75 and the effective velocity pressure is 2/3 of the total static pressure drop.

Solution: We first determine the enthalpy of air leaving the fill matrix by the equation

$$h_{a,e} = h_{a,i} + \frac{L'}{G'}(CR)$$

$$= 38.61 + (1)(18) = 56.61 \text{ Btu/lb}$$

The air at that location is almost saturated. Assuming that it is saturated air, the air temperature is found 90.5 F from Table C-3 and the specific volume is 14.57 ft^3/lb of dry air from Table C-5 (see Appendix C). Then the actual air volume flow rate is

$$ACFM = G' \times v$$

$$= 500,000 \times 14.57/60$$

$$= 121,416 \ ft^3/min$$

From Fig. E-9 for fill type F, we find

$$\overline{\Delta p}_{fill} = 0.0042 \ \text{in. of water/ft}$$

and the static pressure drop due to the tower fill is

$$SP_{fill} = \overline{\Delta p}_{fill} \times (\text{air travel distance})$$

$$SP_{fill} = 0.0042 \times 12 = 0.050 \ \text{in. of water}$$

From Figs. E-10 and E-11 we find the pressure drops across the inlet louvers and drift eliminators as

$$SP_{eliminator} = 0.05 \ \text{in. of water}$$

and

$$SP_{louver} = 0.05 \ \text{in. of water}$$

Then, the total static pressure drop is

$$SP = SP_{fill} + SP_{eliminator} + SP_{louver}$$

$$= 0.050 + 0.05 + 0.05 = 0.150 \ \text{in. of water}$$

and the effective velocity pressure is

$$VP = \tfrac{2}{3}SP = \tfrac{2}{3}(0.150) = 0.100 \ \text{in. of water}$$

Finally, the fan brake horsepower is

$$BHP = \frac{121,416(0.150 + 0.100)}{6356 \times 0.75}$$

$$BHP = 6.37 \ \text{horsepower}$$

To improve the accuracy in estimating the pressure drop in fill matrix, inlet louver, and drift eliminator, we must include the air temperature correction. This can be easily done by multiplying the curve value of pressure loss by the ratio of standard air density to average density. In Kelly's handbook the standard air density used is 0.075 lb/ft^3.

In design the concept of tower cell (i.e. tower modules) is frequently applied. Once the design of typical tower cell is made, the number of tower cells will be determined according to the cooling capacity requirement. In mechanical-draft cooling tower a double-flow arrangement similar to that in Fig. 8-2 is frequently used. The cell length is 1.1 to 1.4 times the fan diameter. The ratio of fill depth to fill height (i.e., configuration parameter) is in the range of 0.4 to 1.0. In natural-draft towers, flow arrangement is either counterflow or crossflow. For economic reasons there are always one or two tower units connected with each generating system. The fill volume and pressure drop can be calculated with the same procedures as those used for mechanical-draft cooling towers. However, the tower body design requires advanced mathematical technique and therefore is not within the range of this text.

EXAMPLE 8-8. A crossflow cooling tower is to be designed for a 225 MW fossil-fuel power plant. The turbine cycle heat balance shows that the maximum heat load is to be 1705 MBtu/hr. The overall cooling system analysis indicates that the tower approach and cooling range are to be 13 and 20 F, respectively. The plant site has ambient air at 90 F and 50% relative humidity. The fill Type F from Kelly's handbook is chosen as the tower fill. The packed height Y and packed depth X are, respectively, 36 ft and 18 ft. The fan used in design has a diameter of 28 ft and mechanical efficiency of 75%. Calculate the volume of tower fill and the fan brake horsepower.

Solution: Using the 13 F approach, the 20 F cooling range, and the 75 F wet-bulb temperature (determined by 90 F dry-bulb temperature and 50% relative humidity), we can determine the tower characteristic curve either by using Kelly's handbook or by using the computer program prepared for this purpose. In this example, let us use the handbook material (Fig. E-4) and reproduce the curve as shown in Fig. 8-20. Next, using the fill specification (Kelly's Type F, $X = 18$ and $Y = 36$), we can identify the fill characteristic curve from Fig. E-8. Transporting the fill curve to Fig. 8-20, we can see the intersection of this curve with the tower characteristics. The intersection of these two curves indicates the overall flow ratio at which the tower characteristics (i.e., required fill characteristics) is equal to the fill performance characteristics. In this example, this ratio is 1.35 and the corresponding KaY/L is 1.69. Since the water flow rate is

$$L' = Q/CR$$

$$= 1705 \times 10^6/20 = 85.25 \times 10^6 \text{ lb/hr}$$

the required air flow rate must be

$$G' = L'/(L'/G') = 85.25 \times 10^6/1.35$$

$$= 63.15 \times 10^6 \text{ lb/hr}$$

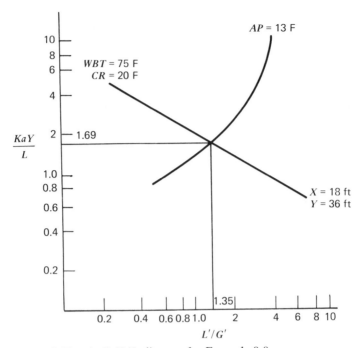

Figure 8-20. A KaY/L diagram for Example 8-8.

Assuming the tower cell length 36 ft (i.e., 1.29 times the fan diameter) and 10 tower cells, the louver face area can be calculated by the equation

$$A_{\text{louver}} = (\text{number of cell})(\text{cell length})(\text{fill height})$$

$$(\text{number of sides})$$

$$= 10 \times 36 \times 36 \times 2 = 25{,}920 \text{ ft}^2$$

and the design air loading G is

$$G = \frac{G'}{A_{\text{louver}}} = \frac{63.16 \times 10^6}{25920} = 2436 \text{ lb/hr/ft}^2$$

Similarly, the design water loading is

$$L = \frac{\text{water flow rate}}{(\text{fill depth})(\text{cell length})(\text{no. of cells})(\text{no. of sides})}$$

or

$$L = \frac{85.25 \times 10^6}{18 \times 36 \times 10 \times 2} = 6577 \text{ lb/hr/ft}^2$$

Evidently both air and water loading are well within the limits. Now we can proceed to calculate the tower fill requirement. The volume of the tower fill is

$$Vol = (\text{fill height})(\text{fill depth})(\text{cell length})$$

$$(\text{number of cells})(\text{number of sides})$$

Substituting the numerical values into this equation, we have

$$Vol = (36)(18)(36)(10)(2)$$

$$= 0.466 \times 10^6 \text{ ft}^3$$

To determine the horsepower for tower fans, we calculate the exit air conditions and air flow rate below:

$$H_{a,e} = H_{a,i} + \frac{L'}{G'}(CR)$$

$$= 38.61 + (1.35)(20) = 65.61 \text{ Btu/lb}$$

Using this exit air enthalpy, the exit air temperature is found 96.4 F from Table C-3. Then, from Table C-5 the air specific volume is found to be 14.878 ft^3/lb of dry air. The actual air flow rate for each fan is

$$ACFM = \frac{G' \times v}{60 \times \text{number of cell}}$$

$$= \frac{63.15 \times 10^6 \times 14.878}{60 \times 10} = 1.57 \times 10^6 \text{ ft}^3/\text{min}$$

To determine the pressure drop along the air path, we find from Fig. E-9

$$\overline{\Delta P}_{\text{fill}} = 0.0134 \text{ in. of } H_2O/\text{ft of air travel}$$

from Fig. E-10

$$SP_{\text{eliminator}} = 0.092 \text{ in. of } H_2O$$

and from Fig. E-11

$$SP_{\text{louver}} = 0.092 \text{ in. of } H_2O$$

The static pressure drop across the tower fill is

$$SP_{\text{fill}} = (\overline{\Delta P}_{\text{fill}})(X)(\text{correction factor})$$

The correction factor in this equation is the ratio of standard air density to the average density in the tower fill. The inlet and exit density are, respectively, 0.0734 and 0.0698 lb/ft^3, with an average value of 0.0716 lb/ft^3. Thus, the correction factor is 0.0750/0.0716. Substituting the numerical values into the above equation, we have the static pressure drop across the fill as

$$SP_{fill} = (0.0134)(18)(0.0750/0.0716)$$

$$= 0.253 \text{ in. of } H_2O$$

Then, the total static pressure drop is

$$SP = 0.092 + 0.092 + 0.253 = 0.437 \text{ in. of } H_2O$$

Approximating the velocity pressure as 2/3 of the static pressure drop, we have the velocity pressure as

$$VP = 2/3 \times 0.437 = 0.291 \text{ in. of } H_2O$$

Finally, the fan brake horsepower is

$$BHP = \frac{(ACFM)(SP + VP)}{6356(EFF)}$$

$$= \frac{(1.57 \times 10^6)(0.437 + 0.291)}{(6356)(0.75)}$$

$$= 239.8 \text{ horsepower (per each fan)}$$

The fan power calculation in the examples is conservative. Fan power has been based on the calculated values of the air flow rate and the pressure drop along the air path. In practice, these two calculated values must be multiplied by some safety factor or margin factor that may be as high as 1.2 to 1.3.

Before this section is concluded, several parameters affecting the tower size are identified and briefly discussed. The first parameter is the design ambient wet-bulb temperature. It has been stated previously that there will be no water cooling unless the air wet-bulb temperature is less than the water temperature. In practice, the ambient wet-bulb temperature is at least 5 F below the water temperature.

The ambient wet-bulb temperature varies seasonally as well as regionally. In power plant design it is customary to set the design value such that only during 5% of summertime would the ambient air temperature exceed the design value. Table 8-7 indicates typical design values for various locations in the United States. Information for other locations can be found in the reference [1].

The design value of wet-bulb temperature has significant impact on the tower size. Consider a base tower designed for 75 F wet-bulb temperature, 20 F cooling range,

Table 8-7
Typical Design Wet-Bulb Temperatures

Locations	Wet-Bulb Temperature (F)
Bismarck, North Dakota	70
Casper, Wyoming	60
Charleston, West Virginia	71
Logan, Utah	63
Fort Worth, Texas	77
Columbus, Ohio	74
Sante Fe, New Mexico	63
Billings, Montana	65
Louisville, Kentucky	76
Springfield, Illinois	76

12 F tower approach, and 1.0 flow rate ratio and use this tower size as a reference point. Fig. 8-21 indicates an increase of tower size with a decrease of design value in the ambient wet-bulb temperature. For instance, the tower size will increase approximately 22% as the design wet-bulb temperature changes from 75 to 70 F. This inverse relationship is mainly because of the reduction of the enthalpy difference in the tower.

The second parameter is tower approach. As indicated previously, the lowest attainable water temperature from a wet cooling tower is the ambient wet-bulb temperature. Reaching a zero approach requires an infinitely large tower size and is therefore obviously impractical. In design the tower approach must be positive and always above 5 F. As seen in Fig. 8-21, the tower approach is inversely related to the tower size if other conditions remain unchanged. For a given condenser pressure a small tower would generally lead to use of a large condenser. The reverse is also true.

The third parameter is cooling range. Figure 8-21 indicates that an increase in cooling range will result in an increase in tower size. In addition, the cooling range

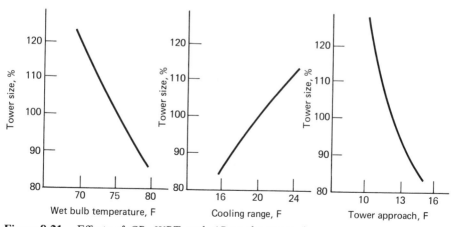

Figure 8-21. Effects of *CR*, *WBT*, and *AP* on the tower size.

will affect the condenser pressure at which a steam turbine operates. In other words the selection of tower cooling range will affect the plant fuel consumption and thus the plant economic performance.

The fourth one is the tower heat load. The design value is equal to the maximum heat load that will occur at the time the steam turbine operates at a maximum capacity. The maximum heat load is generally available from the plant heat balance. It should be pointed out that the tower and condenser heat load are the same. For a given cooling range, the heat load is linearly related to the circulating water flow, and therefore linearly related to the tower size.

8-6 COOLING TOWER PERFORMANCE

A cooling tower does not always operate exactly under the conditions for which it is designed. Ambient weather conditions generally are different from the design values. Heat load may not be exactly equal to the maximum value usually selected for design. Because of these problems, off-design tower performance calculation is always desirable and required for simulation and optimization.

The Merkel's equation indicates that the required tower characteristics is a function of the cooling range, tower approach, flow rate ratio, and ambient wet-bulb temperature. The relationship of these variables can be expressed in a graphical form such as that in the "blue book" published by the Cooling Tower Institute. At the time the tower is designed, the required tower characteristics are determined. When the tower operates at different ambient wet-bulb temperature, (other conditions remain constant) the tower will produce cold water at the temperature predicted by the Merkel's equation. Since the cold water temperature (or tower approach) is not an explicit function of other variables, the method of solution is obviously trial and error in nature. Similarly, when a tower operates at the heat load less than the design value, the tower would again produce a different cold water temperature.

Perhaps the prediction of off-design tower performance can be better illustrated by using the graphs of the "blue book." Let us say a tower designed for a cooling range of 22 F, an ambient wet-bulb temperature 69 F, a tower approach 16 F, and a flow rate ratio of 1.2, the required tower characteristics is 1.20 (see Fig. E-2). When this tower operates at the ambient wet-bulb temperature 60 F (other conditions are identical to the design value), the tower will still have the same tower characteristics (i.e., 1.20). With $KaV/L = 1.20$, the cooling range 22 F, ambient wet-bulb temperature 60 F, and the flow rate ratio 1.2 the graph of the "blue book" (not included in this text) will indicate a tower approach of 20 F. Thus, the cold water temperature is 80 F (20 F + 60 F), which is quite different from the design value of 85 F.

Let us use the same tower, but this time we operate it at a different heat load, say, 18% below the design value. Since the heat load is directly proportional to the cooling range at a given water flow rate, a 18% reduction in heat load means a 18% decrease in the cooling range. With $KaV/L = 1.20$, cooling range 18 F, ambient wet-bulb temperature 69 F, and the flow rate ratio 1.2, the graph of the blue book (Fig. E-3) will set the tower approach at 14 F and, therefore, the cold water temperature at 83 F. It must be pointed out that the fill characteristics remains unchanged only if the flow rate ratio is a constant. The relationship between these two terms is shown by Eq. (8-20). It can be also expressed as

$$\frac{(KaV/L)_{\text{design}}}{(KaV/L)_{\text{current}}} = \frac{(L'/G')^{\beta}_{\text{design}}}{(L'/G')^{\beta}_{\text{current}}} \tag{8-31}$$

In absence of the tower fill performance equation or specific information provided by vendors, the β value is generally assumed to be -0.6.

EXAMPLE 8-9.   A counterflow cooling tower was designed to cool 1000 gpm from 105 F to 83 F at the ambient wet-bulb temperature 69 F. The design L/G is 1.4, and tower fans require 14 horsepower. Now the water flow is increased by 10% in operation, with the air flow, cooling range, and wet-bulb temperature the same as the values in the design. Calculate the temperature of water leaving the cooling tower.

Solution: Using the design values $WBT = 69$ F, $CR = 22$ F, $AP = 14$ F, and $L/G = 1.4$, we find the design value of KaV/L 1.60 from Fig. E-2. Since the water flow rate is increased by 10%, the current value of L/G will increase by 10%, that is, $1.4 \times 1.1 = 1.54$. Using Eq. (8-31), we can calculate the current value of KaV/L as below:

$$\left(\frac{KaV}{L}\right)_{\text{current}} = (1.6)(1.54)^{-0.6}/(1.4)^{-0.6} = 1.51$$

Using the current conditions, $WBT = 69$ F, $CR = 22$ F, $L/G = 1.54$, and $KaV/L = 1.51$, we find the current approach 15 F from Fig. E-2. Finally, the temperature of water leaving the tower is

$$CWT = AP + WBT$$

$$= 15 + 69 = 84 \text{ F}$$

EXAMPLE 8-10.   The tower of Example 8-8 was designed to cool 170,500 gpm $(85.25 \times 10^6 \text{ lb/hr})$ of water from 108 F to 88 F at the ambient wet-bulb temperature 75 F. The total fan power is 2398 hp. The fill matrix is the Kelly's fill, Type F with a packed height of 36 ft and a packed depth of 18 ft. After the tower construction was completed the field test generated the following data:

Water flow rate	165,000 gph
Hot water temperature	108 F
Cold water temperature	90 F
Wet-bulb temperature	75 F
Total fan driver output	2100 hp

Determine the actual cooling tower capability under the design conditions.

Solution: As indicated in Example 8-8, the design value of L'/G' is 1.35 and design KaY/L is 1.7. To determine the actual fill characteristic curve, we calculate the test value of L'/G' by using the equation

$$(L'/G')_t = (L'/G')_d(L_t'/L_d')(G_d'/G_t')$$

Using the fan laws, the above equation becomes

$$(L'/G')_t = (L'/G')_d(L_t'/L_d')(HP_d/HP_t)^{1/3}$$

Substituting numerical values into the equation yields

$$(L'/G')_t = (1.35)\left(\frac{165,000}{170,500}\right)\left(\frac{2398}{2100}\right)^{1/3}$$

$$= 1.37$$

Using the test values $L'/G' = 1.37$, $CR = 108 - 90 = 18$ F, $AP = 90 - 75 = 15$ F, and $WBT = 75$ F, we can find the test KaY/L value at 1.2 from Fig. E-5. With the test ratio $L'/G' = 1.37$ and test $KaY/L = 1.2$, we can now determine the test point for the tower fill and construct the actual fill characteristic curve by drawing a curve through this point and parallel to the curve used in design. This procedure is presented in Fig. 8-22.

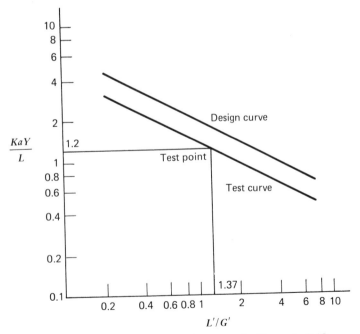

Figure 8-22. Tower fill characteristic curves for Example 8-10.

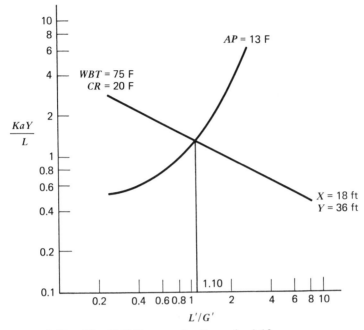

Figure 8-23. The KaY/L curve for Example 8-10.

To determine the actual tower performance under design conditions, we transfer the actual fill characteristics curve to Fig. 8.23 where this curve will intersect with the 13 F approach curve. The intersection of these two curves shown in Fig. 8-23 indicates actual flow ratio L'/G' under the design conditions. That is,

$$(L'/G')\text{actual} = 1.10$$

Finally, the actual tower cooling capability under the design conditions is

$$\text{Cooling capability} = \frac{(L'/G')\text{actual}}{(L'/G')\text{design}} \times 100$$

$$= \frac{1.10}{1.35} \times 100 = 81.48\%$$

As indicated in the foregoing examples, the charts published by the Cooling Tower Institute and Kelly's Associates are undoubtedly convenient in design and performance calculation. This information has been computerized in recent years. Use of these computer programs will be discussed briefly in the next section.

Tower performance information is generally organized and presented in a graphical form. The curves thus prepared are referred to as the tower performance curves.

Figure 8-24 indicates a set of typical performance curves of a mechanical-draft cooling tower. The tower has these design conditions: wet-bulb temperature 79 F, cooling range 23.5 F, tower approach 11.5 F, and water flow rate 145,000 gpm. According to the recommendation of the Cooling Tower Institute, tower performance curves should be prepared for three different water flow rates, that is, 90%, 100%, and 110% of design value. For each water flow rate, three different cooling

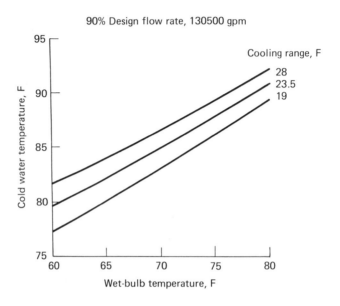

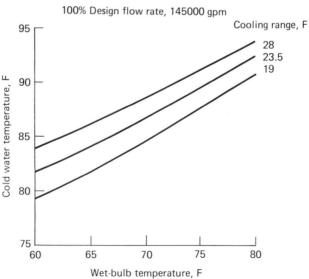

Figure 8-24. Mechanical draft cooling tower performance curves.

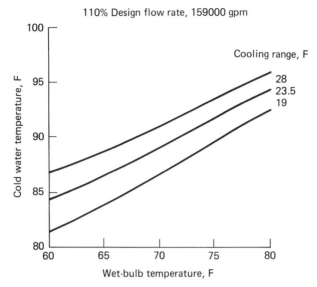

Figure 8-24. (cont.)

ranges are included. These are the design cooling range and $\pm 20\%$ change from the design value.

The performance curves of a cooling tower are generally available from the equipment vendor. These curves are prepared with design and field data. They represent the performance of the tower specified in the vendor's proposal. Evidently, they would vary from one model to another. The performance curves are generally needed for tower evaluation.

For the purpose of tower performance simulation in a digital computer, an equation may be developed from the tower performance data. For the mechanical-draft cooling tower shown in Fig. 8-24 the equation may be in the form such as

$$
\begin{aligned}
CWT = {} & -0.10046 \times 10^2 + 0.22801 \times 10^{-3}(WFR) \\
& + 0.85396(CR) + 0.18617 \times 10^{-5}(CR)(WFR) \\
& + 0.10957 \times 10(WBT) - 0.22425 \times 10^{-5}(WBT)(WFR) \\
& - 0.11978 \times 10^{-1}(WBT)(CR) + 0.14378 \times 10^{-7}(WBT) \\
& \times (CR)(WFR)
\end{aligned}
\tag{8-32}
$$

where

$$CWT = \text{cold water temperature (F)}$$

$$WBT = \text{ambient wet-bulb temperature (F)}$$

$$CR = \text{cooling range (F)}$$

$$WFR = \text{water flow rate (gpm)}$$

The system and component simulations are to be discussed later in this text.

8.7 TOWER BLOWDOWN AND MAKEUP WATER

As water is evaporated in the cooling tower, the concentration of salts or other impurities will increase in the circulating water. To avoid a high concentration level and subsequent scaling of the surface within the tower, it is best to drain off a portion of the water and replace it with the fresh water. This operation is frequently referred to as the tower blowdown. The blowdown is expressed as a percentage of the circulating water flow rate.

Another loss of water in the tower operation is evaporation loss. The evaporation of one pound of water requires approximately 1000 Btu. This heat will cool off 100 lb of water by 10 F. In other words the evaporation loss is approximately 1% of the circulating water flow rate for each 10 F of cooling range.

The third water loss is the tower drift loss. This drift loss results from a carry-off of water droplets by the tower effluent. The drift loss varies with the cooling tower design. The average values are in the range of 0.1 to 0.3% of circulating water flow rate for mechanical-draft towers and 0.3 to 1% for natural-draft towers.

A continuous tower operation is maintained when fresh water is supplied at the same rate as the water losses from the tower. That is,

$$M = B + D + E \tag{8-33}$$

where

$$M = \text{makeup water}$$

$$B = \text{blowdown water}$$

$$D = \text{water drift loss}$$

$$E = \text{evaporation loss}$$

All these terms are generally expressed as a percentage of the circulating water flow rate. For an estimate of the tower blowdown loss, the amount of soluble salts and impurities in blowdown and drift losses must be equal to that in the makeup water. That is,

$$(C_c)(B + D) = (C_m)(M) \tag{8-34}$$

where

$$C_c = \text{solid concentration in circulating water}$$

$$C_m = \text{solid concentration in makeup water}$$

Rearranging the above equation gives

$$R_{cm} = \frac{C_c}{C_m} = \frac{B + D + E}{B + D} \tag{8-35}$$

The term R_{cm} is the ratio of the solid concentration in the circulating water to the solid concentration in the makeup water. In general, this ratio has the average value in the range of 2 to 4. From Eq. (8-35), the tower blowdown is expressed as

$$B = \frac{E}{R_{cm} - 1} - D \tag{8-36}$$

Combining Eq. (8-33) and Eq. (8-35), the tower makeup water is

$$M = \frac{ER_{cm}}{R_{cm} - 1} \tag{8-37}$$

Taking as sample values, $R_{cm} = 2.0$, $D = 0.1$, and $E = 2.0$, we have

$$B = 1.9\%$$

and

$$M = 4\%$$

The water consumption of a cooling tower is mainly due to the evaporation and blowdown losses. The evaporation loss depends not only on the tower heat load, but also on the ratio of the amount of heat carried off by increasing the temperature of the air and by evaporation of the water. Since the tower drift loss is negligibly small, the evaporation loss of a cooling tower can be related to the air flow rate and the air inlet and outlet water vapor contents W_1 and W_2 as

$$\text{Evaporation loss} = G'(W_2 - W_1) \tag{8-38}$$

The air inlet humidity ratio (W_1) is obviously dependent on the local conditions, and the air outlet humidity ratio (W_2) is related to the tower design and operation. In normal operation the air leaving the tower is nearly saturated or completely saturated. As an estimate, one third of the heat transfer in the tower is generally due to the convection while two thirds is due to the evaporation. Thus the evaporation loss will be only about two-thirds of what it would be when the entire heat load is removed through evaporation. The approximation is expressed as

$$E = 0.0667(CR) \tag{8-39}$$

where E is the evaporation loss in terms of the percentage of the circulating water flow rate.

8.8 USE OF COMPUTER PROGRAMS

Very few computer programs are available for classroom use. It is mainly because the program development cost is relatively high. If computer programs are developed by private corporations, they will be generally treated as proprietary. The programs presented here may not be as complete as the industrial application requires, but they serve well as an educational tool [14, 16].

The computer program in the evaporative cooling tower area consists of 5 subprograms. These are

> Tower performance curves (TOWER)
> Counterflow tower
>> Design (TOWER01)
>> Performance simulation (TOWER02)
> Crossflow tower
>> Design (TOWER11)
>> Performance simulation (TOWER12)

As indicated in Sections 8.2 and 8.3, the program TOWER has been developed to generate the tower performance curves. It consists of two parts: one for counterflow and another one for crossflow. This program can essentially generate information as much as those by the Kelly's handbook and Cooling Tower Institute's handbook. The program was prepared in an interactive mode, the typical interaction as seen on the computer terminal screen is reproduced and presented as below:

```
LOAD 1728 TOWERM
READY
RUN
1728 TOWERM 01 / 27 / 83 13:23:05
     ENTER THE TYPE OF COOLING TOWER
"1" FOR COUNTERFLOW; "2" FOR CROSSFLOW
?2
     ENTER THE TYPE OF CALCULATION DESIRED
        "1" FOR KAV / L; "2" FOR APPROACH
?1
     ENTER AP, CR, L / G AND WBT ACCORDINGLY
?12,20,1,75
                AP      =      12.00 F
                CR      =      20.00 F
               L / G    =       1.00
                WBT     =      75.00 F
              KAV / L   =       1.362
     DO YOU WANT ONE MORE COMPUTER RUN?
     ENTER "1" FOR YES; "2" FOR NO
     ?1
     ENTER AP, CR, L / G AND WBT ACCORDINGLY
```

```
?10,20,1,75
            AP      =    10.00 F
            CR      =    20.00 F
            L / G   =     1.00
            WBT     =    75.00 F
          KAV / L   =     1.762
DO YOU WANT ONE MORE COMPUTER RUN?
ENTER "1" FOR YES; "2" FOR NO
?2
STOP
TIME 3.5 SECS
```

The nomenclatures used in this program or others are identical to those in this text. In case that the tower performance curve is needed, one can obtain a series of KaV/L using different values of the flow rate ratio. The calculated results are then used to generate the curve. When the tower performance is to be determined, the tower approach is always desired. If this is the case, the program users should input "2," instructing the computer to calculate the tower approach.

The objective of the design programs (TOWER01 and TOWER11) is to allow the engineer to quickly size the tower fill matrix (packing) and to estimate the fan power. As an illustrative example of using these programs, we consider a crossflow tower for a 225 MW fossil-fuel generating unit. The maximum heat load is 1705.1 MBtu/hr. The tower approach and cooling range are, respectively, 13 and 23.5 F. The plant site conditions are 96 F dry-bulb temperature and 48% relative humidity. The Kelly fill Type F is chosen as the tower fill. Other input information includes water loading 6500 lb/hr/ft^2, air velocity 525 ft/min, and configuration parameter 0.5. The interactive computer output for this design is reproduced and presented in Table 8-8. For reference, the important results are summarized as below:

Tower fill matrix height	36 ft
Tower fill matrix depth	18 ft
Tower fill matrix length	622 ft
Tower fill matrix volume	400,700 ft^3
Total water flow rate	145,000 gpm
Static air pressure drop	0.36 in. of water
Air volume flow rate	12,094,500 cfm
Total fan static brake horsepower	920 hp

The calculational procedures used in this program are similar, but not identical to those presented in the text. More information about this program can be found in the program manual [16]. As indicated in Table 8-8, only static pressure drop is considered. Therefore, the fan horsepower is not the realistic value. In design, the velocity pressure must be taken into account. The safety factor or margin factor should be applied to the air flow rate as well as the pressure drop.

Table 8-8
Crossflow Design Program Interactive Computer Output

```
1728 TOWER11 03/29/84 14:52:57
                                    TOWER11


        THE PROGRAM WAS DESIGNED TO SIZE THE MECHANICAL-DRAFT
        CROSSFLOW COOLING TOWER BY ESTIMATING THE VOLUME OF
        TOWER FILL (PACKING) AND FAN POWER. THE PROGRAM WAS
        DESIGNED ESSENTIALLY ON THE BASIS OF MERKEL THEORY.
        DETAILS ON THE NUMERICAL CALCULATIONS CAN BE FOUND IN
        THE TEXTBOOK POWER PLANT SYSTEM DESIGN BY KAM W. LI
        AND A. PAUL PRIDDY.


        THE PROGRAM NEEDS THE DESIGN WEATHER CONDITIONS, THE
        DESIGN HEAT LOAD (MBTU/HR), COOLING RANGE (DEG F),
        TOWER APPROACH (DEG F), INLET AIR VELOCITY (FPM), AND
        WATER LOADING IN THE TOWER (LB/HR/SQ FT).


        DO YOU STILL WANT TO RUN THIS PROGRAM? ENTER 1
        FOR YES OR ENTER 2 FOR NO.
?1

        PLEASE ENTER THE VALUES WHEN THEY ARE
        ASKED FOR.

        ENTER THE DESIGN DRY BULB TEMPERATURE (DEG F)
?96.
        ENTER THE DESIGN RELATIVE HUMIDITY (%)
?48.
        ENTER THE DESIGN WET BULB TEMPERATURE (DEG F)
?79.
        ENTER THE DESIGN HEAT LOAD (MBTU/HR)
?1705.1
        ENTER THE DESIGN TOWER APPROACH (DEG F)
?13.
        ENTER THE DESIGN COOLING RANGE (DEG F)
?23.5
        ENTER THE DESIGN INLET AIR VELOCITY (FPM)
?525.
        ENTER THE DESIGN TOWER WATER LOADING (LB/HR/SQ FT)
?6500.
        ENTER THE TOTAL FAN EFFICIENCY (%)
?75.
        ENTER THE CONFIGURATION PARAMETER
?.5


        THE PROGRAM NEEDS CONSTANTS C,M, AND N TO DETERMINE
        KA, THE MASS TRANSFER COEFFICIENT PER UNIT ACTIVE
        TOWER VOLUME (LB/HR/CU FT). THE CONSTANTS ARE USED IN
        EQ.(1) SHOWN BELOW:

                KA=C*((L)**M)*((G)**N)        EQ.(1)

        WHERE L IS THE WATER LOADING (LB/HR/SQ FT) AND G IS THE
        AIR LOADING (LB/HR/SQ FT).

        DO YOU HAVE INFORMATION ON THE CONSTANTS?  ENTER
        1 FOR YES OR ENTER 2 FOR NO.
?1
        PLEASE ENTER THE CONSTANTS WHEN THEY ARE ASKED FOR:

        ENTER THE CONSTANT C FOR THE KA EQUATION (EQ.(1))
?.081
        ENTER THE CONSTANT M FOR THE KA EQUATION (EQ.(1))
?.48
        ENTER THE CONSTANT N FOR THE KA EQUATION (EQ.(1))
?.52
```

THE PROGRAM NEEDS CONSTANTS A,B, AND D TO
DETERMINE DP, THE STATIC AIR PRESSURE DROP IN INCHES OF
WATER PER FOOT OF TOWER FILL DEPTH. THE CONSTANTS
ARE USED IN EQ.(2) SHOWN BELOW:

$$DP=A*((L)**B)*((G)**D) \qquad EQ.(2)$$

WHERE L IS THE WATER LOADING (LB/HR/SQ FT) AND G IS THE
AIR LOADING (LB/HR/SQ FT).

DO YOU HAVE INFORMATION ON THE CONSTANTS? ENTER
1 FOR YES, OR ENTER 2 FOR NO.

?1

PLEASE ENTER THE CONSTANTS WHEN THEY ARE ASKED FOR:

ENTER THE CONSTANT A FOR THE DP EQUATION (EQ.(2))
?.3871E-09
ENTER THE CONSTANT B FOR THE DP EQUATION (EQ.(2))
?.94
ENTER THE CONSTANT D FOR THE DP EQUATION (EQ.(2))
?1.16

THE INPUT INFORMATION FOR THE CROSSFLOW TOWER
DESIGN IS SUMMARIZED BELOW:

```
 1.DESIGN DRY BULB TEMPERATURE (DEG F)              96.00
 2.DESIGN RELATIVE HUMIDITY (%)                     48.00
 3.DESIGN WET BULB TEMPERATURE (DEG F)              79.00
 4.DESIGN HEAT LOAD (MBTU/HR)                     1705.10
 5.DESIGN TOWER APPROACH (DEG F)                    13.00
 6.DESIGN COOLING RANGE (DEG F)                     23.50
 7.DESIGN INLET AIR VELOCITY (FPM)                 525.00
 8.DESIGN TOWER WATER LOADING (LB/HR/SQ FT)       6500.00
 9.TOTAL FAN EFFICIENCY (%)                         75.00
10.CONFIGURATION PARAMETER                           0.50
11.CONSTANT C FOR KA EQUATION (EQ.(1))               0.08
12.CONSTANT M FOR KA EQUATION (EQ.(1))               0.48
13.CONSTANT N FOR KA EQUATION (EQ.(1))               0.52
14.CONSTANT A FOR DP EQUATION (EQ.(2))  0.38709991E-09
15.CONSTANT B FOR DP EQUATION (EQ.(2))               0.94
16.CONSTANT D FOR DP EQUATION (EQ.(2))               1.16
```

DO YOU WISH TO MAKE ANY CORRECTIONS ON THE
ABOVE DESIGN VALUES? ENTER 1 FOR YES OR ENTER 2 FOR NO.

?2

CROSSFLOW COOLING TOWER DESIGN CALCULATIONS

```
OVERALL WATER TO AIR FLOW RATIO (DIMENSIONLESS)          1.49
DESIGN VERTICAL NUMBER OF TRANSFER UNITS FOR TOWER       1.65
TOWER FILL MATRIX HEIGHT (FT)                           35.90
TOWER FILL MATRIX DEPTH (FT)                            17.95
TOWER FILL MATRIX LENGTH (FT)                          621.93
TOWER FILL MATRIX VOLUME (CU FT)                    400705.12
TOTAL WATER FLOW RATE (GPM)                         144998.94
TOTAL STATIC PRESSURE DROP IN TOWER (IN OF WATER)        0.36
TOTAL AIR FLOW THROUGH FAN (CFM)                  12094489.0
TOTAL STATIC BRAKE HORSEPOWER FOR FANS (HP)            919.52
```

DO YOU WISH TO CONSIDER DIFFERENT DESIGN
ALTERNATIVES? ENTER 1 FOR YES OR ENTER 2 FOR NO.
?2
STOP
TIME 2.9 SECS

It must be pointed out that this computer program is based on the assumption that the fill configuration parameter is specified and available as an input. In other words, any fill dimension (x, y) is acceptable as long as the ratio of x to y is equal to the value of configuration parameter. In case that the fill dimensions are specified and cannot be changed, caution must be taken in using this program. While these two design computer programs were not developed to accomplish the exact thermal design of cooling tower, these programs do have the capability to predict the effects of parameter variation on the tower design. Once the reference tower is established, one can easily change design parameters right on the computer terminal and immediately see the impact of this change on the tower size and fan horsepower. Through this computer interaction, engineers can proceed to complete the tower thermal design.

The objective of the performance simulation programs (TOWER02 and TOWER12) is to predict the tower exit water temperature under various operating conditions. As indicated in the previous section, the exit water temperature is a function of only three variables: ambient wet-bulb temperature, cooling range, and water flow rate. According to the recommendation of Cooling Tower Institute, the exit water temperature should be plotted against the ambient wet-bulb temperature

Table 8-9
Crossflow Performance Program Output

Crossflow Cooling Tower Performance Calculations	
Design dry-bulb temperature (F)	96.00
Design relative humidity (%)	48.00
Design wet-bulb temperature (F)	79.00
Tower fill matrix height (ft)	35.90
Tower fill matrix depth (ft)	17.95
Tower fill matrix length (ft)	621.93
Current heat load (MBtu/hr)	1705.10
Current water flow rate (gpm)	144,998.94
Current cooling range (F)	23.50
Current air velocity (fpm)	525.00

Crossflow Cooling Tower Exit Water Temperatures for Various Ambient Wet-Bulb Temperatures

Wet-Bulb Temperature (F)	Temperature (F)
90.0	99.1
85.0	95.7
80.0	92.6
75.0	89.5
70.0	86.6
65.0	83.8
60.0	81.2

with cooling range and water flow rate as the parameters. The computer programs use similar format. In one computation run the program generates seven pairs of exit water and ambient wet-bulb temperature. Several computation runs are needed for various combinations of cooling range and water flow rate.

Again, consider the crossflow tower we have just completed in thermal design. The tower performance curve under design cooling range (23.5 F) and water flow rate (145,000 gpm) can be generated by the program TOWER12. The computer output is presented in Table 8-9. It is seen that as the wet-bulb temperature increases, the exit water temperature will increase.

Since the performance calculation involves iterative process, the calculation is usually time-consuming. The computer programs such as TOWER02 and TOWER12 should serve as a convenient tool to engineers.

SELECTED REFERENCES

1. N. P. Cheremisinoff and P. N. Cheremisinoff, *Cooling Tower*, Ann Arbor Science Publishing, 1981.

2. J. D. Gurney and I. A. Cotter, *Cooling Towers*, Maclaren and Sons, Ltd. (London), 1966.

3. N. W. Kelly, *Kelly's Handbook of Crossflow Cooling Tower Performance*, Neil W. Kelly and Associates, Kansas City, Missouri, 1976.

4. Cooling Tower Institute, *Performance Curves*, Houston, Texas, 1967.

5. J. B. Dickey, Jr. and D. W. Dwyer, "Managing Waste Heat with the Water Cooling Tower," published by the Marley Company, Kansas City, Missouri, 1978.

6. N. W. Kelly and L. K. Swenson, "Comparative Performance of Cooling Tower Packing Arrangements," *Chemical Engineering Progress*, Vol. 52, 1956.

7. J. Lichtenstein, "Performance and Selection of Mechanical Draft Cooling Tower," *ASME Transactions*, Vol. 65, 1943.

8. Cooling Tower Institute, "Acceptance Test Procedure for Industrial Water-Cooling Towers, Mechanical Draft Type," Cooling Tower Institute Bulletin ATP-105, Houston, Texas, 1959.

9. Munters Corporation, "Tower Fill Selection Guide," Fort Myers, Florida, January 1974.

10. F. Merkel, "Verdunstungskuehlung," *VDI Forschungsarbeiten* No. 275, Berlin, 1925.

11. S. M. Zivi and B. B. Brand, "An Analysis of the Cross-Flow Cooling Tower," *Refrigerating Engineering*, September 1956.

12. G. F. Hallett, "Performance Curves for Mechanical Draft Cooling Towers," *Journal of Engineering for Power*, ASME Transactions Series A, Vol. 97, October 1975.

13. D. B. Baker and L. T. Mart, "Cooling Tower Characteristics as Determined by the Unit-Volume Coefficient," *Refrigerating Engineering*, September 1952.

14. K. W. Li and B. Oachs, "Computer Modeling of Evaporative Cooling Towers," *The International Journal of Energy Systems*, Vol. 3, No. 1, 1983.

15. D. R. Baker and H. A. Shryock, "A Comprehensive Approach to the Analysis of Cooling Tower Performance," *Journal of Heat Transfer*, ASME Transactions, August 1961.

16. K. W. Li and D. S. Lo, "Computer Programs for Cooling Tower Performance Curves," an internal report, Department of Mechanical Engineering, North Dakota State University, May 1980.

PROBLEMS

8-1. Calculate the crossflow tower characteristics KaY/L for the case where the ambient wet-bulb temperature is 75 F, cooling range 25 F, tower approach 10 F, and fluid ratio 1.25.

8-2. Repeat Problem 8-1 with the ambient wet-bulb temperature changed to 80 F.

8-3. Repeat Problem 8-1 with the tower approach increased to 15 F.

8-4. Repeat Problem 8-1 with the cooling range decreased to 20 F.

8-5. Calculate the counterflow tower characteristics KaV/L for the conditions: $WBT = 80$ F, $CR = 20$ F, $AP = 12$ F, and $L/G = 1.2$.

8-6. Repeat Problem 8-5 with the ambient wet-bulb temperature changed to 70 F.

8-7. The fill characteristic equation for the Kelly's fill Type D (height = 9 ft and depth = 8 ft) is expressed as

$$Ka = 0.103(L)^{0.49} \times (G)^{0.51}$$

Determine α and β if the above equation is changed to the form of Eq. (8-20).

8-8. Repeat Problem 8-7 for the Kelly's fill Type E (height = 24 ft and depth = 20 ft). The fill characteristics equation is

$$Ka = 0.088(L)^{0.42}(G)^{0.58}$$

8-9. The fill information is generally presented in a graphical form. The Kelly's fill (Type F) has the fill characteristics shown in Fig. E-8. Using this data, determine α and β used for Eq. (8-20) for the case $H = 30$ ft and $D = 18$ ft.

8-10. The Kelly's fill (Type F) is selected for tower design. The fill configuration is 30 ft in height and 18 ft in depth. What is the crossflow tower characteristics for the conditions $CR = 30$ F, $WBT = 70$ F, and $AP = 13$ F?

8-11. A crossflow tower system is to be designed for the conditions: heat load = 4000 MBtu/hr, tower approach = 15 F, cooling range = 20 F, $WBT = 75$ F. The fill is the Kelly's fill (Type F) with configuration: height = 36 ft and depth = 18 ft. The fans used in the tower system have a diameter of 28 ft and an overall efficiency of 75%. Determine the volume of tower fill and the fan brake horsepower.

8-12. Repeat Problem 8-11 for the fill with a height of 30 ft and a depth of 12 ft.

8-13. Repeat Problem 8-11 for the case where the tower approach is changed to 13 F.

8-14. A counterflow tower was designed to cool 10,000 gpm from 110 F to 90 F with an 80 F wet-bulb temperature. The fan brake power at design conditions is 140 hp and the design flow rate ratio is 1.3. The following data were obtained from a field test:

Water flow rate	9000 gpm
Hot water temperature	100 F
Cold water temperature	84 F
Wet-bulb temperature	70 F
Total fan power	120 hp

Estimate the tower cooling capability under the design conditions.

8-15. Estimate the cold water temperature at $WBT = 60$ F for the tower designed in Problem 8-11. Other conditions are identical to the design values.

8-16. Estimate the cold water temperature at the reduced heat load for the tower designed in Problem 8-11. The current heat load is 80% of the design value. Other conditions remain unchanged.

8-17. For some reasons the air delivery to the tower designed in Problem 8-11 is reduced by 10% while there is no change in water flow rate, cooling range, and ambient wet-bulb temperature. Estimate the cold water temperature under these new conditions.

CHAPTER NINE

Condensers

As indicated in Chapter 8, the condenser is an important component in a power plant. In the condenser, the latent heat of the turbine exhaust steam is transferred to cooling water and is eventually dissipated to the atmosphere. This waste heat limits the thermal efficiency of modern steam power plants to around 40%. The condenser is generally designed to maintain an economical condenser pressure determined by the available cooling water temperature. The steam condensate is discharged from the condenser at a temperature not lower than the steam saturation temperature. Also, the steam condensate is recuperated from the condenser as pure distillate as required for the feedwater heating system.

The main types of steam condensers are water-cooled surface condensers, water-cooled contact condensers, and air-cooled surface condensers. The most efficient and frequently used is the water-cooled surface condenser, which is discussed in this chapter.

9.1 CONDENSER DESCRIPTION

Figure 9.1 shows the schematic diagram and standard terminology used in this field.

Condensers may be single-pass or two-pass. Two-pass condensers have higher temperature rise and require greater heat transfer surface for equal performance of the single-pass. The two-pass condenser is generally more economical in a cooling tower application while the single-pass is frequently selected in normal river, lake, or seacoast installations. Using more than two passes usually results in an uneconomical operation.

Condensers may be either single pressure or dual pressure. Thermodynamically, the dual-pressure condenser is superior to the single pressure. However, economic factors must be taken into account in the design decision. The dual-pressure condenser has flow and temperature rise characteristics comparable to a two-pass condenser but with an improved heat rate. Therefore, the dual-pressure condenser should be considered in all cooling tower applications or for the plant location where a high water temperature is expected. Also, the dual-pressure operation has a greater improvement in plant heat rate over the single-pressure operation in the application using turbines with heavily loaded exhaust ends.

As shown in Fig. 9-1, the surface condenser consists of a casing or shell with a chamber at each end, frequently referred to as water boxes. Tube sheets separate the water boxes from the center steam space. Banks of tubes connect the water boxes by piercing the tube sheets; the tubes essentially fill the shell or steam space. Circulating pumps force the cooling water through the water boxes and the connecting tubes.

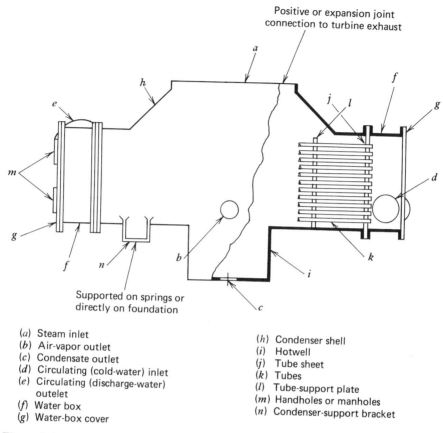

Supported on springs or
directly on foundation

(a) Steam inlet
(b) Air-vapor outlet
(c) Condensate outlet
(d) Circulating (cold-water) inlet
(e) Circulating (discharge-water)
     outelet
(f) Water box
(g) Water-box cover

(h) Condenser shell
(i) Hotwell
(j) Tube sheet
(k) Tubes
(l) Tube-support plate
(m) Handholes or manholes
(n) Condenser-support bracket

Figure 9-1. Surface-condenser and standard terminology.

The tube material falls into two basic groups, freshwater and saltwater. Admiralty, Arsenical Copper, and Type 304 Stainless Steel are common for freshwater applications while Cupronickel, Aluminum Brass, Titanium, and sometimes Type 315 Stainless Steel are used for salt water. The copper alloy tubes are often 18 BWG (Birmingham wire gauge) wall thickness with some 19 BWG applications in recent years. The stainless steel tubes usually have a wall thickness of 22 BWG with few exceptions. The effective tube length and tube diameter must be selected through optimization. More on this subject will be presented later in this text.

The water box may be either divided or nondivided. A divided water box design separates the cooling water path at the condenser. It is recommended for installations where fouling from debris in the water may require removal of one-half of the condenser. In addition this design will facilitate in the locating and plugging of leaking tubes. The water boxes are made of either steel or cast iron. Steel is less expensive and is acceptable for cooling tower and fresh water applications. However, it must be protected against corrosion with a suitable coating material. In general, cast-iron water boxes are used for seawater services.

Tube sheet material should be muntz metal except in some cases in which tubes are to be welded. Tube sheet thickness should be 1/8 inch greater than tube diameter. The various condenser manufacturers have different methods of calculating minimum tube support spacing. In general, the tube support plates should be spaced no longer than 50 tube diameters apart.

The hot well of a condenser is designed to collect the steam condensate. The hot well volume, expressed in minutes of condensate produced at maximum expected turbine throttle flow, should be at least one minute for installations with deaerating heater and storage tank, and three minutes for installations with closed feedwater heating train.

A condenser needs auxiliary equipment to move cooling water through the tubes and to remove air from the steam space and condensate from the hot well. The equipment generally includes a steam-jet air ejector (or mechanical vacuum pumps), an atmospheric relief valve, and circulating and condensate pumps. When the condenser is selected, attention must be given to the initial and operating costs of this auxiliary equipment.

9.2 OVERALL HEAT TRANSFER COEFFICIENT

Consider the cross section of the condenser tube shown in Fig. 9-2. The heat transfer rate is expressed by

$$q = \frac{T_s - T_w}{\dfrac{1}{h_i A_i} + \dfrac{\ln\left(r_o/r_i\right)}{2\pi k L} + \dfrac{1}{h_o A_o}} \qquad (9\text{-}1)$$

where the subscripts i and o pertain to the inside and outside of the condenser tube. The overall heat transfer coefficient U is defined by the relation

$$q = UA(T_s - T_w) \qquad (9\text{-}2)$$

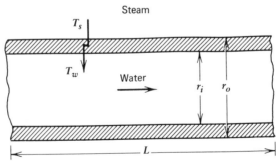

Figure 9-2.   Schematic diagram of a condenser tube.

Combining Eq. (9-1) and Eq. (9-2) would lead to

$$U_i = \frac{1}{\dfrac{1}{h_i} + \dfrac{A_i \ln (r_o/r_i)}{2\pi kL} + \dfrac{A_i}{A_o h_o}} \tag{9-3}$$

or

$$U_o = \frac{1}{\dfrac{A_o}{A_i h_i} + \dfrac{A_o \ln (r_o/r_i)}{2\pi kL} + \dfrac{1}{h_o}} \tag{9-4}$$

Evidently, the overall heat transfer coefficient is based on either the inside or outside area of condenser tube. The U value is theoretically governed by the conductivity of tube material and the inside and outside convective heat transfer coefficients. Since the values of h_o and k for the condenser tube are relatively large as compared with that of h_i, the overall heat transfer coefficient is practically determined by the convective heat transfer coefficient h_i. That is,

$$U_i \doteq h_i \tag{9-5}$$

The Heat Exchange Institute has conducted extensive tests for the purpose of arriving at values that represent maximum overall design limits for the variety of

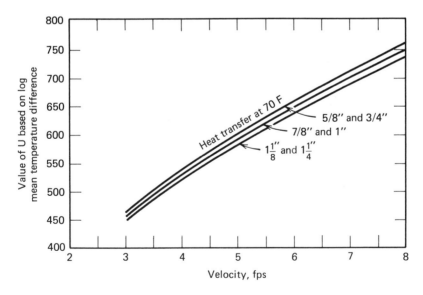

Figure 9-3a. Condenser overall heat transfer coefficient [5]. Reprinted with permission from the Heat Exchange Institute. Standards for Steam Surface Condenser.

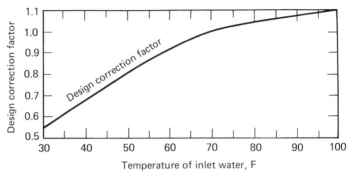

Figure 9-3b. Correction factors for overall heat transfer coefficient [5]. Reprinted with permission from the Heat Exchange Institute. Standards for Steam Surface Condenser.

cooling water specific gravities, specific heats, and salinity ranges. The curves shown in Fig. 9-3a should be used in the design of surface condensers. The curves are based on a new, clean No. 18 BWG Admiralty metal tube with 70 F inlet circulating water temperature. For inlet circulating water temperatures other than 70 F, the basic overall heat transfer coefficient must be multiplied by the corresponding correction factors shown in Fig. 9-3b. For any condenser tube gauge or material other than No. 18 BWG Admiralty, the basic overall heat transfer coefficient should be multiplied by the appropriate correction factors from Table 9-1.

As indicated previously, the overall heat transfer coefficient thus obtained represents the upper limit of its value. After a period of operation, the surfaces of condenser tubes become coated with various deposits present in the steam or cooling

Table 9-1
Correction Factor for Overall Heat Transfer Coefficient [5]

Tube Materials	Tube Wall Gauge (BWG)						
	24	22	20	18	16	14	12
Admiralty metal	1.06	1.04	1.02	1.00	0.96	0.92	0.87
Arsenical copper	1.06	1.04	1.02	1.00	0.96	0.92	0.87
Aluminum	1.06	1.04	1.02	1.00	0.96	0.92	0.87
Aluminum brass	1.03	1.02	1.00	0.97	0.94	0.90	0.84
Aluminum bronze	1.03	1.02	1.00	0.97	0.94	0.90	0.84
Muntz metal	1.03	1.02	1.00	0.97	0.94	0.90	0.84
90–10 Cu–Ni	0.99	0.97	0.94	0.90	0.85	0.80	0.74
70–30 Cu–Ni	0.93	0.90	0.87	0.82	0.77	0.71	0.64
Cold-rolled low carbon steel	1.00	0.98	0.95	0.91	0.86	0.80	0.74
Stainless steels							
Type 410/430	0.88	0.85	0.82	0.76	0.70	0.65	0.59
Type 304/316	0.83	0.79	0.75	0.69	0.63	0.56	0.49
Type 329	0.78	0.76	0.74	0.69	0.65	0.60	0.54
Titanium (tentative)	0.85	0.81	0.77	0.71	—	—	—

water circuits. The heat transfer surface may also become corroded as a result of the interaction between the fluids and the material used for construction of the condenser. In either event, this coating represents an additional resistance to the heat flow and thus results in decreased performance. In the condenser design a cleanliness factor (c) is often used to account for this fouling situation. Design cleanliness can vary from 70 to 90% for the copper alloy tubes with 85% most common and from 80 to 100% for the stainless steel tubes with the majority 90% or higher.

As indicated in Fig. 9-3a, the cooling water velocity inside the tube is a dominating factor in determining the overall heat transfer coefficient. Increasing the water velocity would result in a higher heat transfer coefficient, and therefore in a smaller heat transfer surface. However, the water velocity cannot be increased indefinitely because of high pumping cost. In practice, the water velocity in the condenser tubes will range from 6.0 to 8.0 feet per second (fps) for copper alloy tubes with fresh water, and there is a tendency to limit velocities to a maximum of 6.5 to 7.0 fps with salt water. For stainless steel tubes the velocity has no real maximum limit, and there is some evidence indicating that a minimum velocity may be desirable to maintain better cleanliness. Applications of 7.0 up to 10.0 fps have been made. Although a range of velocities is permissible for a given project, the economic evaluation must be carried out to determine the optimal value.

9.3 HYDRAULIC LOSS

The hydraulic loss of a condenser is always of interest to engineers. It affects not only the initial cost of equipment, but also the operating cost. The Heat Exchange Institute has conducted extensive study in this area and published a series of curves for an estimate of condenser hydraulic loss. These curves and related information are reproduced and presented in Figs. 9-4, 9-5, 9-6, and 9-7.

Figure 9-4 presents the hydraulic loss of condenser tubes. The loss is expressed in terms of feet of water per unit length of condenser tube. These curves have been prepared with the assumption that the tubes must be clean and have a tube gauge of 18 BWG and the water inlet temperature of 70 F. If the actual conditions are different from those, appropriate correction factors from the accompanying table and Fig. 9-5 must be applied.

Figures 9-6 and 9-7 present the head losses to be expected in water boxes and tube entrances and exits. For single-pass condensers, the water box losses should be determined from the curves in Fig. 9-6 using the actual nozzle water velocity. The tube inlet and outlet losses are combined in one curve in Fig. 9-6, and the value for these losses should be taken directly from the curve using the actual water velocity in the tubes. For two-pass condensers, similar procedure should be followed, but using the curves of Fig. 9-7.

The total condenser hydraulic loss is the sum of these four losses: tube loss, tube end loss, water box inlet loss, and water box outlet loss. For condenser performance simulations it may be desirable to express these hydraulic losses in terms of mathematical equations. The following are the mathematical correlations based upon the information provided by the Heat Exchange Institute.

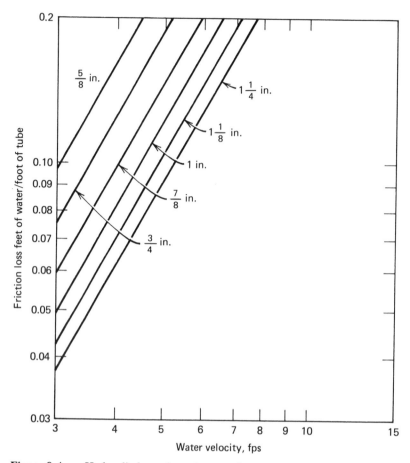

Figure 9-4a. Hydraulic loss of condenser tubes [5]. Reprinted with permission from the Heat Exchange Institute. Standards for Steam Surface Condenser.

Gauge Correction Factor for Pressure Drop							
Tube O.D. In.	12 BWG	14 BWG	16 BWG	18 BWG	20 BWG	22 BWG	24 BWG
0.625	1.38	1.21	1.10	1.00	0.94	0.91	0.89
0.750	1.28	1.16	1.06	1.00	0.95	0.93	0.90
0.875	1.25	1.13	1.06	1.00	0.96	0.94	0.92
1.000	1.19	1.11	1.05	1.00	0.96	0.94	0.93
1.125	1.16	1.09	1.04	1.00	0.97	0.95	0.94
1.250	1.14	1.08	1.04	1.00	0.97	0.96	0.94

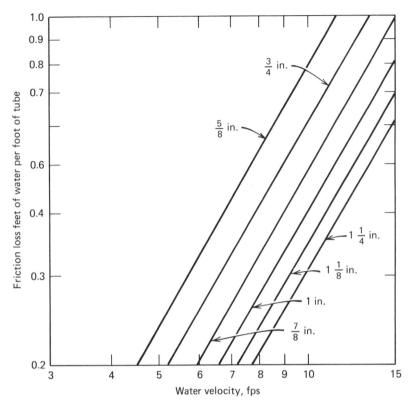

Figure 9-4b. Hydraulic loss of condenser tubes [5]. Reprinted with permission from the Heat Exchange Institute. Standards for Steam Surface Condenser.

Gauge Correction Factor for Pressure Drop							
Tube O.D. In.	12 BWG	14 BWG	16 BWG	18 BWG	20 BWG	22 BWG	24 BWG
0.625	1.38	1.21	1.10	1.00	0.94	0.91	0.89
0.750	1.28	1.16	1.06	1.00	0.95	0.93	0.90
0.875	1.25	1.13	1.06	1.00	0.96	0.94	0.92
1.000	1.19	1.11	1.05	1.00	0.96	0.94	0.93
1.125	1.16	1.09	1.04	1.00	0.97	0.95	0.94
1.250	1.14	1.08	1.04	1.00	0.97	0.96	0.94

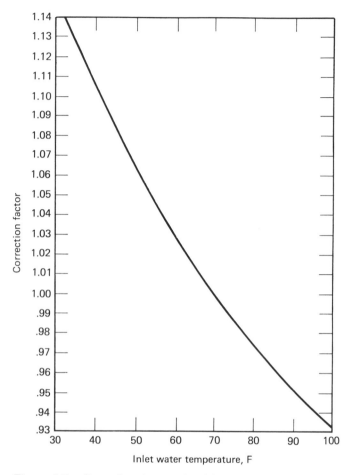

Figure 9-5. Correction factors for friction loss in condenser tubes [5].

Condenser Tube Loss

$$\text{FRLOSS} = 0.0144682 \,(\text{FPSTUB})^{1.74807} \quad \text{for TUBEOD} = 0.625 \qquad (9\text{-}6)$$

$$= 0.0111637 \,(\text{FPSTUB})^{1.74807} \quad \text{for TUBEOD} = 0.750 \qquad (9\text{-}7)$$

$$= 0.00884169 \,(\text{FPSTUB})^{1.74807} \quad \text{for TUBEOD} = 0.875 \qquad (9\text{-}8)$$

$$= 0.00737700 \,(\text{FPSTUB})^{1.74807} \quad \text{for TUBEOD} = 1.00 \qquad (9\text{-}9)$$

$$= 0.00630528 \,(\text{FPSTUB})^{1.74807} \quad \text{for TUBEOD} = 1.125 \qquad (9\text{-}10)$$

$$= 0.00553722 \,(\text{FPSTUB})^{1.74807} \quad \text{for TUBEOD} = 1.250 \qquad (9\text{-}11)$$

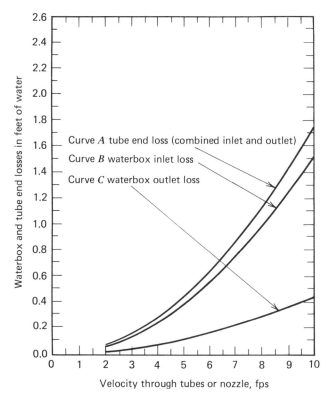

Figure 9-6.   Some hydraulic losses for single-pass
condensers [5].

where FRLOSS is the head loss in feet of water per foot of tube, FPSTUB is the
water velocity through tubes in feet per second, and TUBEOD is the tube outside
diameter in inches. Taking into consideration the effect of inlet water temperature,
the FRLOSS must be multiplied by the correction factor. The temperature
correction factor obtained by the regression method is expressed as follows:

$$\text{CFTEMP} = 1.334 - 0.0076078T + 0.000047097T^2$$

$$- 0.00000012239T^3 \tag{9-12}$$

where CFTEMP is the correction factor and T is the entering cooling water
temperature in degrees F. Taking into consideration the effect of the tube wall
thickness, the FRLOSS should be multiplied by the gauge correction factor shown
in Fig. 9-4.

Condenser Tube End Loss

$$\text{ELOSS1} = 0.01925 - 0.017139 \,(\text{FPSTUB}) + 0.021428 \,(\text{FPSTUB})^2$$

$$- 0.00021038 \,(\text{FPSTUB})^3 \tag{9-13}$$

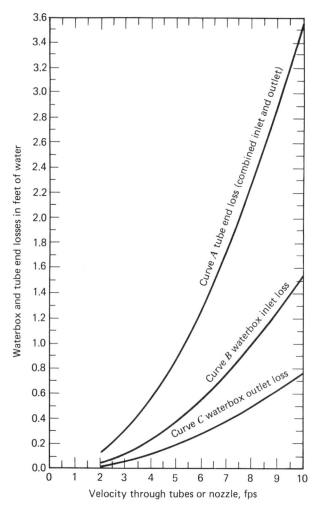

Figure 9-7. Some hydraulic losses for two-pass condensers [5].

for a single-pass condenser, and

$$ELOSS2 = 2 \times ELOSS1 \qquad (9\text{-}14)$$

for a double-pass condenser, where ELOSS1 and ELOSS2 are head losses in feet of water gauge, and FPSTUB is the water velocity through tubes in feet per second.

Waterbox Inlet Loss

$$BILOSS = 0.028196 - 0.029013(FPSNZL) + 0.022216\,(FPSNZL)^2$$

$$- 0.00041926\,(FPSNZL)^3 \qquad (9\text{-}15)$$

for both single-pass and double-pass condensers where BILOSS is the head loss in feet of water gauge, and FPSNZL is the water velocity through the nozzles in feet per second.

Waterbox Outlet Loss

$$BOLOSS1 = -0.017944 + 0.0067363 \, (FPSNZL) + 0.0035301$$

$$\times (FPSNZL)^2 + 0.000058587(FPSNZL)^3 \qquad (9\text{-}16)$$

for a single-pass condenser, and

$$BOLOSS2 = -0.0001297 - 0.0010944(FPSNZL) + 0.0081114$$

$$\times (FPSNZL)^2 - 0.000021328(FPSNZL)^3 \qquad (9\text{-}17)$$

for a double-pass condenser where BOLOSS1 and BOLOSS2 are losses in feet of water gauge.

9.4 CONDENSER THERMAL DESIGN

Consider a typical condenser in which the circulating water passes through the condenser tubes and picks up the latent heat from the steam condensing outside the tubes. Figure 9-8 shows the temperature variation in both fluids. The steam temperature T_s is the saturation temperature at the condenser pressure. The difference between the steam temperature (T_s) and water inlet temperature (T_1) is defined as the initial temperature difference (ITD). The difference between the steam temperature (T_s) and the water exit temperature (T_2) is known as the terminal temperature difference (TTD). The temperature difference varies inside the condenser. To obtain a suitable mean temperature difference across the condenser, we define

$$q = CUA \, \Delta T_m \qquad (9\text{-}18)$$

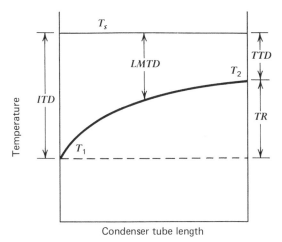

Figure 9-8.   Temperature variation in fluids.

and

$$\Delta T_m = \frac{(ITD) - (TTD)}{\ln\left(\dfrac{ITD}{TTD}\right)} \tag{9-19}$$

where C is the cleanliness factor and ΔT_m is frequently referred to as the log mean temperature difference ($LMTD$). Equation (9-19) can be also expressed as

$$\Delta T_m = \frac{TR}{\ln\left(\dfrac{1}{1 - \dfrac{TR}{ITD}}\right)} \tag{9-20}$$

where TR is the temperature rise in the condenser and defined as $T_2 - T_1$. Equation (9-20) indicates that the log mean temperature difference is simply a function of two variables, temperature rise and initial temperature difference. In addition to Eq. (9-18), the following equations are usually required for the heat transfer surface calculations:

$$q = (500)(TR)(GPM) \tag{9-21}$$

$$(NT) = (GPM)(NP)/(g)(V) \tag{9-22}$$

and

$$(A) = (NT)(L)(m) \tag{9-23}$$

where

$$(GPM) = \text{cooling water flow rate (gpm)}$$

$$(TR) = \text{temperature rise (F)}$$

$$(q) = \text{condenser heat load (Btu/hr)}$$

$$(NT) = \text{total number of condenser tubes}$$

$$(NP) = \text{number of passes of cooling water}$$

$$(g) = \text{gpm per tube at 1.0 fps}$$

$$(m) = \text{square feet per lineal feet of tube}$$

$$(A) = \text{total surface area (ft}^2\text{)}$$

$$(L) = \text{effective tube length (ft)}$$

$$(V) = \text{circulating water velocity (fps)}$$

The tube characteristics such as (g) and (m) can be found in Appendix C.

EXAMPLE 9-1. A condenser is designed for the maximum heat load 1000×10^6 Btu/hr at the condenser pressure 1.5 inches Hg abs. The circulating water has inlet temperature 60 F and outlet temperature 80 F. The decision has been made to select 1 in. OD-22 BWG Type 304 stainless steel for condenser tubes. The water velocity in the tubes is assumed to be 8.00 fps.

Calculate the effective tube length, cooling water flow rate, and total heat transfer area.

Solution: Using Table C-2, we find the steam saturation temperature $T_s = 91.72$ F and calculate the initial temperature difference as

$$ITD = 91.72 - 60 = 31.72 \text{ F}$$

Next, we determine the LMTD by using Eq. (9-20)

$$\Delta T_m = \frac{20}{\ln\left(\dfrac{1}{1 - \dfrac{20}{31.72}}\right)} = 20.08 \text{ F}$$

and the U_o value by using Fig. 9-3 and Table 9-1

$$U_o = 740 \times 0.92 \times 0.79 = 537.8 \text{ Btu/hr-F-ft}^2$$

Then, we use Eq. (9-18) to obtain the required heat transfer area

$$A = 1000 \times 10^6/(537.8 \times 20.08) = 92,600 \text{ ft}^2$$

and use Eq. (9-21) to obtain the cooling water flow rate

$$GPM = 1000 \times 10^6/(500)(20) = 100,000 \text{ gpm}$$

To obtain the effective tube length, we assume the condenser with two passes, and use Eqs. (9-22) and (9-23). That is,

$$NT = (100,000)(2)/(2.182)(8.00) = 11,456$$

and

$$L = (92,600)/(11,456)(0.2618) = 30.88 \text{ ft}$$

Under some circumstances plant designers select the effective tube length and then calculate the temperature rise in the circulating water and terminal temperature difference in condenser. If this is the case, care must be exercised to ensure the condenser terminal temperature difference is not less than 5 F. The following example illustrates this aspect.

EXAMPLE 9-2.   Calculate the condenser surface area and cooling water requirement for the following design conditions.

Heat load (q)	1000×10^6 Btu/hr
Condenser pressure (p)	2.00 in. Hg abs.
Cooling water inlet temperature	75 F
Cooling water velocity (V)	7.00 fps
Number of passes (NP)	Single
Tube diameter (D)	1 in. OD
Tube gauge and material	18 BWG Admiralty
Tube length (L)	50 ft
Tube cleanliness (C)	0.85

Solution: Since the tube length is specified, the calculations involve solving
several working equations simultaneously. First, combining Eq. (9-22) with Eq.
(9-23), we obtain

$$A = (GPM)(NP)(L)(m)/(V)(g) \tag{a}$$

Then, substituting the expression (a) into Eq. (9-18), we get

$$q = \frac{(C)(U)(GPM)(NP)(L)(m)(TR)}{(g)(V)\ln\left(\cfrac{1}{1 - \cfrac{TR}{ITD}}\right)} \tag{b}$$

Also from Eq. (9-21), we have

$$TR = q/(500)(GPM) \tag{c}$$

Finally, we combine (b) and (c) to eliminate TR and have

$$GPM = q \Big/ \left[(500)(ITD)\left(1 - \frac{1}{e^k}\right) \right] \tag{d}$$

where

$$K = \frac{(C)(U)(NP)(L)(m)}{(500)(g)(V)} \tag{e}$$

Using

$$U = 695 \times 1.02 \times 1.0 = 708.9, \; ITD = 101.14 - 75 = 26.14,$$

$$C = 0.85, \qquad NP = 1, \qquad L = 50, \qquad m = 0.2618,$$

$$g = 1.992, \qquad V = 7.0 \text{ and } q = 1000 \times 10^6$$

we have $K = 1.13$ and

$$GPM = \frac{1000 \times 10^6}{500 \times 26.14 \times (1 - e^{-1.13})} = 112,900 \text{ gpm}$$

To determine the condenser heat transfer area, we calculate the temperature rise using Eq. (9-21)

$$TR = 1000 \times 10^6 / 500 \times 112,900 = 17.7 \text{ F}$$

and then use Eq. (9-18) to obtain

$$A = 1000 \times 10^6 / 0.85 \times 708.9 \times 15.65$$

$$A = 105,900 \text{ ft}^2$$

If we wish, we can check to see whether the terminal temperature difference would be less than 5 F. In this case we have

$$TTD = ITD - TR = 26.14 - 17.7$$

$$= 8.4 \text{ F}$$

This kind of calculation can be easily carried out in a digital computer. Table 9-2 presents the computer output for Example 9-2 by using one of the condenser programs available [8]. Like the computer programs for evaporative cooling towers, this condenser program has been developed in an interactive mode.

The thermal design of condenser is not completed unless the pump power is determined. To evaluate the condenser design, power needed for circulating water pump must be known. Using the first law of thermodynamics, we have the following expression for hydraulic horsepower:

$$\text{Hydraulic hp} = \frac{W}{33,000} DH \tag{9-24}$$

or

$$\text{Hydraulic hp} = \frac{(GPM)(DH)(sp\ gr)}{3960} \tag{9-25}$$

where

$$W = \text{water flow rate (lb/min)}$$

$$DH = \text{developed head (ft)}$$

352   Condensers

Table 9-2
Computer Outputs for Example 9-2

```
LOA CONDO1
READY
RUN
CONDO1 03/27/84 15:35:20


     The program was designed to size the single pressure condenser.
     The program requires the water inlet temperature (F), design
condenser pressure (in.Hg. abs) and other inputs.
     Do you still want to run this program?
Enter 1 for yes; 2 for no.
?1
Enter the water inlet temperature (F).
?75
Enter the design condenser pressure (in.Hg. abs).
The maximum value is 5.0.
?2
Enter the design heat load (Btu per hour).
?1000000000
Enter the tube length (feet).
?50
Enter the tube outside diameter (inches).
The choices are: 0.625(5/8), 0.750(3/4), 0.875(7/8),
1.0, 1.125(1 1/8) and 1.25(1 1/4).
?1
Enter the tube material.
     1 = Admiralty metal
     2 = Arsenical Copper
     3 = Aluminum
     4 = Aluminum Brass
     5 = Aluminum Bronze
     6 = Muntz metal
     7 = 90-10 Cu-Ni
     8 = 70-30 Cu-Ni
     9 = Cold-Rolled low Carbon Steel
    10 = Stainless Steel: Type 410/430
    11 = Stainless Steel: Type 304/316
    12 = Stainless Steel: Type 329
?1
Enter the tube wall gauge (Ranging from 12 to 24 with the increment
equal to 1).
?18
Enter the cleanliness of tube (fraction of 1.0).
     1.0 = perfectly clean
     0.0 = absolutely dirty
?.85
Enter the number of passes (either 1 or 2).
?1
Enter the water velocity inside the tube (feet per second).
?7


     The inputs and computed results are summarized as below:
          Water inlet temperature     =          75.00 F
          Design condenser pressure   =           2.00 in.Hg. abs
          Design heat load            =  1000000000.   Btu per hour
          Tube length                 =          50.00 Ft.
          Tube diameter               =           1.000 Inches
          Tube material               = Admiralty metal
          Tube gauge                  =          18.
          Cleanliness                 =           0.85
          Number of passes            =           1.
          Water velocity              =           7.00 Ft./sec.


     TEMPERATURE RISE IN CONDENSER     =          17.76 F
     WATER OUTLET TEMPERATURE          =          92.76 F
     TERMINAL TEMPERATURE DIFFERENCE   =           8.38 F
     WATER FLOW RATE                   =      112639.25 GPM
     NUMBER OF TUBES                   =        8078.
     TUBE SURFACE AREA                 =      105740.62 SQ.FT.


     Do you want another calculation?
Enter 1 for yes; 2 for no.
?2
STOP
TIME 2.9 SECS
```

In condenser calculation the specific gravity of water is frequently assumed as 1.0. The shaft horsepower (brake horsepower) is obtained by dividing the hydraulic horsepower by pump efficiency. That is,

$$\text{Shaft horsepower} = \frac{\text{hydraulic hp}}{\text{pump efficiency}} \qquad (9\text{-}26)$$

EXAMPLE 9-3.   Calculate the hydraulic horsepower for the condenser described in Example 9-2. The external friction loss is 80 ft.

Solution: We first calculate the hydraulic loss of the condenser CL. The term CL consists of condenser tube loss L_1, condenser tube end loss L_2, waterbox inlet loss L_3, and waterbox outlet loss L_4. Using Figs. 9-4, 9-5, and 9-6, we have

$$L_1 = 0.22 \times 50 = 11 \text{ ft}$$

$$L_2 = 0.88 \text{ ft}$$

$$L_3 = 0.78 \text{ ft}$$

$$L_4 = 0.22 \text{ ft}$$

the condenser hydraulic loss is then

$$CL = L_1 + L_2 + L_3 + L_4 = 12.88 \text{ ft}$$

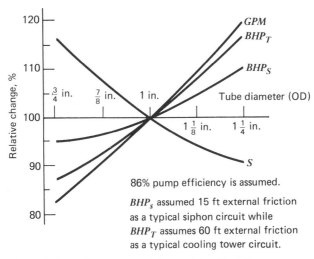

Figure 9-9.   Influence of condenser tube diameters.

The total hydraulic loss is the sum of the condenser loss and its external loss. That is,

$$\text{Total loss} = 12.88 + 80 = 92.88 \text{ ft}$$

Finally, the hydraulic horsepower is obtained by using Eq. (9-25):

$$\text{Hydraulic hp} = 112,900 \times 92.88/3960$$

$$= 2648 \text{ hp}$$

Three fundamental figures make up the major economic influence on any condenser's thermal design. They are the total square feet of condensing surface, the cooling water flow rate, and the brake horsepower required to move the cooling water through the specific circuit. In general, these three terms are in conflict with each other. When the heat transfer surface is reduced, the required water flow will increase and so will the pump power.

Figure 9-9 shows the influence of different tube diameters for a given heat transfer function using a constant tube length and cooling water velocity. With increasing tube diameters, there will be less heat transfer surface and more cooling water required. The relative change in cooling water pumping is also shown for an assumed siphon water circuit and cooling tower water circuit, both of which are commonly used today.

Figure 9-10 shows the effect of changing the velocity of the cooling water within a given tube diameter for a fixed tube length. With higher velocities, the required heat transfer surface is reduced, but the cooling water is increased and along with it the condenser hydraulic loss and pump power. Similarly, Fig. 9-11 depicts the relative

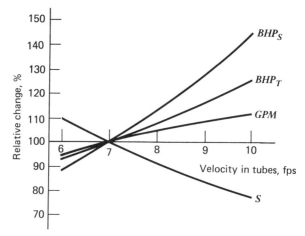

Figure 9-10. Influences of water velocity inside the condenser tubes.

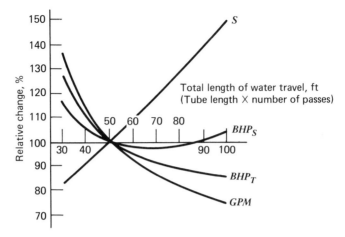

Figure 9-11. Influences of the total length of water travel.

changes of the three basic parameters as the total length of cooling water travel (tube length times the number of passes) is increased.

9.5 CONDENSER PERFORMANCE

The condenser is usually sized to perform a heat transfer function under a given set of conditions. These conditions are called the design conditions or the conditions at the design point. In operation the condenser may not always be at the design point. This is mainly due to the variation of condenser heat load or a change of circulating water inlet temperature. Because of these, the condenser evaluation must be based on the design as well as off-design performances.

The condenser performance can be easily predicted showing that the condenser absolute pressure is a function of condenser heat load and circulating water inlet temperature. The derivation of this relationship is presented as below:

Starting with Eq. (9-18)

$$q = CUA \frac{(T_s - T_1) - (T_s - T_2)}{\ln\left(\dfrac{T_s - T_1}{T_s - T_2}\right)} \tag{9-27}$$

and the modified expression of Eq. (9-21)

$$q = WC_p(T_2 - T_1) \tag{9-28}$$

we combine these two equations to obtain

$$\ln \frac{T_s - T_2}{T_s - T_1} = \frac{-CUA}{WC_p} \tag{9-29}$$

or

$$T_s = \frac{T_2 - T_1 \exp\left(-\dfrac{CUA}{WC_p}\right)}{1 - \exp\left(-\dfrac{CUA}{WC_p}\right)} \tag{9-30}$$

Substituting T_2 from Eq. (9-28) into Eq. (9-30) will give

$$T_s = \frac{\dfrac{q}{WC_p} + T_1\left[1 - \exp\left(\dfrac{-CUA}{WC_p}\right)\right]}{1 - \exp\left(-\dfrac{CUA}{WC_p}\right)} \tag{9-31}$$

or

$$T_s = C_1 q + T_1 \tag{9-32}$$

where

$$C_1 = \frac{1}{WC_p}\left[1 - \exp\left(-\frac{CUA}{WC_p}\right)\right]^{-1} \tag{9-33}$$

It is noted from Eq. (9-32) that the condenser temperature (T_s) is a function of condenser heat load (q) and circulating water inlet temperature (T). The condenser temperature has a linear relationship with the heat load. In practice, the condenser pressure is frequently used instead of condenser temperature. It must be pointed out that in the above derivation no attention is given to the effects of noncondensable gases in the condenser and the limitation of gas-removal equipment. In operation, the condenser temperature (or pressure) cannot be such that the condenser terminal difference is less than 5 F. If this should occur in calculations, the Heat Exchange Institute recommends that the equation $ITD = TR + 5$ be used for an estimate of condenser temperature.

EXAMPLE 9-4.   Calculate the condenser pressure for the condenser described in Example 9-2. The operating conditions are identical to those at the design point except that the condenser heat load is increased to 1200×10^6 Btu/hr.

Solution: Since the operating conditions are identical to those at the design point except for the condenser heat load, we obtain from Example 9-2.

$$CU = 602.6 \text{ Btu/hr-ft}^2\text{-F}$$

$$GPM = 112,900 \text{ gpm}(w = 56.45 \times 10^6 \text{ lb/hr})$$

Then, we calculate C_1 by Eq. (9-33)

$$C_1 = \frac{1}{56.45 \times 10^6}\left[1 - \exp\left(-\frac{602.6 \times 105,900}{56.45 \times 10^6}\right)\right]^{-1}$$

$$C_1 = 0.026 \times 10^{-6} \text{ hr-F/Btu}$$

Using Eq. (9-32), the condenser temperature is

$$T_s = (0.026 \times 10^{-6})(1200 \times 10^6) + 75 = 106.4 \text{ F}$$

Since the terminal temperature difference (*TTD*) in operation cannot be less than 5 F, the *TTD* must be checked in calculation. The *TTD* for this case is

$$TTD = ITD - TR$$

$$= 31.4 - 21.2 = 10.2 \text{ F}$$

and, therefore, the *TTD* is acceptable. The condenser pressure at the saturation temperature 106.4 is found 2.35 in. Hg abs. from Table C-2.

   If we use the condenser programs available [8], Example 9-4 is repeated. The computer output is reproduced and shown in Table 9-3.

EXAMPLE 9-5.   Calculate the condenser pressure for the following conditions:

Condenser surface area (A)	300,000 ft^2
Water flow rate (*GPM*)	175,000 gpm (87.5×10^6 lb/hr)
Tube diameter (d)	7/8 in.
Tube gauge and material	18 BWG, Aluminum Brass
Number of passes (*NP*)	2
Effective tube length (L)	36 ft
Condenser heat load (q)	2500×10^6 Btu/hr
Circulating water inlet temperature (T_1)	80 F
Tube cleanliness factor (C)	1.00

Table 9-3
Computer Output for Example 9-4

```
LOA COND02
READY
RUN
COND02 03/27/84 15:30:28


       The program was designed to simulate an operation performance of
single-pressure condenser.
       the program requires the operation heat load (Btu/Hr), operation
water inlet temperature (F), design temperature rise (f) and other
inputs.
       Do you still want to run this program?
Enter 1 for yes; 2 for no.
?1
Enter the heat load under consideration (Btu/Hr).
?1200000000
Enter the water inlet temperature (F).
?75
Enter the design temperature rise for the condenser (F).
?17.76
Enter the design heat load (Btu per hour).
?1000000000
Enter the tube length (feet).
?50
Enter the tube outside diameter (inches).
The choices are: 0.625(5/8), 0.75(3/4), 0.875(7/8),
1.0, 1.125(1 1/8) and 1.25(1 1/4).
?1
Enter the tube material.
        1 = Admiralty metal
        2 = Arsenical copper
        3 = Aluminum
        4 = Aluminum Brass
        5 = Aluminum Bronze
        6 = Muntz metal
        7 = 90-10 Cu-Ni
        8 = 70-30 Cu-Ni
        9 = Cold-Rolled low Carbon Steel
       10 = Stainless Steel: Type 410/430
       11 = Stainless Steel: Type 304/316
       12 = Stainless Steel: Type 329
?1
Enter the tube wall gauge (Ranging from 12 to 24 with the increment
equal to 1).
?18
Enter the cleanliness of tube (fraction of 1.0).
       1.0 = perfectly clean
       0.0 = absolutely dirty
?.85
Enter the number of passes (either 1 or 2 ).
?1
Enter the water velocity inside the tube (feet per second).
?7


       The inputs and computed results are summarized as below:
          Operation heat load       =  1200000000.  Btu per Hour
          Water inlet temperature   =       75.00 F
          Design temperature rise   =       17.76 F
          Design heat load          =  1000000000.  Btu per hour
          Tube length               =       50.00 Ft.
          Tube diameter             =        1.000 Inches
          Tube material             = Admiralty metal
          Tube gauge                =       18.
          Cleanliness               =        0.85
          Number of passes          =        1.
          Water velocity            =        7.00 Ft./sec.


       WATER OUTLET TEMPERATURE    =      96.31 F
       CONDENSER STEAM TEMPERATURE =     106.37 F
       CONDENSER PRESSURE          =       2.35 IN.HG. ABS


       Do you want another calculation?
Enter 1 for yes; 2 for no.
?2
STOP
TIME 2.0 SECS
```

Solution: Starting with Eq. (9-23), we calculate the number of condenser tubes

$$NT = 300,000/36 \times 0.2291 = 36374$$

and with Eq. (9-22) we determine the water velocity inside the tubes

$$V = \frac{175,000 \times 2}{1.478 \times 36374} = 6.51 \text{ fps}$$

Using this water velocity, the overall heat transfer coefficient found from Fig. 9-3 is

$$U = 670 \times 1.04 \times 0.97 = 675 \text{ Btu/hr-ft}^2\text{-F}$$

The C_1 defined by Eq. (9-33) is calculated as follows:

$$C_1 = \frac{1}{87.5 \times 10^6}\left[1 - \exp\left(-\frac{675 \times 300,000}{87.5 \times 10^6}\right)\right]^{-1}$$

$$C_1 = 0.01268 \times 10^{-6} \text{ hr-F/Btu}$$

Finally, from Eq. (9-32), the condenser temperature is

$$T_s = (0.01268 \times 10^{-6})(2500 \times 10^6) + 80 = 111.7 \text{ F.}$$

The corresponding condenser pressure is found 2.73 in. Hg abs. from the steam table (Table C-2). The terminal temperature difference is

$$TTD = ITD - TR$$

$$= 31.7 - 28.5 = 3.12 \text{ F}$$

Since the calculation results in a *TTD* less than 5 F, according to the recommendation of the Heat Exchange Institute, the initial temperature difference *ITD* must be calculated again by

$$ITD = TR + 5$$

$$ITD = 28.5 + 5 = 33.5 \text{ F}$$

Then, the condenser temperature (steam temperature) becomes

$$T_s = ITD + T_1$$

$$= 33.5 + 80 = 113.5 \text{ F}$$

From Table C-2, the corresponding condenser pressure is found 2.87 in. Hg abs.

The condenser performance is frequently expressed in a graphical form. Figure 9-12 indicates typical performance curves showing absolute pressures for varying condenser duties and inlet circulating water temperatures. These performance curves are provided by the equipment vendors at the time the equipment proposal is submitted. The performance information is usally based on design calculations as well as field testing data.

To simulate the condenser performance under various conditions in the digital computer, the performance equation must be prepared with the equipment data. For instance, using the data such as those in Fig. 9-12, we can present the performance equation as

$$CP = 0.16302 \times 10$$

$$-0.50095 \times 10^{-1}(CIRC)$$

$$+0.55796 \times 10^{-3}(CIRC)^2$$

$$+0.32946 \times 10^{-3}(HL)$$

$$-0.10229 \times 10^{-4}(HL)(CIRC)$$

$$+0.16253 \times 10^{-6}(HL)(CIRC)^2$$

$$+0.42658 \times 10^{-6}(HL)^2$$

$$-0.92331 \times 10^{-8}(HL)^2(CIRC)$$

$$+0.71265 \times 10^{-10}(HL)^2(CIRC)^2 \tag{9-34}$$

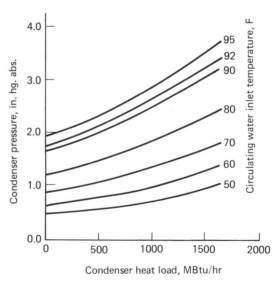

Figure 9-12. Condenser performance curves.

for the range

$$250 \leq HL \leq 1250 \text{ MBtu/hr}$$

and

$$70 \leq CIRC \leq 95 \text{ F}$$

More on the equipment performance simulations will be presented in Chapter 10.

9.6 DUAL-PRESSURE CONDENSER

A dual-pressure condenser may be designed with either separate condenser shells or with a single shell divided into two separate compartments. In either case, the cold inlet water produces a lower pressure in compartment "1" than the same, but warmer water produces in compartment "2." The net result of this operation is always a lower average condenser pressure as compared with that obtained with the same water flow rate and heat transfer surface area in a conventional single-pressure condenser. The improved condenser performance is mainly due to the fact that heat is transferred from the steam to the circulating water in a more uniform temperature difference.

Figure 9-13 shows temperature variations in both single-pressure and dual-pressure condensers. In both condensers the water flow rates and heat transfer surface areas are the same. The vertical line $x - x$ represents a physical partition in the steam side of dual-pressure condenser. The dotted lines in Fig. 9-13b are the conditions of the single-pressure condenser and are reproduced from Fig. 9-13a. It is

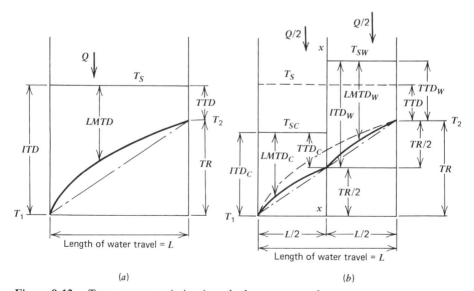

Figure 9-13. Temperature variation in a dual-pressure condenser.

seen that the temperature T_s of the single-pressure condenser is always higher than that of the first compartment, but lower than that of the second compartment in the dual-pressure. For any condenser the temperature difference $T_s - T_{s1}$ is always greater than that of $T_{s2} - T_s$. Therefore, the average temperature in dual-pressure condenser is always a value less than that in the single-pressure. These relationships are derived as below:

Applying Eq. (9-32) to the first compartment of the dual-pressure condenser, we have

$$T_{s1} = C_{11}\frac{q}{2} + T_1 \tag{9-35}$$

where

$$C_{11} = \frac{1}{WC_p}\left[1 - \exp\left(\frac{-CUA}{2WC_p}\right)\right]^{-1} \tag{9-36}$$

Letting $R = TR/ITD$, it is easily proved that

$$\exp\left(-CUA/WC_p\right) = 1 - R \tag{9-37}$$

Substituting Eq. (9-37) into Eq. (9-36), we have

$$C_{11} = \frac{1}{WC_p}\left[1 - (1 - R)^{1/2}\right]^{-1} \tag{9-38}$$

Then, we combine Eq. (9-35) and Eq. (9-38) to generate the temperature of the first compartment as

$$T_{s1} = T_1 + \tfrac{1}{2}(TR)\left[1 - (1 - R)^{1/2}\right]^{-1} \tag{9-39}$$

Similarly, we can derive the temperature of the second compartment as

$$T_{s2} = T_1 + \frac{(TR)}{2} + \frac{(TR)}{2}\left[1 - (1 - R)^{1/2}\right]^{-1} \tag{9-40}$$

Since the temperature of single-pressure condenser is simply $T_s = ITD + T_1$, we have

$$T_s - T_{s1} = ITD\left\{1 - \tfrac{1}{2}R\left[1 - (1 - R)^{1/2}\right]^{-1}\right\} \tag{9-41}$$

$$T_{s2} - T_s = ITD\left\{\frac{R}{2} + \frac{R}{2}\left[1 - (1 - R)^{1/2}\right]^{-1} - 1\right\} \tag{9-42}$$

and

$$T_s - \frac{T_{s1} + T_{s2}}{2} = ITD\left\{1 - \frac{R}{4} - \frac{R}{2}\left[1 - (1 - R)^{1/2}\right]^{-1}\right\} \qquad (9\text{-}43)$$

Equation (9-43) indicates the condenser temperature difference between the single-pressure and the dual-pressure when both have the same water flow rate and heat transfer area. If Eq. (9-43) is extended to multiple-pressure condensers, Eq. (9-43) will become

$$T_s - \frac{T_{s1} + T_{s2} + \cdots + T_{sn}}{n} = ITD\left\{1 - \frac{R}{n}\left(\frac{n-1}{2}\right) - \frac{R}{n}\left[1 - (1 - R)^{1/2}\right]^{-1}\right\}$$

$$(9\text{-}44)$$

where n is the number of compartments in the multiple-pressure condenser. In calculation, each compartment in a multiple-pressure condenser should be treated as a separate unit, the water temperature correction factor and consequently the heat transfer coefficient will be increased for all compartments that follow the first. Once the condenser compartment measure (i.e., steam temperature) is determined, the turbine heat rate can be either predicted by the heat balance method shown in Chapter 7 or approximated by using a simplified diagram such as that in Fig. 9-14. For most modern turbine designs and steam conditions, Fig. 9-14 will be sufficiently accurate in predicting heat rate change due to the use of dual pressure condenser. This can best be illustrated in the following examples.

EXAMPLE 9-6. The condenser designed in Example 9-2 is modified by adding a partition in the steam side. Both compartments thus created have the same water flow rate and equal heat transfer surface area. Other conditions are identical to those in Example 9-2. Calculate the pressures of these condenser compartments.

Solution: From Example 9-2 we have

$$TR = 17.7 \text{ F}$$

$$T_s = 101.14 \text{ F}$$

and

$$ITD = 26.14 \text{ F}$$

First, we calculate the ratio R as

$$R = TR/ITD$$

$$= 17.7/26.14$$

$$= 0.677$$

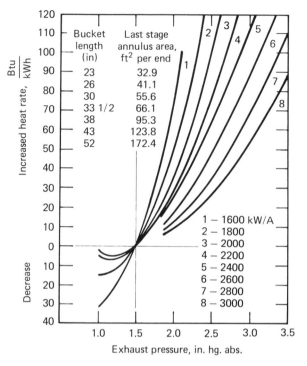

Figure 9-14.   Turbine heat rate correction [4].

Using Eq. (9-41), we have

$$T_{s1} = T_s - ITD\left[1 - \frac{R}{2}\left(1 - (1 - R)^{1/2}\right)^{-1}\right]$$

Substituting numerical values into the equation gives

$$T_{s1} = 95.56 \text{ F}$$

Similarly, we obtain the temperature of the second compartment by using Eq. (9-42)

$$T_{s2} = 104.04 \text{ F}$$

The condenser pressures are therefore

$$P_{s1} = 1.69 \text{ in. Hg abs.}$$

$$P_{s2} = 2.18 \text{ in. Hg abs.}$$

EXAMPLE 9-7. Consider a TC4F steam turbine with nominal output of 350,000 kW. The last-stage bucket length is 26 in. When a single-pressure condenser is applied, the turbine exhaust pressure is expected to be 2.0 in. Hg abs. With a dual-pressure condenser, the cold-end exhaust pressure would be improved to 1.69 in. Hg abs. and the warm-end becomes 2.18 in Hg abs. Estimate the change of turbine heat rate.

Solution: The annulus area provided by one row is found 41.1 ft² from Fig. 9-14. The turbine last-stage loading is therefore

$$\text{Loading} = \frac{350,000}{4 \times 41.1}$$

$$= 2129 \text{ kW/ft}^2$$

Entering Fig. 9-14 with this loading and the condenser pressure 2.0 in. Hg abs., we obtain a heat rate increase of 32 Btu compared with the arbitrary base performance at 1.5 in. Hg abs. Reentering Fig. 9-14 with condenser pressures 1.69 and 2.18 in. Hg abs., we have the readings 10 Btu/kWh, and 48 Btu/kWh. Evidently, there is improvement of 22 Btu/kWh on one half of the turbine flow and a performance loss of 16 Btu/kWh on another half. Thus, the next change of turbine heat rate is reduced by 6 Btu/kWh.

It is seen from the above examples that a dual-pressure condenser is thermodynamically superior to the single-pressure condenser. When both condensers have the same amount of water flow and heat transfer surface, the dual-pressure condenser will operate at a lower pressure than the single-pressure condenser and therefore will result in an improvement of turbine net heat rate. It must be pointed out, however, that the equations derived above are based on the assumption that the overall heat transfer coefficient is the same in both compartments of dual-pressure condenser. To improve accuracy in design and performance calculations, each compartment in a dual-pressure condenser should be treated as a separate unit. The temperature correction factor should be estimated for each compartment and applied accordingly.

Referring back to Example 9-6 and considering the difference of overall heat transfer coefficient in both compartments, we found the average condenser pressure 1.92 in. Hg as compared with 1.94 in. Hg previously calculated. This amounts to a 4% difference. These and other results are summarized in Table 9-4. It is also seen from the table that the average condenser pressure will decrease as the number of compartments increases. However, the rate at which the condenser pressure decreases will become diminishingly small.

To maintain the same condenser pressure and water flow rate, the use of a dual-pressure arrangement will result in a reduction of heat transfer surface.

Table 9-4
Average Condenser Pressure in a Multipressure Condenser

Number of Compartments	Compartment Number	Overall Heat Transfer Coefficient (Btu / hr-ft²-F)	Compartment Pressure (in. Hg abs.)	Average Condenser Pressure (in. Hg abs.)
1	1	708.9	2.00	2.00
2	1	708.9	1.68	1.92
	2	736.7	2.16	
3	1	708.9	1.60	1.91
	2	729.7	1.89	
	3	740.2	2.24	
4	1	708.9	1.56	1.90
	2	724.2	1.77	
	3	734.6	2.01	
	4	743.6	2.28	

Referring back again to Example 9-6 and keeping the condenser pressure at 2.00 in. Hg, we found the heat transfer surface reduced to 92% of the original value calculated for the single-pressure condenser. In case that the condenser pressure and condenser surface are to be maintained constant, the water flow through the condenser is expected to be lower for dual-pressure arrangement. An approximate reduction of 7% is given for Example 9-6. Dual-pressure condensers have been used in a number of recent plant installations.

9.7 PARAMETERS AFFECTING THE CONDENSER SELECTION

In previous sections, we identified the design parameters affecting the condenser. These include

    Condenser heat load
    Condenser pressure
    Cold water temperature
    Water velocity inside condenser tubes
    Condenser tube diameter
    Condenser tube length
    Condenser tube cleanliness factor and tube materials

The first design parameter is the condenser heat load. In the closed-water-circuit system, the heat load is approximately equal to the cooling tower heat load. As indicated in Chapter 8, the design value must be the maximum heat load that would occur at the time the steam turbine operates at a maximum capacity. This information is generally available from the plant heat balance analysis.

The design condenser pressure is another important parameter. As the design condenser pressure is increased, the condenser surface area will decrease. Also, the amount of cooling water required will decrease with an increase in condenser pressure. However, it does not mean that a high condenser pressure is always the best choice.

The selection of cold water temperature has a significant influence on the condenser size and cost. For a given design condenser pressure, increasing the cold water temperature means an increase in the condenser surface and thus an increase in the initial investment. However, it does not hold that a low cold water temperature should always be used. As indicated in Chapter 8, this cold water temperature also has a significant impact on the size and cost of the cooling tower.

Three tube diameters normally considered for condenser use are 3/4, 7/8, and 1.0 in. The choice among them is not easy. The general tendency is to utilize the large diameter as condensers become larger. Figure 9-9 shows a comparison of condensers, using several different tube diameters designed for the same conditions. As the tube diameter increases, there will be less condenser surface required, but more cooling water needed. With increasing tube diameters, the cooling water pumping power would also increase. Evidently certain compromises must be made in the system design.

Condenser tube length is another parameter needed for preparation of equipment specification. In general, the selection of a longer tube will result in less cooling water required and a decrease in the pumping power. Also, the condenser surface will increase as the tube length increases. In selecting the tube length, consideration must be given to space required to pull tubes, and to possibly a deeper basement requirement, since the longer condenser may take an additional neckpiece diversion in order to obtain proper steam distribution. This latter is sometimes a problem overlooked in foundation design, but it is obvious that steam must be distributed to the entire tube length in order to get good performance. Generally, longer lengths are recommended provided that sufficient space is available.

Cooling water velocity is certainly one of the prime considerations in condenser design. The normal velocity range is from 6 to 10 fps. Heat transfer is improved with the higher velocity, but the friction loss also increases. Generally, higher velocities result in large quantities of cooling water, and less condenser surface while other design conditions remain unchanged.

Figure 9-10 shows the influences of cooling water velocity. As the velocity increases from 7 to 9 fps, the condenser surface will be reduced approximately by 18% and at the same time the cooling water required increases by 9%. In addition to considerations of heat transfer performance and pump power, selecting a design velocity value is generally contingent upon many other factors including the possible effect on condenser tube life, due to impossible increased erosion or corrosion with high velocities.

The most commonly used cleanliness factor for condenser design is around 85%. This means that only 85% of condenser tube surface is assumed to be effective. Thus, the cleanliness factor amounts to a sort of insurance against the necessity for cleaning the tubes too frequently. Selecting different design values for the cleanliness factor will affect the size and cost of a condenser. For instance, the difference between using a cleanliness factor of 70% rather than 100%, will result in an increased cost of approximately 25%. Therefore, this design parameter should also deserve attention.

There are many different tube materials for selection. In recent years the materials indicated below have been quite popular.

- Admiralty metal
- Arsenical copper
- Aluminum
- Aluminim brass
- Aluminum bronze
- 90-10 Cu-Ni
- 70-30 Cu-Ni
- Cold-rolled low carbon steel
- Stainless steel type 410/430
- Stainless steel type 304/316
- Stainless steel type 329
- Titanium

Probably the most common material used for condenser tubes is Admiralty metal with a 70% copper, 29% zinc, and 1% tin. This alloy has the highest heat transfer and has proved best suited for the largest number of installations. However, there is no formula for selection of tube materials. Each installation should be evaluated on an individual basis. Attention must be paid to the initial investment as well as operating costs.

As indicated in this section, some of condenser design parameters are also those of cooling tower. The selection of these parameters affects not only the condenser design, but also the tower design. In most cases the condenser size is inversely related to the cooling tower size if other conditions remain unchanged. Obviously, the selection of condenser parameters cannot be made unless an analysis is made on the overall system, which may consist of condenser, cooling tower, low-pressure turbine and other auxiliary equipment. Topics on system simulation and optimization are presented in the next two chapters.

SELECTED REFERENCES

1. D. Q. Kern, *Process Heat Transfer*, McGraw-Hill, 1950.
2. A. P. Fraas, and M. N. Ozisik, *Heat Exchanger Design*, Wiley, 1965.
3. P. J. Marto and R. H. Nunn (editors), *Power Condenser Heat Transfer*

Technology, McGraw-Hill, 1981.

4. W. E. Palmer and E. H. Miller, "Why Multipressure Condenser Turbine Operation?" *Proceedings of the American Power Conference*, Chicago, 1965.

5. Heat Exchange Institute, *Standards for Steam Surface Condensers*, Sixth Edition, New York, 1970.

6. N. H. Afgan and E. V. Schlunder (editors), *Heat Exchangers: Design and Theory Sourcebook*, McGraw-Hill, 1974.

7. P. J. Pottor, *Power Plant Theory and Design*, Second Edition, Ronald Press, 1959.

8. K. W. Li., "Computer-Aided Design of Condenser," *Computer in Engineering 1984*, ASME publication, 1984.

PROBLEMS

9-1. Calculate the condenser surface area for a 150,000 kW tandem compound, double-flow reheat turbine-generator using 1800 psia, 1000 F steam, with 735,000 lb/hr flow to the condenser and an assumed 950 Btu heat rejection per pound. The following design parameters should be used:

Cold Water Temperature	80 F
Condenser pressure	2.00 in. Hg abs.
Number of pass	single
Tube diameter	7/8 in.
Tube length	30 ft
Cold water velocity	7 fps
Tube cleanliness	1.00
Tube gauge and materials	18 BWG Admiralty

9-2. Repeat Problem 9-1 with the same design parameters except that the water velocity increased to 9 fps.

9-3. Repeat Problem 9-1 with the same design conditions except that the tube diameter changed to 1 in. OD.

9-4. Repeat Problem 9-1 with the same design conditions except that the tube length increased to 40 ft.

9-5. Calculate the hydraulic horsepower for the condenser described in Problem 9-1. The total friction except that of the condenser is equal to 100 ft of water.

9-6. Prepare and draw the performance curves for the condenser described in Problem 9-1. The cold water temperatures used are 70 F, 80 F, and 90 F.

9-7. Calculate the condenser pressure for the condenser described in Problem 9-1 at the time when the tube cleanliness is no better than 75%.

9-8. A condenser is designed for a maximum heat load 1000×10^6 Btu/hr at the condenser pressure 2.0 in. Hg abs. The parameters used for this design are

Cold water temperature	70 F
Number of passes	2
Water velocity	7 fps
Condenser temperature rise	20 F
Tube outside diameter	1 in. OD
Tube gauge and material	22 BWG stainless steel Type 304
Tube cleanliness factor	0.85

Calculate the effective tube length, cooling water flow rate, and total heat transfer surface area.

9-9. Prepare and draw the performance curves for the condenser described in Problem 9-8. The cold water temperatures used are 60 F, 70 F, and 80 F.

9-10. A condenser is designed for a maximum heat load 1500×10^6 Btu/hr at the condenser pressure of 2.0 in. Hg abs. The water inlet temperature is 66.7 F. The condenser Admiralty tube (60 feet long) has a 1 in. OD-18 BWG and a cleanliness factor of 85%. The water velocity in the tube is 6.9 fps. Calculate the condenser heat transfer surface area and pump horsepower.

Assume that the condenser has a single pass and the total external hydraulic loss is 80 ft.

9-11. The condenser designed in Problem 9-10 is changed into a dual-pressure condenser by adding a partition on the steam side. Other conditions are identical to those in Problem 9-10. Estimate the reduction in condenser pressure.

9-12. The condenser designed in Problem 9-10 is changed into a dual-pressure condenser by adding a partition inside the condenser. To maintain the average condenser pressure at 2.0 in. Hg abs., the water flow must be reduced accordingly if other conditions remain unchanged. Estimate the reduction in cooling water requirements.

9-13. The condenser designed in Problem 9-10 is changed to a dual-pressure by adding a partition inside the condenser. If the average condenser pressure is desired to be maintained at 2.0 in. Hg abs. the tube length must be reduced accordingly (other conditions are unchanged too). Estimate the reduction in the condenser heat transfer surface area.

9-14. A dual-pressure condenser is designed for the maximum heat load 2040 MBtu/hr. The design condenser pressure is 3.70/4.30 in. Hg abs. Estimate the heat transfer surface area, the number of tubes and the cooling water flow rate in gallons per minute for the conditions listed below:

Cooling water inlet temperature	95 F
Temperature rise	20 F
Cooling water velocity	7.2 fps
Tube diameter	1 in. OD
Tube gauge and materials	18 BWG Admiralty
Tube cleanliness	0.85

CHAPTER TEN

Simulation

Most problems in the power plant system design and operation can be reduced in the final analysis to the problems of optimization. Optimization is a collective process of finding the set of conditions required to achieve the best result from a given situation. The methods of optimization must be brought to bear on every system design. The subject, seemingly simple, is in fact a complex one and can best be mastered by reviewing the components of an optimization study.

One major component of an optimization study is system simulation. Before anything is done, the system must be clearly defined for the problem under consideration. The system generally consists of one or more components related to each other to perform one particular task. Figure 10-1 shows a typical system structure and the interrelationship of components in the system. The system is usually characterized by receiving certain inputs and producing certain outputs.

System simulation is the process in which the system performance characteristics are determined under various operating conditions. Most systems are designed for maximum load, but operate most of the time at loads less than the design value. Because of this fact, the designers must employ system simulation and carry out an economic and optimization analysis throughout the operating range. System simulation is also employed for improving an existing system. For example, the power plant condenser will start to deteriorate as the condenser tubes become fouled with various deposits. The question of when the tube should be cleaned will naturally arise. Obviously, it is not economical, if not impossible, to simulate this situation in the real system and to make detailed measurements. However, a simple computer simulation will provide the needed information for the maintenance decision.

This chapter will first discuss the component simulation in general. Various techniques are presented to formulate the component performance equations. Next discussed is the system simulation. In this aspect the information-flow diagram and various methods of solving simultaneous nonlinear algebraic equations are presented in detail.

10.1 COMPONENT SIMULATION

Referring back to Fig. 10-1, the system generally consists of one or more components interacting with each other to perform a certain task. Before the system simulation can be executed, the model for each component of the system must be available. The model will define the outputs of that component for a given set of inputs. The model is either analytical or black-box in nature. An analytical model is one which is expressed in analytical form, whereas in a black-box model, the output

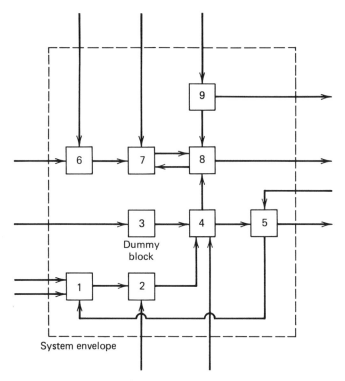

Figure 10-1. A typical system structure.

of the model to a particular input is determined by a suitable experiment. The two models are quite interchangeable.

In the analytical approach the physical laws (such as principles of conservation) are used to develop an analytical equation (or a set of equations) that will define uniquely the outputs for a specified set of inputs. In order that these equations are suitable for use in conjunction with the analytical methods of optimization, these equations must have certain features such as differentiability. If not, simplification will be necessary. This simplification can result from valid physical or mathematical approximation. In general, the analytical approach is used for the case where the component under consideration does not exist or experimentation on an existing equipment is not feasible because of high cost or other practical reasons. The analytical model can be converted into a black-box scheme through a series of numerical computations. This route is usually taken when the analytical model is complex and cannot be simplified and changed to a desirable form.

In a black-box approach the component (equipment) under consideration usually exists. Experimentations on the existing equipment must be so conducted that unique outputs can be measured for a given set of inputs. These outputs may be used directly in numerical search, or they can be converted into an analytical expression by surface fitting techniques. The analytical expression thus developed may not be much different from that developed in the mechanistic approach. For optimization purposes, they are quite indistinguishable.

10.2 SURFACE FITTING TECHNIQUES

Surface fitting is defined here as the procedure of representing a set of black-box outputs (such as experimental data or numerical results from a complex analytical model) in terms of the inputs by a mathematical relationship. In most cases this mathematical expression will yield a good approximation.

The most useful form of representation is probably a generalized polynomial approximation. It is

$$y = \beta_0 + \beta_1 x_1 + \beta_2 x_2 + \cdots + \beta_n x_n$$

$$+ \beta_{1,1} x_1^2 + \beta_{1,2} x_1 x_2 + \cdots + \beta_{n,n} x_n^2 \tag{10-1}$$

$$+ \beta_{1,1,1} x_1^3 + \beta_{1,1,2} x_1^2 x_2 + \cdots + \beta_{n,n,n} x_n^3 + \cdots$$

Obviously, the accuracy of the representation depends to some extent upon the number of terms included in the polynomial and on the choice of the coefficient β. In the simulation of power plant components such as condenser, cooling tower, and steam turbine system, the polynomial representation of equipment performance is much simplified. In general, it contains no more than two or three independent variables, and each variable has a degree no higher than the third. Because of this, the remainder of the section will only discuss the surface fitting problems frequently encountered in the power plant component simulation.

10.2.1 Linear Approximation

Suppose that a set of data points is to be approximated by the first-degree polynomial

$$y = a + bX \tag{10-2}$$

The deviation of the data point from that calculated from this equation is defined as the residual at that point. Mathematically it is

$$R_i = a + bx_i - y_i \qquad i = 1, 2, \ldots, n$$

The sum of the squares of these deviations is

$$E = \sum_{i=1}^{n} R_i^2 = \sum_{i=1}^{n} (a + bx_i - y_i)^2$$

If we wish to select the a and b so as to minimize E, then we will differentiate E with respect to a and b separately and set the derivatives equal to zero. Performing the operation results in

$$na + b \sum_{i=1}^{n} x_i = \sum_{i=1}^{n} y_i \qquad (10\text{-}3)$$

$$a \sum_{i=1}^{n} x_i + b \sum_{i=1}^{n} x_i^2 = \sum_{i=1}^{n} x_i y_i \qquad (10\text{-}4)$$

Solving Eqs. (10-3) and (10-4) simultaneously gives

$$a = \frac{\sum x_i^2 \sum y_i - \sum x_i \sum x_i y_i}{n \sum x_i^2 - \left(\sum x_i \right)^2} \qquad (10\text{-}5)$$

$$b = \frac{n \sum x_i y_i - \sum x_i \sum y_i}{n \sum x_i^2 - \left(\sum x_i \right)^2} \qquad (10\text{-}6)$$

EXAMPLE 10-1. From the following data points obtain y as a linear function of x.

y_i	x_i
5.90	5
1.90	1
5.10	4
3.10	2
4.05	3

Solution: We seek an equation of the form

$$y = a + bx$$

We first calculate the quantities indicated below:

$$\sum x_i = 15.0$$

$$\sum y_i = 20.05$$

$$\sum x_i y_i = 70.15$$

and

$$\sum x_i^2 = 55.0$$

Then, we calculate the value of a and b by using Eqs. (10-5) and (10-6)

$$a = \frac{(55)(20.05) - (15)(70.15)}{(5)(55) - (15)^2} = 1.01$$

$$b = \frac{(5)(70.15) - (15)(20.05)}{(5)(55) - (15)^2} = 1.0$$

Thus the required linear equation is

$$y = 1.01 + 1.0x$$

The method of least squares just presented has a broad application. In fact, it can be used to obtain the best straight line in other coordinate systems.

EXAMPLE 10-2. The cost information of cast iron cascade cooler is given below:

Effective External Surface (ft)^2	Estimate Cost ($)
40	100
60	230
80	300
100	370
150	550
200	770

From this data express the cost estimate as a function of the external surface.

Solution: We assume that the correlation has the form $y = cx^d$ where x is the external surface area (ft^2) and y is the cost estimate (dollars). The equation can be expressed as

$$\ln y = \ln c + d \ln x$$

It is seen that the term $\ln y$ is linearly related to the term $\ln x$. The constants, $\ln c$ and d, can be determined by the previous method. We first calculate $\ln y$ and $\ln x$ as below:

$X = \ln x$	$Y = \ln y$
3.688	4.605
4.094	5.438
4.382	5.703
4.605	5.913
5.011	6.309
5.298	6.646

Then, we calculate the quantities indicated below:

$$\sum X_i = 27.078$$

$$\sum Y_i = 34.614$$

$$\sum X_i Y_i = \sum (\ln x_i)(\ln y_i) = 158.291$$

and

$$\sum x_i^2 = \sum (\ln x_i)^2 = 123.949$$

Using Eq. (10-5) and (10-6), we have

$$a = \frac{(123.949)(34.614) - (27.078)(158.291)}{(6)(123.949) - (27.078)^2} = 0.3978$$

$$b = \frac{(6)(158.291) - (27.078)(34.614)}{(6)(123.949) - (27.078)^2} = 1.1902$$

Therefore, the required equation is (i.e., $a = \ln c$, $b = d$)

$$\ln y = 0.3978 + 1.1902 \ln x$$

or

$$y = 1.488 x^{1.1902}$$

A plot of this relation and its associated data points is shown in Fig. 10-2.

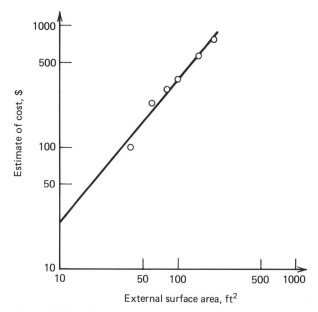

Figure 10-2. Cost relationship for Example 10-2.

10.2.2 Quadratic Approximation

Let a set of data points be approximated by the second-degree polynomial

$$y = a + bx + cx^2 \tag{10-7}$$

where the coefficients (a, b, and c) are to be determined. The procedures are similar to those used for a linear approximation. First, the residual for the observation i is defined as

$$R_i = a + bx_i + cx_i^2 - y_i \tag{10-8}$$

The sum of the squares of all deviations is

$$E = \sum_{i=1}^{n} R_i^2 = \sum_{i=1}^{n} \left(a + bx_i + cx_i^2 - y_i \right)^2$$

The quantity E can be minimized by setting the following derivatives equal to zero:

$$\frac{\partial E}{\partial a} = 0, \qquad \frac{\partial E}{\partial b} = 0, \qquad \text{and} \qquad \frac{\partial E}{\partial c} = 0 \tag{10-9}$$

Solving these three equations simultaneously would yield the values of a, b, and c. The derivation is more tedious than the linear case. Here the results are simply presented as

$$a = \frac{\left[\left(\sum x_i^2 \right)^2 - \sum x_i \sum x_i^3 \right]\left(\sum y_i x_i^2 \right) + \left(\sum x_i \sum x_i^4 - \sum x_i^2 \sum x_i^3 \right)\left(\sum y_i x_i \right) + \left[\left(\sum x_i^3 \right)^2 - \sum x_i^2 \sum x_i^4 \right]\left(\sum y_i \right)}{\left[\left(\sum x_i^2 \right)^2 - \sum x_i \sum x_i^3 \right]\left(\sum x_i^2 \right) + \left(n \sum x_i^3 - \sum x_i \sum x_i^2 \right)\left(\sum x_i^3 \right) + \left[\left(\sum x_i \right)^2 - n \sum x_i^2 \right]\left(\sum x_i^4 \right)}$$

$$\tag{10-10}$$

$$b = \frac{\left(n \sum x_i^3 - \sum x_i \sum x_i^2 \right)\left(\sum y_i x_i^2 \right) + \left[\left(\sum x_i^2 \right)^2 - n \sum x_i^4 \right]\left(\sum y_i x_i \right) + \left(\sum x_i^4 \sum x_i - \sum x_i^2 \sum x_i^3 \right)\left(\sum y_i \right)}{\left[\left(\sum x_i^2 \right)^2 - \sum x_i \sum x_i^3 \right]\left(\sum x_i^2 \right) + \left(n \sum x_i^3 - \sum x_i \sum x_i^2 \right)\left(\sum x_i^3 \right) + \left[\left(\sum x_i \right)^2 - n \sum x_i^2 \right]\left(\sum x_i^4 \right)} \tag{10-11}$$

$$c = \frac{\left[\left(\sum x_i \right)^2 - n \sum x_i^2 \right]\left(\sum y_i x_i^2 \right) + \left(n \sum x_i^3 - \sum x_i \sum x_i^2 \right)\left(\sum y_i x_i \right) + \left[\left(\sum x_i^2 \right)^2 - \sum x_i \sum x_i^3 \right]\left(\sum y_i \right)}{\left[\left(\sum x_i^2 \right)^2 - \sum x_i \sum x_i^3 \right]\left(\sum x_i^2 \right) + \left(n \sum x_i^3 - \sum x_i \sum x_i^2 \right)\left(\sum x_i^3 \right) + \left[\left(\sum x_i \right)^2 - n \sum x_i^2 \right]\left(\sum x_i^4 \right)} \tag{10-12}$$

10.2.3 Multivariable Linear Approximation

In component simulation there are many cases where the component output is a multivariable function. For example, the condenser pressure is a function of condenser heat load and inlet water temperature. As indicated in Eq. (10-1), the multivariable function can be very complex. In this section we discuss only the simplest form of this function, that is, the multivariable linear equation

$$y = \beta_0 + \beta_1 x_1 + \beta_2 x_2 + \beta_3 x_3 + \cdots + \beta_n x_n \tag{10-13}$$

To obtain symmetry with respect to the coefficient, we rewrite this equation as

$$y = \beta_0 x_0 + \beta_1 x_1 + \cdots + \beta_n x_n \tag{10-14}$$

or

$$y = \sum_{i=0}^{n} \beta_i x_i$$

where x_0 is unity by definition. The coefficients β_i defining the plane surface are again determined by the method of least squares. The value of E for this case is

$$E = \sum_{j=1}^{r} R_j^2 = \sum_{j=1}^{r} \left(y_j - \sum_{i=0}^{n} \beta_i x_{ij} \right)^2 \tag{10-15}$$

If we wish to minimize the quantity E, we must differentiate E with respect to various β_i and set the derivatives equal to zero. Mathematically, the necessary conditions for a minimum value of E are

$$\frac{\partial E}{\partial \beta_i} = 0 \qquad \text{for} \qquad i = 0, 1, 2, \ldots, n \tag{10-16}$$

In other words there will be $n + 1$ equations for $n + 1$ unknown coefficients. These equations can be derived and shown as

$$\begin{aligned}
C_{0,0}\beta_0 + C_{0,1}\beta_1 + C_{0,2}\beta_2 + \cdots + C_{0,n}\beta_n &= D_0 \\
C_{1,0}\beta_0 + C_{1,1}\beta_1 + C_{1,2}\beta_2 + \cdots + C_{1,n}\beta_n &= D_1 \\
&\cdots \cdots \cdots \cdots \cdots \cdots \cdots \\
C_{n,0}\beta_0 + C_{n,1}\beta_1 + C_{n,2}\beta_2 + \cdots + C_{n,n}\beta_n &= D_n
\end{aligned}$$

or

$$\sum_{k=0}^{n} C_{i,k}\beta_k = D_i \qquad \text{for} \qquad i = 0, 1, 2, \ldots, n \tag{10-17}$$

where

$$C_{i,k} = C_{k,i} = \sum_{j=1}^{r} x_{i,j} x_{k,j} \tag{10-18}$$

$$D_i = \sum_{j=1}^{r} x_{i,j} y_j \tag{10-19}$$

Solving these equations simultaneously would generate the coefficients β_i.

EXAMPLE 10-3. From the following data derive the equation $y(x_1, x_2)$ in the form of $y = \beta_0 + \beta_1 x_1 + \beta_2 x_2$.

Data Point	Input		Output
j	$x_{1,j}$	$x_{2,j}$	y_j
1	0	0	-25
2	0	1	-16
3	0	2	-9
4	0	3	-4
5	1	1	3
6	1	2	10
7	1	3	15
8	2	2	27

Solution: We first use Eqs. (10-18) and (10-19) to calculate the terms $C_{i,k}$ and D_i. The results are

$$C_{0,0} = \sum_{j=1}^{8} x_{0,j} x_{0,j} = 8$$

$$C_{0,1} = \sum_{j=1}^{8} x_{0,j} x_{1,j} = 5$$

$$C_{0,2} = \sum_{j=1}^{8} x_{0,j} x_{2,j} = 14$$

$$C_{1,1} = \sum_{j=1}^{8} x_{1,j}^2 = 7$$

$$C_{1,2} = \sum_{j=1}^{8} x_{1,j} x_{2,j} = 10$$

$$C_{2,2} = \sum_{j=1}^{8} x_{2,j}^2 = 32$$

$$D_0 = \sum_{j=1}^{8} x_{0,j} y_j = 1.0$$

$$D_1 = \sum_{j=1}^{8} x_{1,j} y_j = 82$$

$$D_2 = \sum_{j=1}^{8} x_{2,j} y_j = 76$$

Substituting $C_{i,k}$ and D_i into Eq. (10-17), we have

$$8\beta_0 + 5\beta_1 + 14\beta_2 = 1$$

$$5\beta_0 + 7\beta_1 + 10\beta_2 = 82$$

$$14\beta_0 + 10\beta_1 + 32\beta_2 = 76$$

Solving these three equations simultaneously would give

$$\beta_0 = -23.47 \qquad \beta_1 = 18.82 \qquad \beta_2 = 6.76$$

Therefore, the desired equation is

$$y = -23.47 + 18.82x_1 + 6.76x_2$$

Equation (10-13) is frequently applied in engineering and other fields. Several computer packages have been prepared for determination of coefficients β_i. One of these packages is by International Business Machines [8]. The program is written in Fortran and will run a regression analysis to find the best fit in the form of

$$y = \beta_0 + \sum_{i=1}^{n} \beta_i x_i \qquad \text{where} \qquad n < 40$$

It should be pointed out that the maximum number of variables, dependent and independent, is 40. To illustrate the use of this IBM package, the mechanical draft cooling tower data shown in Fig. 8-24 is used for regression analysis. The dependent variable is the cold water temperature, while the independent variables are ambient wet-bulb temperature, cooling range, and cooling water flow rate. The input information is summarized in Table 10-1. As indicated, there are only 45 observations for this regression.

The best-fit equation generated by this program is

$$CWT = 0.54078(WBT) + 0.43889(CR)$$

$$+ 0.00014(WFR) + 18.57 \qquad (10\text{-}20)$$

where

$$CWT = \text{Cold water temperature (F)}$$

$$WBT = \text{Ambient wet-bulb temperature (F)}$$

$$CR = \text{Cooling range (F)}$$

$$WFR = \text{Cooling water flow rate (gpm)}$$

Table 10-1
Input to IBM Regression Package on Mechanical Draft Cooling Tower Performance

Cold Water Temperature (F)	Wet-Bulb Temperature (F)	Cooling Range (F)	Circulatory Water Flow Rate (gpm)
81.80	60.	28.0	130500.
84.15	65.	28.0	130500.
86.75	70.	28.0	130500.
89.45	75.	28.0	130500.
92.30	80.	28.0	130500.
79.90	60.	23.5	130500.
82.25	65.	23.5	130500.
85.00	70.	23.5	130500.
87.95	75.	23.5	130500.
91.10	80.	23.5	130500.
77.30	60.	19.0	130500.
80.15	65.	19.0	130500.
83.20	70.	19.0	130500.
86.30	75.	19.0	130500.
89.60	80.	19.0	130500.
84.00	60.	28.0	145000.
86.25	65.	28.0	145000.
88.60	70.	28.0	145000.
91.25	75.	28.0	145000.
94.00	80.	28.0	145000.
81.80	60.	23.5	145000.
84.30	65.	23.5	145000.
86.80	70.	23.5	145000.
89.65	75.	23.5	145000.
92.60	80.	23.5	145000.
79.20	60.	19.0	145000.
81.80	65.	19.0	145000.
84.70	70.	19.0	145000.
87.75	75.	19.0	145000.
90.90	80.	19.0	145000.
86.85	60.	28.0	159500.
88.80	65.	28.0	159500.
91.15	70.	28.0	159500.
93.60	75.	28.0	159500.
96.20	80.	28.0	159500.
84.40	60.	23.5	159500.
86.70	65.	23.5	159500.
89.20	70.	23.5	159500.
91.75	75.	23.5	159500.
94.60	80.	23.5	159500.
81.60	60.	19.0	159500.
84.10	65.	19.0	159500.
86.80	70.	19.0	159500.
89.75	75.	19.0	159500.
92.75	80.	19.0	159500.

Another example involving a use of the IBM computer program is the regression analysis of the condenser data shown in Fig. 9-12. Table 10-2 shows the exact data inputed into the program. The best-fit equation generated in this regression analysis is

$$CP = 0.00087(HL) + 0.05642(CIRC) - 3.40 \qquad (10\text{-}21)$$

where

$$CP = \text{Condenser pressure (in. Hg abs.)}$$

$$HL = \text{Condenser heat load (MBtu/hr)}$$

$$CIRC = \text{Circulating cold water temperature (F)}$$

It must be pointed out that this kind of linear regression may not always produce satisfactory results. It is mainly because all curvilinear terms are omitted from the regression analysis. The condenser regression equation just presented has a maximum residual around 21.5%, while the tower equation around 1.3%.

Table 10-2
Input to IBM Regression Package on Condenser Performance

Condenser Pressure (in. Hg abs.)	Condenser Heat Load (MBtu / hr)	Circulating Water Inlet Temperature (F)
0.980	250.	70.
1.100	500.	70.
1.230	750.	70.
1.380	1000.	70.
1.545	1250.	70.
1.355	250.	80.
1.520	500.	80.
1.695	750.	80.
1.880	1000.	80.
2.090	1250.	80.
1.845	250.	90.
2.055	500.	90.
2.280	750.	90.
2.530	1000.	90.
2.800	1250.	90.
1.965	250.	92.
2.180	500.	92.
2.425	750.	92.
2.680	1000.	92.
2.970	1250.	92.
2.145	250.	95.
2.380	500.	95.
2.640	750.	95.
2.925	1000.	95.
3.230	1250.	95

10.2.4 Regression Computer Program For Power Plant Cooling Systems

As indicated in the last section, the multivariable linear approximation cannot be always successfully applied to the performance of power plant cooling systems. In this section measures have been taken to include some curvilinear terms [4, 5]. The functional relationships are modified and expressed in a matrix form:

$$
Y = \begin{bmatrix} X_1^0 & X_1^1 & X_1^2 & X_1^3 \end{bmatrix}
\begin{bmatrix}
\begin{bmatrix} X_2^0 & X_2^1 & X_2^2 & X_2^3 \end{bmatrix}
\begin{bmatrix}
b_{000} & b_{001} & b_{002} & b_{003} \\
b_{010} & b_{011} & b_{012} & b_{013} \\
b_{020} & b_{021} & b_{022} & b_{023} \\
b_{030} & b_{031} & b_{032} & b_{033}
\end{bmatrix}
\begin{bmatrix} X_3^0 \\ X_3^1 \\ X_3^2 \\ X_3^3 \end{bmatrix} \\[3em]
\begin{bmatrix} X_2^0 & X_2^1 & X_2^2 & X_2^3 \end{bmatrix}
\begin{bmatrix}
b_{100} & b_{101} & b_{102} & b_{103} \\
b_{110} & b_{111} & b_{112} & b_{113} \\
b_{120} & b_{121} & b_{122} & b_{123} \\
b_{130} & b_{131} & b_{132} & b_{133}
\end{bmatrix}
\begin{bmatrix} X_3^0 \\ X_3^1 \\ X_3^2 \\ X_3^3 \end{bmatrix} \\[3em]
\begin{bmatrix} X_2^0 & X_2^1 & X_2^2 & X_2^3 \end{bmatrix}
\begin{bmatrix}
b_{200} & b_{201} & b_{202} & b_{203} \\
b_{210} & b_{211} & b_{212} & b_{213} \\
b_{220} & b_{221} & b_{222} & b_{223} \\
b_{230} & b_{231} & b_{232} & b_{233}
\end{bmatrix}
\begin{bmatrix} X_3^0 \\ X_3^1 \\ X_3^2 \\ X_3^3 \end{bmatrix} \\[3em]
\begin{bmatrix} X_2^0 & X_2^1 & X_2^2 & X_2^3 \end{bmatrix}
\begin{bmatrix}
b_{300} & b_{301} & b_{302} & b_{303} \\
b_{310} & b_{311} & b_{312} & b_{313} \\
b_{320} & b_{321} & b_{322} & b_{323} \\
b_{330} & b_{331} & b_{332} & b_{333}
\end{bmatrix}
\begin{bmatrix} X_3^0 \\ X_3^1 \\ X_3^2 \\ X_3^3 \end{bmatrix}
\end{bmatrix}
$$

$$(10\text{-}22)$$

for the case of three independent variables involved;

$$
Y = \begin{bmatrix} X_1^0 & X_1^1 & X_1^2 & X_1^3 \end{bmatrix}
\begin{bmatrix}
b_{00} & b_{01} & b_{02} & b_{03} \\
b_{10} & b_{11} & b_{12} & b_{13} \\
b_{20} & b_{21} & b_{22} & b_{23} \\
b_{30} & b_{31} & b_{32} & b_{33}
\end{bmatrix}
\begin{bmatrix} X_2^0 \\ X_2^1 \\ X_2^2 \\ X_2^3 \end{bmatrix}
\tag{10-23}
$$

for the case of two independent variables; and

$$
Y = \begin{bmatrix} b_0 & b_1 & b_2 & b_3 \end{bmatrix}
\begin{bmatrix} X_1^0 \\ X_1^1 \\ X_1^2 \\ X_1^3 \end{bmatrix}
\tag{10-24}
$$

for the case of one independent variable. These equations are established on the basis of two assumptions. First, there is no need of special transformations such as logarithm and square root for representation of performance data. Second, each independent variable can be only raised up to the third power. It should be noticed that Eqs. (10-22), (10-23), and (10-24) are similar to Eq. (10-13) if all generated terms such as X_1^2, X_1^3, and $X_1 X_2$ are treated as independent variables. In other words, the regression to be considered here is still of multiple linear regression model.

Equations (10-22), (10-23), and (10-24) can be used as general functions for preparation of a generalized regression computer program for power plant cooling system. Equation (10-22) is selected for cases of three independent variables involved. Since each variable is assumed to go as high as the third power, there are 27

Table 10-3
Possible Combinations for the Function of Three Independent Variables

Arrangement	Highest Power of Independent Variable			Number of Terms in Multiple Linear Regression
	X_1	X_2	X_3	
1	1	1	1	8
2	1	1	2	12
3	1	1	3	16
4	1	2	1	12
5	1	2	2	18
6	1	2	3	24
7	1	3	1	16
8	1	3	2	24
9	1	3	3	32
10	2	1	1	12
11	2	1	2	18
12	2	1	3	24
13	2	2	1	18
14	2	2	2	27
15	2	2	2	36
16	2	3	1	24
17	2	3	2	36
18	2	3	3	48
19	3	1	1	16
20	3	1	2	24
21	3	1	3	32
22	3	2	1	24
23	3	2	2	36
24	3	2	3	48
25	3	3	1	32
26	3	3	2	48
27	3	3	3	64

different arrangements as shown in Table 10-3. Each arrangement has a different number of terms. For example, there are eight terms in the arrangement no. 1 where each independent variable has the highest power of one. For this arrangement, Eq. (10-22) becomes

$$
Y = \begin{bmatrix} 1 & X_1 \end{bmatrix}
\begin{bmatrix}
\begin{bmatrix} 1 & X_2 \end{bmatrix} \begin{bmatrix} b_{000} & b_{001} \\ b_{010} & b_{110} \end{bmatrix} \begin{bmatrix} 1 \\ X_3 \end{bmatrix} \\
\begin{bmatrix} 1 & X_2 \end{bmatrix} \begin{bmatrix} b_{100} & b_{101} \\ b_{110} & b_{111} \end{bmatrix} \begin{bmatrix} 1 \\ X_3 \end{bmatrix}
\end{bmatrix}
\tag{10-25}
$$

or

$$
\begin{aligned}
Y = b_{000} \quad & + b_{001}X_3 \\
+ b_{010}X_2 \quad & + b_{011}X_2X_3 \\
+ b_{100}X_1 \quad & + b_{101}X_1X_3 \\
+ b_{110}X_1X_2 & + b_{111}X_1X_2X_3
\end{aligned}
\tag{10-26}
$$

In arrangement no. 23, the first variable has the highest power of three, while the second and third variables have the highest power of two. Because of this power combination, there are 36 terms involved. The regression model becomes

$$
Y = \begin{bmatrix} X_1^0 & X_1^1 & X_1^2 & X_1^3 \end{bmatrix}
\begin{bmatrix}
\begin{bmatrix} X_2^0 & X_2^1 & X_2^2 \end{bmatrix} \begin{bmatrix} b_{000} & b_{001} & b_{002} \\ b_{010} & b_{011} & b_{012} \\ b_{020} & b_{021} & b_{022} \end{bmatrix} \begin{bmatrix} X_3^0 \\ X_3^1 \\ X_3^2 \end{bmatrix} \\
\begin{bmatrix} X_2^0 & X_2^1 & X_2^2 \end{bmatrix} \begin{bmatrix} b_{100} & b_{101} & b_{102} \\ b_{110} & b_{111} & b_{112} \\ b_{120} & b_{121} & b_{122} \end{bmatrix} \begin{bmatrix} X_3^0 \\ X_3^1 \\ X_3^2 \end{bmatrix} \\
\begin{bmatrix} X_2^0 & X_2^1 & X_2^2 \end{bmatrix} \begin{bmatrix} b_{200} & b_{201} & b_{202} \\ b_{210} & b_{211} & b_{212} \\ b_{220} & b_{221} & b_{222} \end{bmatrix} \begin{bmatrix} X_3^0 \\ X_3^1 \\ X_3^2 \end{bmatrix} \\
\begin{bmatrix} X_2^0 & X_2^1 & X_2^2 \end{bmatrix} \begin{bmatrix} b_{300} & b_{301} & b_{302} \\ b_{310} & b_{311} & b_{312} \\ b_{320} & b_{321} & b_{322} \end{bmatrix} \begin{bmatrix} X_3^0 \\ X_3^1 \\ X_3^2 \end{bmatrix}
\end{bmatrix}
$$

$$
\tag{10-27}
$$

or

$$
\begin{aligned}
Y = b_{000} \quad &+ b_{001} X_3 \quad &+ b_{002} X_3^2 \\
+ b_{010} X_2 \quad &+ b_{011} X_2 X_3 \quad &+ b_{012} X_2 X_3^2 \\
+ b_{020} X_2^2 \quad &+ b_{021} X_2^2 X_3 \quad &+ b_{022} X_2^2 X_3^2 \\
+ b_{100} X_1 \quad &+ b_{101} X_1 X_3 \quad &+ b_{102} X_1 X_3^2 \\
+ b_{110} X_1 X_2 \quad &+ b_{111} X_1 X_2 X_3 \quad &+ b_{112} X_1 X_2 X_3^2 \\
+ b_{120} X_1 X_2^2 \quad &+ b_{121} X_1 X_2^2 X_3 \quad &+ b_{122} X_1 X_2^2 X_3^2 \\
+ b_{200} X_1^2 \quad &+ b_{201} X_1^2 X_3 \quad &+ b_{202} X_1^2 X_3^2 \\
+ b_{210} X_1^2 X_2 \quad &+ b_{211} X_1^2 X_2 X_3 \quad &+ b_{212} X_1^2 X_2 X_3^2 \\
+ b_{220} X_1^2 X_2^2 \quad &+ b_{221} X_1^2 X_2^2 X_3 \quad &+ b_{222} X_1^2 X_2^2 X_3^2 \\
+ b_{300} X_1^3 \quad &+ b_{301} X_1^3 X_3 \quad &+ b_{302} X_1^3 X_3^2 \\
+ b_{310} X_1^3 X_2 \quad &+ b_{311} X_1^3 X_2 X_3 \quad &+ b_{312} X_1^3 X_2 X_3^2 \\
+ b_{320} X_1^3 X_2^2 \quad &+ b_{321} X_1^3 X_2^2 X_3 \quad &+ b_{322} X_1^3 X_2^2 X_3^2
\end{aligned}
\tag{10-28}
$$

In the regression computer program, a regression analysis is made for each arrangement when a sufficient number of observations or data points is available. After these regression analyses are done, the best fit is selected for representation of performance data. Again, the criteria used is the least sum of squares of deviation.

Tables 10-4 and 10-5 indicate various arrangements, respectively, for the cases involving two variables and one variable. The regression process and selection of the best arrangement are similar to those for the case involving three independent variables.

The generalized regression computer program was prepared according to the mathematical approximations indicated by Eqs. (10-22), (10-23), and (10-24). The background information about the regression programs were presented in references [4, 5].

To illustrate a use of this generalized regression computer program, we again use the mechanical-draft cooling tower data shown in Table 10-1 for regression analysis.

Table 10-4
Possible Combinations for the Function of Two Independent Variables

Arrangement	Highest Power of Independent Variable of		Number of Terms in Multiple Linear Regression
	X_1	X_2	
1	1	1	4
2	1	2	6
3	1	3	8
4	2	1	6
5	2	2	9
6	2	3	12
7	3	1	8
8	3	2	12
9	3	3	16

Table 10-5
Possible Combinations for the Function of One
Independent Variable

Arrangement	Highest Power of Independent Variable X_1	Number of Terms in Multiple Linear Regression
1	1	2
2	2	3
3	3	4

The regression equation thus obtained is

$$CWT = \begin{bmatrix} 1 & WBT & WBT^2 & WBT^3 \end{bmatrix}$$

$$\times \begin{bmatrix} [1\ CR] \begin{bmatrix} -0.10046 \times 10^2 & 0.22801 \times 10^{-3} \\ 0.85396 & 0.18617 \times 10^{-5} \end{bmatrix} \begin{bmatrix} 1 \\ WFR \end{bmatrix} \\ [1\ CR] \begin{bmatrix} 0.10957 \times 10 & -0.22425 \times 10^{-5} \\ -0.11978 \times 10^{-1} & 0.14378 \times 10^{-7} \end{bmatrix} \begin{bmatrix} 1 \\ WFR \end{bmatrix} \\ [1\ CR] \begin{bmatrix} 0 & 0 \\ 0 & 0 \end{bmatrix} \begin{bmatrix} 1 \\ WFR \end{bmatrix} \\ [1\ CR] \begin{bmatrix} 0 & 0 \\ 0 & 0 \end{bmatrix} \begin{bmatrix} 1 \\ WFR \end{bmatrix} \end{bmatrix}$$

$$(10\text{-}29)$$

or

$$CWT = -0.10046 \times 10^2 + 0.22801 \times 10^{-3}(WFR)$$

$$+ 0.85396(CR) + 0.18617 \times 10^{-5}(CR)(WFR)$$

$$+ 0.10957 \times 10(WBT) - 0.22425 \times 10^{-5}(WBT)(WFR)$$

$$- 0.11978 \times 10^{-1}(WBT)(CR) + 0.14378 \times 10^{-7}(WBT)(CR)(WFR)$$

$$(10\text{-}30)$$

The nomenclatures of this equation are identical to those used in Eq. (10-20). For this regression the maximum deviation (residual) is approximately 0.6%.

When the condenser performance data shown in Table 10-2 is used in conjunction with this generalized regression program, the performance equation generated is

$$CP =$$

$$
\begin{bmatrix} 1 & HL & HL^2 & HL^3 \end{bmatrix}
$$

$$
\times
\begin{bmatrix}
0.16302 \times 10 & -0.50095 \times 10^{-1} & 0.55796 \times 10^{-3} & 0 \\
0.32946 \times 10^{-3} & -0.10229 \times 10^{-4} & 0.16253 \times 10^{-6} & 0 \\
0.42658 \times 10^{-6} & -0.92331 \times 10^{-8} & 0.71265 \times 10^{-10} & 0 \\
0 & 0 & 0 & 0
\end{bmatrix}
\begin{bmatrix}
1 \\
CIRC \\
CIRC^2 \\
CIRC^3
\end{bmatrix}
$$

$$(10\text{-}31)$$

In an algebraic form Eq. (10-31) becomes

$$CP = 0.16302 \times 10$$

$$-0.50095 \times 10^{-1}(CIRC)$$

$$+0.55796 \times 10^{-3}(CIRC)^2$$

$$+0.32946 \times 10^{-3}(HL)$$

$$-0.10229 \times 10^{-4}(HL)(CIRC)$$

$$+0.16253 \times 10^{-6}(HL)(CIRC)^2$$

$$+0.42658 \times 10^{-6}(HL)^2$$

$$-0.92331 \times 10^{-8}(HL)^2(CIRC)$$

$$+0.71265 \times 10^{-10}(HL)^2(CIRC)^2 \qquad (10\text{-}32)$$

A comparison of Eq. (10-21) with Eq. (10-32) produced a significant statistical difference. Both equations are performance equations of the same condenser, but generated with two different approximations. When the curvilinear terms are taken into consideration, the maximum deviation is approximately 1.8% and is much less that using only linear terms.

10.3   HEAT EXCHANGER PERFORMANCE EQUATIONS (ANALYTICAL APPROACH)

The analytical approach used to develop the component performance equation is frequently selected in the engineering simulation. This approach is characterized by utilizing certain physical principles. It is the process of finding an equation or equations that represent performance characteristics of equipment. The heat exchanger is one of the items of equipment frequently encountered in power plant design. The heat exchanger simulation in a typical case is to determine the outlet temperature of working substances and the rate of heat transfer. The performance equation of a heat exchanger is best formulated by using the concept of heat exchanger effectiveness. The definition of heat exchanger effectiveness is

$$E = \frac{\text{actual heat transfer rate}}{\text{maximum possible heat transfer rate}} \qquad (10\text{-}33)$$

The actual heat transfer is either equal to the energy lost by the hot substance or the energy gained by the cold substance. The maximum possible heat transfer rate is the ideal rate that could be attained if one of the substances were to experience a maximum temperature difference in the exchanger. Mathematically, it is

$$q_{max} = C_{min}(T_{h,i} - T_{c,i}) \qquad (10\text{-}34)$$

where

$$T_{h,i} = \text{inlet temperature of hot substance}$$

$$T_{c,i} = \text{inlet temperature of cold substance}$$

$$C_{min} = \dot{m}_h C_{p,h} \quad \text{if} \quad \dot{m}_h C_{p,h} < \dot{m}_c C_{p,c}$$

or

$$C_{min} = \dot{m}_c C_{p,c} \quad \text{if} \quad \dot{m}_c C_{p,c} < \dot{m}_h C_{p,h}$$

In heat transfer literature the term $C = \dot{m}C_p$ (product of mass flow rate and its specific heat) is usually referred to as the capacity flow rate. The ratio of hot fluid capacity rate to the cold fluid capacity rate, that is, C_h/C_c, has a significant influence on the heat exchanger effectiveness. For example, the effectiveness of counterflow heat exchanger can be expressed as

$$E = \frac{1 - \exp\left[-NTU\left(1 - \dfrac{C_{min}}{C_{max}}\right)\right]}{1 - \dfrac{C_{min}}{C_{max}}\exp\left[-NTU\left(1 - \dfrac{C_{min}}{C_{max}}\right)\right]} \qquad (10\text{-}35)$$

In a heat exchanger the fluid capacity flow rate must be either C_{max} or C_{min}, depending on the relative magnitude. The NTU is the number of transfer units indicating the relative heat transfer surface area. Mathematically, the number of transfer units (NTU) is defined as

$$NTU = UA/C_{\min}$$

Kay and London have presented the effectiveness for various heat exchangers [3]. Some of their results are summarized in Table 10-6.

To formulate the performance equation for a heat exchanger, we use the symbols as shown in Fig. 10-3 and present the actual heat transfer rate as

$$q = EC_{\min}(T_{h,i} - T_{c,i}) \tag{10-36}$$

Also, the actual heat transfer rate can be expressed as

$$q = \dot{m}_h C_{p,h}(T_{h,i} - T_{h,e}) \tag{10-37}$$

Combining Eq. (10-36) with Eq. (10-37) will give

$$T_{h,e} = T_{h,i} - \frac{EC_{\min}}{\dot{m}_h C_{p,h}}(T_{h,i} - T_{c,i}) \tag{10-38}$$

Table 10-6
Heat-Exchanger Effectiveness Relations [3]

$$N = NTU = \frac{UA}{C_{\min}} \qquad C = \frac{C_{\min}}{C_{\max}}$$

Flow Geometry	Relation
Double pipe:	
Parallel flow	$\varepsilon = \dfrac{1 - \exp[-N(1 + C)]}{1 + C}$
Counterflow	$\varepsilon = \dfrac{1 - \exp[-N(1 - C)]}{1 - C\exp[-N(1 - C)]}$
Cross flow:	
Both fluids unmixed	$\varepsilon = 1 - \exp\left\{\dfrac{C}{n}[\exp(-NCn) - 1]\right\}$ where $n = N^{-0.22}$
Both fluids mixed	$\varepsilon = \left[\dfrac{1}{1 - \exp(-N)} + \dfrac{C}{1 - \exp(-NC)} - \dfrac{1}{N}\right]^{-1}$
$C_{\max}$ mixed, $C_{\min}$ unmixed	$\varepsilon = (1/C)\{1 - \exp[C(1 - e^{-N})]\}$
$C_{\max}$ unmixed, $C_{\min}$ mixed	$\varepsilon = 1 - \exp\{(1/C)[1 - \exp(-NC)]\}$
Shell and tube:	
One shell pass, 2, 4, 6 tube passes	$\varepsilon = 2\left\{1 + C + (1 + C^2)^{1/2}\dfrac{1 + \exp[-N(1 + C^2)^{1/2}]}{1 - \exp[-N(1 + C^2)^{1/2}]}\right\}^{-1}$

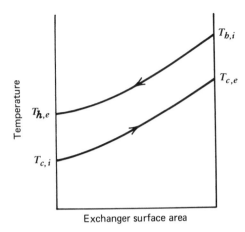

$T_{b,i}$

$T_{c,e}$

$T_{h,e}$

$T_{c,i}$

Temperature

Exchanger surface area

Figure 10-3. Temperature variation in a counterflow heat exchanger.

Similarly, the outlet temperature of cold substance is

$$T_{c,e} = T_{c,i} + \frac{EC_{min}}{\dot{m}_c C_{p,c}}(T_{h,i} - T_{c,i}) \tag{10-39}$$

Equations (10-38) and (10-39) are the performance equations for the heat exchanger. They are used to calculate the temperature of the working substances under various operating conditions. The following examples will illustrate the use of the heat exchanger effectiveness in simulation.

EXAMPLE 10-4. A drain cooler in the feedwater heating system was designed to raise the feedwater temperature from 150 to 152 F at the water flow rate 10^6 lb/hr. Under the design conditions the drain has a temperature 190 F and flow rate 10^5 lb/hr. Estimate the outlet temperature of feedwater and the rate of heat transfer when the drain temperature changes to 210 F for some reasons.

Solution: We first calculate the drain cooler effectiveness under design conditions. Since the actual heat transfer rate is

$$q_{act} = \dot{m}_c C_{p,c}(T_{c,e} - T_{c,i})$$

$$= (10^6)(1)(152 - 150)$$

$$= 2 \times 10^6 \text{ Btu/hr}$$

and the maximum heat transfer is

$$q_{max} = C_{min}(T_{h,i} - T_{c,i})$$

$$= (10)^5(1)(190 - 150)$$

$$= 4 \times 10^6 \text{ Btu/hr}$$

the cooler effectiveness is by Eq. (10-33)

$$E = \frac{q_{act}}{q_{max}} = \frac{2 \times 10^6}{4 \times 10^6}$$

$$E = 0.5$$

It is noted that the cooler effectiveness in this problem will remain constant even though the drain temperature changes from 190 to 210 F. Making use of Eq. (10-39), we have

$$T_{c,e} = 150 + \frac{(0.5)(10^5)}{(10^6)}(210 - 150)$$

$$= 153 \text{ F}$$

It is seen that this feedwater temperature is quite insensitive to the change of the drain water temperature. Under these new operating conditions, however, the heat transfer rate becomes

$$q = \dot{m}_c C_{p,c}(T_{c,e} - T_{c,i})$$

$$= (10^6)(1.0)(153 - 150)$$

$$= 3 \times 10^6 \text{ Btu/hr}$$

EXAMPLE 10-5. For the same drain cooler as described in Example 10-4 estimate the outlet temperature of feedwater when the drain water flow rate is reduced to a 75% of the design value. Other conditions are identical to those of design values.

Solution: We calculate the design value of overall conductance UA by using the equation

$$q = UA(LMTD) \tag{a}$$

where

$$LMTD = \frac{(T_{hi} - T_{ce}) - (T_{he} - T_{ci})}{\ln \dfrac{T_{hi} - T_{ce}}{T_{he} - T_{ci}}}$$

Using the following temperatures,

$$T_{hi} = 190 \text{ F}, \qquad T_{ci} = 150 \text{ F}, \qquad T_{ce} = 152 \text{ F}$$

and

$$T_{he} = T_{hi} - \frac{m_c}{m_h}(T_{ce} - T_{ci})$$

$$= 190 - \frac{10^6}{10^5}(152 - 150)$$

$$= 170 \text{ F}$$

we have the log-mean-temperature difference as

$$LMTD = 28.04 \text{ F}$$

At design conditions, the heat transfer rate between the drain and feedwater is equal to the heat gained by the feedwater, which is

$$q = \dot{m}_c C_{p,c}(T_{c,e} - T_{c,i})$$

$$= (10^6)(1.0)(152 - 150)$$

$$= 2 \times 10^6 \text{ Btu/hr}$$

Substituting these values into Eq. (a), we have the overall conductance as

$$UA = 2 \times 10^6 / 28.04$$

$$= 71326 \text{ Btu/hr} - \text{F}$$

In determining the drain cooler effectiveness under the operating conditions, the term UA is assumed unchanged as the drain water is reduced to 75% of the design value. Since

$$C_c = 0.75 \times 10^5 \text{ Btu/hr} - \text{F}$$

$$C_h = 10^6 \text{ Btu/hr} - \text{F}$$

we have

$$C_{min}/C_{max} = C_c/C_h = 0.075$$

and

$$UA/C_{min} = 71326/0.75 \times 10^5$$

$$= 0.951$$

Using Eq. (10-35) we obtain $E = 0.604$. Finally, the outlet temperature of the feedwater is by using Eq. (10-39).

$$T_{c,e} = 150 + 0.604 \frac{0.75 \times 10^5}{10^6}(190 - 150)$$

$$= 151.8 \text{ F}$$

The outlet temperature of the feedwater decreases slightly. This happens mainly because the effects of drain flow reduction are almost offset by the improvement of the heat exchanger effectiveness.

Equations (10-38) and (10-39) are applicable not only to counterflow heat exchanger, but also to the heat exchanger with other arrangements such as parallel-flow and crossflow. As long as the heat exchanger effectiveness is available, these two equations can be used to predict the performance under various operating conditions.

Equations (10-38) and (10-39) are also applicable to the condenser, which is an important element in the power plant. For the condenser, the heat exchanger effectiveness can be obtained by setting $C_{min}/C_{max} = 0$ in Eq. (10-35). That is,

$$E = 1 - \exp(-NTU) \qquad (10\text{-}40)$$

To determine the steam temperature in the condenser, one should combine Eq. (10-40) with Eq. (10-39); then the resultant equation is expressed as

$$T_{h,i} = T_{c,i} + \frac{T_{c,e} - T_{c,i}}{1 - \exp(-NTU)} \qquad (10\text{-}41)$$

The $T_{h,i}$ is the steam temperature, which will remain unchanged during a condensation process. Since

$$q = \dot{m}_c C_{p,c}(T_{c,e} - T_{c,i})$$

and

$$NTU = UA/\dot{m}_c C_{p,c}$$

we can also change Eq. (10-41) to the following form

$$T_{h,i} = T_{c,i} + Kq \tag{10-42}$$

where

$$K = \frac{1}{\dot{m}_c C_{p,c}} \left[1 - \exp\left(\frac{-UA}{\dot{m}_c C_{p,c}} \right) \right]^{-1} \tag{10-43}$$

Equation (10-42) derived here by using the concept of heat exchanger effectiveness is identical to Eq. (9-32).

When the ratio $C_{min}/C_{max} = 1.0$, the effectiveness of counterflow heat exchanger can be also modified and expressed as

$$E = (NTU)/(1 + NTU)$$

This situation frequently arises in the regenerator of gas turbine system where the air specific heat is approximately equal to the gas specific heat of combustion products.

10.4 COOLING TOWER PERFORMANCE EQUATIONS (ANALYTICAL APPROACH)

A cooling tower is a heat exchanger. Like other heat exchangers, cooling tower simulation in a typical situation is to calculate the cold water temperature. However, the tower performance equation developed by an analytical approach will be quite different from the equations developed in the last section. It will be an implicit equation of which the solution can be obtained only by an iterative method. In the following is a brief review of counterflow tower performance equation by an analytical method.

First, we start with the fill characteristic equation in the form

$$\frac{KaV}{L} = \alpha \left(\frac{L}{G} \right)^{\beta} \tag{10-44}$$

where α and β are constants to be determined experimentally. Generally they are related to the tower fill and the air velocity through the fill. In the analytical method this equation will be treated as a function of available tower characteristics in terms of the flow ratio. Once the term L/G is specified, the tower characteristics can be calculated according to Eq. (10-44).

Second, we use the Merkel's equation to approximate the counterflow tower operation. The Merkel's equation is generally in the form of

$$\frac{KaV}{L} = \int \frac{dt_w}{H_w - H_a} \tag{10-45}$$

Equation (10-45) was derived in Chapter 8, using the principles of heat and mass transfer. Also indicated there is the functional relationship among these five variables: required tower characteristics KaV/L, cooling range CR, tower approach AP, and fluid flow ratio L/G, and ambient wet-bulb temperature WBT. According to Eq. (10-45), the required tower characteristic KaV/L is an explicit function of other four variables. For simulation purposes, however, the required tower characteristic is always known and the tower approach AP is to be calculated. Therefore, the tower simulation will simply be treated in Eq. (10-45) as an implicit function of tower approach of which the solution can be obtained by an iterative method.

Counterflow cooling tower performance is represented by Eqs. (10-44) and (10-45). These equations can be applied directly in the optimization process. As an alternative, they can be used to generate performance curves through a series of numerical computations. These performance curves are similar to those in Fig. 8-24. These curves are then used directly in the optimization scheme (i.e., numerical search), or they can be converted back into a mathematical expression by surface fitting techniques. It must be pointed out that the mathematical expression thus developed will be of an explicit form and therefore much more convenient in application.

10.5 METHODS OF SOLVING SIMULTANEOUS LINEAR ALGEBRAIC EQUATIONS

Solving simultaneous linear equations is frequently encountered in the field of simulation. As indicated in the previous section, the surface fitting technique used to develop the component performance equation often requires solving a set of algebraic equations. Similar problems will also appear in the system simulation. A large number of methods have been proposed to perform this task and various computer programs are now available for applications [9].

Two methods are presented in this section. They are Gauss's elimination method and Gauss-Seidel iteration method. Both methods are well adapted to use on the computer.

10.5.1 Gauss's Elimination Method

A system of linear equations is generally in the form

$$\begin{aligned}
A_{11}X_1 + A_{12}X_2 + \cdots + A_{1n}X_n &= C_1 \\
A_{21}X_1 + A_{22}X_2 + \cdots + A_{2n}X_n &= C_2 \\
\cdots\cdots\cdots\cdots\cdots\cdots\cdots\cdots\cdots\cdots\cdots\cdots \\
A_{n1}X_1 + A_{n2}X_2 + \cdots + A_{nn}X_n &= C_n
\end{aligned} \tag{10-46}$$

For convenience of presentation, we can express this linear system in the tabular form as below:

$$
\begin{array}{ccccccc}
X_1 & X_2 & X_3 & \cdots & X_n & C_i \\
A_{11} & A_{12} & A_{13} & \cdots & A_{1n} & C_1 \\
A_{21} & A_{22} & A_{23} & \cdots & A_{2n} & C_2 \\
\cdots\cdots\cdots\cdots\cdots\cdots\cdots \\
A_{n1} & A_{n2} & A_{n3} & \cdots & A_{nn} & C_n
\end{array}
\tag{10-47}
$$

Gauss's elimination method first reduces the system of the above table to a triangular form by eliminating one unknown at a time. That is, the unknown X_1 is eliminated from all equations except the first one. This can be done by dividing the first equation by A_{11} and subtracting this equation multiplied by A_{i1} ($i = 2, 3, \ldots, n$) from the remaining $n - 1$ equations. The unknown X_2 is then eliminated from $n - 2$ of the $n - 1$ equations not containing X_1. This elimination process is continued until only X_n appears in the last equation. At that time the tabular form would become a triangular form as below:

$$
\begin{array}{ccccccc}
X_1 & X_2 & X_3 & \cdots & X_n & C_i \\
1 & t_{12} & t_{13} & \cdots & t_{1n} & K_1 \\
 & 1 & t_{23} & \cdots & t_{2n} & K_2 \\
\cdots\cdots\cdots\cdots\cdots\cdots \\
 & & & & 1 & K_n
\end{array}
\tag{10-48}
$$

The second step of Gauss's elimination method is to use backward substitution to solve all unknown X_i ($i = 1, 2, 3, \ldots, n$). That is, X_n is obtained from the nth equation in the triangular form. Substituting X_n into the $(n - 1)$th equation will give X_{n-1}. This process is continued until all unknowns are determined. The following example illustrates this method solving a set of linear algebraic equations.

EXAMPLE 10-6.   Solve the following linear system of equations.

$$X_1 + X_2 + X_3 + X_4 = 1$$

$$X_1 + 2X_2 + X_3 + 2X_4 = 2$$

$$X_1 + 2X_2 + 3X_3 + 6X_4 = 5$$

$$X_1 + X_2 + 3X_3 + 3X_4 = 2$$

Solution: For convenience, the linear system is expressed in the tabular form.

Eqs.	X_1	X_2	X_3	X_4	C
a	1	1	1	1	1
b	1	2	1	2	2
c	1	2	3	6	5
d	1	1	3	3	2

To reduce the above form into a triangular form, we first eliminate X_1 from Eqs. b, c, and d by subtracting Eq. a from these equations. The immediate results are

Eqs.	X_1	X_2	X_3	X_4	C
a	1	1	1	1	1
b		1	0	1	1
c		1	2	5	4
d		0	2	2	1

Next, we eliminate X_2 from Eq. c. This is done by subtracting Eq. b from Eq. c. The linear system at this moment would become

Eqs.	X_1	X_2	X_3	X_4	C
a	1	1	1	1	1
b		1	0	1	1
c			2	4	3
d			2	2	1

Finally, we eliminate X_3 from Eq. d by subtracting Eq. c from Eq. d. With this step completed the triangular form (matrix) is obtained and expressed as follows:

Eqs.	X_1	X_2	X_3	X_4	C
a	1	1	1	1	1
b		1	0	1	1
c			2	4	3
d				−2	−2

Using the backward substitution we get $X_4 = 1$ from Eq. d. Substituting X_4 into Eq. c gives $X_3 = -0.5$. Continuing this backward substitution, we have $X_2 = 0$ and $X_1 = 0.5$.

10.5.2 Gauss-Seidel Iteration Method

A set of algebraic equations such as those given in Eq. (10-46) may be solved by the method of successive approximation. Of many techniques, the Gauss-Seidel iterative method presents the great advantage of simplicity. To apply the Gauss method, solve each equation of the system for the unknown with the largest coefficient. The resultant expression becomes

$$\begin{aligned}
X_1 &= b_{12}X_2 + b_{13}X_3 + \cdots + b_{1n}X_n + K_1 \\
X_2 &= b_{21}X_1 + b_{23}X_3 + \cdots + b_{2n}X_n + K_2 \\
&\cdots\cdots\cdots\cdots\cdots\cdots\cdots\cdots\cdots\cdots\cdots \\
X_n &= b_{n1}X_1 + b_{n2}X_2 + \cdots + b_{n,n-1}X_{n-1} + K_n
\end{aligned} \qquad (10\text{-}49)$$

In the Gauss method the initial values $X_j^{(0)}$ (where $j = 1, 2, \ldots, n$) are first substituted into the right-hand side of Eq. (10-49) to determine improved values $X_j^{(1)}$. Next, the values $X_j^{(1)}$ are substituted into the right-hand side of Eq. (10-49) to obtain further improved values of $X_j^{(2)}$. This process is repeated until $X_j^{(m)}$ is close to

$X_j^{(m+1)}$ within the required accuracy. The $X_j^{(m+1)}$ (where $j = 1, 2, \ldots, n$) are the roots of the linear algebraic system.

In the Gauss-Seidel iteration method the initial value $X_j^{(0)}$ is usually set equal to zero for $j = 2, 3, \ldots, n$ so that $X_1^{(1)} = k_1$. In the second equation of Eq. (10-49) $X_j^{(0)}$ is set equal to zero for $j = 3, 4, \ldots, n$, but X_1 is taken equal to $X_1^{(1)}$; the equation is then solved for $X_2^{(1)}$. Similarly, in all other equations the latest available value of the X_j is used at each step. Evidently, the method would produce a faster convergence than the simple Gauss method.

EXAMPLE 10-7.   Solve the following system by Gauss-Seidel iteration method.

$$10X_1 + 2X_2 + X_3 = 9$$

$$2X_1 + 20X_2 - 2X_3 = -44$$

$$-2X_1 + 3X_2 + 10X_3 = 22$$

Solution: We first express this system as

$$X_1 = 0.9 - 0.2X_2 - 0.1X_3 \tag{a}$$

$$X_2 = -2.2 - 0.1X_1 + 0.1X_3 \tag{b}$$

$$X_3 = 2.2 + 0.2X_1 - 0.3X_2 \tag{c}$$

Then, assuming the initial values $X_2^{(0)} = X_3^{(3)} = 0$, we calculate $X_1^{(1)} = 0.90$ by Eq. (a). With $X_1^{(1)} = 0.90$ and $X_3^{(0)} = 0$, we have $X_2^{(1)} = -2.29$ from Eq. (b). Now, using the latest available values of the X_j ($X_1^{(1)} = 0.90$ and $X_2^{(1)} = -2.29$), we determine $X_3^{(1)} = 3.07$ from Eq. (c). Repeating this iteration process again and again, the final results will be obtained and expressed as

$$X_1 = 1.00 \qquad X_2 = -2.00 \qquad X_3 = 3.00$$

The following table indicates all intermediate results in each iteration process.

m	0	1	2	3	4
$X_1^{(m)}$	–	0.90	1.05	1.00	1.00
$X_2^{(m)}$	0	-2.29	-2.00	-2.00	-2.00
$X_3^{(m)}$	0	3.07	3.01	3.00	3.00

Both methods presented here for solving simultaneous linear algebraic equations are well known and easily adapted to use in the computer.

10.6 METHODS OF SOLVING NONLINEAR ALGEBRAIC EQUATIONS

Nonlinear algebraic equations are frequently encountered in the simulation and optimization. One of the methods for this kind of problem is called the Newton-

Raphson Method. This section will present this method and its associated procedures.

A typical nonlinear equation is the nth-degree algebraic equation

$$y(x) = a_n x^n + a_{n-1} x^{n-1} + \cdots + a_1 x + a_0 = 0 \tag{10-50}$$

For equations of degree four or less the solutions may be obtained by formula. For example, the solutions of the quadratic equation

$$y(x) = a_2 x^2 + a_1 x + a_0 = 0 \tag{10-51}$$

are given by

$$x_{1,2} = \frac{-a_1 \pm \left(a_1^2 - 4a_2 a_0\right)^{1/2}}{2a_2} \tag{10-52}$$

However, for equations of higher degree numerical methods are always desirable. The method introduced here is the Newton-Raphson method that is based on the Taylor series expansion. Let $x = x_c$ is one of the solutions for Eq. (10-50); the Taylor series expansion of $y(x)$ at the point $x = x_c$ is

$$y(x) = y(x_c) + \frac{dy(x_c)}{dx}(x - x_c) + \frac{1}{2}\frac{d^2 y(x_c)}{dx^2}(x - x_c)^2 + \cdots \tag{10-53}$$

When we omit the terms involving the second order or higher, Eq. (10-53) becomes

$$y(x) = y(x_c) + \frac{dy(x_c)}{dx}(x - x_c) \tag{10-54}$$

If we let $x = x_t$ where x_t is the trial solution of Eq. (10-50), we can further change Eq. (10-54) to

$$y(x_t) = y(x_c) + \frac{dy(x_c)}{dx}(x_t - x_c) \tag{10-55}$$

Since $y(x_c) = 0$, Eq. (10-55) becomes

$$y(x_t) = \frac{dy(x_c)}{dx}(x_t - x_c) \tag{10-56}$$

To make the above equation convenient in use, we approximate the first derivative at x_c by the first derivative at x_t. That is,

$$y(x_t) = \frac{dy(x_t)}{dx}(x_t - x_c) \tag{10-57}$$

or

$$x_c = x_t - \frac{y(x_t)}{y^1(x_t)} \tag{10-58}$$

In calculation the solution of the nonlinear equation is first assumed. The assumed value will obviously not satisfy the equation. Then, Eq. (10-58) is used to generate a new trial value. Repeat this iterative process until this equation is satisfied with a desirable accuracy.

EXAMPLE 10-8.   Calculate the roots of the nonlinear equation

$$x^3 + 2x^2 + 2.2x + 0.4 = 0$$

Solution: First we let

$$y(x) = x^3 + 2x^2 + 2.2x + 0.4 \tag{a}$$

and

$$y^1(x) = 3x^2 + 4x + 2.2 \tag{b}$$

Second, we assume an initial trial solution

$$x_{t1} = 0$$

Then it gives

$$y(x_{t1}) = 0.4$$

and

$$y^1(x_{t1}) = 2.2$$

Using Eq. (10-58), we find the second trial solution as

$$x_{t2} = 0 - \frac{0.4}{2.2}$$

$$x_{t2} = -0.182$$

The value of $x = -0.182$ is a more desirable value and should be used for the next iteration. The entire iterative process is summarized as follows:

$x_{t,i}$	$y(x_t)$	$y^1(x_t)$	$x_{t,i+1}$	
0	0.40	2.20	−0.182	
−0.182	0.06	1.57	−0.222	
−0.222	-7.7×10^{-4}	1.46	−0.222	($x_1 = -0.222$)

Thus the first root of Eq. (a) is approximately -0.222. To determine the second and third roots, we obtain the reduced equation from Eq. (a) as follows:

$$x^2 + 1.778x + 1.79 = 0$$

Using Eq. (10-52), we have

$$x_{2,3} = -0.889 \pm i$$

EXAMPLE 10-9.   Carry out the numerical calculation of Example 4-5 by using the Newton-Raphson method.

Solution: Combining Eqs. (a), (b) and (c) of Example 4-5 into one equation of one variable, we have

$$Y(L_A) = 0.4818 \times 10^{-7}L_A^4 - 0.9089 \times 10^{-4}L_A^3 + 0.6842 \times 10^{-1}L_A^2$$

$$-0.2107 \times 10^2 L_A + 9860 - 0.9592 \times 10^{-7}(L - L_A)^4$$

$$+0.7811 \times 10^{-4}(L - L_A)^3 - 0.2625 \times 10^{-1}(L - L_A)^2$$

$$+0.2189 \times 10(L - L_A) - 9003 \tag{d}$$

Note that the value of load (L) is given 1000 MW. First, we assume the initial trial value $L_A = 700$ MW. This trial value will obviously not satisfy Eq. (d). Using Eq. (10-58), we can determine the new trial value for L_A. The calculations are

$$Y(700) = -337.3$$

$$Y^1(700) = 11.3$$

and

$$L_{A,\text{new}} = L_A - \frac{Y(700)}{Y^1(700)}$$

$$= 700 - \frac{-337.3}{11.3} = 730$$

The new trial value for L_A is better than the previous and should be used in next iteration. For simplicity, the first three iterative processes are summarized below:

$L_{A,i}$	$Y(L_{A,i})$	$Y^1(L_{A,i})$	$L_{A,i+1}$
700	-337.3	11.3	730
730	-90.0	11.8	737
737	51.4	10.9	732.3

Therefore, the unit A should have an approximate load 732.3 MW and the unit B have $L_B = L - L_A = 267.7$ MW. The corresponding incremental heat rate can be obtained by substituting L_A into Eq. (a) of Example 4-5. The result is approximately 9292 Btu/kWh.

The Newton-Raphson method can be extended to solve a set of multiple nonlinear equations. Suppose that a set of three nonlinear equations shown below is to be considered

$$y_1(x_1, x_2, x_3) = 0 \qquad (10\text{-}59)$$

$$y_2(x_1, x_2, x_3) = 0 \qquad (10\text{-}60)$$

$$y_3(x_1, x_2, x_3) = 0 \qquad (10\text{-}61)$$

First, we expand the first equation y_1 in a Taylor series about the point (x_{1c}, x_{2c}, x_{3c}). If we omit all terms involving second or higher orders, this series becomes

$$y_1(x_1, x_2, x_3) = y_1(x_{1c}, x_{2c}, x_{3c})$$

$$+ \frac{\partial y_1(x_{1c}, x_{2c}, x_{3c})}{\partial x_1}(x_1 - x_{1c})$$

$$+ \frac{\partial y_1(x_{1c}, x_{2c}, x_{3c})}{\partial x_2}(x_2 - x_{2c})$$

$$+ \frac{\partial y_1(x_{1c}, x_{2c}, x_{3c})}{\partial x_3}(x_3 - x_{3c}) \qquad (10\text{-}62)$$

Following the same procedures used for the derivation of Eq. (10-57), we can modify Eq. (10-62) and express it as

$$y_1(x_{1t}, x_{2t}, x_{3t}) = \frac{\partial y_1(x_{1t}, x_{2t}, x_{3t})}{\partial x_1}(x_{1t} - x_{1c})$$

$$+ \frac{\partial y_1(x_{1t}, x_{2t}, x_{3t})}{\partial x_2}(x_{2t} - x_{2c})$$

$$+ \frac{\partial y_1(x_{1t}, x_{2t}, x_{3t})}{\partial x_3}(x_{3t} - x_{3c}) \qquad (10\text{-}63)$$

Similarly, we have, respectively, for the second and third equations

$$y_2(x_{1t}, x_{2t}, x_{3t}) = \frac{\partial y_2(x_{1t}, x_{2t}, x_{3t})}{\partial x_1}(x_{1t} - x_{1c})$$

$$+ \frac{\partial y_2(x_{1t}, x_{2t}, x_{3t})}{\partial x_2}(x_{2t} - x_{2c})$$

$$+ \frac{\partial y_2(x_{1t}, x_{2t}, x_{3t})}{\partial x_3}(x_{3t} - x_{3c}) \qquad (10\text{-}64)$$

$$y_3(x_{1t}, x_{2t}, x_{3t}) = \frac{\partial y_3(x_{1t}, x_{2t}, x_{3t})}{\partial x_1}(x_{1t} - x_{1c})$$

$$+ \frac{\partial y_3(x_{1t}, x_{2t}, x_{3t})}{\partial x_2}(x_{2t} - x_{2c})$$

$$+ \frac{\partial y_3(x_{1t}, x_{2t}, x_{3t})}{\partial x_3}(x_{3t} - x_{3c}) \qquad (10\text{-}65)$$

It is seen that there are three equations for three unknown variables: $(x_{1t} - x_{1c})$, $(x_{2t} - x_{2c})$, and $(x_{3t} - x_{3c})$. Once, these variables are available, the new trial values can be determined by

$$x_{1,\text{new}} = x_{1t} - (x_{1t} - x_{1c}) \qquad (10\text{-}66)$$

$$x_{2,\text{new}} = x_{2t} - (x_{2t} - x_{2c}) \qquad (10\text{-}67)$$

$$x_{3,\text{new}} = x_{3t} - (x_{3t} - x_{3c}) \qquad (10\text{-}68)$$

In application the solutions of a nonlinear set are first assumed. The assumed values will probably not satisfy the nonlinear system. Then, Eqs. (10-63), (10-64), and (10-65) are simultaneously solved to provide the information by which the new trial values are determined. This iterative process will be repeated again and again until the nonlinear system is reasonably satisfied. The following example will illustrate the Newton-Raphson method in solving a set of nonlinear equations.

EXAMPLE 10-10.   Calculate x_1 and x_2 that will satisfy the nonlinear system

$$500x_1^2 + x_2 - 320 = 0$$

$$3x_1 + 21 \times 10^{-6}x_2^{1.8} - 0.9 = 0$$

Solution: Let

$$y_1 = 500x_1^2 + x_2 - 320$$

and

$$y_2 = 3x_1 - 21 \times 10^{-6}x_2^{1.8} - 0.9$$

Differentiating these equations will give

$$\frac{\partial y_1}{\partial x_1} = 1000x \qquad \frac{\partial y_1}{\partial x_2} = 1.0$$

$$\frac{\partial y_2}{\partial x_1} = 3.0 \quad \text{and} \quad \frac{\partial y_2}{\partial x_2} = -37.8 \times 10^{-6}x_2^{0.8}$$

Also, let the initial trial solutions be

$$x_1 = 0 \quad \text{and} \quad x_2 = 0$$

Applying Eqs. (10-63) and (10-64) and omitting the third term of the right-hand side, we have

$$-320 = (0)(x_{1t} - x_{1c}) + (x_{2t} - x_{2c}) \tag{a}$$

$$-0.9 = 3(x_{1t} - x_{1c}) - (0)(x_{2t} - x_{2c}) \tag{b}$$

Solving Eq. (a) and Eq. (b) simultaneously gives

$$(x_{1t} - x_{1c}) = -0.30$$

$$(x_{2t} - x_{2c}) = -320$$

Using Eqs. (10-66) and (10-67), we obtain the new trial solutions as

$$x_{1,\text{new}} = x_{1t} - (x_{1t} - x_{1c})$$

$$= 0 - (-3.0) = 0.3$$

and

$$x_{2,\text{new}} = x_{2t} - (x_{2t} - x_{2c})$$

$$= 0 - (-320) = 320$$

These new values are evidently better than the previous. Using these new values and the same procedure, we have another set of algebraic equations.

$$45 = 300(x_{1t} - x_{1c}) + (x_{2t} - x_{2c}) \tag{c}$$

$$-0.68 = 3(x_{1t} - x_{1c}) - 3.816 \times 10^{-3}(x_{2t} - x_{2c}) \tag{d}$$

Solving these two equations gives

$$(x_{1t} - x_{1c}) = -0.122$$

$$(x_{2t} - x_{2c}) = 81.671$$

Therefore, the latest trial solutions are

$$x_{1,\,new} = 0.3 - (-0.122) = 0.422$$

$$x_{2,\,new} = 320 - (81.671) = 238.32$$

Repeating the above procedures three more times, we have the solutions

$$x_1 = 0.425$$

$$x_2 = 229.84$$

which practically satisfy the nonlinear system.

10.7 SYSTEM SIMULATION

As indicated at the beginning of this chapter, system simulation is the process in which the system performance characteristics are determined under various operating conditions. A general procedure used for power plant system simulation can be summarized as follows:

1. Define the system for study.

2. Prepare and examine the system structure diagram similar to that in Fig. 10-1.

3. Formulate a model for each system component.

4. Construct a model for the system as an integrated unit.

5. Examine and define the internal and external restrictions imposed upon the system.

6. Carry out the system simulation by determining the system performance under various operating conditions.

In the following is a brief discussion of these steps with some reference to a particular system. But these notions have a much wider field of application.

Figure 10-4 indicates a turbine cold end system that mainly consists of turbine system, condensers, and cooling towers. Steam exhausted from the turbine first enters the condenser and releases heat by condensation. The condensate then returns to the feedwater heating circiut that is part of the turbine cycle. In the meantime, the cooling water (sometimes referred to as the circulating water) receives heat in the condenser and then transfers heat away in cooling towers. In operation these three components will affect each other in performance. The purpose of the system simulation is to determine the performance of each component in the system environment and the performance of the system as an integrated unit.

The second step is to prepare the system structure diagram, which is sometimes referred to as the information-flow diagram. In general, the diagram is used to show the arrangement and interrelationship of the various components constituting the total system. As indicated in Fig. 10-1, the component is represented by a block, which will define the outputs of that component for a given set of inputs. In other words, the block can be treated as a component model. The interconnecting lines show the system streams that are nothing more than lines of information flow. The system streams are frequently classified into three types: input, output, and interlinking. The input streams enter the system from the outside. The output streams are those leaving the system and transmitting the simulation results from the system. Those originating and terminating within the system are so-called interlinking streams.

The preparation of a system structure diagram is an important step in a system simulation. It will indicate whether the simulation is sequential or simultaneous. The sequential simulations are those in which it is possible to start with input information and proceed through the entire calculation in sequence. This kind of simulation is simple and straightforward. However, most simulations encountered in power plant design are not of the sequential type, but of the simultaneous type. The

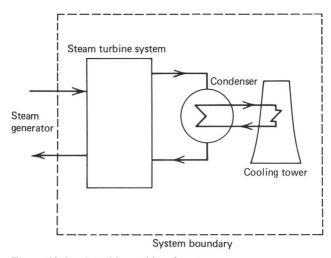

Figure 10-4. A turbine cold-end system.

simultaneous simulations are characterized by a physical coupling of various components in the system. In mathematical language the simultaneous simulation is the problem of solving a set of equations simultaneously.

Figure 10-5 shows the system structure diagram for the turbine cold-end system. Examination of this diagram reveals that there is no component where the calculation can begin and follow through in sequence with the others. Because of this, the simulation is of the simultaneous type. In fact, there are five blocks representing three major components (condenser, turbine, and cooling tower), and two couplings. As indicated later, there will be five equations for the five unknowns. Solving these equations with different inputs will generate various operation information.

Step 3 in system simulation is to formulate a model for each system component. The model may be either in the form of an analytical equation or in the form of graphical data. For an illustration, the components of the turbine cold-end system

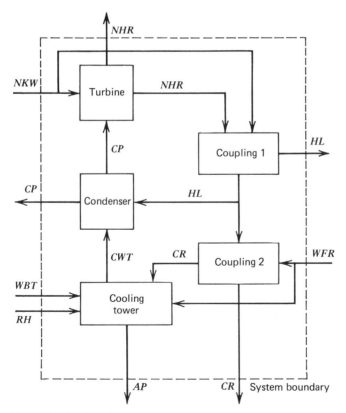

Figure 10-5. Simulation structure diagram for a turbine cold-end system.

should have the following functional relationships:

$$F_1(NKW, CP, NHR) = 0 \qquad \text{for steam turbine} \qquad (10\text{-}69)$$
$$F_2(CWT, HL, CP, WFR) = 0 \qquad \text{for condenser} \qquad (10\text{-}70)$$
$$F_3(WBT, RH, CR, WFR, CWT) = 0 \qquad \text{for cooling tower} \qquad (10\text{-}71)$$

where

NKW = turbine net output
NHR = turbine net heat rate
CP = condenser pressure
CWT = cold water temperature
HL = heat load
WBT = ambient wet-bulb temperature
RH = ambient relative humidity
CR = cooling range
WFR = cooling water flow rate

To complete a mathematical model for the system in step 4, additional coupling equations will be needed. For the cooling system under consideration the first equation is to couple the turbine with the condenser while the second is to couple the condenser with the cooling tower. Their formulations are based upon the principle of energy conservation. The functional relationships of these two couplings are

$$F_4(NHR, NKW, HL) = 0 \qquad (10\text{-}72)$$

and

$$F_5(WFR, HL, CR) = 0 \qquad (10\text{-}73)$$

The model for the turbine cold-end system is completed by combining the three component equations and two coupling equations. These five equations contain five unknowns: CP, NHR, CWT, HL, and CR. For this system simulation the required inputs are WBT, RH, WFR, and NKW.

Before the system model is simulated, attention must be paid to the internal and external restrictions imposed upon the system. Without proper care in this area, the results of system simulation may be meaningless or misleading. For instance, the steam turbine generally operates at the exhaust pressure (or condenser pressure) 5 in. Hg abs. or below. This restriction must be observed in system simulation. In other words, any system simulation involving the condenser pressure higher than 5 in. Hg abs. is not compatible with reality.

It should be pointed out that the restriction on each component of the system is also a restriction to the system. Therefore, the restrictions imposed on the system are usually more severe than those on the components.

Finally, the system model is simulated by simultaneously solving the component equations and related coupling equations. The methods of solution will vary from case to case. Two methods introduced in Section 10.5 are frequently used in thermal system simulations. Other methods can be found in the references [1, 10].

EXAMPLE 10-11. A condenser-cooling tower system operates at a constant heat load 1000 MBtu/hr. The condenser performance is expressed by the equation

$$CP = 0.00087(HL) + 0.05642(CWT) - 3.40 \qquad \text{(a)}$$

for WFR = 145,000 gpm and the cooling tower performance is by

$$CWT = 0.54078(WBT) + 0.43889(CR) + 0.00014(WFR) + 18.57 \qquad \text{(b)}$$

The units are CP in inches Hg abs., HL in MBtu/hr, CWT and WBT in F. Calculate the condenser pressures for the ambient wet-bulb temperatures 65, 75, and 80 F.

Solution: The heat received by cooling water in the condenser is equal to the heat removed in the cooling tower. In equation form the heat load in the cooling tower is

$$(HL) \times 10^6 = (\dot{m})(C_p)(CR)$$

where

$$HL = \text{heat load in the tower or condenser (MBtu/hr)}$$

$$\dot{m} = \text{water flow rate (lb/hr)}$$

$$C_p = \text{specific heat for water (Btu/lb-F)}$$

$$CR = \text{cooling range (F)}$$

When the water flow rate is expressed in gallons per minute (1 gpm = 500 lb/hr), the above equation becomes

$$CR = \frac{10^6(HL)}{(500)(1)(WFR)}$$

$$CR = 2000\frac{(HL)}{(WFR)}$$

For the heat load 1000 MBtu/hr and flow rate 145,000 gpm in this example, the cooling range must be

$$CR = 13.79 \text{ F}$$

To determine the condenser pressure, one should solve Eqs. (a) and (b). Since the system simulation is of the sequential type, one can simply start with Eq. (b) and then with Eq. (a). The results are summarized as follows:

WBT (F)	65	75	80
CWT (F)	80.07	85.48	88.19
CP (in. Hg abs.)	1.99	2.29	2.48

The above calculation is based on the assumption that the system heat load is constant regardless of the condenser pressure. In practice, this must be treated as an approximation. When the steam turbine performance data are available, this assumption should be avoided. The following example will demonstrate a simulation of the system consisting of steam turbine, condenser, and cooling tower.

EXAMPLE 10-12. The turbine "cold-end" system generally consists of turbine exhaust end, condenser, and cooling tower. For a 250 MW unit, the turbine performance data with the maximum steam throttle flow can be approximated by the two fourth degree polynomial equations as follows:

$$NHR = -45.19(CP)^4 + 420(CP)^3$$

$$-1442(CP)^2 + 2248(CP) + 6666 \qquad (a)$$

and

$$NKW = 4883(CP)^4 - 44,890(CP)^3$$

$$+152,600(CP)^2 - 231,500(CP) + 383,400 \qquad (b)$$

The condenser and mechanical-draft cooling tower have the performance equations, respectively,

$$CP = 1.6302 - 0.50095 \times 10^{-1}(CWT)$$

$$+0.55796 \times 10^{-3}(CWT)^2 + 0.32946 \times 10^{-3}(HL)$$

$$-0.10229 \times 10^{-4}(HL)(CWT) + 0.16253 \times 10^{-6}$$

$$\times (HL)(CWT)^2 + 0.42658 \times 10^{-6}(HL)^2$$

$$-0.92331 \times 10^{-8}(HL)^2(CWT) + 0.71265 \times 10^{-10}$$

$$\times (HL)^2(CWT)^2 \qquad \text{for} \qquad WFR = 145,000 \text{ GPM} \qquad (c)$$

and

$$CWT = -0.10046 \times 10^2 + 0.22801 \times 10^{-3}(WFR)$$

$$+0.85396(CR) + 0.18617 \times 10^{-5}(CR)(WFR)$$

$$+0.10957 \times 10(WBT) - 0.22425 \times 10^{-5}(WBT)(WFR)$$

$$-0.11978 \times 10^{-1}(WBT)(CR)$$

$$+0.14378 \times 10^{-7}(WBT)(CR)(WFR) \tag{d}$$

The circulating water flow is assumed 145,000 gpm. Estimate: (1) heat load, (2) condenser pressure, (3) turbine net heat rate, (4) turbine net output, (5) tower approach, and (6) tower cooling range under various ambient wet-bulb temperatures.

Solution: In addition to the given performance equations, we need two coupling equations to complete the mathematical model for this cooling system. The first equation is the coupling between the condenser and the cooling tower. The heat load for these two must be the same. As derived in Example 10-11, the coupling equation is

$$CR = 2000\frac{(HL)}{(WFR)} \tag{e}$$

The second equation is the coupling between the steam turbine and the condenser. When minor heat losses from the turbine system are neglected, the turbine waste heat rejection must take place entirely in the condenser. That is, the waste heat rejection is also the amount of heat removed from the condenser. In equation form, it is

$$HL = (NHR - 3412)(NKW)/10^6 \tag{f}$$

These four performance equations and two coupling equations constitute the mathematical model for the turbine "cold-end" system. There are six equations for six unknowns (CP, CR, CWT, NHR, NKW, and HL). Since the system is not of sequential type, one must solve these equations simultaneously by the numerical method such as those in Section 10.5. The results are summarized as

below:

	Ambient Wet-Bulb Temperatures		
	60 F	70 F	80 F
Heat load (MBtu/hr)	1153	1154	1154
Condenser pressure			
(in. Hg abs.)	1.89	2.27	2.71
Net heat rate			
(Btu/kWh)	8023	8051	8089
Net output (kw)	250,217	248,825	246,717
Tower approach (F)	17.7	13.8	9.9
Cooling range (F)	15.9	15.9	15.9

10.8 CIRCULATING WATER SYSTEM SIMULATION

Figure 10-6 shows a schematic diagram of typical circulating water system. The system consists of pumps, condenser, and cooling tower connected by the associated piping. The system is characterized by low-pressure drop and high water flow rate. For an economic and reliable operation, the system frequently has more than one pump. Two pumps may be operated in parallel to satisfy the high demand, with just one pump used for the low demands. A third pump is required as a stand-by unit. For small generating units a two-pump rather than a three pump arrangement may be adopted.

The circulating water pumps always run against a low head and handle a large water flow. Typical performance characteristics for this kind of pump are shown in Fig. 10-7. It is seen that the pump head and input vary with the pump capacity. When pumps are operated in parallel, the performance of the combined system is obtained by adding the capacities at the same head. For pumps connecting in series, the performance is obtained by adding the pump heads at the same capacity. The latter is seldom used in the circulating water system.

The pump characteristic curves are very useful in system simulation and optimization. These curves are generally provided by the equipment vendors. In system design it is sometimes desirable to express the pump performance characteristics in terms of algebraic equations by using the curve fitting technique. For instance, the circulating water pump with a capacity of 145,000 gpm at the pump head 90 ft may have the performance data:

Performance of One Pump at a Constant Speed

Capacity (10^3 gpm)	Head (ft of water)	Pump Input (hp)
174	72.9	3938
145	90.0	4020
116	108.0	4020
87	118.8	3859
58	126.9	3778

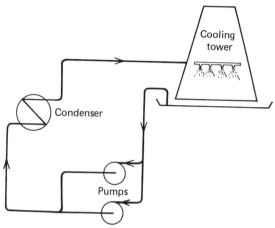

Figure 10-6. Schematic diagram for a circulating water system.

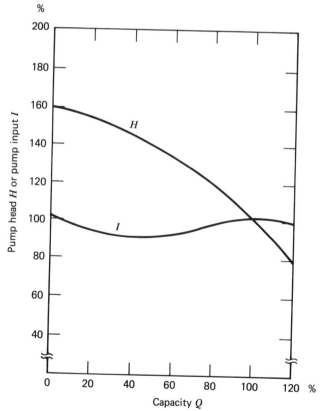

Figure 10-7. Typical performance curve of a circulatory water pump.

The head-capacity (H-C) correlation can be expressed as

$$H = 0.7423 \times 10^{-6}(C)^4 - 0.3321 \times 10^{-3}(C)^3 + 0.5074 \times 10^{-1}(C)^2$$

$$-0.3507 \times 10(C) + 216.0 \tag{10-74}$$

Similarly, the input-capacity (I-C) correlation is

$$I = 0.2545 \times 10^{-3}(C)^4 - 0.1118(C)^3 + 0.1746 \times 10^2(C)^2$$

$$-0.1145 \times 10^4(C) + 30420 \tag{10-75}$$

Since the pump hydraulic horsepower can be estimated by Eq. (9-25), combining this hydraulic power with the pump input will give the pump efficiency. Obviously, the pump efficiency will vary with the pump capacity as it is expected.

In addition to the pump performance, the circulating water system friction is needed in the system design. The system head is the sum of all frictions including pipe friction and static head. The pipe friction loss in a pumping system is a function of pipe size, length, number, and type of fittings. For a turbulent flow this friction loss varies roughly as the square of the water flow in the system. For laminar flow the friction loss varies linearly with the water flow rate. The static head is due to different elevations in the system and will remain constant regardless of water flow rate.

Figure 10-8 shows a typical water system head curve. The system head loss equals the static head at the zero flow rate. As the water flow increases, the head loss will increase. Superimposing the pump H-C curve on the system head curve gives the point at which a particular pump will operate in the system. If the resistance of a given piping system is changed by partially closing a valve or making some design modification, the system head curve will change accordingly. In Fig. 10-8, partially closing a valve in the discharge line produces the artificial system head curve, shifting the operation point to a higher head, but a lower capacity. Opening a valve wider will have an opposite effect. In simulation the system head loss is preferably expressed in a mathematical form such as

$$\text{Head loss} = a + bQ + cQ^2 \tag{10-76}$$

where

$$Q = \text{water flow rate}$$

The constants a, b, c in Eq. (10-76) are to be determined by the conditions of circulating water system. In application, the equation for the system head loss is frequently established by using an analytical approach. For instance, a pump delivers water from a lower reservoir to a higher reservoir. The elevation differs by 300 ft. In addition, the friction loss in the connecting pipe is related to the water flow

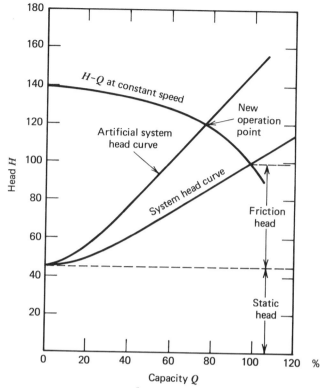

Figure 10-8. System head curve superimposed on pump head capacity curve.

rate by $15Q^{1.8}$. Then the total system head loss will become

$$\text{Head loss} = 300 + 15Q^{1.8} \tag{10-77}$$

Pump selection for a given application is a frequent task in power plant design. To select a circulating water pump properly will affect not only the initial plant cost, but also the plant operating cost. In fact, the circulating water pumps generally consume 1 to 2% of the plant net output. For a 800 MW generating unit, it means 8 to 16 MW used for the pump operation. Like other equipment selection, consideration must be given to full-load operation as well as part-load operation. Therefore, simulation of pumping system is extremely important. The following examples will illustrate this aspect of power plant system design.

EXAMPLE 10-13. A circulating water system has two pumps connected in parallel. The system has an arrangement similar to that in Fig. 10-6. The water system has the total head loss expressed as

$$H_s = 51.2 + 1.25 \times 10^{-3}(C)^2 \tag{a}$$

where H_s is in feet of water and C is in thousands of gallons per minute. Both pumps have the same performance data, which are approximated by the equation

$$H_p = 0.7654 \times 10^{-5}(C_p)^4 - 0.1699 \times 10^{-2}(C_p)^3 + 0.1257(C_p)^2$$
$$- 0.4247 \times 10(C_p) + 0.1615 \times 10^3 \tag{b}$$

After an overhaul, the second pump has the performance equation changed to

$$H_p' = 0.9053 \times 10^{-5}(C_p')^4 - 0.2076 \times 10^{-2}(C_p')^3 + 0.1608(C_p')^2$$
$$- 0.5548 \times 10(C_p') + 0.1782 \times 10^3 \tag{c}$$

Determine the flow distribution between these two pumps before and after the pump overhaul.

Solution: Before the pumps are overhauled, both pumps are expected to deliver the same amount of the water in the system. This means that the total water flow rate (C) must be equal to two times the flow rate per pump. That is,

$$C = 2C_p \tag{d}$$

In a steady-state operation, the pump head also equals the system head loss. Therefore, combining Eqs. (a), (b), and (d) gives

$$51.2 + 1.25 \times 10^{-3}(2C_p)^2 = 0.7654 \times 10^{-5}(C_p)^4 - 0.1699 \times 10^{-2}(C_p)^3$$
$$+ 0.1257(C_p)^2 - 0.4247 \times 10(C_p) + 0.1615 \times 10^3$$

Arranging all these terms on one side of the equation, we have

$$Y = 0.7654 \times 10^{-5}(C_p)^4 - 0.1699 \times 10^{-2}(C_p)^3 + 0.1157(C_p)^2$$
$$- 0.4247 \times 10(C_p) + 0.1103 \times 10^3$$

Solving the equation by the Newton-Raphson method gives

$$C_p = 61.8 \times 10^3 \text{ gpm}$$

After the pump overhaul, the second pump characteristics are changed. Under steady-state and steady-flow conditions, the following relationships must hold.

$$C = C_p + C_p' \tag{e}$$

$$H_s = H_p = H_p' \tag{f}$$

Combining these two equations with Eqs. (a), (b), and (c), we have

$$51.2 + 1.25 \times 10^{-3}\left(C_p + C_p'\right)^2 = 0.7654 \times 10^{-5}\left(C_p\right)^4 - 0.1699 \times 10^{-2}\left(C_p\right)^3$$

$$+ 0.1257\left(C_p\right)^2 - 0.4247 \times 10\left(C_p\right)$$

$$+ 0.1615 \times 10^3$$

and

$$51.2 + 1.25 \times 10^{-3}\left(C_p + C_p'\right)^2 = 0.9053 \times 10^{-5}\left(C_p'\right)^4 - 0.2076 \times 10^{-2}\left(C_p'\right)^3$$

$$+ 0.1608\left(C_p'\right)^2 - 0.5548 \times 10\left(C_p'\right)$$

$$+ 0.1782 \times 10^3$$

Arranging all these terms on one side of the equation gives

$$Y_1 = 0.7654 \times 10^{-5}\left(C_p\right)^4 - 0.1699 \times 10^{-2}\left(C_p\right)^3 + 0.124\left(C_p\right)^2$$

$$- 1.25 \times 10^{-3}\left(C_p'\right)^2 - 2.5 \times 10^{-3}\left(C_p C_p'\right)$$

$$- 0.4247 \times 10\left(C_p\right) + 0.1103 \times 10^3$$

and

$$Y_2 = 0.9053 \times 10^{-5}\left(C_p'\right)^4 - 0.2076 \times 10^{-2}\left(C_p'\right)^3 + 0.1598\left(C_p'\right)^2$$

$$- 1.25 \times 10^{-3}\left(C_p\right)^2 - 2.50 \times 10^{-3}\left(C_p C_p'\right)$$

$$- 0.5548 \times 10\left(C_p'\right) + 0.1270 \times 10^3$$

Now the Newton-Raphson method is used to find C_p and C_p' that reduce Y_1 and Y_2 to zero. The final results are

$$C_p = 70.8 \times 10^3 \text{ gpm} \quad \text{and} \quad C_p' = 75.6 \times 10^3 \text{ gpm}$$

Details of applying the Newton-Raphson method, including an illustrative example, are in Section 10.6

EXAMPLE 10-14. Two design alternatives of a circulating water system are to be considered. Both alternatives have the same design flow rate 160,000 gpm and the same head loss as

$$H_s = 57 + 4 \times 10^{-3}(C)^{1.8} \tag{a}$$

where (C) is in thousands of gallons per minute. In the first design alternative one pump is normally used while another identical pump is used as a stand-by. In the second alternative three identical pumps are provided. Two of them are normally used and the third one is for emergency. The manufacturer provides the pump performance data as

$$H_a = 0.7423 \times 10^{-6}(C)^4 - 0.3321 \times 10^{-3}(C)^3 + 0.5074 \times 10^{-1}(C)^2$$

$$- 0.3507 \times 10(C) + 0.2160 \times 10^3 \tag{b}$$

for the first alternative and

$$H_b = 0.1469 \times 10^{-4}(C)^4 - 0.3414 \times 10^{-2}(C)^3 + 0.2769(C)^2$$

$$- 0.1002 \times 10^2(C) + 0.2558 \times 10^3 \tag{c}$$

for the second alternative. Both pumps have the same efficiency as below:

gpm for Pump A	gpm for Pump B	Efficiency (%)
145,000	72,500	82
108,750	54,375	71
72,500	36,250	60

Calculate the annual energy consumption for these two design alternatives using the operating information listed below:

System Loading (gpm)	Annual Operation Hour (hr)
145,000	2,800
108,750	1,500
72,500	1,500

Solution: To determine the power consumption for the first arrangement, we first estimate the pump head for various flow rates by using (b). Then we calculate the hydraulic power by using Eq. (9-25). Combining this hydraulic power with the given pump efficiency will give the required pump power. The results are summarized as follows:

Flow Rate (gpm)	Pump Head (ft)	Hydraulic Power (hp)	Pump Power (hp)	System Head Loss (ft)
145,000	89.98	3294	4018	88.08
108,750	111.39	3059	4308	75.52
72,500	122.92	2250	3750	65.93

Also included in the above table is the system head loss for various flow rates. It is seen that they are lower than the pump head developed. This means that the pumping system will have a throttling control for part-load operation. To determine the annual energy consumption, we use

$$\text{annual consumption} = \sum_{i=1}^{n} (\text{pump power})_i (\text{annual operation hour})_i$$

where the subscript i indicates the number of loadings. Substituting the numerical values into this equation gives

$$= (4018 \times 0.746 \times 2800) + (4308 \times 0.746 \times 1500)$$

$$+ (3750 \times 0.746 \times 1500) = 17.41 \times 10^6 \text{ kWh/yr}$$

A similar procedure is used for the second alternative. It is assumed that both pumps will be operated for the first two loadings while only one pump is used for the third loading (i.e., 72,500 gpm). The results are summarized as follows:

Flow Rate (gpm)	Pump Head (ft)	Hydraulic Power (hp)	Pump Power (hp)	System Head Loss (ft)
72,500	89.66	1641.5	2001.8	88.08
54,375	109.21	1499.6	2112.1	75.52
72,500[a]	89.66	1641.5	2001.8	65.93

[a]*Note.* One pump in operation

and the annual energy consumption is

$$= (2)(2001.8 \times 0.746 \times 2800) + (2)(2112.1 \times 0.746 \times 1500)$$

$$+ (1)(2001.8 \times 0.746 \times 1500)$$

$$= 15.33 \times 10^6 \text{ kWh/yr}$$

Thus, the three-pump arrangement is superior to the two-pump arrangement in terms of annual energy consumption.

The simulation technique for a circulating water pump can be applied to other pumps such as boiler feed, condensate, and heater drain. However, it must be noted that these pumps are quite different from the circulating water pump. For example, the boiler feed pump usually operates against a high head and in a medium capacity. For condensate the pump head and capacity are always medium. The pump used with the feedwater heater is referred to as the heater drain pump. The general feature of this pump is a high pump head and low capacity.

Similar to the pump-water pipe system is the fan-duct system. In power plant design the fan-duct system is frequently encountered. These may include boiler induced-draft fan and forced-draft fan. The fan performance data are presented in the same manner as that for the pumps. The simulation techniques presented in this section are generally applicable. In fan simulation, attention must be given to the fan speed. In case the fan speed is not constant, the fan head (equivalent to pump head) will be a function of flow capacity as well as fan speed.

SELECTED REFERENCES

1. W. F. Stoecker, *Design of Thermal Systems*, McGraw-Hill, 1971.

2. C. Daniel and F. S. Wood, *Fitting Equations to Data*, Second Edition, Wiley, 1980.

3. W. M. Kays and A. L. London, *Compact Heat Exchangers*, Second Edition, McGraw-Hill, 1964.

4. K. W. Li, K. Matusuoka, and J. Cashman, "Application of Multiple Linear Regression to Power Plant Cooling System Simulation," *Electric Power System Research*, Vol. 2, No. 4, 1980.

5. K. W. Li, "Simulation of Steam Turbine, Condenser, and Cooling Tower in Power Plant Operation," *Computer in Engineering 1982*, ASME publication, 1982.

6. W. H. Li, *Engineering Analysis*, Prentice-Hall, 1960.

7. J. D. Finn, *A General Model for Multivariate Analysis*, Holt, Rinehart and Winston, 1974.

8. International Business Machines, *Scientific Subroutine Package Version III, IBM* 360A-CM-03X, 1969.

9. J. M. McCormick and M. G. Salvadori, *Numerical Methods in FORTRAN*, Prentice-Hall, 1964.

10. G. S. G. Beveridge and R. S. Schechter, *Optimization: Theory and Practice*, McGraw-Hill, 1970.

PROBLEMS

10-1.   The enthalpies of saturated air at the atmospheric pressure are given

below:

T (F)	h (Btu / lb)
90	55.93
92	58.78
94	61.77
96	64.92
98	68.23
100	71.73
102	75.42
104	79.32
106	83.42
108	87.76
110	92.34

Obtain a linear approximation in form of $h = a + bT$.

10-2. Using the same data as indicated in Problem 10-1, obtain a quadratic approximation in form of $h = a + bT + cT^2$. Also compare this approximation with the previous linear approximation. Which one would you recommend?

10-3. The heat transfer information of fluid flow across a specific tube bank is given in the following table:

Nusselt Number (Nu)	Reynolds Number (Re)
54	6,000
60	7,000
64	8,000
70	9,000
74	10,000
89	15,000
120	20,000
140	25,000

The correlation is expected to be in form of $(Nu) = a(Re)^b$. Estimate the numerical values of a and b.

10-4. A condenser manufacturer presents the following field data for one of their condenser models:

Cold Water Temperature (F)	Heat Load (MBtu / hr)	Condenser Pressure (in. Hg abs.)
80	3,000	2.15
80	4,000	2.25
80	5,000	2.40
90	3,000	2.85
90	4,000	2.95
90	5,000	3.15
95	3,000	3.25
95	4,000	3.45
95	5,000	3.60

Obtain a linear approximation for the condenser pressure in terms of heat load and inlet cold water temperature. Identify the restrictions imposed on the correlation equation.

10-5. A regenerator of gas turbine system is designed to heat the air with the exhaust gas from the turbine. The regenerator has the geometric configuration and heat transfer surface specification as such that UA is equal to 1500 Btu/hr-F. The air enters at the temperature 400 F and at the flow rate of 6000 lb/hr. The exhaust gas is available at the temperature 1000 F. Determine the temperature of the air leaving the regenerator.

Assume that there is no leakage from both sides and that the specific heat of air and gas are equal.

10-6. **(a)** Repeat Problem 10-5 for the same conditions except the exhaust gas temperature changed to 900 F.

(b) Repeat Problem 10-5 for the same conditions except the air inlet temperature changed to 450 F.

(c) For the regenerator described in Problem 10-5, develop a mathematical model by which the temperature of air leaving the exchanger can be predicted under various possible conditions.

10-7. Solve the following system by the Gauss-Seidel iteration method.

$$10X_1 - 2X_2 - X_3 - X_4 = 3$$

$$-2X_1 + 10X_2 - X_3 - X_4 = 15$$

$$-X_1 - X_2 + 10X_3 - 2X_4 = 27$$

$$-X_1 - X_2 - 2X_3 + 10X_4 = -9$$

10-8. Solve the following system by the Gauss-Seidel iteration method:

$$-6X_1 + X_2 + X_3 = -1133$$

$$X_1 - 6X_2 + X_3 = -3200$$

$$X_1 + X_2 - 6X_3 = -4200$$

10-9. Solve the following system by Gauss's elimination.

(a) $X_1 + 2X_2 + X_3 = 5$

$2X_1 + 2X_2 + X_3 = 6$

$X_1 + 2X_2 + 2X_3 = 7$

(b) $2X_1 + 2X_2 + X_3 + 2X_4 = 7$

$-X_1 + 2X_2 + X_4 = -2$

$-3X_1 + X_2 + 2X_3 + X_4 = -3$

$-X_1 + 2X_4 = 0$

10-10. Solve the following equations by the Newton-Raphson method.
(a) $4 \sin x - e^x = 0$
(b) $x^3 + 12.1x^2 + 13.1x + 22.2 = 0$
(c) $x^4 + x^3 + 0.56x^2 - 1.44x - 2.88 = 0$

10-11. Solve the following sets of equations by the Newton-Raphson method
(a) $x_2^2 + 2x_1x_2 + 13x_1^2 - 60x_1 - 200 = 0$

$x_1^2 + 2x_1x_2 + 25x_2^2 - 40x_2 - 300 = 0$

(b) $0.7654x_1^4 - 169.9x_1^3 + 12400x_1^2 - 125x_2^2 - 250x_1x_2 - 424700x_1$

$+ 11.03 \times 10^6 = 0$

$0.9053x_2^4 - 207.6x_2^3 + 15980x_2^2 - 125x_1^2 - 250x_2x_1 - 554800x_2$

$+ 12.7 \times 10^6 = 0$

10-12. The feedwater pump at 3350 rpm has its performance characteristics as below:

Flow Rate (gpm)	Developed Head (Ft)	Efficiency (%)
0	7440	0
800	7420	27
1600	7330	46
2400	7200	58
3200	7010	67
4000	6730	74
5000 (design point)	6200	77
5600	5800	77

(a) Calculate the shaft horsepower for water at the rate of 3500 gpm.
(b) Obtain the correlation equation for the pump H-C curve.
(c) Obtain the correlation equation for the pump I-C curve.

10-13. A boiler feedwater pump receives 5000 gpm of saturated water at 360 F and discharge the water at the pressure 2760 psia with an efficiency of 75%. Calculate:

(a) The temperature and enthalpy of the water at the pump discharge.

(b) The pump power in kilowatts.

10-14. A water pump has the performance equation as

$$H = 12Q^{-2} + 20Q^{-1} + 300$$

Determine the operation point when the water piping system has the total friction loss as

$$H = 200 + 1.2Q^{1.8}$$

where H is in feet of water and Q is in gallons per minute.

10-15. Determine the operation point of a fan-duct system. The fan performance equation at a constant speed is

$$H = -0.1653 \times 10^{-7}\left(\frac{Q}{10^3}\right)^4 + 0.7800 \times 10^{-5}\left(\frac{Q}{10^3}\right)^3 - 0.1367 \times 10^{-2}$$

$$\times \left(\frac{Q}{10^3}\right)^2 + 0.1065\left(\frac{Q}{10^3}\right) + 8.520$$

and the duct has the total head loss as

$$H = 2.9 \times 10^{-4}\left(\frac{Q}{10^3}\right)^2$$

where H is the total pressure (or total head loss) in inches of water and Q is the air flow in cfm.

10-16. Repeat Problem 10-15 when the fan speed is reduced by 20%. The fan laws expressed below should be used to change the fan performance equation.

$$\frac{Q_1}{Q_2} = \frac{N_1}{N_2}, \qquad \frac{H_1}{H_2} = \left(\frac{N_1}{N_2}\right)^2, \qquad \frac{P_1}{P_2} = \left(\frac{N_1}{N_2}\right)^3$$

where P and N are, respectively, the fan power and fan speed.

10-17. Calculate and plot the efficiency, head and horsepower for the data of Problem 10-12 for a speed of 2840 rpm. The relationships of fan laws are applicable for this calculation.

10-18. A boiler forced-draft fan at 875 rpm has the performance characteristics as below.

Air Flow (cfm)	Static Pressure (in. of water)	Static Efficiency (%)
250,000	14.4	75
200,000	15.1	69
150,000	15.3	61
100,000	15.4	49
50,000	15.4	30
0	15.2	0

Calculate the shaft horsepower, and obtain the correlation equations for the fan H-C curve and I-C curve. The average air temperature is 100 F.

10-19. Calculate the annual energy consumption for the fan described in Problem 10-18. The fan will operate for 3000 hours per year at 200,000 cfm,

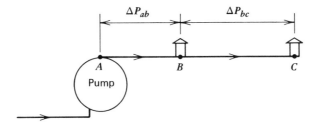

Figure 10-9.   A pump-hydrant system for Problem 10-20.

3000 hours per year at 150,000 cfm, and 1500 hours per year at 100,000 cfm.

The system head curve is a square curve from the design point of 250,000 cfm. The motor efficiency is 92%.

10-20.   A pump delivers water to two hydrants as shown in Fig. 10.9. The water flow rate through the hydrant is expressed as $Q = $ constant $(P)^{1/2}$ where P is the pressure at the hydrant location. Prepare the information flow diagram for this pump-hydrant system, indicating whether the system simulation is sequential or simultaneous.

10-21.   The effectiveness of counterflow heat exchanger is expressed by Eq. (10-35). For the special case $C_{min}/C_{max} = 1$, Eq. (10-35) becomes indeterminate. Show that the effectiveness for this case can be derived and expressed as

$$E = \frac{NTU}{1 + NTU}$$

CHAPTER ELEVEN

Optimization

11.1 INTRODUCTION

Optimization is a collective process of finding the set of conditions required to achieve the best result from a given situation. Optimization has always played an important part in the design. Basic to any optimization process are the definition of the system to be investigated and the selection of criterion to be optimized. In a power plant design the system may be very simple, such as a steam pipe between the superheater outlet and the turbine inlet or very complex such as the turbine cold-end system, which mainly consists of the turbine exhaust end, condenser, and cooling tower (or cooling pond and other similar devices). The criterion most frequently used in power plant design is the minimum owning cost, which usually means a sum of initial investment and operating cost. For instance, the optimal insulation thickness of steam pipe is obtained by considering the annual capital cost based on the investment required for the pipe insulation and the operating cost (mainly the fuel cost for this case) corresponding to the heat loss. The optimal thickness is so determined that the sum of annual expenses is a minimum. Another example is the optimal velocity in a water pipe. It is well known that as the water velocity increases, the pipe diameter will decrease for a given water flow rate and therefore so will the annual capital and maintenance cost. However, with increasing water velocity, the pumping cost including both capital and operation will also increase. Evidently, there are two competing influences in selecting the water velocity. The optimization is the process of finding the water velocity that will result in a minimum annual cost.

The criterion used for optimization is always expressed as a mathematical function. This objective function is generally in terms of the system variables. Using the pipe insulation as an example, the objective function is the annual owning cost and the system independent variable (only one variable for this case) is the thickness of the insulation. It must be pointed out that the criterion used for optimization is not always monetary. In design there may be some nonmonetary criteria such as safety and reliability. Also, there may be some other restrictions imposed on the objective function.

Even though the optimization process may vary from one case to another, the general procedure used is quite similar. The procedure is summarized below:

1. Define the system for optimization study.
2. Select a suitable criterion for evaluation.
3. Develop the system structure diagram as explained in the last chapter.

428

4. Construct a mathematical model (i.e., objective function) using the system structure diagram.

5. Identify all restrictions imposed upon the model.

6. Carry out the simulation process by expressing the objective function in terms of the system variables.

7. Carry out the optimization process by finding optimal values for all system variables.

In this chapter we will focus our attention on the construction of mathematical models and various methods of optimization.

11.2 MATHEMATICAL MODEL CONSTRUCTION

The mathematical model for optimization generally consists of an objective function and its associated restrictions. The objective function y is a function of system variables $x_1, x_2, \ldots, x_n$. The restrictions imposed on the objective may be of either the equality or inequality type. In mathematical form the objective function is

$$y = y(x_1, x_2, \ldots, x_n) \qquad (11\text{-}1)$$

and the m restrictions are

$$g_i = g_i(x_1, x_2, \ldots, x_n) = 0 \qquad (11\text{-}2)$$

or

$$g_i = g_i(x_1, x_2, \ldots, x_n) \geq 0 \qquad (11\text{-}3)$$

where

$$i = 1, 2, \ldots, m$$

Construction of mathematical model for a specific engineering case is not easy in general. It involves a good understanding of the engineering system and reasonable skills in mathematics. As is indicated later, experiences play an important part in this particular step. The following examples will illustrate general aspects of model construction.

EXAMPLE 11-1. Develop a mathematical model for optimizing the dimension of a liquid storage tank. The tank is cylindrical in shape and has a fixed volume V. The unit-area costs of the side wall and end wall are, respectively, equal to S and E. For some reasons, the radius of the storage tank should be larger than R_1.

Solution: The total cost of the storage tank can be expressed as

$$C_t = EA_e + SA_s$$

where the end wall surface A_e and the side-wall surface A_s are, respectively,

$$A_e = 2\pi R^2$$

$$A_s = 2\pi RL$$

Substituting A_e and A_s into the equation for the total cost, we have

$$C_t = 2\pi(R^2E + RLS)$$

Since the storage volume is fixed and equal to $\pi R^2 L$, the above equation becomes

$$C_t = 2\pi\left(R^2E + \frac{VS}{\pi R}\right) \tag{a}$$

Equation (a) is the objective function for optimization. It indicates that the total cost of the storage tank is a function of the tank radius. In an optimization the radius of the tank is so determined that the total cost will be a minimum.

The objective function is subject to the restriction

$$R \geq R_1$$

which is specified in the problem statement.

EXAMPLE 11-2.   Construct a mathematical model for optimizing the insulation thickness for a steam pipe. The pipe has been designed with L in length and r_1 in outside radius, carrying a superheated steam of temperature T_1 and pressure P_1 from the steam generator outlet to the turbine inlet. The annual minimum owning cost is used as the criterion of this optimization.

Solution: The annual owning cost (C_t) is the sum of annual capital cost (C_{inv}) based on the investment required for the insulation and the annual fuel cost (C_{fuel}) based on the heat loss from the steam pipe. That is,

$$C_t = C_{inv} + C_{fuel}$$

or

$$C_t = (V)(UC)(AFCR) + (\text{Loss})(UF)(n) \tag{a}$$

where

$$V = \text{Volume occupied by insulation materials}$$

$$AFCR = \text{Annual fixed charge rate}$$

$$UC = \text{Unit cost for insulation materials}$$

$$\text{Loss} = \text{heat loss from the steam pipe}$$

$$UF = \text{Levelized unit cost for boiler fuel}$$

$$n = \text{Number of operation hours per year}$$

To simplify the calculation for the heat loss, we assume that the thermal resistance in the heat transfer path is concentrated in the insulation layer and at the outer surface. Then, the heat loss per unit time is

$$\text{Loss} = \frac{T_1 - T_0}{\dfrac{\ln(r_0/r_1)}{2\pi KL} + \dfrac{1}{2\pi r_0 L h_0}}$$

or

$$\text{Loss} = \frac{2\pi KL(T_1 - T_0)}{\ln\left(\dfrac{r_0}{r_1}\right) + \dfrac{K}{h_0 r_0}} \tag{b}$$

where K is thermal conductivity of insulation materials, h_0 is thermal convective conductance on the insulation surface, and r_0 the outside radius of insulation layer. It must be pointed out that the addition of insulation to the outside of the steam pipe does not always reduce the heat loss.

Also, the V representing the amount of insulation materials is

$$V = \left(\pi r_0^2 - \pi r_1^2\right)L$$

or

$$V = \pi r_1^2 \left(\frac{r_0^2}{r_1^2} - 1\right)L \tag{c}$$

Substituting Eqs. (b) and (c) into Eq. (a), we have

$$C_t = \left(\pi r_1^2 L\right)(UC)(AFCR)\left(\left(\frac{r_0}{r_1}\right)^2 - 1\right) + \frac{(2\pi KL)(UF)(n)(T_1 - T_0)}{\ln\left(\dfrac{r_0}{r_1}\right) + \dfrac{k}{h_0 r_0}}$$

or

$$C_t = C_1\left[\left(\frac{r_0}{r_1}\right)^2 - 1\right] + C_2\left[\ln\left(\frac{r_0}{r_1}\right) + \frac{k}{h_0 r_0}\right]^{-1} \tag{d}$$

where

$$C_1 = \left(\pi r_1^2 L\right)(UC)(AFCR)$$

and

$$C_2 = (2\pi KL)(UF)(n)(T_1 - T_0)$$

Equation (d) is the objective function for this example. It indicates that the annual owning cost C_t is a function of the insulation thickness (or the outside radius of insulation layer). There is only one inequality constraint imposed on the objective function. That is

$$r_0 \geq r_1 \geq 0 \qquad \text{(e)}$$

EXAMPLE 11-3.   A three-stage air compression system is designed to receive air at one atmospheric pressure and discharge at 27 atm. Between each stage the air passes through an intercooler that brings the temperature back to the inlet condition.

Construct a mathematical model for selecting the intermediate pressures such that the total compression work is at a minimum.

Solution: Assuming that the compression process is isentropic and the air behaves as an ideal gas, we can express the work for each stage as

$$W = C_p(T_i - T_e)$$

or

$$W = C_p T_i \left(1 - \frac{T_e}{T_i} \right)$$

In terms of the pressure ratio across the compressor stage, the above equation can be further changed to

$$W = \frac{kRT_i}{k-1} \left[1 - \left(\frac{P_e}{P_i} \right)^{(k-1/k)} \right] \qquad \text{(a)}$$

where k is the specific heat ratio and R is the gas constant of air. For a three-stage compression system such as that in Fig. 11-1, the compression work is therefore

$$W = \frac{k}{k-1} RT_1 \left[1 - \left(\frac{P_2}{P_1} \right)^{(k-1/k)} \right] + \frac{k}{k-1} RT_1 \left[1 - \left(\frac{P_3}{P_2} \right)^{(k-1/k)} \right]$$

$$+ \frac{k}{k-1} RT_1 \left[1 - \left(\frac{P_4}{P_3} \right)^{(k-1/k)} \right] \qquad \text{(b)}$$

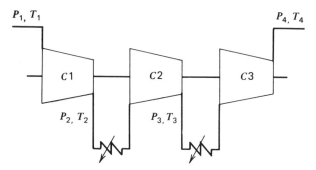

Figure 11-1. Schematic diagram for a three-stage compression system.

or

$$W = \frac{k}{k-1}RT_1\left(3 - \left(\frac{P_2}{P_1}\right)^{(k-1/k)} - \left(\frac{P_3}{P_2}\right)^{(k-1/k)} - \left(\frac{P_4}{P_3}\right)^{(k-1/k)}\right) \quad \text{(c)}$$

Eq. (c) is the objective function. In the optimization process this function will be minimized. The restriction for the optimization is

$$\left(\frac{P_2}{P_1}\right)\left(\frac{P_3}{P_2}\right)\left(\frac{P_4}{P_3}\right) = 27 \quad \text{(d)}$$

At the conclusion of this section it must be pointed out again that the construction of mathematical models is generally not easy. Setting up the problem to the point where an optimization method can take over represents more than half of the total effort in the optimization study. The above examples represent some of the simplest mathematical models. For the system involving more than one component, the mathematical model would be much more complicated. The model should not be formulated unless the system structure diagram is completely defined. These will be illustrated later.

11.3 COMBINED SYSTEM OF COOLING TOWER AND CONDENSER

In Chapters 8 and 9 the cooling tower and condenser are discussed on an individual basis. However, in practice, these items of equipment always operate as a part of a cooling system. In the cooling system design, one of the problems frequently faced by the consulting engineers is the matching of cooling tower and condenser. As shown later, the general relationship indicates the condenser size will increase as the tower size decreases. Under these conditions the engineers must optimize the cooling system by considering the tower and condenser at the same time.

In the cooling system the condenser inlet water temperature is equal to the tower exit temperature. To match the tower with the condenser, these two components must be subjected to the same conditions. That is, the tower cooling range must equal the temperature rise in the condenser, the water flow rates through the condenser and tower must be the same, and the temperature of the water leaving the condenser must equal the tower inlet water temperature. Let us consider a 225 MW fossil-fuel power plant cooling system. The system is designed with the following parameters:

Condenser pressure	3.5 in. Hg abs.
Heat load	1705.1 MBtu/hr
Ambient air dry-bulb temperature	96 F
Ambient air relative humidity	48%

Table 11-1 summarizes three different designs with the tower approaches 11 F, 13 F, and 15 F. It is seen that as the tower approach decreases, the tower size will increase, and at the same time the size of corresponding condenser will decrease. Evidently, the tower approach must be optimized in the cooling system design. In these calculations all other parameters are kept as the constants. These include the

Table 11-1
Two-Pass Single-Pressure Condenser and Tower Matching

Case Study Item	System Design Alternative		
	1	2	3
Cooling System Design Conditions			
Condenser pressure (in. Hg abs.)		3.5	
Heat load (MBtu/hr)		1705.1	
Air wet-bulb temperature (F)		79	
Air dry-bulb temperature (F)		96	
Air relative humidity (%)		48	
Cooling Tower Design			
Tower approach (F)	11	13	15
Cooling range (F)	25.1	23.5	12.8
Water flow rate (gpm)	136,000	145,000	156,000
Tower fill matrix height (ft)	52	36	26
Tower fill matrix depth (ft)	26	18	13
Tower fill matrix length (ft)	400	622	920
Tower fill matrix volume (ft^3)	547,200	400,700	313,700
Total static air pressure drop (in. of water)	0.45	0.36	0.31
Total air flow through fans (cfm)	11,397,600	12,094,500	12,965,600
Total fan static brake (hp)	1083	920	838
Condenser Design			
Number of tubes	19,509	20,850	22,370
Surface area (ft^2)	183,900	196,500	210,800
Pump work (hp)	2800	2406	2200

condenser tube length of 36 ft, tube diameter 1 in. and tube material, 18 BWG admiralty. Also the water velocity is assumed at 7 fps and the tube cleanliness factor is 0.85. In the cooling tower, the tower fill is assumed to be Type F from the Kelly's handbook. The water loading, air velocity, and fill configuration factor are chosen, respectively, to be 6500 lb/hr/ft², 525 ft/min, and 0.5.

As indicated in Chapter 9, the dual-pressure condenser is frequently considered for the system with a cooling tower. It has been pointed out that the use of a dual-pressure condenser would result in a reduction of cooling water flow rate if the condenser pressure and surface area are kept constant. The reduction of water flow rate means a reduction of power consumption for the circulating water pump. It also has some positive influences on the cooling tower design.

Let us consider the same power plant cooling system, that is, condenser pressure 3.5 in. Hg abs. and heat load 1705.1 MBtu/hr under the design ambient conditions 96 F and 48% (relative humidity) RH. Let us use the following data:

Inlet water temperature (i.e., tower approach = 13 F)	92 F
Tube length	44 ft
Tube diameter	1 in.
Tube material	18 BWG Admiralty metal
Cleanliness factor	0.85
Water velocity	7 ft/sec
Number of passes	single
Pressure drop external to the condenser	60 ft of H_2O

Using the calculational procedures presented in Chapter 9, we found the surface required for the single-pressure condenser to be 151,600 ft², the cooling water flow at 183,000 gpm, and its pumping power 3300 hp. When a dual-pressure arrangement was used, the condenser water flow was reduced to 176,000 gpm and its pump power is cut down to 3160 hp.

Different condenser design will lead to the different cooling tower design. To match the above single-pressure and dual-pressure condensers, two different cooling towers were designed using the procedures presented in Chapter 8. The results are summarized in Table 11-2. The tower matching the dual-pressure condenser is seen to have essentially the same height and depth (in terms of the tower fill) as the tower for the single-pressure condenser, but a shorter length, smaller fill volume, and fan power. Therefore, the tower designed for the dual-pressure condenser is expected to be cheaper than that for the single-pressure condenser. This will partially compensate for the additional condenser cost. In the cooling system design, cost optimization must be carried out before a decision is made about whether the dual-pressure condenser should be used.

Table 11-2
Tower Design for Single- and Dual-Pressure Condensers

Design Calculations	Tower Design for Single-Pressure Condenser	Tower Design for Dual-Pressure Condenser
Tower fill height (ft)	28.3	29.4
Tower fill depth (ft)	14.15	14.7
Tower fill length (ft)	997.3	921
Tower fill volume (ft^3)	398,880	397,140
Water flow rate (gpm)	183,197	175,643
Total static air pressure drop (in. of water)	0.32	0.33
Total air flow through fans (cfm)	15,091,827	14,499,609
Total fan static brake (hp)	1015	993
Pump work (hp)	3300	3160

11.4 METHODS OF OPTIMIZATION FOR SINGLE-VARIABLE FUNCTIONS

Methods used for optimizing the single-variable functions can be broadly classified into two groups: analytical and numerical. The analytical method requires the continuity of the objective function and its derivatives. It also requires as a condition that the first derivative will vanish at the location at which the optimal value of the function occurs. Because of these restrictive conditions numerical rather than analytical search methods are frequently used for optimizing the thermal systems including those of power plants.

The numerical search methods are generally based upon the same approach. First a reference point (or a set of reference conditions) is chosen. Then each search method selects a new set of independent variables and tests the objective function to see if the new set gives a better value of the objective function. Based upon this comparison, another set of variables is chosen, the acceptable choice being such that the value of the objective function must be improved. As indicated later in this section, the various search methods differ essentially only in their choice of a new search location.

The numerical search method is generally characterized by its inability to determine the exact optimal location. All it can do is specify the range of the independent variable in which the optimal value of the objective function will occur. The range frequently called the interval of uncertainty can always be reduced to a desired value (but not to the zero value) by increasing the number of trials in the search process.

The numerical search methods presented in this section include:

1. Direct search with regular-sized intervals.

2. Direct search with irregular-sized intervals.

The function to be optimized must be defined in a specific range. In addition, the function must be unimodal (*unimodal* means there is only one optimal value in the range of interest).

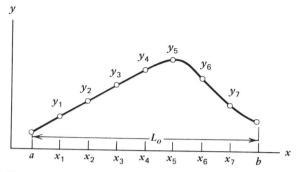

Figure 11-2. Preplanned uniform search.

11.4.1 Direct Search with Regular-Sized Intervals

Consider the objective function $y = f(x)$, which is unimodal and defined in the range $a \leq x \leq b$. The most obvious search method with regular-sized interval is that in which the original interval $(L_o = b - a)$ is divided into $n + 1$ intervals and the values of the function $y = f(x)$ will be calculated at the n intermediate points:

$$x_i = a + \frac{i(b - a)}{n + 1} \qquad \text{where} \qquad i = 1, 2, \ldots, n \qquad (11\text{-}4)$$

Comparison of these values will indicate the optimal value of the function and the location where it occurs. The optimal value in this text means either the maximum value or the minimum value of the objective function. Figure 11-2 indicates the search process with seven computational experiments.* Obviously the optimal location x is either in the interval between x_4 and x_5 or in the interval between x_5 and x_6. Therefore the interval of uncertainty for this search process must be equal to the regular interval multiplied by 2. In general form the uncertainty interval is

$$L_n = 2\left(\frac{b - a}{n + 1}\right) \qquad (11\text{-}5)$$

To measure the efficiency of a search method, the ratio of the uncertainty interval to the original interval is frequently used. For the above search method this ratio is

$$\alpha = \frac{L_n}{L_o} = \frac{2}{n + 1} \qquad (11\text{-}6)$$

The method described above is frequently referred to as a preplanned uniform search. In general, this method has a low efficiency because not all computational experiments done are needed in the search. Now let us consider a more efficient

---

*A computational experiment is designed to generate the value of the objective function under a specific set of conditions.

method. In this method, instead of doing all computational experiments in advance, we will look at the results of some experiments (two or three) before planning the next set of computational experiments. Here we shall call this method the sequential uniform search.

Again consider the objective function $y = f(x)$, which is unimodal and defined in the region $a \leq x \leq b$. In this method we use two computational experiments at a time to eliminate the region in which the optimal value of the function will never exist. First, calculate the two values of the function at positions at one-third and two-thirds of the original interval, that is

$$x_1 = a + \tfrac{1}{3}L_o \qquad \text{and} \qquad x_2 = a + \tfrac{2}{3}L_o \qquad (11\text{-}7)$$

Figure 11-3 indicates the two possibilities (the case of equal values of the function is excluded here for the time-being). Comparison of y_1 and y_2 will reduce the search interval from L_o to $\tfrac{2}{3}L_o$. The new search interval is then treated in the same manner, adding two computational experiments and making the choice of a latest search interval. After the second search, the interval of uncertainty will become $L_4 = (\tfrac{2}{3})^2 L_o$. This procedure is repeated until the interval of uncertainty is within a reasonable limit. In a more general form the interval of uncertainty is

$$L_n = L_{2m} = \left(\tfrac{2}{3}\right)^m L_o \qquad (11\text{-}8)$$

or

$$L_n = \left(\frac{2}{3}\right)^{n/2} L_o \qquad \text{where} \qquad n = 2, 4, 6, \ldots \qquad (11\text{-}9)$$

In the above equations m is the number of trials in the search process and n is the number of computational experiments. Since there are two experiments in one trial,

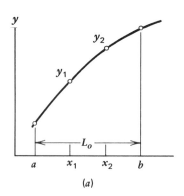 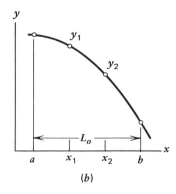

(a)   (b)

Figure 11-3.   Sequential uniform search.

the relation $n = 2m$ must exist. The α ratio of this search method is then

$$\alpha = \frac{L_n}{L_o} = \left(\frac{2}{3}\right)^m \tag{11-10}$$

or

$$\alpha = \frac{L_n}{L_o} = \left(\frac{2}{3}\right)^{n/2} \qquad \text{where} \qquad n = 2, 4, 6, \ldots \tag{11-11}$$

The efficiency of this search method will be improved slightly by using three computational experiments at one trial. The procedures are identical to those in the case of two experiments. For reference the interval of uncertainty is simply presented as

$$L_n = L_o\left(\frac{1}{2}\right)^{\frac{n-1}{2}} \qquad \text{where} \qquad n = 3, 5, 7, \ldots \tag{11-12}$$

and the α ratio is

$$\alpha = \left(\frac{1}{2}\right)^{\frac{n-1}{2}} \qquad \text{where} \qquad n = 3, 5, 7, \ldots \tag{11-13}$$

In conclusion, the preplanned method is always less efficient than the sequential methods. In spite of these characteristics, the preplanned uniform search is still very popular because of simplicity and therefore frequently used by engineers. Especially at the time when computers are available, a large number of computational experiments can be performed easily and economically.

11.4.2 Direct Search with Irregular-Sized Intervals

To eliminate a certain region in the search for an optimum, at least two computational experiments are required. In the methods presented in the previous section, the computational experiments were done at uniformly spaced locations. It may be of interest to see if any advantage can be gained by using irregular-sized intervals.

The direct search method utilizing irregular-sized intervals again have two different approaches: preplanned and sequential search. Since the preplanned approach has proven less efficient than the sequential, the preplanned is omitted from this presentation. The sequential methods presented here include the dichotomous search and Fibonacci search.

Dichotomous Search Method

Consider the objective function $y = f(x)$, which is unimodal and defined in the region $a \le x \le b$. We first carry out a pair of computational experiments around the midpoint of the search region, but separated by a distance δ. The δ is a small positive number and so chosen that the two experiments give significantly different results. More specifically, the first two locations are

$$x_1 = a + \frac{b - a}{2} - \frac{\delta}{2} \tag{11-14}$$

and

$$x_2 = a + \frac{b - a}{2} + \frac{\delta}{2} \tag{11-15}$$

Then we determine that the new search region is that between a and x_2 or that between x_1 and b, depending upon whether $y(x_1) > y(x_2)$ or $y(x_2) > y(x_1)$. At the end of the first cycle (or the end of the first two experiments) the search region is approximately half of the original one. This search region can be further reduced by carrying out a second pair of computational experiments around the midpoint of the search region. In this fashion we can repeat the cycle, doing two experiments in each cycle and reducing the search region by half each time. The procedure is not completed until the accuracy is within the desirable limit. If the original search region is $L_o = b - a$, the search region after the first cycle will become

$$L_2 = \frac{L_o}{2} + \frac{\delta}{2} = \frac{1}{2}(L_o + \delta) \tag{11-16}$$

Similarly, the search region after the second cycle is

$$L_4 = \tfrac{1}{2}(L_2 + \delta) = \tfrac{1}{4}(L_o + 3\delta) \tag{11-17}$$

In a general form, the search region after the $n/2$ cycles is

$$L_n = \frac{L_o + \delta(2^{(n/2)} - 1)}{2^{(n/2)}} \tag{11-18}$$

It must be pointed out that in this search method the number of cycles used is exactly equal to $n/2$ where n is the number of computation experiments taken. The α ratio for this search method is

$$\alpha = \frac{L_n}{L_o} = \left(\frac{1}{2}\right)^{(n/2)} + \left[1 - \left(\frac{1}{2}\right)^{(n/2)}\right]\frac{\delta}{L_o} \tag{11-19}$$

If we neglect the effect of δ, the α ratio would become

$$\alpha = \left(\frac{1}{2}\right)^{(n/2)} \tag{11-20}$$

Fibonacci Search Method

One of the most efficient methods for optimizing a simple variable function is the Fibonacci method. This method is sequential in nature, but in this method the number of computational experiments is determined in advance. The locations at which the objective function is to be evaluated are controlled by calculations involving the use of Fibonacci numbers. The definition of the Fibonacci series is

$$F_0 = 1 \tag{11-21}$$

$$F_1 = 1 \tag{11-22}$$

and

$$F_n = F_{n-1} + F_{n-2} \quad \text{for} \quad n > 1 \tag{11-23}$$

Some values of the series are given in Table 11-3. Here is the step-by-step procedure of the Fibonacci search method:

Step 1. Decide the number of computational experiments n used in this search.

Table 11-3
Some Numbers of the Fibonacci Series

n	F_n
0	1
1	1
2	2
3	3
4	5
5	8
6	13
7	21
8	34
9	55
10	89
11	144
12	233
13	377
14	610
15	987
16	1,597
17	2,584
18	4,181
19	6,765
20	10,946

Step 2. Place the first two experiments at the locations.

$$x_1 = a + l_1 \qquad (11\text{-}24)$$

$$x_2 = b - l_1 \qquad (11\text{-}25)$$

where

$$l_1 = \frac{F_{n-2}}{F_n} L_1 \qquad (11\text{-}26)$$

Evidently l_1 is the distance from both ends of the original search region L_1. The objective function is still $y = f(x)$, defined for $a \le x \le b$.

Step 3. Eliminate the interval (a, x_1) from further search if $y(x_1) < y(x_2)$ or the interval (x_2, b) if $y(x_2) < y(x_1)$. After this step, the new search region is approximately two-thirds of the original region. That is,

$$L_2 = L_1 - l_1 \qquad (11\text{-}27)$$

or

$$L_2 = L_1 - L_1 \frac{F_{n-2}}{F_n} = L_1 \left(1 - \frac{F_{n-2}}{F_n} \right) = L_1 \frac{F_{n-1}}{F_n} \qquad (11\text{-}28)$$

Step 4. Place the second two experiments in the search region L_2. The experiments are at the locations with the distance l_2 away from both ends of the search region. The distance l_2 is

$$l_2 = \frac{F_{n-3}}{F_{n-1}} L_2 \qquad (11\text{-}29)$$

Step 5. Apply the rule indicated in Step 3 and further reduce the region from L_2 to L_3 where $L_3 = L_2 - l_2$.

Step 6. Continue the process of placing two experiments at a time and reducing the search region by one-third. It must be pointed out that one of the two experiments used in each cycle is always identical to the experiment from the previous cycle. Therefore, only one new experiment is actually needed in each cycle (except the first cycle where two experiments must be initiated).

Step 7. Place the last computational experiment (i.e., the nth one) at the location closest to the midpoint of the search region L_{n-1}. Since the $(n-1)$th experiment is exactly at the midpoint of the region L_{n-1}, one can apply the dichotomous concept, and with the last two experiments select the optimal location of the region $L_n = \frac{1}{2} L_{n-1}$.

For reference, the generalized relationships for Fibonacci search are summarized as follows:

$$l_k = \frac{F_{n-(k+1)}}{F_{n-(k-1)}} L_k \tag{11-30}$$

$$L_m = \frac{F_{n-(m-1)}}{F_n} L_1 \tag{11-31}$$

and

$$\alpha = \frac{L_n}{L_1} = \frac{1}{F_n} \tag{11-32}$$

Note that L_1 is the original search region defined in $a \le x \le b$.

11.4.3 Numerical Demonstrations

In order to demonstrate a use of various direct search methods presented in the last section and to compare their efficiencies on a common basis, we shall consider the objective function $y = 10x - x^2$, defined in the region $1 \le x \le 10$. If we assume that the function is complicated enough, we apply the direct search methods rather than the analytical to estimate the maximum of this function. The details are presented as below.

By Preplanned Uniform Search Method

In this method we will use eight computational experiments (i.e., $n = 8$) evenly distributed in the interval of interest. The locations of these experiments are shown in Fig. 11-4. The computational results are presented in Table 11-4. It is seen that the maximum value is 25 (it also happens to be the exact maximum value). Using Eq. (11-5), we have the interval of uncertainty

$$L_8 = 2\frac{10 - 1}{8 + 1} = 2$$

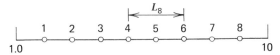

o = locations of computational experiements.

Figure 11-4. Locations of computational experiments determined by the preplanned uniform search.

Table 11-4
Computational Results by Preplanned Uniform Search

Exp. No. n	x_n	y_n
1	2	16
2	3	21
3	4	24
4	5	25
5	6	24
6	7	21
7	8	16
8	9	9

For comparison, the α ratio calculated with Eq. (11-6) is

$$\alpha = \frac{2}{8 + 1} = 0.222$$

By Sequential Uniform Search Method

Again, we use eight computational experiments to search the maximum value of the objective function. However, we run two experiments at a time. The results of these two experiments will determine the locations for the next set of experiments. Figure 11-5 indicates all locations in the search process. Since the search region is reduced once for every two experiments, this change is also presented in Fig. 11-5.

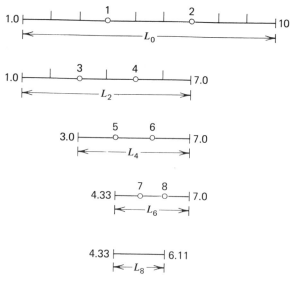

Figure 11-5. Locations of computational experiments determined by the sequential uniform search method.

Table 11-5
Computational Results by the Sequential Uniform Search Method

Search Cycle No. m	Experiment No. n	x_n	y_n	Search Range After Cycle m	Interval of Uncertainty L_{2m}
1	1	4	24.00		
	2	7	21.00	$1 \leq x \leq 7.0$	6.0
2	3	3	21.00		
	4	5	25.00	$3.0 \leq x \leq 7.0$	4.0
3	5	4.33	24.55		
	6	5.66	24.56	$4.33 \leq x \leq 7.0$	2.67
4	7	5.22	24.95		
	8	6.11	23.76	$4.33 \leq x \leq 6.11$	1.78

The computational result by this search method is summarized in Table 11-5. For comparison we also include the interval of uncertainty and the α ratio as follows:

$$L_8 = \left(\tfrac{2}{3}\right)^{8/2}(10 - 1) = 1.78$$

and

$$\alpha = \left(\tfrac{2}{3}\right)^{8/2} = 0.1975$$

The maximum value of this function is approximately equal to 24.95. It will exist at the interval $4.33 \leq x \leq 6.11$.

By Dichotomous Search Method

The dichotomous search method is quite similar to the sequential uniform search in the sense that both methods would take two computational experiments in one search cycle and then decide the locations for the experiments in the next search cycle. However, the difference between the two methods lies in the manner by which the locations of experiments are determined. Figure 11-6 shows all locations by the dichotomous search method. The distance between two experiments, δ, is taken as 0.2. As indicated, two experiments are always located around the midpoint of the search region, which is reduced by half in every search cycle.

Table 11-6 shows the details of this search process. At the end the maximum value is estimated as 24.99 in the interval between 4.94 and 5.50. Again, for

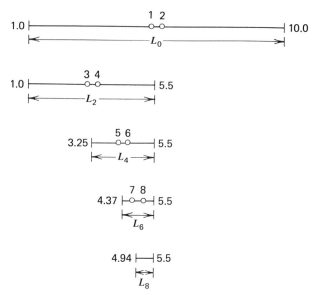

Figure 11-6. Locations of experiments determined by the dichotomous search method.

reference, the interval of uncertainty and the α ratio are calculated as

$$L_8 = \frac{(9 - 1)}{2^{8/2}} = 0.563$$

and

$$\alpha = \left(\tfrac{1}{2}\right)^{8/2} = 0.0625$$

By Fibonacci Search Method
Again, use eight computational experiments to determine the position of the optimum. The solution is presented in Table 11-7. Clearly, the last experiment (No. 8) is to be located at the same position as the experiment No. 6. Therefore, one must apply the dichotomous concept and place the last experiment very close to the remaining valid experiment (i.e., experiment No. 6). This permits the last experiment to be used to select the location of the optimum to within $\frac{1}{2}L_{n-1}$. In this case the interval of uncertainty is $L_8 = \frac{1}{2}L_7 = 0.26$. This can be also confirmed by using Eq. (11-31).

Using the positive δ or negative δ will determine whether the valid final interval is (4.70, 4.97) or (4.97, 5.23). Figure 11.7 indicates the search intervals at various stages and the locations for these eight computational experiments.

Table 11-6
Computational Results by The Dichotomous Search Method

Search Cycle No. m	Experiment No. n	x_n	y_n	Search Range After Cycle m	Interval of Uncertainty L_{2m}
1	1	5.40	24.84		
	2	5.60	24.64	$1.0 \leq x \leq 5.50$	4.50
2	3	3.15	21.58		
	4	3.35	22.27	$3.25 \leq x \leq 5.50$	2.25
3	5	4.27	24.47		
	6	4.47	24.72	$4.37 \leq x \leq 5.5$	1.125
4	7	4.84	24.97		
	8	5.04	24.99	$4.94 \leq x \leq 5.5$	0.536

For comparison, we use Eq. (11-32) and calculate the α ratio as

$$\alpha = \frac{1}{F_n} = \frac{1}{34} = 0.029$$

In regard to these four numerical demonstrations, two observations should be made. The first is that all direct search methods provide essentially the same maximum value of the objective function. However, the associated interval of

Table 11-7
Computational Results by the Fibonacci Search Method

Search Cycle No. m	Experiment No. n	Distance l_m	x_n	y_n	Search Range After Cycle m	Interval of Uncertainty L_{m+1}
1	1	3.44	4.44	24.68		
	2		6.55	22.59	$1 \leq x \leq 6.55$	5.55
2	3	2.11	3.11	21.43		
	1*		4.44	24.68	$3.11 \leq x \leq 6.55$	3.44
3	1*	1.32	4.44	24.68		
	4		5.23	24.94	$4.44 \leq x \leq 6.55$	2.11
4	4*	0.79	5.23	24.94		
	5		5.76	24.42	$4.44 \leq x \leq 5.76$	1.32
5	6	0.528	4.97	24.99		
	4*		5.23	24.94	$4.44 \leq x \leq 5.23$	0.79
6	7	0.263	4.70	24.87		
	6*		4.97	24.99	$4.70 \leq x \leq 5.23$	0.53
7	8	0.263	4.97	24.99	$4.97 \leq x \leq 5.23$	0.26

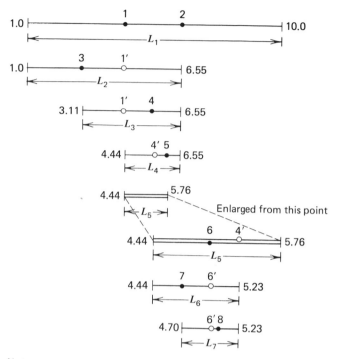

Notes:
  • Position of new experiment.
  ○ Position of old experiment, that is, experiment a is
    identical to a'.

Figure 11-7. Locations of computational experiments by the Fibonacci search method.

uncertainty varies greatly from the value 2 by the preplanned uniform search to the value 0.26 by the Fibonacci search method. The second observation is that while the preplanned uniform search method has a lowest efficiency (i.e., a highest reduction ratio), this method is simplest in application. It is this simplicity that contributes so much to the popularity of the preplanned uniform search. In general, other efficient search methods should not be used unless the computational experiment of objective function is very laborious and lengthy.

To evaluate these four direct search methods, it may be significant to compare the number of computational experiments required to achieve a given accuracy or a given reduction ratio. Table 11-8 summarizes the results in this respect. It is seen that the Fibonacci method has the greatest computational advantage, especially at the time when a low reduction ratio α is desired. For instance, for $\alpha = 10^{-4}$, the Fibonacci requires $n = 20$ while the next best, the dichotomous, needs $n = 28$. However, it must be pointed out that in the Fibonacci method, the number of computational experiments must be decided in advance and some of these experiments may be located at rather odd values of the independent variables.

Table 11-8
A Comparison of the Number of Computational Experiments for a Given α Ratio

Search Methods	Reduction Ratio α			
	0.1	0.01	0.001	0.0001
Preplanned uniform search	19	199	1,999	19,999
Sequential uniform search				
(two per cycle)	12	24	36	46
Dichotomous search	8	14	20	28
Fibonacci search	6	11	16	20

11.5 METHODS OF OPTIMIZATION FOR MULTIVARIABLE FUNCTIONS

In thermal system design the optimization of multivariable functions is frequently encountered. Although many of these can be reduced by direct eliminations to a problem involving just a single independent variable, some cases do arise in which the optimization of multivariable functions cannot be avoided. There are many numerical search methods for this purpose. But the ones to be presented below are: (1) univariate search, (2) pattern search, and (3) the steepest ascent (descent) method.

11.5.1 Univariate Search

In the univariate search, the objective function is optimized with respect to one variable at a time. The basic procedure is to substitute trial values of all but one independent variable into the function and optimize the resulting function in terms of one remaining variable. Then, the optimal value is substituted into the function, and the function is optimized with respect to another variable. Repeating the procedure, the function would be optimized with respect to all variables one at a time. After all these steps, the first phase is completed, and then the entire process of sequential optimization is repeated. The calculation would cease only when the successive change of the function value is less than a specified tolerance. For example, let us consider a function of two variables:

$$Z = f(x, y) \tag{11-33}$$

The geometrical representation of this function is, of course, a surface. For convenience of this presentation, the lines of constant Z are drawn on the x-y plane, as depicted in Fig. 11-8. To obtain the minimum value of this function we start with a trial value for y (i.e., $y = y_a$) and optimize the function with respect to x. At the end of this optimization process, we find the optimal value for x (i.e., $x = x_b$). Then, we substitute this optimal x into the function and optimize the resulting expression with respect to y. After this step, we find the point at $x = x_c$ and $y = y_c$. Repeating these processes, we would reach the point D and eventually the point M, where the function would have a minimum value.

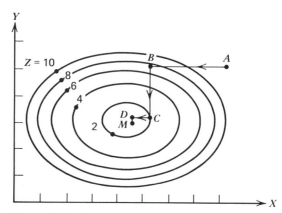

Figure 11-8. Contour lines for $Z = f(x, y)$.

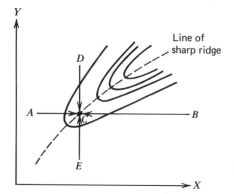

Figure 11-9. Failure of the univariate search at a ridge.

The method selected for performing the single-variable optimization may be either the exhaustive search technique or Fibonacci search technique. As indicated in the last section, these techniques have been found very adaptable to the digital computer. It should be pointed out that there are several conditions under which the univariate search will fail. For example, the objective function has a sharp change as shown in Fig. 11-9. Once the sharp ridge is reached, such as the point C, the process of optimizing the function with respect to x or y would cease. Therefore, the nonoptimal value may be regarded as the optimal. Figure 11-10 gives another example. When the interval for independent variables is too large, the optimizing process may also cease even though the optimum has not been reached. Fortunately, these conditions are not usually found in the thermal system designs.

11.5.2 Pattern Search

One of the pattern search methods is the sequential simplex method developed by Spendly, Hext, and Himsworth [1]. The method takes a regular geometric figure as a basis. Computational experiments are placed at the locations of vertices of the geometric figure. Then, the vertex with the worst value of objective function is rejected and a new vertex is found for replacement. The new vertex must be so located that the shape of the geometric figure will be preserved, and the value of

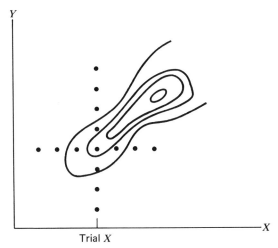

Figure 11-10. Failure of the univariate search with too large an interval.

objective function at that point will become better. Thus, the method proceeds in the fashion of rejecting a vertex and generating a new vertex. The search process is completed when the geometric figure contains the optimal region.

As an illustration, we consider a function of two variables $y = f(x_1, x_2)$. The lines of constant y are drawn on the x_1-x_2 plane as depicted in Fig. 11-11. Three initial points $(1, 2, 3)$ are so selected that they must form an equilateral triangle. Computational experiments are then carried out at the locations of these vertices. The vertex (1) is immediately rejected because of the worst value of function and the vertex (4) is selected to form a second equilateral triangle (i.e., 2-3-4). As indicated in Fig. 11-11, the vertex (4) is the mirror image of the rejected vertex (1) in the mirror

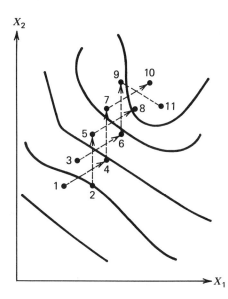

Figure 11-11. Progression of the simplex toward the optimum.

2-3. The second equilateral triangle (2-3-4) is next used as the basis for fresh moves. Continuing these repetitive operations, the sequence of points shown in Fig. 11-11 will be obtained. It is seen that this pattern search will follow the path that zigzags in the favorable direction and eventually reaches the region when the optimum value of objective function can be found.

The concept just presented can be extended to the function of n variables. Below we present the generalized relationships that could be used for computer programming. The derivations can be found in the reference [1].

To start a simplex, a set of $n + 1$ points is required for the function of n variables. The coordinates of these $n + 1$ initial points are presented in Table 11-9. These values are based on the assumption that the search is started at the origin. Using a function of two variables for illustration (i.e., $n = 2$) we have

$$p = \frac{a(1 + \sqrt{3})}{2\sqrt{2}} = 0.9657a \tag{11-34}$$

and

$$q = \frac{a(\sqrt{3} - 1)}{2\sqrt{2}} = 0.2587a \tag{11-35}$$

Then, the coordinates of the initial three points are written as $(0, 0)$, $(0.9657a, 0.2587a)$, and $(0.2587a, 0.9657a)$, according to the Table 11-9. The value of a is the side of the equilateral triangle and corresponds to the step size in the pattern search. The smaller the value of a is, the slower the search process will progress. But at the end the region of uncertainty is reduced. In general, the value of a is selected by experiences. If it were desired to start the search at a base point (C_1, C_2) rather than

Table 11-9
Coordinates for a Set of Simplex Vertices

Point j	\multicolumn{7}{c}{n coordinates of each point}
	$\xi_{1,j}$	$\xi_{2,j}$	$\xi_{3,j}$	$\xi_{4,j}$	$\cdots$	$\xi_{n-1,j}$	$\xi_{n,j}$
1	0	0	0	0	$\cdots$	0	0
2	p	q	q	q	$\cdots$	q	q
3	q	p	q	q	$\cdots$	q	q
n	q	q	q	q	$\cdots$	p	q
$n + 1$	q	q	q	q	$\cdots$	q	p

where

$$p = \frac{a}{n\sqrt{2}} (\sqrt{n + 1} + n - 1)$$

$$q = \frac{a}{n\sqrt{2}} (\sqrt{n + 1} - 1)$$

the origin $(0, 0)$, then the coordinates of the starting simplex would become

$$(C_1, C_2),$$
$$(C_1 + 0.9657a, \qquad C_2 + 0.2587a) \qquad \text{and}$$
$$(C_1 + 0.2587a, \qquad C_2 + 0.9657a)$$

To generate a new simplex (i.e., a new geometric figure) in the pattern search, we must find a new vertex to replace the vertex at which the objective function has a worse value. The new vertex and the n unchanged ones combine to form a new simplex figure. The coordinates of the new vertex are

$$X_{i,n} = \left[\frac{2}{n} \left(\sum_{j=1}^{n+1} X_{i,j} - X_{i,R} \right) \right] - X_{i,R} \qquad \text{for} \qquad i = 1, 2, \ldots, n \quad (11\text{-}36)$$

where

$X_{i,n}$ = coordinates of new vertex

$X_{i,R}$ = coordinates of replaced vertex

$X_{i,j}$ = coordinates of all vertices in the original simplex figure

n = number of variables in the objective function

For a function of two variables Eq. (11-36) is simplified and expressed as

$$X_{1,n} = (X_{1,1} + X_{1,2} + X_{1,3} - X_{1,R}) - X_{1,R}$$
$$X_{2,n} = (X_{2,1} + X_{2,2} + X_{2,3} - X_{2,R}) - X_{2,R} \qquad (11\text{-}36a)$$

EXAMPLE 11-4.   Locate the minimum of the function represented by $y = 2/X_1 + 8X_1^2 X_2 + X_2^2$ and determine the region of uncertainty.

Solution: We arbitrarily start the search at a reference point $(1, 1)$ and use the equations at Table 11-9 to calculate p and q. The value of p and q are

$$p = 0.9657a$$

$$q = 0.2587a$$

If the value of a is set at 0.1, the coordinates of the starting simplex will become $(1, 1)$, $(1.0966, 1.0259)$ and $(1.0259, 1.0966)$. All these points and corresponding values of the objective function are given by points 1 to 3 in Table 11-10.

The pattern search now starts according to the procedures presented above. In the initial simplex (1-2-3) we eliminate point 2 because it has the largest value of objective function, we then generate point 4. By using Eq. (11-36a) we find that the replacement point 4 has three coordinates:

$$X_{i,4} = X_{i,1} + X_{i,3} - X_{i,2} \qquad \text{for} \qquad i = 1, 2$$

or

$$X_{1,4} = 1.0 + 1.0259 - 1.0966 = 0.9293$$

and

$$X_{2,4} = 1.0 + 1.0966 - 1.0259 = 1.0707$$

At the new point 4, the value of the objective function y_4 is found to be equal to 8.40, as indicated in Table 11-10. Now, we have rejected point 2 from the original triangle (1-2-3) and formed a new triangle (1-4-3). This new triangle is then used as a basis for fresh moves. All subsequent calculations are recorded in Table 10-11. It is seen that the minimum value of the function occurs at point 13. The region of uncertainty is

$$X_1 = 0.517 \pm a = 0.517 \pm 0.1$$

and

$$X_2 = 0.870 \pm a = 0.870 \pm 0.1$$

Table 11-10
Pattern Search in Example 8-4

Point j	Coordinates of Point j		Value of Function at Point j	Points in Simplex	Point Rejected
	$x_{1,j}$	$x_{2,j}$	y_j		
1	1.0	1.0	9.0		
2	1.0966	1.0259	10.64		
3	1.0259	1.0966	9.97	1-2-3	2
4	0.9293	1.0707	8.40	1-4-3	3
5	0.9034	0.9741	7.62	1-4-5	1
6	0.8327	1.0448	7.05	6-4-5	4
7	0.8068	0.9842	6.51	6-7-5	5
8	0.7361	1.0189	6.09	6-7-8	6
9	0.7102	0.9221	5.68	9-7-8	7
10	0.6395	0.9926	5.39	9-10-8	8
11	0.6136	0.8960	5.14	9-10-11	9
12	0.5429	0.9665	5.02	12-10-11	10
13	0.5170	0.8700	4.97	12-13-11	11
14	0.4460	0.9405	5.09	12-13-14	14
15	0.6136	0.8960	5.14	12-13-15	15

11.5.3 Steepest Ascent Method

The method of steepest ascent (or descent for minimization) is a cyclic search technique. It involves two basic steps in search of an optimum. The first is to determine the search direction at which the objective function is expected to change at the most favorable rate. The second is to select the search step size (or search interval). For the convenience of this presentation, we first examine a two-dimensional case and then extend the results to the case of n dimensions.

Let $y = f(x_1, x_2)$ be the function of two variables having continuous partial derivatives and a well-defined value for each point (x_1, x_2). Figure 11-12 indicates the lines of constant value y on the x_1-x_2 plane. At an arbitrary point 0, the rate change of y in a given direction such as α is given by

$$\frac{dy}{ds} = \left(\frac{\delta y}{\delta x_1}\right)_0 \left(\frac{dx_1}{ds}\right) + \left(\frac{\delta y}{\delta x_2}\right)_0 \left(\frac{dx_2}{ds}\right) \tag{11-37}$$

or

$$\frac{dy}{ds} = \left(\frac{\delta y}{\delta x_1}\right)_0 \cos \alpha + \left(\frac{\delta y}{\delta x_2}\right)_0 \sin \alpha \tag{11-38}$$

where $(\delta y/\delta x_1)_0$ and $(\delta y/\delta x_2)_0$ are the partial derivatives evaluated at the point 0. Evidently its directional derivative dy/ds will change as the angle α changes. To optimize the directional derivative, we differentiate this term with respect to α and set the resultant equation equal to zero, that is,

$$\frac{d}{d\alpha}\left(\frac{dy}{ds}\right) = 0 \tag{11-39}$$

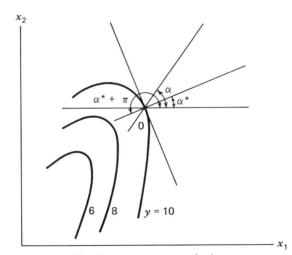

Figure 11-12. Steepest ascent method.

or

$$\left(\frac{\delta y}{\delta x_1}\right)_0 (\sin \alpha) - \left(\frac{\delta y}{\delta x_2}\right)(\cos \alpha) = 0 \qquad (11\text{-}40)$$

Then, the angle α^* optimizing dy/ds would become

$$\tan \alpha^* = \frac{\left(\dfrac{\delta y}{\delta x_2}\right)_0}{\left(\dfrac{\delta y}{\delta x_1}\right)_0} \qquad (11\text{-}41)$$

As indicated in Fig. 11-12, there are two angles that will satisfy the above expression. They can be presented as

$$\cos \alpha^* = \frac{\left(\dfrac{\delta y}{\delta x_1}\right)_0}{\pm \sqrt{\left(\dfrac{\delta y}{\delta x_1}\right)_0^2 + \left(\dfrac{\delta y}{\delta x_2}\right)_0^2}} \qquad (11\text{-}42)$$

$$\sin \alpha^* = \frac{\left(\dfrac{\delta y}{\delta x_2}\right)_0}{\pm \sqrt{\left(\dfrac{\delta y}{\delta x_1}\right)_0^2 + \left(\dfrac{\delta y}{\delta x_2}\right)_0^2}} \qquad (11\text{-}43)$$

The positive sign is used for maximization and the negative sign is for minimization. In the physical sense, the angle α^* is the direction perpendicular to the counter of constant y. For every (x_1, x_2), there must be a definite value for α^*. This local property of α^* must be taken into consideration when the search process is carried out.

Once the search direction is decided, the search interval must be determined. For the two-dimensional case, the search interval (step size) s is related to the coordinates of the starting and final points by

$$x_{1,f} - x_{1,0} = s \cos \alpha^* = \frac{s \left(\dfrac{\delta y}{\delta x_1}\right)_0}{\pm \sqrt{\left(\dfrac{\delta y}{\delta x_1}\right)_0^2 + \left(\dfrac{\delta y}{\delta x_2}\right)_0^2}} \qquad (11\text{-}44)$$

and

$$x_{2,f} - x_{2,0} = s \sin \alpha^* = \frac{s \left(\frac{\delta y}{\delta x_2} \right)_0}{\pm \sqrt{\left(\frac{\delta y}{\delta x_1} \right)_0^2 + \left(\frac{\delta y}{\delta x_2} \right)_0^2}} \qquad (11\text{-}45)$$

If we extend these results to the function of n variables, we have

$$x_{i,f} - x_{i,0} = s m_{i,0} \qquad i = 1, 2, \ldots, n \qquad (11\text{-}46)$$

where

$$m_{i,0} = \frac{\left(\frac{\delta y}{\delta x_i} \right)_0}{\pm \sqrt{\sum_{i=1}^{n} \left(\frac{\delta y}{\delta x_i} \right)_0^2}} \qquad (11\text{-}47)$$

The subscripts 0 and f denote the initial and final locations, respectively. The term $m_{i,0}$ indicates the direction evaluated at the point 0. Again, the positive sign should be taken for maximization while the negative sign for minimization.

Since the angle α^* or the directional term m_i is not constant, the search interval s must be infinitely small and, therefore, results in a very slow search process. Obviously, such an infinitely small s is not a practical choice. In practice, the search interval s is arbitrarily assigned or the step size for one variable such as Δx, is selected and then other Δx_i are determined with Eq. (11-46). Figure 11-13 indicates these movements for a two-dimensional case. The search starts at point 0 and moves in the direction of steepest ascent to point 1. The process continues to point 2, point

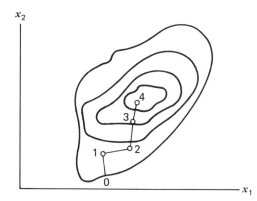

Figure 11-13. Steepest ascent method with uniform step size.

3, and eventually reaches point 4. At point 4, we start a new search with a smaller interval. We repeat these searches until the region of uncertainty is within the desirable range.

In contrast to the single step of predetermined length used in the above method, another method is to proceed along the direction of steepest ascent from the starting point until the optimal value is reached in this direction. Figure 11-14 indicates computational experiments in this search and the optimal point 1 in the first direction. At point 1, the most favorable direction is again determined and the search starts in the second direction. The subsequent movements are presented in Fig. 11-14.

EXAMPLE 11-5. Estimate the minimum value of the function $y = 3x_1^2 + 4x_2^2$ by the method of steepest descent. Assume that the starting point O is $(1, 1)$.

Solution: The gradient at any point is defined by the two derivatives $\delta y/\delta x_1 = 6x_1$ and $\delta y/\delta x_2 = 8x_2$. At the starting point O, we have

$$\left(\frac{\delta y}{\delta x_1}\right)_O = 6 \quad \text{and} \quad \left(\frac{\delta y}{\delta x_2}\right)_O = 8$$

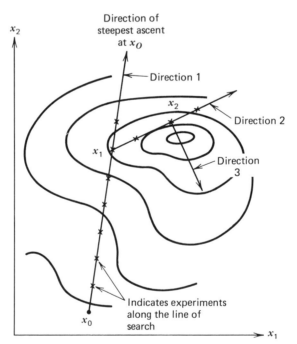

Figure 11-14. Search based on direction of steepest ascent.

From Eq. (11-47) the direction of steepest descent at that point can be found as

$$m_{1,0} = \frac{-6}{\sqrt{(6)^2 + (8)^2}} = -\frac{3}{5}$$

$$m_{2,0} = \frac{-8}{\sqrt{(6)^2 + (8)^2}} = -\frac{4}{5}$$

The parametric representation of the straight line in the direction of steepest descent is then given by

$$x_1 = 1 - \tfrac{3}{5}s; \qquad x_2 = 1 - \tfrac{4}{5}s$$

Substituting these two expressions into the objective function, we have

$$y = 3\left(1 - \tfrac{3}{5}s\right)^2 + 4\left(1 - \tfrac{4}{5}s\right)^2$$

Since this equation is one dimensional, any of the search techniques in Section 11.4 can be used to determine the optimal s. The result is $s = 4/3$ and $y = 0.13$ at that point. The coordinates of the new point (referred to as point 1) are

$$x_{1,1} = 1 - \left(\tfrac{3}{5}\right)\left(\tfrac{4}{3}\right) = \tfrac{1}{5}$$

$$x_{2,1} = 1 - \left(\tfrac{4}{5}\right)\left(\tfrac{4}{3}\right) = -\tfrac{1}{15}$$

Now, we can start a new search cycle and use point 1 as the basis. The gradient at this new point has components

$$\left(\frac{\delta y}{\delta x_1}\right)_1 = \frac{6}{5} \quad \text{and} \quad \left(\frac{\delta y}{\delta x_2}\right)_1 = \frac{-8}{15}$$

The direction of steepest descent is given by

$$m_{1,1} = -0.9139 \quad \text{and} \quad m_{2,1} = 0.4061$$

Then, we search along a line defined by

$$x_1 = 0.2 - 0.9139s$$

and

$$x_2 = -0.0666 + 0.4061s$$

Again, substituting these two expressions into the objective function, we have

$$y = 3(0.2 - 0.9139s)^2 + 4(-0.0666 + 0.4061s)^2$$

The distance of movement is quite small in this second search. We find that $s = 0.2$ and $y = 0.00173$ at that point. The coordinates of this point are then

$$x_{1,2} = 0.2 - 0.9131(0.2) = 0.017$$

$$x_{2,2} = -0.0666 + 0.4061(0.2) = 0.014$$

It is seen that some improvements on y have been made in the second search cycle (from 0.13 changed to 0.00173). If the search is continued, the optimal value of y is expected to approach zero.

11.6 A SIMPLIFIED COST OPTIMIZATION

Equipment cost estimates are not always available in the regular literature mainly because equipment cost is determined by many time-dependent variables, such as labor and material costs. The commercial competitive environment will also affect the equipment cost. The method of cost estimating varies from one company to another. Some methods are empirical, while others are semi-analytical. The accuracy of cost estimates is determined by the need and desire. For equipment proposals, the cost estimate must be carefully made and must be as accurate as possible. On the other hand, when the cost estimates are needed in a comparative study, the accuracy may become less important.

Taking a crossflow mechanical draft cooling tower as an example, the Marley Company published a set of curves for a cost estimate [4]. Some of these curves are reproduced and presented in Fig. 11-15. It is seen that the so-called rating factor is determined by the ambient wet-bulb temperature, tower cooling range, and tower approach. When the rating factor is determined, the required tower unit is calculated by

$$TU = (RF)(WFR) \tag{11-48}$$

where

$$TU = \text{required tower units}$$

$$RF = \text{rating factor}$$

$$WFR = \text{cooling water flow rate (gpm)}$$

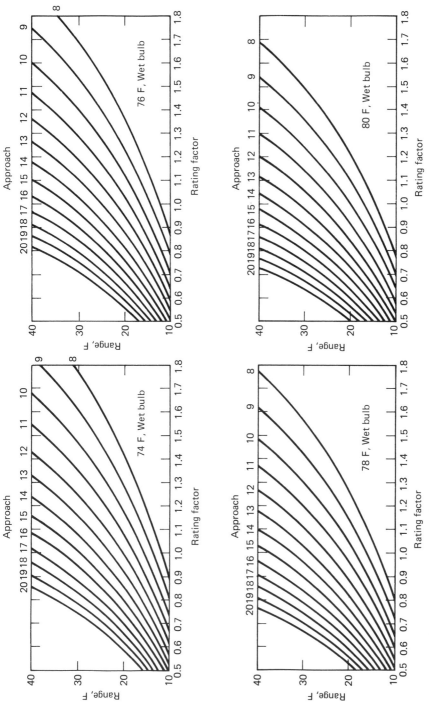

Figure 11-15.   The rating factors charts. (Courtesy of the Marley Company)

The rating factor indicates a relative degree of difficulty for the cooling process. The required tower units represent a relative tower size and are therefore related to the tower cost. For computer simulation and optimization, the rating factor of the cross flow mechanical-draft cooling tower may be expressed in the following form:

$$RF = (BRF)(FW)(FC)(FA) \qquad (11\text{-}49)$$

where

$$BRF = \text{basic rating factor}$$

$$FW = \text{wet-bulb temperature correction factor}$$

$$FC = \text{cooling range correction factor}$$

$$FA = \text{approach correction factor}$$

The basic rating factor and other correction factors are summarized in Table 11-11. To estimate the cooling tower cost, one must determine the required tower units by Eq. (11-48) and then utilize the tower unit cost.

The tower unit cost is usually in terms of dollars per tower unit. The unit cost information is either provided by the equipment vendor or determined by experiences in similar projects. From 1976 to 1978, the tower unit cost was around $12.00 per tower unit according to the Marley publication [4]. It must be emphasized that these cost estimates are used only for a comparative study.

Based on the previous experiences, the tower fan power has been found to be linearly related to the number of tower units required. The equation below may be used for a quick estimate:

$$FP = 0.011 \text{ (number of tower unit)} \qquad (11\text{-}50)$$

where FP represents the horsepower required for tower fan operation.

In cooling system design the relative condenser cost is frequently needed. This can be estimated by the equation

$$\text{Condenser cost} = (Es)[a + b(A) + c(N)] \qquad (11\text{-}51)$$

where

$$A = \text{heat transfer surface area } (\text{ft}^2)$$

$$N = \text{number of condenser tubes}$$

$$Es = \text{escalation factor}$$

$$a, b, c = \text{empirical constants}$$

In Eq. (11-51) the coefficient b represents the unit cost per unit of heat transfer

Table 11-11
Correction Factors for the Mechanical-Draft Cooling Tower Rating Factor

Range	BRF	Correction Factors
$WBT > 72$ F	0.82	$FW = 1.043 - 0.01475(WBT - 74) - 0.004(WBT - 74)^2 + 0.0003125(WBT - 74)^3$
		$FC = 0.72 + 0.0324(CR - 15) - 0.00036(CR - 15)^2 - 0.000008(CR - 15)^3$
$AP > 12$ F		$FA = 1.134 - 0.06816(AP - 14) + 0.00013(AP - 14)^2 + 0.00023(AP - 14)^3$
$WBT > 72$ F	1.36	$FW = 1.05 - 0.0192(WBT - 74) - 0.00374(WBT - 74)^2 + 0.000415(WBT - 74)^3$
		$FC = 0.635 + 0.0465(CR - 10) - 0.0012(CR - 10)^2 + 0.00002(CR - 10)^3$
$AP \leq 12$ F		$FA = 1.225 - 0.1257(AP - 6) + 0.00675(AP - 6)^2 - 0.0000625(AP - 6)^3$
65 F $\leq WBT \leq 72$ F	0.96	$FW = 1.05 - 0.0289(WBT - 68) + 0.002745(WBT - 68)^2 - 0.000395(WBT - 68)^3$
		$FC = 0.73 + 0.03606(CR - 15) - 0.00128(CR - 15)^2 + 0.0000373(CR - 15)^3$
$AP \geq 15$ F		$FA = 1.151 - 0.08806(AP - 14) + 0.007145(AP - 14)^2 - 0.0004325(AP - 14)^3$
65 F $\leq WBT \leq 72$ F	1.34	$FW = 1.06 - 0.026(BT - 68) - 0.002624(WBT - 68)^2 + 0.0002915(WBT - 68)^3$
		$FC = 0.623 + 0.0485(CR - 10) - 0.00128(CR - 10)^2 + 0.0002(CR - 10)^3$
$AP < 15$ F		$FA = 1.21 - 0.1185(AP - 8) + 0.007(AP - 8)^2 - 0.000125(AP - 8)^3$
35 F $\leq WBT \leq 65$ F	1.25	$FW = 1.26 - 0.028(WBT - 35) + 0.0002(WBT - 35)^2$
		$FC = 0.58 + 0.1172(CR - 10) - 0.01216(CR - 10)^2 + 0.000464(CR - 10)^3$
		$FA = 1.128 - 0.0713(AP - 18) + 0.004(AP - 18)^2 - 0.0001667(AP - 18)^3$

surface area and is usually expressed in terms of dollars per square foot. The coefficient c represents the installation cost per condenser tube and is expressed in dollars per tube. The constant a is related to the cost of various items such as condenser shells and condenser hot well.

The relative pumping cost consists of the cost for the pump and the electric motor that is used to drive the pump. The relative pump cost is linearly proportional to the water flow rate for a given pump head. Since the pump head varies in a narrow range, the relative pump cost is usually approximated by

$$\text{Pump cost} = C_1(WFR) \tag{11-52}$$

where (WFR) represents the circulating water flow rate through the pump and is given in units of gallons per minute. The coefficient C_1 is the empirical constant and is obviously dependent on the market situation. Similarly, the pump motor cost is estimated by

$$\text{Motor cost} = C_2(\text{motor horsepower}) \tag{11-53}$$

where

$$C_2 = \text{empirical constant (dollars/hp)}$$

Table 11-12
Cost Estimates for Three Different Combined Systems

Line No.	Variables and Basis of Estimates	System 1	System 2	System 3
1	Ambient wet-bulb temperature (F)	79	79	79
2	Tower approach (F)	11	13	15
3	Cooling range (F)	25.1	23.5	21.8
4	Water flow rate (gpm)	136,000	145,000	156,000
5	Rating factor	1.07	0.87	0.74
6	Tower units ($\times 10^6$)	0.1455	0.1261	0.1155
7	Relative tower cost (millions of dollars)	1.746	1.513	1.386
8	Condenser surface area (ft^2)	183,900	196,500	210,800
9	Number of condenser tubes	19,500	20,850	22,370
10	Relative condenser cost (millions of dollars)	0.9585	1.0242	1.0987
11	Pump work (hp)	4293	4577	4924
12	Relative pump-motor cost (millions of dollars)	0.2575	0.2746	0.2954
13	Relative water-pump cost (millions of dollars)	0.4080	0.4350	0.4680
14	Total relative equipment cost (millions of dollars)	3.37	3.23	3.25

Referring back to the three combined systems shown in Table 11-1, we can determine the total relative equipment cost by using the following information:

Tower unit cost	$12/TU
Tube installation cost	$2/tube
Tube material cost	$5/ft^2
Pump head	100 ft
Pump efficiency	80%
Pump motor cost	$60/hp
Pump cost	$3/gpm

The results are presented in Table 11-12. The total relative cost of the combined system is the sum of condenser, cooling tower, and pump cost. In these calculations the system relative cost becomes minimum for the case in which the design tower approach is 13 F. It is also shown that the cost of cooling tower is inversely related to that of the condenser.

More on the cooling system optimization is presented in Chapter 12.

11.7 HOT WATER GENERATING SYSTEM

Hot water is frequently produced in a power plant. Hot water may be used within the plant or supplied to the outside for various purposes. The hot water temperature varies in a wide range, depending on the nature of water usage. For an industrial use the water temperature can be as high as 300 to 400 F. The hot water generation system may be a waste heat boiler fueled by the exhaust gas from gas turbines or a series of heat exchangers heated by the extraction steam. Or it is simply an industrial boiler producing nothing except hot water at a desired temperature. The selection of hot water generating system is obviously determined by many variables such as type of fuels and availability of process steam. In addition, economic factors must be taken into consideration. Even when a particular system is determined, there are still many decisions that will require some engineering optimizations.

Figure 11-16 is a schematic diagram of a hot water system. The system consists of two heaters connected in a series. The water flows through these heaters and its temperature is raised to a desired level. The heating media are two sources of process steam. Because of different steam pressures, steam costs are not the same. The higher the steam pressure, the higher will be the steam cost. Also, the higher-pressure heater is expected to cost a little more than the low-pressure heater. The question that arises in design is how to distribute the heat load between these two heaters. At first glance, greater heat load should be put on the low-pressure heater. But attention must be paid to the fact that the average temperature difference may be too low in the low-pressure heater and thus result in a large heat transfer area. The objective function to be optimized is derived in the following way.

Let the annual cost be the sum of initial cost and steam cost. Since the hot water system consists of two heaters, the annual cost must involve two terms as

$$C = \left(C_i + C_s \right)_h + \left(C_i + C_s \right)_l \qquad (11\text{-}54)$$

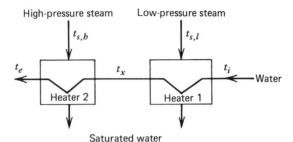

Figure 11-16. A schematic diagram of hot water system.

where

$$C_i = \text{annual initial cost (dollars/yr)}$$

$$C_s = \text{annual steam cost (dollars/yr)}$$

The subscripts h and l denote the high pressure and low pressure, respectively. The steam cost is determined by the heat transfer rate in the heater, which is also the amount of heat the water will receive. Using the notations shown in Fig. 11-16, we present these expressions as

$$\dot{Q}_l = \dot{m}C_p(t_x - t_i) \tag{11-55}$$

and

$$\dot{Q}_h = \dot{m}C_p(t_e - t_x) \tag{11-56}$$

The annual steam costs for both heaters are then

$$C_{s,l} = \frac{(a_l)}{(LH)_l}(\dot{Q}_l)(\tau) \tag{11-57}$$

and

$$C_{s,h} = \frac{(a_h)}{(LH)_h}(\dot{Q}_h)(\tau) \tag{11-58}$$

where

$$a = \text{steam cost (dollars/lb)}$$

$$\tau = \text{operating hours per year}$$

$$LH = \text{latent heat, approximately 890 Btu/lb}$$

To determine the annual initial cost, we must calculate the heat transfer surface area in the heater. Using the concept of overall heat transfer coefficient, the low-pressure heater surface is

$$A_l = \frac{\mathring{Q}_l}{U_l (LMTD)_l} \tag{11-59}$$

where

$$(LMTD)_l = \frac{(t_{s,l} - t_x) - (t_{s,l} - t_i)}{\ln \dfrac{t_{s,l} - t_x}{t_{s,l} - t_i}} \tag{11-60}$$

Then, the relative low-pressure heater cost is simply the product of the heat transfer surface area and the unit cost. If we apply the annual fixed charge rate, the annual relative initial cost becomes

$$C_{i,l} = (b_l)(A_l)(AFCR) \tag{11-61}$$

where

$$b_l = \text{low-pressure heater surface cost, dollar/ft}^2$$

Combining Eq. (11-59) and Eq. (11-61) gives

$$C_{i,l} = \frac{(b_l)(AFCR)(\mathring{Q}_l)}{U_l} \frac{\ln \dfrac{t_{s,l} - t_x}{t_{s,l} - t_i}}{(t_{s,l} - t_x) - (t_{s,l} - t_i)} \tag{11-62}$$

Similarly, the annual relative initial cost for high-pressure heater is

$$C_{i,h} = \frac{(b_h)(AFCR)(\mathring{Q}_h)}{U_h} \frac{\ln \dfrac{t_{s,h} - t_e}{t_{s,h} - t_x}}{(t_{s,h} - t_e) - (t_{s,h} - t_x)} \tag{11-63}$$

It is seen that both steam cost and relative initial cost are functions of the intermediate temperature t_x. As this temperature increases, more heat load will be placed on the low-pressure heater. The reverse is also true. For instance, let us use

Table 11-13
Heat Load Distributions and Heater Sizes

Intermediate Temperature T_x (F)	Heat Load $\dot{Q}$ (MBtu / hr)		Heater Surface Area A (ft²)	
	HP Heater	LP Heater	HP Heater	LP Heater
150	60.00	0	2082.0	0
175	50.00	10.00	1948.1	169.4
200	40.00	20.00	1786.9	384.6
225	30.00	30.00	1584.8	679.9
250	20.00	40.00	1313.8	1152.9
275	10.00	50.00	900.8	2466.7

the following data:

Water flow rate	400,000 lb/hr
Water inlet temperature	150 F
Water outlet temperature	300 F
Steam pressures	50 psia
	80 psia
Steam cost	
Low-pressure	$6.0/$10^6$ Btu
High-pressure	$6.04/$10^6$ Btu
Heater surface cost	
Low-pressure	$10.1/ft²
High-pressure	$10.2/ft²
Overall heat transfer coefficient	500 Btu/hr-ft²-F
Annual fixed charge rate	20%
Operation hours	2000 hr/yr

We vary the intermediate temperature t_x and determine the optimal value at which the annual relative cost will become minimum. The calculations are summarized in Table 11-13 and 11-14.

It is seen that as the intermediate temperature t_x increases, the heat load supported by the low-pressure heater will increase. Because of large temperature differences, however, the high-pressure heater always has advantage over the low-pressure heater in terms of heat transfer surface required. For instance, when the heat load is solely supported by the high-pressure heater, the heater surface is approximately 2082 ft² and, when supported mostly by the low-pressure, it becomes 2466 ft².

Table 11-14 presents some interesting facts. If the initial cost is a sole variable to be considered in optimization, a single high-pressure heater should be utilized for the hot water production. As indicated in Table 11-14, this arrangement will result in a minimum cost. On the other hand, if steam cost is also taken into consideration, the optimal result will be entirely different. For this case the optimal t_x is around 250 F.

Table 11-14
Comparative Hot Water System Costs

Intermediate Temperature T_x (F)	Steam Cost C_s ($/yr)			Relative Heater Cost C_i ($/yr)			Total Relative Cost ($/yr)
	HP Heater	LP Heater	Subtotal	HP Heater	LP Heater	Subtotal	
150	724,800	0	724,800	4,247	0	4,247	729,047
175	604,000	120,000	724,000	3,974	342	4,316	728,316
200	483,200	240,000	723,200	3,645	777	4,422	727,622
225	262,400	360,000	722,400	3,233	1,374	4,607	727,007
250	241,600	480,000	721,600	2,680	2,329	5,009	726,609
275	120,000	600,000	720,800	1,838	4,983	6,821	727,621

That is, the high-pressure heater will take one-third of the total heat load while the low-pressure heater takes care of another two-thirds. Steam cost is an important variable in this optimization. When the cost for low-pressure steam is substantially lower than that for high-pressure steam, it will definitely favor the use of a low-pressure heater. This is particularly true when the hot water system is operated on a continuous basis.

In optimization certain constraints must be met. These may include in the above case the minimum heater size and the heater terminal temperature. The heater terminal temperature is defined as the difference between the saturated temperature of steam and the temperature of water leaving the heater. This value is usually 5 F or higher. Since the heater size does not increase continuously, but changes like a step function, this, too, must be taken into account in optimization.

SELECTED REFERENCES

1. G. S. G. Beveridge and R. S. Schechter, *Optimization*: *Theory and Practice*, McGraw-Hill, 1970.

2. B. Carnaham, H. A. Luther, and J. O. Wikes, *Applied Numerical Methods*, Wiley, 1969.

3. P. J. Wilde, *Optimum Seeking Methods* Prentice-Hall, 1964.

4. J. B. Dickey, Jr. and D. W. Dwyer, *Managing Waste Heat with the Water Cooling Tower*, Marley Company, Kansas City, Missouri, 1978.

PROBLEMS

11-1. A reheat gas turbine is designed to receive the gas of 2000 F and 40 atmospheric pressure. Construct a mathematical model for selecting the reheat pressure such that the total turbine work is a maximum. Assume that the turbine inlet temperatures for both turbine cylinders are the same and equal to 2000 F.

11-2. Obtain a numerical solution of Problem 11-1 by using one of the direct search methods.

11-3. A power company has several simple-cycle gas turbine systems for peaking power generation. A proposal is made to install a waste heat boiler using the exhaust gas from the turbines to generate hot water. Construct a mathematical model for selecting the optimal size of waste heat boiler and obtain a numerical solution for the following data:

Exhaust gas flow rate	10^6 lb/hr
Exhaust gas temperature	1000 F
Specific heat of exhaust gas	0.25 Btu/lb/F

Hot water conditions

Outgoing	360 F
Return	110 F
Overall heat transfer coefficient based on gas-side area	8 Btu/hr/ft^2/F
Initial cost of waste heat boiler	$25/ft^2
Economic life	20 years
Annual fixed charge rate	15%
Levelized O & M cost (for waste heat boiler only)	$0.05/10^6 Btu
Number of operation hours	2000 hr
Levelized market value of hot water	$1.50/10^6 Btu

Be advised that measures must be taken to ensure that the stack gas temperature of waste heat boiler is not lower than 300 F.

11-4. Repeat Problem 11-3, but change the system operating hours to 5000 hr/yr.

11-5. An insulated steam pipe is to be designed. The pipe will deliver the steam at the rate 50,000 lb/hr and at a distance of 30 ft. For simplicity, no consideration is given to the difference in elevation. The steam inlet conditions are 100 psia and 650 F. Optimize the pipe diameter and insulation thickness for the following conditions:

Pipe Cost:

$$C_p = L \exp \left[0.72(\ln D)^2 - 2.25(\ln D) + 7.92 \right]$$

Insulation Cost:

$$C_I = L \{ (0.04933D + 0.20875) [\exp (0.9472 \ln Q + 1.4804)] + 8.0 \}$$

Thermal energy cost:

$$C_{th} = \$3.00/10^6 \text{ Btu}$$

Pump energy cost:

$$C_e = 50 \text{ mills/kWhr}$$

where

C_p = capital cost of pipe, dollars

C_I = capital cost of pipe insulation, dollars

C_{th} = unit thermal energy cost, dollars/10^6 Btu

C_e = unit electrical energy cost, mills/kWhr

L = pipe length, ft

D = pipe nominal diameter, in.

Q = pipe insulation thickness, in.

Interest rate: 9%

Annual fixed charge rate: 16%

Annual operation time: 5000 hr

11-6. A heat exchanger system (See Fig. 11-16) is used to heat 1000 gpm of water from 150 to 300 F by two process steams. These two steams are saturated and at the pressures 50 and 100 psia respectively. The overall heat transfer coefficient of these two heaters is approximately 500 Btu/hr-ft²-F. The low-pressure and high-pressure steam cost, respectively, $2.0 and $2.25 per thousand pounds. The initial cost of low-pressure heater is expressed in form of $10(A)$ where A is the heater area in square feet. The high-pressure heater costs 10% more than the low-pressure. Estimate the optimal heat load distribution between these two heaters. The conditions below should be used for evaluation:

Operation time	5000 hr/yr
Annual fixed charge rate	15%
Economic life	20 yr

11-7. Estimate by the Fibonacci method the maximum value of the objective function

$$Y = 12X - 2X^2 + 10 \qquad \text{for} \qquad 0 \le X \le 10$$

and determine the region of uncertainty.

11-8. Repeat Problem 11-7 by the dichotomous method.

11-9. Repeat Problem 11-7 by the sequential uniform search method.

11-10. Determine by the method of steepest descent the minimum value of the function

$$Y = X_1^2 + 3X_1^2X_2^2 - 2X_1X_2X_3 + 5X_3^2$$

Assume that the starting point is $(-2, 3, 4)$.

11-11. The objective function is given below

$$Y = 3X_1^2 - 6X_1X_2 + 4X_2^2$$

Determine by the univariate search method the minimum value of the function.

11-12. Search for the minimum of the objective function $Y = 6X_1^2 + 3X_2^2$ by the univariate search method.

11-13. Repeat Problem 11-12 by the pattern search method.

11-14. Air is to be compressed from 25 to 2500 psia. The compression will be carried out by a centrifugal compressor in series with a reciprocating compressor. The intercooler between them will return the temperature of the air to the inlet condition 120 F. The equations for the initial cost of the

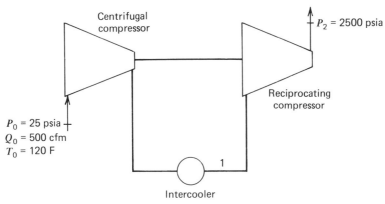

Figure 11-17.   Compression system for Problem 11-14.

compressors are

$$C_{\text{cent}} = 2(Q_c) + 1600\left(\frac{P_1}{P_0}\right)$$

$$C_{\text{recip}} = 6(Q_r) + 800\left(\frac{P_2}{P_1}\right)$$

where Q_c and Q_r are the air flow rate in cubic feet per minute, respectively, for centrifugal and reciprocating compressor. Fig. 11-17 presents the compression system and its information. Construct the mathematical model for optimizing the total initial cost. Also estimate the corresponding pressure ratio.

11-15. A hot-water tank (10×10 ft) is to be insulated. Determine the optimum insulation thickness using the following data:

Average water temperature	200 F
Average ambient temperature	75 F
Conductivity of insulation	0.02 Btu/hr/ft/F
Cost of heat	$0.02/10^3$ Btu
Cost of insulation	$15/ft^3$
Economic life	10 yr
Interest rate	12%

The criteria in the optimization is the minimum total cost of the insulation and the standby heating.

CHAPTER 12

Cooling System Design

12.1 INTRODUCTION

It has been well established by the second law of thermodynamics that heat rejection at the low temperature of steam power cycle is necessary to the power cycle's operation. The magnitude of waste heat is significant. For every kilowatt hour of electricity generated, there are approximately two kilowatt hours of waste heat emitted to the atmosphere. It must be pointed out that not all waste heat will be rejected through the cooling system. In fact, about 10% goes through the boiler stack. The waste heat from cooling system is either in the form of hot-dry air or in the form of hot-humid air.

There are many methods available to deal with the thermal discharge of power plant cooling systems. One of the fundamental approaches is to improve the plant thermal efficiency. Changes in the plant efficiency have a marked effect on the ratio of heat rejection to power generation. Figure 12-1 indicates that this ratio decreases from 2.33 to 1.0 when the plant efficiency improves from 30 to 50%.

Another approach to the reduction of thermal discharge is to utilize the waste heat. In theory, the idea is simple and attractive. In practice, however, there are many technical and nontechnical problems associated with an utilization of waste heat. One of the technical problems is the low temperature at which the waste heat is available. The matching between the waste heat supply and demand is a serious obstacle. The nontechnical problems are obviously in the area of development planning and cooperation among the institutions. For these reasons, utilization of waste heat has been carried out only on a small scale in the nation.

The third fundamental approach to the thermal discharge is to dispose of the waste heat. While this approach does not eliminate the waste heat, it would disperse the waste heat in such a manner that the impacts on the environment become tolerable. Ever since 1882 when the Pearl Street Station was operated in New York, the power industry has developed a variety of cooling systems for dissipating waste heat from thermal power plants. These cooling systems are once-through condenser cooling, evaporative cooling towers (sometimes referred to as wet towers), cooling ponds, and dry-type cooling towers.

In a once-through cooling system the waste heat from power plants is transferred directly to the cooling water, which would raise the temperature in the receiving water body. In some cases this temperature rise may adversely affect the ecology of water body. In addition, the once-through cooling system continuously requires a significant amount of water supply during operation. Because of this dependency on the water availability, the plant sites are generally limited to the locations near natural water bodies such as large rivers, lakes and estuaries.

474

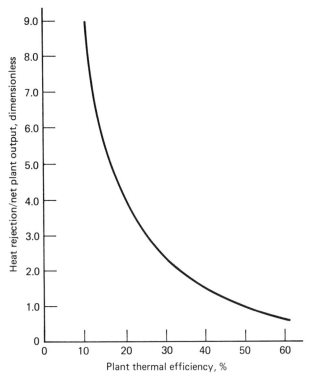

Figure 12-1. Heat rejected per unit of output versus the plant thermal efficiency.

Evaporative cooling towers, as described in Chapter 8, provide a new freedom in locating power plants. In the evaporative cooling towers the water is cooled by contact with air. The water and air flows are directed in such a way as to provide maximum heat transfer. The heat is transferred to the air primarily by evaporation (about 75%) while the remainder is accomplished by convection. The cooling water requirement for evaporative cooling towers is much smaller than that for the once-through cooling. For this reason, power plants with evaporative cooling towers can be located at a distance from the water body.

For economic reasons a cooling pond may also be used as a cooling system for power plants. The heat transfer mechanisms in the cooling pond are similar to those of cooling towers. That is, the water is cooled by contact with air. The lowest water temperature achievable by this method is equal to the ambient wet-bulb temperature. The cooling water requirement generally referred to as the makeup water, is little higher than that for the cooling towers. However, the land needed for pond construction is significant. In general, it takes one to four acres for one megawatt of generating facility.

As an alternative method for removing waste heat from power plants, the dry-type cooling towers have recently received some attention. In dry cooling towers waste heat is rejected directly to the atmosphere through a system of heat exchangers. Within these heat exchangers the circulating water flows inside the finned tubes while it transfers the heat to the moving air outside by conduction and convection. In contrast to the wet cooling towers, there is no evaporation in the dry cooling towers. Because of these features, the thermal discharge to water body, the tower evaporation and vapor plumes will not exist in the dry cooling system. It should be pointed out that the minimum water temperature attainable by dry cooling is higher than that by the wet cooling and is always greater than the ambient dry-bulb temperature. Since the heat transfer mechanism is not so effective as compared with the evaporative heat transfer, the condenser pressure will generally result in a higher level and therefore, the plant performance will be adversely affected.

To have a good understanding of a cooling system, one should not isolate the cooling system from other components in the power generating system. Figure 12-2 shows a typical arrangement. Cooling water first receives the heat in condensers and then transfers it away in the cooling system by various heat transfer mechanisms. Except once-through cooling, cooling water generally goes through a closed circuit. There are many interactions among the components. In power plant design, engineers do not look for the best performance of one component, but look for the optimal overall system performance.

It is of interest to know the amount of cooling water required for power plant operation. The estimation chart shown in Fig. 12-3 has been prepared with the following equations:

$$\dot{Q} = 500 \dot{m} C_p (TR) \tag{12-1}$$

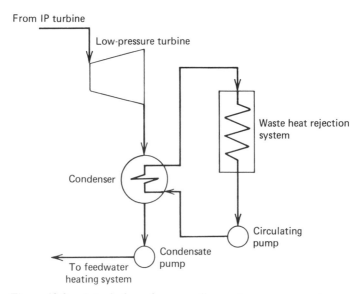

Figure 12-2. A typical condenser-cooling system arrangement.

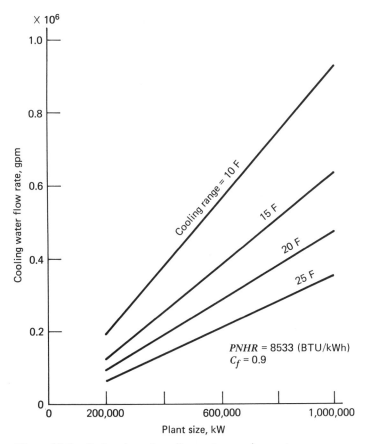

Figure 12-3. Estimation of cooling water requirement.

and

$$\overset{\circ}{Q} = (PNHR - 3413) \times C_f \times (PSIZE)$$ (12-2)

where

$\overset{\circ}{Q}$ = waste heat from cooling system (Btu/hr)

$\overset{\circ}{m}$ = cooling water flow rate rate (gpm)

TR = temperature rise in condenser (F)

C_p = specific heat of cooling water (Btu/lb/F)

$PNHR$ = plant net heat rate (Btu/kWh)

C_f = correction factor (dimensionless)

$PSIZE$ = plant size (kW)

For example, the cooling water will amount to 368,000 gpm for a 1000 MW power plant when the temperature rise is taken at 25 F. It should be pointed out that this cooling water is not entirely conserved in the process. There will be some minor losses in the system such as evaporation, drift and blowdown. But these losses will constitute about four to five percent of the total cooling water.

The following sections present a brief introduction of various heat rejection systems. These include once-through cooling, cooling ponds and dry-type cooling towers. The evaporative cooling tower is presented in Chapter 8. Also in the following sections are the evaluation of various cooling systems, taking into consideration some important economic and technical factors.

12.2 ONCE-THROUGH COOLING

Once-through cooling was frequently used as the heat rejection method in the past. The main problem of this method is to design the intake and outfall channels to properly and efficiently handle the appropriate cooling water flow. The temperature rise of cooling water in the condenser may be calculated from Eq. (12-1). Rearranging the equation gives

$$TR = \mathring{Q}/500 \mathring{m} C_p \tag{12-3}$$

The warm water is then discharged back to the original water body. The added heat is dissipated to the earth's atmosphere and the water temperature will eventually return to normal. In once-through cooling system design, the flow and temperature patterns around the discharge outlet must be determined and are used to estimate the level of thermal impacts. This can be done either by mathematical modelling or by experiments. These topics are beyond the scope of this text.

In recent years the federal and local governments have established a number of water quality regulations. In general, they are very stringent and have made the once-through cooling system very difficult, if not impossible, to be implemented in power plant construction. Another reason for its declining use is the huge cooling water requirement. This water dependency will limit plant sites only to the locations near large rivers, lakes and estuaries.

It is of interest to note from Eq. (12-3) that the temperature rise through the condenser is inversely proportional to the water flow rate. It means that a reduction in water flow rate will result in an increase of temperature rise. The larger the temperature rise, the higher the steam condensing temperature will be for a given inlet temperature. This will lead to a higher back pressure and thus, a higher turbine heat rate and a loss of generating capability. In other words, a saving in pump work (both initial and operating costs) is often offset by a deteriorating turbine performance.

When the turbine exhaust pressure is determined in advance, a reduction in water flow rate will mean an increase of condenser temperature rise and therefore an increase of condenser heat transfer surface. In this case designers must balance the saving in pumping against the increase of condenser cost.

12.3 COOLING POND

Cooling ponds are increasingly used as a source of circulating water for power plant operation. This is especially true when land is available at a reasonable cost and where geographic conditions are suitable for pond construction. In general, a cooling pond is the simplest, cheapest, and least expensive method. Also, this cooling method makes it possible to develop the power plant near the fuel source or the system load center. However, cooling ponds have low heat transfer rates, thereby requiring large surface area.

A cooling pond is a body of water in which the circulating water will dissipate its heat into the atmosphere and reduce its temperature. The heat transfer mechanisms between the water surface and the atmosphere are complicated. They involve radiation, convection, conduction and evaporation. The components of heat transfer rate on the water surface may be written as follows:

$$\phi_n = \phi_s - \phi_{sr} + \phi_a - \phi_{ar} - \phi_{br} - \phi_e - \phi_c \qquad (12\text{-}4)$$

where

ϕ_n = net heat input to a water body

ϕ_s = short-wave radiation incident to the water surface

ϕ_{sr} = reflected short-wave radiation

ϕ_a = incoming long-wave radiation from the atmosphere

ϕ_{ar} = reflected long-wave radiation

ϕ_{br} = long-wave back radiation emitted by the water body

ϕ_e = energy utilized by evaporation

ϕ_c = energy conducted from the water body as sensible heat

The incoming short-wave radiation is the solar radiation transmitted directly from the sun to the earth. Its magnitude generally depends on the altitude of the sun, which varies daily and seasonally for a given location on the earth. It is also influenced by the conditions of cloud cover. The incoming long-wave radiation is the atmospheric radiation which is originated from the gases in the atmosphere, such as water vapor and carbon monoxide. The magnitude of this long-wave radiation depends on the ambient conditions as well as the cloud conditions. Not all incoming radiation reaching a body of water will pass through the water surface. In fact, a fraction of this radiation is always reflected away from the water surface. The major mechanisms of heat loss from a water surface are radiation, evaporation, and conduction. The magnitude of these are dependent on the temperature of the water. From a standard heat transfer textbook it is well known that the long-wave back radiation emitted from the water surface is proportional to the fourth power of the absolute surface temperature. The amount of heat lost by evaporation is proportional to the difference in saturation vapor pressure at the water surface temperature and the water vapor pressure in the ambient air. The heat conducted from the surface is simply related to the temperature difference between the water and the air. It should be pointed out again that all these heat transfers are surface phenomena, the magnitude depending upon the water surface temperature directly or indirectly. Typical values of these terms are presented in Fig. 12-4.

In practice, all the incident and reflected radiation terms are frequently grouped in a net radiation term ϕ_r. In doing so, Eq. (12-4) becomes

$$\phi_n = \phi_r - (\phi_{br} + \phi_e + \phi_c) \tag{12-5}$$

ϕ_r is a function of meterological variables at the plant site and the quantity $\phi_{br} + \phi_e + \phi_c$ depends in part on the water surface temperature. Below some formulas are suggested for these estimates.

The back radiation (ϕ_{br}) emitted from the water surface is approximated by the Stephan-Boltzman equation

$$\phi_{br} = \varepsilon\sigma(T_s + 460)^4 \tag{12-6}$$

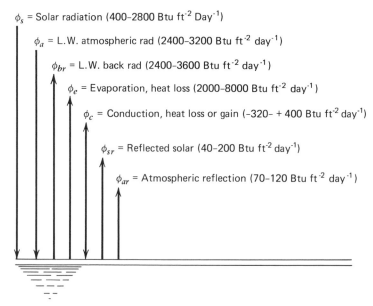

ϕ_s = Solar radiation (400–2800 Btu ft^{-2} Day^{-1})

ϕ_a = L.W. atmospheric rad (2400–3200 Btu ft^{-2} day^{-1})

ϕ_{br} = L.W. back rad (2400–3600 Btu ft^{-2} day^{-1})

ϕ_e = Evaporation, heat loss (2000–8000 Btu ft^{-2} day^{-1})

ϕ_c = Conduction, heat loss or gain (–320– + 400 Btu ft^{-2} day^{-1})

ϕ_{sr} = Reflected solar (40–200 Btu ft^{-2} day^{-1})

ϕ_{ar} = Atmospheric reflection (70–120 Btu ft^{-2} day^{-1})

Figure 12-4.   Typical values of heat inputs across a water surface [6].

where

ε = emissitivity of water surface, approximately 0.97

σ = Stephan-Boltzman constant 4.15×10^{-8} Btu/ft²-day-R⁴

T_s = water surface temperature, F

In calculation, linearization of Eq. (12-6) is sometimes very desirable. To do that, we first express Eq. (12-6) by using a binomial expansion series. That is,

$$\phi_{br} = \varepsilon\sigma(460)^4\left[1 + 4\left(\frac{T_s}{460}\right) + 6\left(\frac{T_s}{460}\right)^2 + 4\left(\frac{T_s}{460}\right)^3 + \cdots\right] \qquad (12\text{-}7)$$

When we omit the second and higher order terms, Eq. (12-7) becomes

$$\phi_{br} = \varepsilon\sigma(460)^4\left(1 + 4\frac{T_s}{460}\right) \qquad (12\text{-}8)$$

or

$$\phi_{br} = 1800 + 15.7T_s \qquad (12\text{-}9)$$

The heat loss by evaporation (ϕ_e) is the product of the water latent heat and the evaporation loss from the water surface. The evaporation loss from the water surface to the atmosphere is usually represented empirically by the equation

$$m_{ev} = \rho f(v)(e_s - e_a) \qquad (12\text{-}10)$$

where

m_{ev} = evaporation loss, (lb/day-ft^2)

ρ = water density (lb/ft^3)

e_a = water vapor pressure at T_a (in. Hg abs.)

e_s = saturated water vapor pressure at T_s (in. Hg abs.)

$f(v)$ = wind empirical function to be defined later (ft/day-in. Hg)

The latent heat for water depends on the water temperature. In the temperature range of interest, the relationship is approximately

$$L = 1087 - 0.54T_s \qquad (12\text{-}11)$$

where L = latent heat of water, Btu/lb. Combining Eqs. (12-10) and (12-11) will yield an expression for the heat loss by evaporation. That is,

$$\phi_e = Lm_{ev}$$

$$\phi_e = (1087 - 0.54T_s)(\rho)f(v)(e_s - e_a)$$

$$\phi_e = (68,000 - 34T_s)f(v)(e_s - e_a) \tag{12-12}$$

The empirical wind function $f(v)$ is mainly related to the wind velocity. However, the exact function is still unknown. Various forms have been proposed by many investigators. One of these is the Brady's equation [6]

$$f(v) = 0.0254 + 0.000254V^2 \tag{12-13}$$

where

$$V = \text{wind velocity (mph)}$$

Equation (12-12) is used to estimate the heat loss by evaporation, if the conditions T_s, T_a, and V are given. As indicated in the next section, it is sometimes desirable to replace the vapor pressure differential in Eq. (12-12) by some temperature differential. Using the thermodynamic properties of water vapor, we approximate $(e_s - e_a)$ in terms of temperature [6]:

$$e_s - e_a = \beta(T_s - T_d) \tag{12-14}$$

where T_d is the dew point temperature of the air. The proportionality factor β is a function of temperature and expressed as

$$\beta = 0.01 - 0.000335T^* + 0.0000080T^{*2} \tag{12-15}$$

where

$$T^* = (T_s + T_d)/2 \tag{12-16}$$

The heat loss by conduction (ϕ_c) is equal to the product of a heat transfer coefficient and the temperature differential between the air and water surface. In general, this heat loss is an order of magnitude less than other losses. Bowen suggested that the conduction loss is proportional to the heat loss by evaporation [6]. That is,

$$\phi_c = \phi_e R_B \left(\frac{T_s - T_a}{e_s - e_a} \right) \tag{12-17}$$

where

$$R_B = \text{the Bowen ratio, numerically equal to } 0.0102 \text{ in. Hg/F}$$

Substituting Eq. (12-12) into Eq. (12-17) yields

$$\phi_c = (68,000 - 34T_s)f(v)[0.0102(T_s - T_a)] \tag{12-18}$$

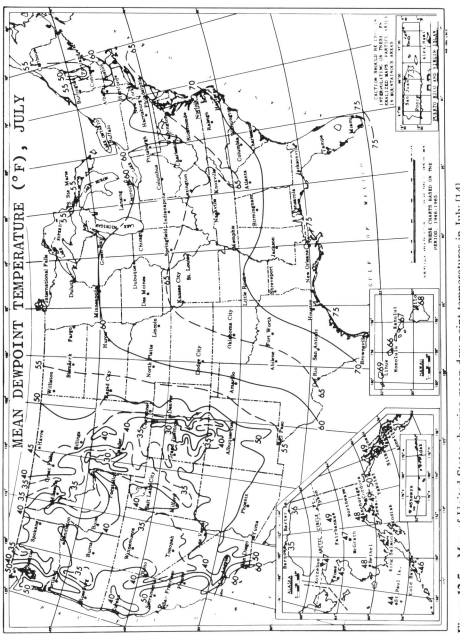

Figure 12-5. Map of United States showing mean dew point temperatures in July [14].

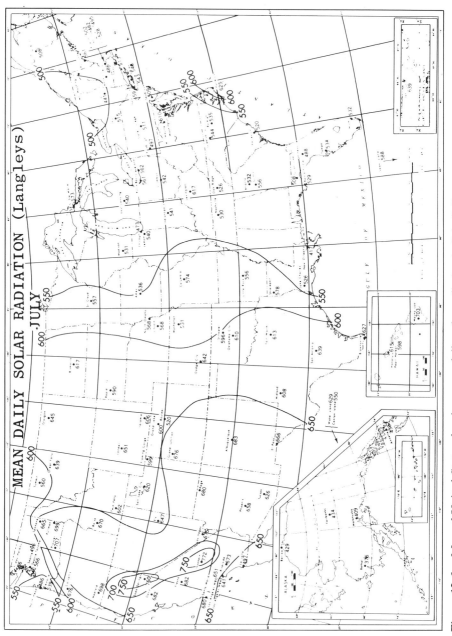

Figure 12-6.   Map of United States showing mean daily solar radiation in July [14].

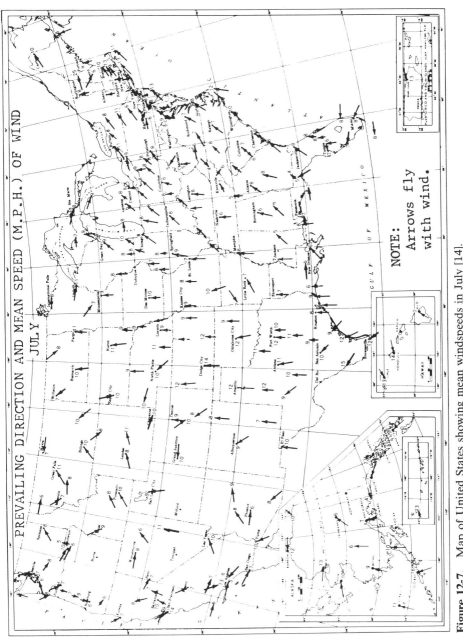

Figure 12-7. Map of United States showing mean windspeeds in July [14].

The radiation absorption (ϕ_r) is obviously a site-related function. The value of this function should be determined by using the climatic data for the particular site. Figures 12-5 through 12-7 are typical maps indicating the average monthly values for the dew point temperature, gross solar radiation, and wind speed in the nation. More climatic data can be found in the publication *Climatic Atlas of the United States* [14].

Using the formulas just derived, the net input to a body of water will become

$$\phi_n = \phi_r - 4 \times 10^{-8}(T_s + 460)^4 - (68{,}000 - 34T_s)f(v)$$
$$\times [(e_s - e_a) + 0.0102(T_s - T_a)] \tag{12-19}$$

It should be noted the evaporation term is valid only for $e_s > e_a$. During the time period when $e_s < e_a$, the evaporation term should be simply set equal to zero.

12.3.1 Equilibrium Temperature and Surface Heat Exchange Coefficient

In cooling pond design the concept of equilibrium temperature T_e is frequently utilized. The equilibrium temperature is defined as the water surface temperature at which there is no net heat transfer between the water surface and its surroundings. In other words when a water body has temperature below equilibrium temperature, it will approach the equilibrium temperature by receiving heat from the ambient. When a water body has temperature above equilibrium temperature, it will approach the equilibrium temperature by transferring heat away. From this definition, the equation used for the calculation of equilibrium temperature is obtained by setting ϕ_n to zero and T_s to T_e in Eq. (12-19). That is,

$$\phi_r = 4 \times 10^{-8}(T_e + 460)^4 + (68{,}000 - 34T_e)f(v)$$
$$\times [(e_e - e_a) + 0.0102(T_e - T_a)] \tag{12-20}$$

Equation (12-20) is an implicit equation for equilibrium temperature. For a given net radiation ϕ_r, wind empirical function $f(v)$, partial vapor pressure at the atmosphere e_a, the equilibrium temperature T_e can be calculated by a trial and error method.

Using the approximations developed in the last section and letting the quantity $(68{,}000 - 34T_e)$ be equal to 66,300 (assume $T_e = 50$ F), the equilibrium temperature is given by

$$T_e = \frac{\phi_r + 66{,}300 f(v)(\beta T_d + 0.0102 T_a) - 1800}{15.7 + 66{,}300 f(v)(\beta + 0.0102)} \tag{12-21}$$

Another term frequently used in the cooling pond design is the surface heat exchange coefficient K. This term K relates the change of heat transfer rate to the change of water surface temperature. That is,

$$K = -\frac{\partial \phi_n}{\partial T_s} \tag{12-22}$$

The negative sign results from the fact that the net heat input to a water body will decrease as the surface temperature T_s increases. Substituting Eq. (12-19) into Eq.

(12-22) and again, assuming $68{,}000 - 34T_s = 66{,}300$, we have

$$K = (4 \times 10^{-8})(4)(T_s + 460)^3 + 66{,}300f(v)(\beta + 0.0102) \qquad (12\text{-}23)$$

To further simplify this equation, we let

$$(T_s + 460)^3 \doteq 460^3$$

and then, have the approximation equation as

$$K = 15.7 + 66{,}300f(v)(\beta + 0.0102) \qquad (12\text{-}24)$$

Finally, using the empirical wind function suggested by Brady, Eq. (12-24) becomes

$$K = 15.7 + 1680(1 + 0.01V^2)(\beta + 0.0102) \qquad (12\text{-}25)$$

Figure 12-8 presents a chart for quick estimate of the surface heat exchange coefficient.

EXAMPLE 12-1. Estimate the equilibrium temperature and surface heat exchange coefficient for the conditions: wind speed 10 mph, ambient dew point temperature 72 F, ambient air temperature 92 F, and net radiation absorption 3000 Btu/day-ft^2.

Solution: The calculational procedures are as follows:

1. Assume $T_e = 74$ F.

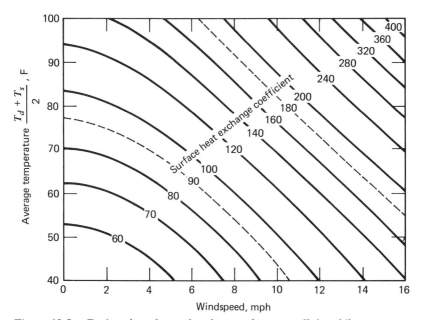

Figure 12-8. Design chart for surface heat exchange coefficient [6].

2. Find K by using Eq. (12-25)

$$T^* = \tfrac{1}{2}(T_s + T_d) = \tfrac{1}{2}(74 + 72) = 73 \text{ F}$$

$$\beta = 0.01 - 0.000335T^* + 0.000008T^{*2}$$

$$\beta = 0.0282$$

$$K = 15.7 + 1680(1 + 0.01V^2)(\beta + 0.0102)$$

$$= 15.7 + 1680(1 + 0.01 \times 10^2)(0.0282 + 0.0102)$$

$$= 144.7 \frac{\text{Btu}}{\text{day-ft}^2\text{-F}}$$

3. Find T_e by using Eq. (12-21)
   Since $\phi_r = 3000 \text{ Btu/day-ft}^2$, $T_a = 92 \text{ F}$

   $$\beta = 0.0282, \quad T_d = 72 \text{ F}$$

   and $f(v) = 0.0254 + 0.000254V^2$

   $$= 0.0508$$

   we have

$$T_e = \frac{3{,}000 + 66{,}300 \times 0.0508(0.0282 \times 72 + 0.0102 \times 92) - 1800}{15.7 + 66{,}300 \times 0.0508(0.0282 + 0.0102)}$$

$$T_e = 77.2 \text{ F}$$

Since the calculated T_e is different from the assumed value (74 F), we repeat this calculational process by using the new assumption $T_e = 76$ F. The final results are

$$K = 147 \frac{\text{Btu}}{\text{day-ft}^2\text{-F}}$$

and

$$T_e = 76.2 \text{ F}$$

12.3.2 Pond Size Estimate

Cooling ponds are used in areas where river flows are not adequate to meet the demand for cooling water. The cooling pond layout is determined by topography, available land, and ease of construction at a particular plant site. To be more specific, the required pond surface area is definitely related to the circulation pattern and temperature distribution in the pond, which are in turn determined by the pond geometry (shape and depth), placement of intake and discharge. In addition, the meteorological factors, such as wind speed, will play an important role.

The general objectives in cooling pond design are to dissipate the heat rejected by the power plant and to yield the lowest possible intake temperature of the cooling water. These two objectives are not the same. It is possible to satisfy the first objective without achieving the lowest intake temperature. For a given set of meteorological conditions, the intake temperature is largely determined by the degree of longitudinal mixing and the uniformity of flow distribution between the discharge and intake locations. Both longitudinal mixing and nonuniform flow distribution tend to convey warmer water from the discharge region to the intake region, and thereby increase the intake temperature. In design, these situations should be avoided or eliminated as much as possible.

There have been no general method reported in literature relating to the cooling pond design. Here we plan to discuss only two simple ponds: completely mixed and flow-through pond. For these two, mathematical models have been well formulated and accepted.

The Completely Mixed Pond

A completely mixed pond is the pond in which the water surface temperature is almost uniform except in a small region near the plant discharge. This condition seems to exist in ponds where the surface area is large compared to the plant pumping rate, and the pond temperature is uniform and several degrees above the equilibrium temperature. In other words, the induced temperature rise above the ambient is small and in the order of 5 to 10 F. Since the water surface temperature (T_s) is uniform in the pond, the heat transfer from the water surface using the concept of the equilibrium temperature is

$$\mathring{Q} = KA(T_s - T_e) \qquad (12\text{-}26)$$

where

$\mathring{Q}$ = heat dissipation from the pond Btu/day

K = exchange coefficient $(\text{Btu/day-ft}^2\text{-F})$

A = pond surface area (ft^2)

T_s = surface temperature (F)

T_e = equilibrium temperature (F)

In a steady-state, steady-flow condition this heat dissipation rate is equal to the difference between the heat into the pond and the heat out of the pond. This net heat input is also approximately the amount of heat removed from the condenser. In terms of the plant intake and discharge temperature, it is

$$\mathring{Q} = \rho C_p \mathring{Q}_p (T_{dis} - T_i) \qquad (12\text{-}27)$$

where

ρ = water density (lb/ft^3)

C_p = water specific heat (Btu/lb-F)

$\mathring{Q}_p$ = water flow rate (ft^3/day)

T_{dis} = water discharge temperature (F)

T_i = water intake temperature (F)

In a completely mixed pond, the intake temperature is equal to the pond surface temperature. Combining Eq. (12-26) and Eq. (12-27) gives

$$\rho C_p \mathring{Q}_p (T_{dis} - T_s) = KA(T_s - T_e) \tag{12-28}$$

Then, the pond surface temperature (or intake temperature) is expressed as

$$T_s = \frac{T_{dis} + rT_e}{r + 1} \tag{12-29}$$

where

$$r = \frac{KA}{\rho C_p \mathring{Q}_p} \tag{12-30}$$

Equation (12-29) can also be expressed as

$$\frac{T_s - T_e}{T_{dis} - T_e} = \frac{1}{1 + r} \tag{12-31}$$

In both Eq. (12-29) and Eq. (12-31) the parameter r is dimensionless. At the location near the plant discharge (i.e., $r = 0$), the pond temperature is equal to the plant discharge temperature T_{dis}. As the parameter r increases (i.e., the pond surface area increases), the pond surface temperature will decrease and approach the equilibrium temperature as a limit. This is clearly shown in Fig. 12-9.

The parameter r is sometimes expressed in terms of the average pond depth and detention time. Let the average detention time of the pond be t_d, and the pond

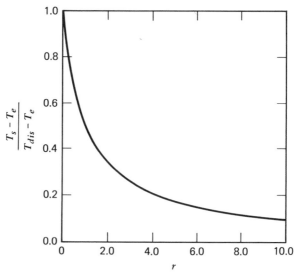

Figure 12-9. Temperature excess for a completely mixed cooling pond.

volume be V, the relationship between them will be

$$t_d = \frac{V}{\mathring{Q}_p}$$ (12-32)

Substituting Eq. (12-32) into (12-30) gives

$$r = \frac{Kt_d}{\rho d C_p}$$ (12-33)

where

$$d = \text{the average pond depth, ft}$$

The Flow-Through Pond

A flow-through pond is the pond in which the condenser discharge moves like a plug flow, with no mixing with the receiving water. As indicated later, the water surface temperature decreases exponentially from the plant discharge. The time required for water to pass through the pond depends on the rate of pumping and the pond volume while the temperature drop from the plant discharge to the plant intake depends on the rate of pumping and the pond surface area. Since the water surface temperature varies, the heat transfer rate from the pond surface is not expected to be uniform.

The equation predicting the temperatures for a flow-through pond is derived in a manner similar to that for the completely mixed pond. The heat transfer rate from the pond surface is

$$\mathring{Q} = \int_0^A K(T_s - T_e)\,dA$$ (12-34)

In a steady-state, steady-flow condition, this heat transfer rate is equal to the amount of heat removed from the plant condenser. That is,

$$\rho C_p \mathring{Q}_p (T_{dis} - T_i) = \int_0^A K(T_s - T_e)\,dA$$ (12-35)

To obtain the equation for the pond surface temperature, we relate the heat dissipation rate from the pond surface area dA to the pond surface temperature drop by

$$-\rho C_p \mathring{Q}_p\,dT_s = K(T_s - T_e)\,dA$$ (12-36)

or

$$-\rho C_p \mathring{Q}_p \frac{dT_s}{dA} = K(T_s - T_e)$$ (12-37)

This is the first-order differential equation with the boundary condition $T_s = T_{dis}$ at $A = 0$. The solution for this differential system is

$$\frac{T_s - T_e}{T_{dis} - T_e} = e^{-r}$$ (12-38)

where

$$r = \frac{KA}{\rho C_p \mathring{Q}_p}$$ (12-39)

Figure 12-10 presents a plot of $(T_s - T_e)/(T_{dis} - T_e)$ versus the dimensionless parameter r. It should be noted that the surface temperature (T_s) in Eq. (12-38) is not constant and must be treated as the surface temperature at the location where the pond surface area is A between this point and the plant discharge.

EXAMPLE 12-2.   A plant discharges cooling water into a cooling pond at the rate of 3×10^6 ft^3/hr. The discharge and intake temperatures are, respectively, 100 F and 85 F. A previous study indicates that the equilibrium temperature at the site is 72 F and the surface exchange coefficient is 150 Btu/day-ft^2-F. Estimate the required pond surface area if the pond is of completely mixed type or of flow-through type.

Solution: First, we used Eq. (12-31) for the completely mixed pond and found

$$\frac{85 - 72}{100 - 72} = \frac{1}{1 + r}$$

or

$$r = 1.15$$

Using Eq. (12-30), we estimate the required pond surface area as

$$A = \frac{r\rho C_p \mathring{Q}_p}{K}$$

$$A = \frac{(1.15)(62.4)(1.0)(24 \times 3 \times 10^6)}{150}$$

$$A = 34.44 \times 10^6 \text{ ft}^2 \quad \text{or} \quad 790.9 \text{ acres}$$

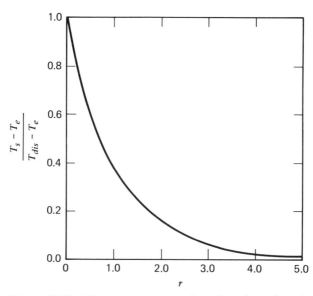

Figure 12-10.   Temperature excess for a flow-through cooling pond.

Next we used Eq. (12-38) for the flow-through pond and found

$$e^{-r} = \frac{85 - 72}{100 - 72}$$

$$r = 0.77$$

Substituting $r = 0.77$ into Eq. (12-39) gives

$$A = \frac{(0.77)(62.4)(1.0)(24 \times 3 \times 10^3)}{150}$$

$$A = 23.06 \times 10^6 \text{ ft}^2 \quad \text{or} \quad 529.6 \text{ acres}$$

EXAMPLE 12-3. A 1100 acre cooling pond is available for a coal-fired power plant. The plant has two 800 MW generating units with total condenser heat load 2×4080 MBtu/hr. A previous analysis recommended that the condenser temperature rise should be 28 F. Estimate the condenser pressure for these two units if the pond is of the completely mixed type.

Assume that the equilibrium pond temperature is 83 F and the surface heat exchange coefficient 161 Btu/day-ft²-F.

Solution: To calculate the parameter r, we modify Eq. (12-30) as

$$r = \frac{KA\Delta T_p}{\mathring{Q}} \tag{a}$$

where ΔT_p is the condenser temperature rise, and $\mathring{Q}$ is the total condenser heat load in terms of Btu/day. Using

$$\mathring{Q} = 2 \times 4080 \text{ MBtu/hr} = 195.8 \times 10^9 \text{ Btu/day}$$

and

$$A = 1100 \text{ acres} = 47.9 \times 10^6 \text{ ft}^2$$

we have

$$r = \frac{(161)(47.9 \times 10^6)(28)}{195.84 \times 10^9} = 1.1$$

To determine the discharge and intake temperatures, we utilize the two relationships as follows

$$\frac{T_i - T_e}{T_{dis} - T_e} = \frac{1}{1 + r} \tag{b}$$

and

$$T_{dis} - T_i = 28 \qquad \text{(c)}$$

Solving Eq. (b) and Eq. (c) gives

$$T_{dis} = 136.8 \text{ F}$$

$$T_i = 108.8 \text{ F}$$

Assuming the condenser terminal temperature difference is 5 F, the condenser steam temperature will be 141.8 F. Finally, the condenser pressure is 6.03 in. Hg abs.

In the above example the condenser pressure estimate was based on the assumption that the cooling pond is a completely mixed pond. If the cooling pond is a flow-through pond, the condenser pressure is expected to be lower and in fact approximately equal to 4.54 in. Hg abs.

It is of interest to compare the performance characteristics of a completely mixed pond with that of a flow-through pond. When these two ponds have the same discharge and intake temperature, the completely mixed pond will require a greater surface area than a flow-through pond. This was numerically confirmed in Example 12-2. A derivation of the general relationship is reserved as a student excercise. When these two ponds have the same water surface area, the intake temperature of a completely mixed pond will be greater than that of a flow-through pond.

Occasions may arise that the cooling pond is neither a completely mixed pond nor a flow-through pond. For other types of cooling pond, readers should consult the literatures in this field [5].

12.3.3 Evaporation Loss

In this section the equations are presented for predicting the evaporation loss from a cooling pond. Attention is focused on the completely mixed pond and flow-through pond. For both ponds the evaporation rate from the water surface is assumed to be proportional to the difference between the saturation vapor pressure at the water surface temperature and the water vapor pressure in the ambient air. The constant of proportionality between the evaporation rate and the vapor pressure differential is the wind speed function defined by Eq. (12-13). Applying this principle to the differential water surface area gives

$$dQ_{ev} = f(v)(e_s - e_a)\,dA \qquad (12\text{-}40)$$

where

$$Q_{ev} = \text{evaporation rate, ft}^3/\text{day}$$

For a completely mixed pond the water surface temperature is essentially constant. Therefore, Eq. (12-40) can be integrated over the entire water surface and expressed as

$$Q_{ev} = f(v)(e_s - e_a)A \qquad (12\text{-}41)$$

Equation (12-41) involves a calculation of water vapor pressures. For an approximation, the vapor pressure differential is frequently replaced by the temperature and expressed as

$$Q_{ev} = f(v)\beta(T_s - T_d)A \qquad (12\text{-}42)$$

Making use of Eq. (12-24), Eq. (12-42) will become

$$Q_{ev} = \frac{K - 15.7 - 675f(v)}{66,300}(T_s - T_d)A \qquad (12\text{-}43)$$

It is seen that the evaporation rate is expressed in terms of the water surface temperature for a given set of plant site conditions. Eq. (12-43) may be less accurate than (12-41), but much more convenient in use. The temperature T_s is the water surface temperature and also the intake temperature to the plant. When there is no plant operation or no artificial heat source, the temperature T_s must be treated as the natural water surface temperature. This natural water temperature is closely related to the equilibrium temperature. Studies indicated that the annual cycle of natural water temperature has the same amplitude as the equilibrium temperature but lags behind about 15 days.

To determine the evaporation loss from a completely mixed cooling pond, one must first determine the evaporation loss from the water surface at its natural temperature. Since this loss exists even before the plant is operated, this loss is subtracted from the actual evaporation, which is calculated with the actual water surface temperature during the plant operation. The resultant evaporation loss is the loss induced by the plant operation. For convenience of calculation the equation for the plant induced evaporation loss is derived as follows.

Let us define the forced temperature rise ΔT_f as the water temperature increase above the natural level. The forced temperature rise is related to the plant condenser heat load by the equation

$$\Delta T_f = \frac{\mathring{Q}}{KA} \qquad (12\text{-}44)$$

where

$$\mathring{Q} = \rho C_p \mathring{Q}_p \Delta T_p \qquad (12\text{-}45)$$

The term ΔT_p is the condenser temperature rise. Combining Eq. (12-43), Eqs.

(12-44) and (12-45) will yield the plant-induced evaporation loss as

$$\frac{Q_{ev}}{\Delta T_p \mathring{Q}_p} = \frac{[K - 15.7 - 675f(v)]\rho C_p}{66,300K} \tag{12-46}$$

where

$$Q_{ev} = \text{plant-induced evaporation loss, ft}^3/\text{day}$$

Equation (12-46) indicates the fraction of plant condenser flow lost to the evaporation for every degree of condenser temperature rise. This fraction is a function of wind speed as well as the surface temperature. In general, it varies from 0.03 to 0.07 in percentage per degree of cooling. When the natural evaporation loss is taken into account, this value will become much higher.

In a flow-through pond the evaporation rate per unit surface area is expected to change with the distance from the plant discharge. Because of this, integration must be performed to determine the evaporation loss per day. Integrating Eq. (12-40) over the entire surface area gives

$$Q_{ev} = \int_A f(v)(e_s - e_a)\, dA \tag{12-47}$$

Following the same approach used for a completely mixed pond, Eq. (12-47) can be modified and expressed as

$$Q_{ev} = \frac{K - 15.7 - 675f(v)}{66,300} \int_A (T_s - T_d)\, dA \tag{12-48}$$

It is seen that the evaporation loss from a flow-through pond is in an integration form. The temperature distribution in the pond must be available before the evaporation loss can be numerically determined. Presented in the following is a numerical example for evaporation loss calculation.

EXAMPLE 12-4. The temperature distribution for a 755 acre flow-through pond has been determined and expressed as

$$T_s = 65 + 15 \exp(-33.4 \times 10^{-9} A) \tag{a}$$

where T_s is the water surface temperature in F and A is the water surface area in square feet. At the pond site the wind speed is 5 mph and the water vapor pressure in the atmosphere is 0.6 in. Hg. Estimate the evaporation loss from this cooling pond.

Table 12-1
Evaporation Loss Calculation

Location from the Plant Discharge i	Water Surface Temperature at Location i (F)	Surface Area Between the Discharge and Location i (ft²)	Surface Area Division j (ft²)	Average Water Surface Temperature in Division j (F)	Average Saturation Vapor Pressure in Division j (in. Hg)	Value of $(e_s \Delta A)_j$ (in. Hg-ft²)
1	80	0				
2	78	4.286×10^6	4.286×10^6	79	1.00	4.286×10^6
3	76	9.288×10^6	5.002×10^6	77	0.94	4.702×10^6
4	74	15.299×10^6	6.011×10^6	75	0.88	5.289×10^6
5	72	22.826×10^6	7.527×10^6	73	0.82	6.172×10^6
6	70	32.905×10^6	10.079×10^6	71	0.77	7.760×10^6

$$\sum_j (e_s \Delta A)_j = 28.2 \times 10^6$$

Solution: We first determine the value of the wind speed function as

$$f(v) = 0.0254 + 0.000254(5)^2$$

$$= 0.03175 \frac{\text{ft}}{\text{day-in. Hg}}$$

Then, we change the integration of Eq. (12-47) into the form

$$Q_{ev} = f(v)\left[\sum_j (e_s \Delta A)_j - e_a A\right] \tag{b}$$

In this calculation the pond surface is divided into five divisions. For each division we determine the product of the surface area and the saturation vapor pressure at the water surface temperature. The summary of the calculation is presented in Table 12-1. The summation of these five terms $\sum_j (e_s \Delta A)_j$ is equal to 28.2×10^6. Substituting this term into Eq. (b) will give the evaporation loss as

$$Q_{ev} = 0.03175\left[28.2 \times 10^6 - (0.6)(32.905 \times 10^6)\right]$$

$$= 268{,}759 \text{ ft}^3/\text{day}$$

The evaporation loss is a major item for the water balance of a cooling pond, but other items can not be neglected. Attention must be paid to the water loss from a cooling pond by outflow as well as seepage. Also, rainfall and runoff must be taken into consideration. For most cooling ponds the rainfall may be equal to a substantial fraction of the water loss by evaporation.

12.4 DRY-TYPE COOLING TOWER

The concept of direct cooling with air is simple and not new to industries. In petrochemical and chemical industries, air cooling has been widely used for more than four decades. Other examples are aircraft engine oil coolers, radiators in automobiles, air preheaters in steam generators, and regenerators in gas turbine systems. However, the use of air-cooled condensers or air-cooled heat exchangers for removing the waste heat from power plants is still in an infant stage. Table 12-2 shows a list of generating plants with dry-type cooling towers.

The dry cooling systems may be classified on the basis of the configuration of the fluid flow path through the air-cooled heat exchangers. The three most common types of flow path configuration are: (1) parallel flow, (2) counter flow, (3) crossflow. The most important difference between these three types lies in the relative amounts of heat transfer surface area required for a given heat load under the same conditions. The dry cooling systems can also be classified according to the types of

Table 12-2
Generating Plants with Dry-Type Cooling Towers [15]

Location	Rating (MW)	Type of Dry Tower	Year Commissioned
Rugeley, England	120	Heller	1962
Ibbenburen, Germany	150	Heller	1967
Wolfsburg, Germany	3–50	GEA Direct	1961–67
Grootvlei, South Africa	200	MAN/Birwilco (indirect)	1971
Gyongyos, Hungary	2–100	Heller	1969
	2–200	Heller	Under construction
Razdan, USSR	3–200	Heller	1970–72
Wyodak, Wyoming	22	GEA Direct	1969
	3	Direct	1962
Utrillas, Spain	160	GEA Direct	1970
Quetta, Pakistan	7.5	Baldwin–Lima–Hamilton (Direct)	1964
Bavaria	40	GEA Direct	1960
Windhok, South Africa	3–30	GEA Direct	1971
Switzerland	4.3	GEA Direct	1969
Luxemburg	13	GEA Direct	1956
Rome, Italy	2–20	GEA Direct	1957
Cologne, Germany	28	GEA Direct	1958
Sindelfingen, Germany	11 & 15	GEA Direct	1960–61
Worms, Germany	5	GEA Direct	1962
Chile	3.6	GEA Direct	1963
Ludwigshafen, Germany	38	GEA Direct	1966
Eilenburg, Germany	5.3	Heller	NA
Duaujvarus, Hungary	16	Heller	1961

condensers used. Figure 12-11 illustrates three types of dry cooling system. They are:

1. Direct system.
2. Indirect system with jet condensers.
3. Indirect system with tube condensers.

In industry, the direct system is often referred to as GEA (*Gesellschaft für Luftkondensation*) system, and the indirect system with jet condensers is named as Heller's system after the inventor Professor L. Heller. As shown in Fig. 12-11, these dry cooling systems are similar to each other except in the area of condensation. In the direct system the air-cooled heat exchanger also acts as a condenser. The exhausted steam condenses within the heat exchangers and then returns to the boiler feedwater circuit. Obviously, the pressure in the direct system is always below atmospheric pressure and air leakage may become serious. Therefore, the air leakage must be carefully considered in both design and construction. In the indirect system there are

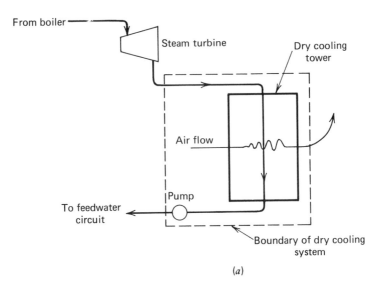

(a)

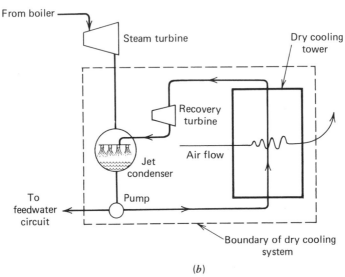

(b)

Figure 12-11. Types of dry cooling systems. (a) Direct system. (b) Indirect system with jet condensers. (c) Indirect system with tube-type condensers.

no vacuum conditions and, thus, the problem of air leakage does not exist. Compared with the direct system, indirect systems are more complex because of more pump facilities and condensers.

In all these dry cooling systems, the hot fluid (either circulating water or the condensing steam) flows inside the finned tubes while it transfers the heat to moving air outside by conduction and convection. In contrast to the wet cooling towers,

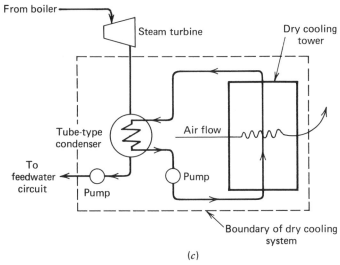

Figure 12-11. (*cont.*)

there is no evaporation in the dry cooling towers; therefore it needs practically no water supply. The air motion in the air-cooled heat exchangers is induced either by mechanical-draft fans or by natural-draft towers.

12.4.1 Air-Cooled Heat Exchanger Design

The main component of dry-type cooling system is the air-cooled heat exchanger. The heat transfer phenomena is quite simple and easily understood. Figure 12-12 illustrates a cross section of the heat exchanger tube. The hot fluid, hot water or condensing steam, flows inside the tube while the coolant (air) moves outside. The heat is transferred from the hot fluid to the inner surface of the tube by the mechanism of forced convection. The rate of heat flow q_i is

$$q_i = h_i A_i (T_f - T_i) \tag{12-49}$$

In the steady state conditions, heat is conducted through the tube wall at the same rate and is

$$q_w = \frac{K_w A_w}{\Delta w} (T_i - T_o) \tag{12-50}$$

After passing through the wall, the heat flows to the coolant by forced convection and/or free convection. The rate of heat flow q_o is

$$q_o = h_o A_o (T_o - T_a) \tag{12-51}$$

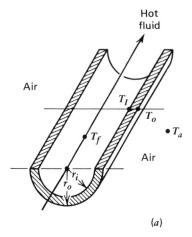

(a)

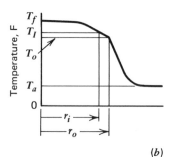

(b)

Figure 12-12.   Cross section of circular tube in the air-cooled heat exchangers. (*a*) Cross section. (*b*) Temperature distribution.

Combining Eq. (12-49), Eq. (12-50), and Eq. (12-51) produces

$$q = U_o A_o (T_f - T_a) \qquad (12\text{-}52)$$

where

$$\frac{1}{U_o A_o} = \frac{1}{h_i A_i} + \frac{\Delta w}{K_w A_w} + \frac{1}{h_o A_o} \qquad (12\text{-}53)$$

The term U_o is the overall heat transfer coefficient based on the outside surface area. It should be noticed that the fouling factors have been omitted in the derivation of Eq. (12-53). The term $\Delta w / K_w A_w$ is the thermal resistance of the tube and is generally small compared with other terms. For this reason, we may rewrite Eq. (12-53) as

$$\frac{1}{U_o A_o} = \frac{1}{h_i A_i} + \frac{1}{h_o A_o} \qquad (12\text{-}54)$$

Equation (12-54) indicates that the total thermal resistance along the heat path is the sum of two thermal resistances. One resistance is due to convection at the inner surface of the tube, while the other is the convection resistance at the outside surface. In air-cooled heat exchangers the heat transfer coefficient on the air side (h_o) is quite small, and thus the associated thermal resistance is quite large. To overcome this difficulty, we could employ the finned tube in the air-cooled heat exchanger. Accordingly, Eq. (12-54) will be modified and changed to

$$\frac{1}{U_o A_o} = \frac{1}{h_i A_i} + \frac{1}{\eta_o h_o A_o} \tag{12-55}$$

The term η_o is the surface efficiency applied to A_o which is now the total outside surface area of finned tubes. The surface efficiency is calculated by the equation

$$\eta_o = 1 - \frac{A_f}{A_o}(1 - \eta_f) \tag{12-56}$$

where A_f denotes the surface area of fins and η_f is the fin efficiency, which is defined as a ratio of heat transfer rate from the fin surface to the heat transfer rate if the entire fin surface were at the base temperature. Generally, the fin efficiency is a function of fin length, fin thickness, and fin materials. For straight and circular fins, curves in Fig. 12-13 are used for an estimate of fin efficiency. Information for other types of fins can be found in reference [13].

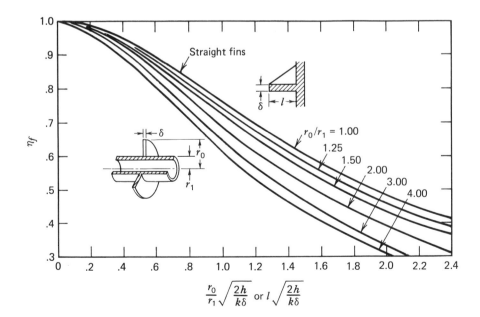

Figure 12-13. Heat transfer effectiveness of straight and circular fins [13].

To determine the overall heat transfer coefficient, convective heat transfer coefficients must be known. In general, this information is arranged in the dimensionless form as below:

$$N_{st}N_{pr}^{2/3} = F(N_R) \qquad (12\text{-}57)$$

where

$$N_{st} = \text{Stanton number}$$
$$N_{pr} = \text{Prandtl number}$$
$$N_R = \text{Reynolds number}$$

All these terms are defined and well explained in standard heat transfer textbooks. For illustration, we examine three types of finned tubes as indicated in Fig. 12-14. The technical descriptions of these finned tubes including heat transfer coefficients and friction factors are presented in Figs. 12-15, 12-16, and 12-17. Using these curves, we can easily determine the heat transfer coefficient and friction factor on the air side. While the heat transfer coefficient and friction factor on the air side are

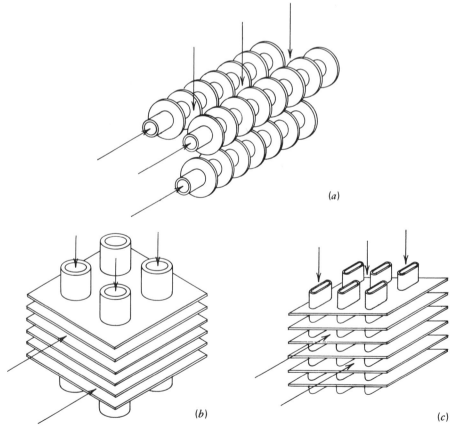

(a)

(b)

(c)

Figure 12-14. Three banks of finned tubes. (a) Circular fins and tubes. (b) Circular tubes and continuous fins. (c) Flat tubes and continuous fins.

highly dependent on the geometrical conditions of finned tubes, those on the water side are quite insensitive to these factors. In other words the heat transfer coefficient and friction factor on the water side can be determined with the generalized equations. We may use Moody's friction chart or the equation below for determination of friction factor

$$f = 0.184N_R^{-0.2} \tag{12-58}$$

For heat transfer coefficient on the water side, we use

$$N_n = 0.023N_R^{0.8}N_{Pr}^{0.33} \tag{12-59}$$

where

$$N_n = \text{Nusselt number}$$

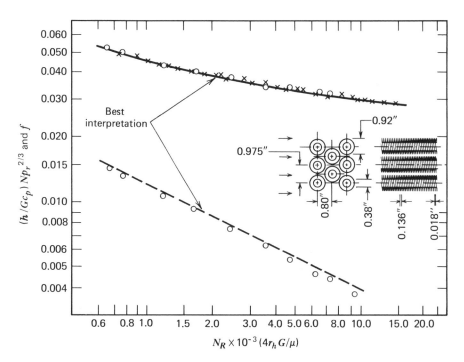

Tube outside diameter = 0.38 in.
Fin pitch = 7.34/in.
Flow passage hydraulic diameter, $4r_h$ = 0.0154 ft.
Fin thickness (average)* = 0.018 in., aluminum
Free-flow area/frontal area, σ = 0.538
Heat transfer area/total volume, α = 140 ft^2/ft^3
Fin area/total area = 0.892

Note: Experimental uncertainty for heat transfer results possibly somewhat greater than the nominal $\pm 5\%$ quoted for the other surfaces because of the necessity of estimating a contact resistance in the bi-metal tubes.
*Fins slightly tapered.

Figure 12-15. Finned circular tubes, surface CF-7.34 [13]. Reproduced with permission from McGraw-Hill.

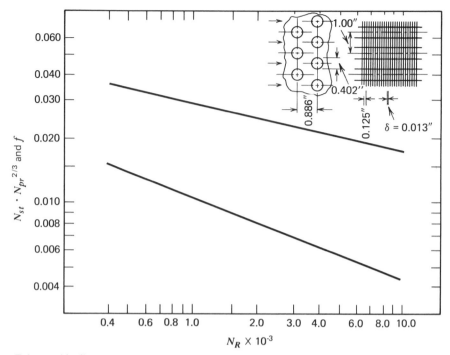

Tube outside diameter = 0.402 in.
Fin pitch = 8.0/in.
Flow passage hydraulic diameter, $4r_h$ = 0.01192 ft.
Fin thickness = 0.013 in.
Free-flow area/frontal area, σ = 0.534
Heat transfer area/total volume, α = 179 ft^2/ft^3
Fin area/total area = 0.913

Note: Minimum free-flow area in spaces transverse to flow.

Figure 12-16. Finned circular tubes, surface 8.0—3/8 T [13]. Reproduced with permission from Trane Company.

Evidently, the selection of heat transfer surface is one of the major decisions in air-cooled heat exchangers design. It affects not only the initial investment, but also the heat transfer performance as well as power consumption in operation. The optimal balance of these conflicting influences must therefore be determined.

Air-cooled heat exchanger design involves several trial-and-error procedures. The design inputs are generally the heat load, inlet temperatures, and hot water flow rate. The initial step is to assume an air temperature rise in the heat exchanger and calculate the air capacity rate (i.e., the product of air mass flow rate and its specific heat). The second step is to determine the water capacity rate. After these, we can estimate the effectiveness of the heat exchanger. As indicated in Chapter 10, the heat exchanger effectiveness is a function of the capacity-rate ratio and the number of transfer units for a given flow arrangement. When the effectiveness and capacity-rate ratio are given, the number of transfer units can be determined. This will in turn yield the product of overall heat transfer coefficient and heat transfer surface area.

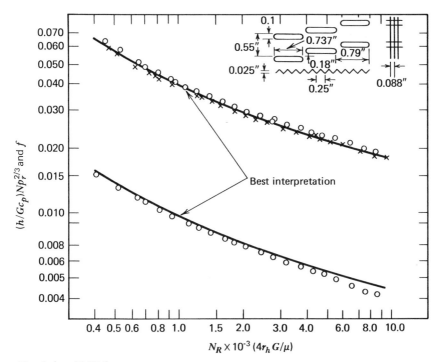

Fin pitch = 11.32/in.
Flow passage hydraulic diameter, $4r_h$ = 0.01152 ft.
Fin metal thickness = 0.004 in., copper
Free-flow area/frontal area, σ = 0.780
Total heat transfer area/total volume, α = 270 ft^2/ft^3
Fin area/total area = 0.845

Figure 12-17. Finned flat tubes, surface 11.32-0.737-SR [13]. Reproduced with permission from McGraw-Hill.

Assuming the value of U_o, based on previous experience with similar situation, the heat transfer surface area can be calculated.

A trial arrangement of this surface area is made in the most economical way as to length of tube, number of tube bundles, and depth of tube rows. At this point the air temperature rise and overall heat transfer coefficient must be calculated using this particular surface arrangement. Unless the match between the calculated and the assumed values is within the desirable limit, the above procedures must be repeated.

In air-cooled heat exchanger design, pressure drops in both streams are important. It affects not only power consumption in operation, but also the size of equipment (pumps and fans) and thus the initial cost.

12.4.2 Effects on Power Generation

The performance equation of air-cooled heat exchanger can be set up as follows:

$$ITD = aQ^b \tag{12-60}$$

where the initial temperature difference (ITD) is defined as the temperature difference between the inlet hot water and inlet air. The values of a and b are determined by field measurements. For example, the Ibbenburen and Rugeley plants, which have been in operation, have, respectively [15],

$$ITD = 0.501 \times Q^{0.717} \tag{12-61}$$

and

$$ITD = 0.247 \times Q^{0.793} \tag{12-62}$$

where Q is the exchanger heat load with a unit of 1 million Btu per hour.

The performance equation shown by Eq. (12-60) indicates that as the ambient air temperature increases, the inlet hot water temperature or condenser pressure will increase for a given heat load. At a design ITD 70 F, for example, the ambient air temperature 80 F results at the inlet hot water temperature of 150 F. This temperature will correspond to the condenser pressure 7.57 in. Hg if we neglect the terminal temperature difference in the condenser. Evidently, this condenser pressure is much higher than that allowed for conventional turbines. There are two alternatives for this situation. One is to reduce the design value of initial temperature difference by using a large air-cooled heat exchanger. Another alternative is to modify the conventional turbine so that it can operate at a high back pressure. Based on the results of several studies [9, 10], the optimal ITD for most parts of this country is in the range of 60 to 75 F. Therefore, the steam turbine for a dry-type cooling tower plant must be carefully selected so that it can operate at a back pressure higher than 5 in. Hg.

It is well known that the ambient dry-bulb temperature will vary at a much wider range than the wet-bulb temperature. In dry cooling, the turbine exhaust pressure is dependent on the ambient dry-bulb temperature and therefore changes over a wide range. If the dry-type towers are sized for summer peak, the turbine exhaust pressure could be as low as 2.0 to 2.5 in. Hg abs. in winter months. In other words, the turbine in a dry-tower plant must be capable of operating at a high back pressure as well as at a wide pressure range.

Because of high back pressure in the dry tower power plant, the turbine cycle efficiency is expected to be reduced as compared with the conventionally cooled power plant. Also, the turbine generating capability will be reduced if the throttle steam flow to the turbine remains unchanged. It is generally true in the dry-tower plant design that a relatively efficient turbine cycle when supported by a large dry tower would require a smaller balance of plant. Conversely, a less efficient turbine cycle resulting from a small dry tower installation would require a bigger balance of plant. Evidently, the dry-tower plant optimization is extremely important.

There is no doubt that the dry cooling system has several advantages over the wet-type cooling. These include no evaporation loss and its associated environmental impacts. The dry-tower plant can be located in a more desirable area, which may result in possible savings in transmission, fuel transportation and other savings.

12.4.3 Dry / Wet Combined Cooling

Throughout the years a variety of cooling systems have been developed and used for dissipating waste heat from thermal power plants. As indicated in the previous sections, these include once-through cooling, cooling ponds, evaporative cooling towers (sometimes called wet towers), and dry-type cooling towers. In addition to these basic systems, various combinations are often proposed. For instance, a wet tower is incorporated in the once-through cooling system and used as a helper tower. In recent years interest has been increasingly shown in another combination which consists of wet towers and dry towers. This combination is possible because these two towers have different characteristics and can supplement each other in the performance.

There are many possible combinations of wet and dry cooling towers. On the basis of the connection of fluid flow circuits, they can be classified as follows:

1. *S-S* combined system. Series connection on the water side and series connection on the air side.

2. *P-P* combined system. Parallel connection on the water side and parallel connection on the air side.

3. *P-S* combined system. Parallel connection on the water side and series connection on the air side.

4. *S-P* combined system. Series connection on the water side and parallel connection on the air side.

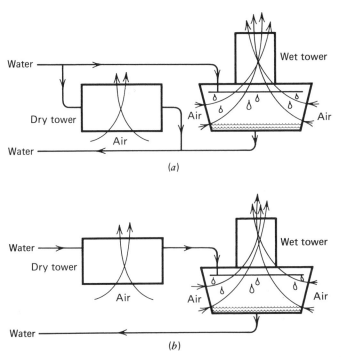

Figure 12-18. Combined cooling systems. (*a*) *P-P* combined system. (*b*) *S-P* combined system. (*c*) *P-S* combined system. (*d*) *S-S* combined system.

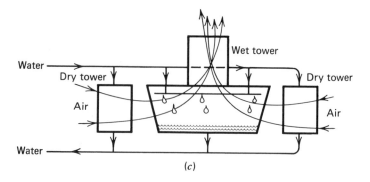

(c)

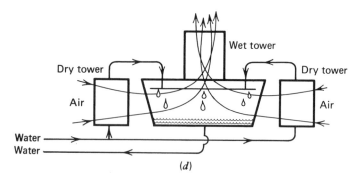

(d)

Figure 12-18. (*cont.*)

Figure 12-18 shows schematic diagrams for each combination. For simplicity, the bypass piping needed for flexibility in operation is omitted. It should be pointed out that the combined cooling could be a physically integrated unit, or simply a combination of wet towers and dry towers. In the parallel water-side arrangement, water distribution between these two sections can vary in a wide range. Depending on the objective formulated for the combined cooling system, either wet towers or dry towers could be treated as the helper tower, which will operate occasionally or at the time it is needed.

The selection of combined cooling systems is a major decision in design. It is generally influenced by the design objective. There are several possible objectives for using the combined cooling system:

1. To incorporate dry towers in the wet cooling system and reduce or eliminate the vapor plume from towers.

2. To incorporate wet towers in the dry cooling system and reduce the condenser pressure.

3. To construct power plants at the location where not enough water is available for the wet cooling system.

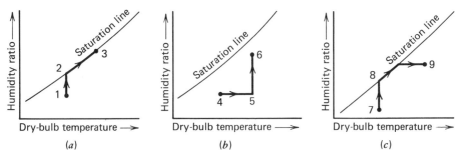

Figure 12-19. Processes in the combined cooling system.

Considering these objectives in order, the first objective is best illustrated in the simplified psychrometric charts shown in Fig. 12-19. The process 1–2–3 represents the process of cooling air in the wet towers. It is seen that the exiting air, denoted by point 3, is fully saturated and its relative humidity is 100%. It is this 100% relative humidity in combination with other climatological conditions that will give rise to the vapor plume from wet towers. Reducing this relative humidity of exiting air will certainly lead to a reduction of vapor plume. In general, the reduction of relative humidity can be realized by raising the air temperature through heating. The process 4–5–6 in Fig. 12-19 represents the case in which the air is heated in the dry towers before it enters the wet towers. Similarly, the process 7–8–9 represents the case in which the air is heated before leaving the wet towers. Both processes indicate that the relative humidity at the tower exit will be less than 100%. The cooling capacity of dry towers for this system is comparatively small and they may operate only occasionally.

As described previously, the condenser pressure in the dry tower plant is generally high. This situation becomes more serious at a high ambient air temperature. To overcome this undesirable feature, wet towers can be incorporated in the dry cooling system and operate only at certain ambient air temperatures. Several studies have been carried out showing the feasibility of this combined cooling system [12].

The third possible objective for using the combined cooling is self-explanatory. The combined cooling system of this kind will consist of wet towers and dry towers. Both towers are comparable in size and operate on a continuous basis. Of course, the exact arrangement of this combined cooling will depend upon the results of technical as well as economical evaluations for a particular plant site.

In summary, the objective for using the combined cooling system is different for each power plant project. The different objective will lead to different selection of combined cooling system. Thus, it is necessary and important for engineers to clearly define their objective at the very beginning.

12.5 COMPARISON OF WASTE HEAT REJECTION SYSTEMS

Discussions have been made so far on once-through cooling, wet-type cooling tower, cooling pond and dry-type cooling tower. Each of these systems has its own design and performance features. In this section we discuss their differences and similarities.

Figure 12-20 indicates a general relationship between the cooling system temperature and turbine exhaust pressure. During typical summer operation, once-through cooling systems generally operate with turbine exhaust pressures of 1.5 to 2.5 in. Hg. Wet cooling tower systems will yield exhaust pressures of 2.5 to 4.5 in. Hg while dry-type cooling towers will result in turbine exhaust pressures in excess of 6 to 8 in. Hg. Exhaust pressures with cooling ponds are in the same general range as wet-type cooling towers.

Water consumption is also different. Figure 12-21 shows a typical example of water consumption in the northeastern region. These curves were prepared by assuming a relative humidity of 60%, cloud cover of 70%, wind speed at 8 mph, and a cooling range of 20 F. Water consumption with once-through cooling is in the same general range as cooling ponds, while for dry-type cooling towers it is practically equal to zero.

Operation of all waste heat rejection systems will have some impact on the environment. These impacts may be in the form of temperature rise in the receiving water body or vapor plume that spreads over the planet's surrounding area. While the dry-type cooling towers have no vapor plumes, they still reject a significant amount of heat to the atmosphere. Naturally, there are some questions as to the

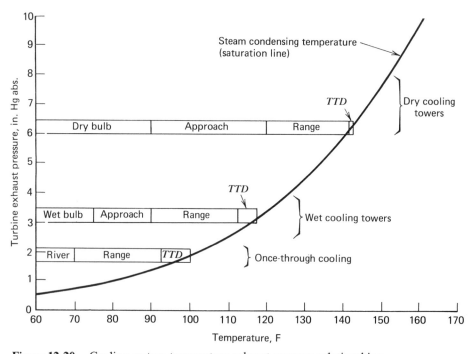

Figure 12-20. Cooling system temperature-exhaust pressure relationships.

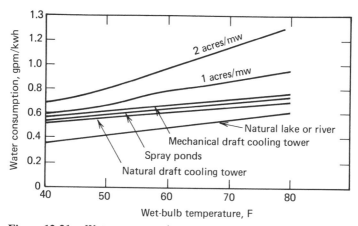

Figure 12-21. Water consumption versus wet-bulb temperature [15].

effects the waste heat from power plant operation will have on the environment. At what level will these impacts be acceptable to the public? To safeguard people's health and protect the environment, the federal and local governments have established a number of regulations for plant design and operation.

While there are impacts on the environment resulting from the cooling system operation, the environment in turn has some influences on the cooling system performance. Figure 12-22 shows the basic environmental factors affecting cooling system performance. For instance, the once-through cooling performance is mainly determined by the intake water temperature, which is a complex function of ambient air temperature, relative humidity, wind speed, cloud cover, day of the year, and

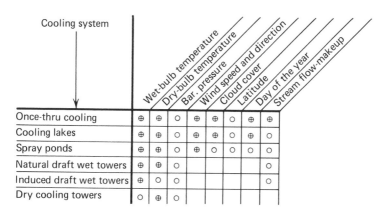

Cooling system	Wet-bulb temperature	Dry-bulb temperature	Bar. pressure	Wind speed and direction	Cloud cover	Latitude	Day of the year	Stream flow-makeup
Once-thru cooling	⊕	⊕	○	⊕	⊕	○	⊕	⊕
Cooling lakes	⊕	⊕	○	⊕	⊕	○	⊕	○
Spray ponds	⊕	⊕	○	⊕	○	○	○	○
Natural draft wet towers	⊕	⊕	○					○
Induced draft wet towers	⊕	○	○					○
Dry cooling towers	○	⊕	○					

⊕ = major variable
○ = secondary variable

Figure 12-22. Environmental influence on cooling system performance [3].

various stream flow effects. In design of cooling systems, therefore, local conditions must be taken into consideration.

The comparison of various cooling systems is not complete without including an economic aspect. Economic evaluation should at least consist of capital cost, operation, and maintenance cost. These costs are unfortunately subject to change at almost any moment. Also, these costs vary widely according to the conditions of the region in which the plant is built. Therefore, we make no attempt to compare the cooling systems in a numerical fashion. Rather, we offer several general observations.

Once-through cooling is the least expensive method. This method should be used provided that the site is suitable and governmental regulations can be met at a reasonable cost.

Dry-type cooling is the most expensive method. This method should be generally avoided unless cooling water is not available at the site or is very expensive in the semidesert area. The high cost of dry-type cooling is mainly due to the high cost of air-cooled heat exchanger and the penalty resulting from the generating capability loss.

In recent years wet-type cooling towers have frequently been selected as a waste heat rejection system. Its total cost is somewhere between the once-through cooling and the dry cooling. A cooling pond becomes competitive to the wet-type cooling tower when the site conditions are favorable and the land is not expensive.

12.6 SYSTEM DESIGN CONSIDERATIONS

The cooling system mainly consists of condenser, heat rejection equipment, circulating water pump and its associated piping and accessories. The cooling system may be thought of as a complete package that will interface to the turbine exhaust end and accomplish the condensation of steam.

For a selected steam turbine, the objective of a cooling system design is to match the condenser with a cooling tower (or waste heat rejection equipment in general) in such a manner that the combination will result in a minimum evaluated cost. The costs included in the evaluation should at least consist of:

1. Construction cost.
2. Plant fuel cost resulting from different condenser pressures.
3. Cooling system operating cost.
4. Cooling system maintenance cost.
5. Generating capability loss penalty.
6. Generating energy loss penalty.

The design decision may be then made on the basis of the present worth of lifetime total evaluated cost. In the following are some explanations for each cost item.

Construction Cost

Construction costs mainly consist of expenses for planning, engineering design, equipment materials, and erection. Since a power plant project takes several years for completion, interest paid during construction time must be taken into consideration.

The construction cost of the cooling system should include those of the condenser, waste heat rejection equipment, circulating water pump, and its associated pipings. Other costs essentially constant with various designs may be excluded for simplicity. The construction cost is not always available especially at the time when the cooling system is in the design stage, or when the equipment specifications are being prepared. For these cases cost estimates should be made on the basis of past experiences in similar projects. The cost information can also be derived from the literature provided by equipment vendors. This practice is entirely acceptable because the cost differential, rather than the absolute cost, is needed in the system design.

Plant Fuel Cost

The design of cooling system affects not only the cooling system performance, but also the overall plant performance. It is well known that as the design condenser pressure is decreased by using a large cooling system, the plant will use less fuel for a given power output. Conversely, a high condenser pressure resulting from a small cooling system will require more fuel. It is for this reason that the plant fuel cost must be taken into consideration in the cooling system design.

Calculation of plant fuel cost is quite lengthy. The fuel consumption depends upon the loading patterns during the entire plant economic life. It also depends upon the weather conditions under which various load demands are imposed on the unit. In other words, means must be found to correlate the weather conditions with future anticipated loading on the turbine-generator unit. Depending upon the desired accuracy and the availability of input data, various approaches have been developed and used in practice. One approach is simply to select the number of weather set conditions for study. The weather conditions, which mainly consist of ambient dry-bulb temperature and relative humidity, can be given as the average on a seasonal, monthly, or some other basis. For each set of weather conditions the number of operating hours is specified for various unit loadings. Table 12-3 indicates the operating conditions for a typical year with 12 sets of weather conditions and four different loadings. It is seen that there are 48 sets of conditions under which the cooling system will operate. Corresponding to each set of conditions, the cooling system will generate one condenser pressure and, thus, one plant net heat rate. Combination of the plant net heat rate and other data information will yield the annual fuel consumption.

It should be pointed out that the loading pattern of a turbine-generator may vary every year and, therefore, the fuel consumption must be calculated on the annual basis. Because of changing fuel price during the unit's economic life, the escalation factor must be taken into account.

Table 12-3
Example of Cooling System Operation Conditions

Month	Weather Conditions Dry-Bulb Temperature (F) Relative Humidity(%)	Turbine-Generator Loading			
		100%	80%	60%	40%
		Operation Hours			
January	47				
	50	100	125	125	300
February	59				
	48	100	125	125	300
March	69				
	50	100	125	125	300
April	72				
	51	200	150	150	200
May	80				
	51	200	150	150	200
June	84				
	50	200	150	150	200
July	86				
	50	100	125	125	300
August	90				
	48	100	125	125	300
September	70				
	49	100	125	125	300
October	62				
	50	100	150	100	300
November	49				
	40	100	150	100	300
December	48				
	50	100	150	150	300

Operating Cost

The operating cost is primarily the cost of operating circulating water pumps and cooling tower fans. It may also include the costs for makeup water and water treatment.

Power requirements for cooling systems are first estimated using the information of unit loading pattern. Then, they are converted into the operating cost. The costs for makeup water and water treatment are highly sensitive to the local conditions. In some sites, pumping power for makeup water may be significant and therefore, must be taken into account.

Maintenance Cost

Like other equipment the cooling system needs regular maintenance for smooth operation. The maintenance cost generally consists of two parts. The first is related to the size of cooling system and is usually constant regardless the number of hours that the system is in operation; the second part is dependent upon the unit's output and usually expressed in terms of dollars per kilowatt hour.

The maintenance cost varies significantly with different heat rejection systems. For instance, the cooling pond has a maintenance cost of around $1.25/kW-yr, while the mechanical-draft cooling tower is around $2.50/kW-yr. However, for the same kind of heat rejection system, the maintenance cost generally does not change very much.

Generating Capability Loss Penalty

In assessing the merits of cooling system for a given turbine-generator unit, the maximum generating capability must be calculated and compared. Any difference should be made up by additional generating facility. This added generating facility requires both capital expenditure and operating revenues. These costs are, respectively, referred to as the penalty cost due to the loss of generating capability and the penalty cost due to the loss of energy generation. They may constitute a significant portion of total evaluated cost for a cooling system.

There are different approaches used by consulting engineers for calculation of generating capability loss penalty. One approach is first to determine the generating capability at the time when the ambient air temperature is the historical maximum and the turbine operates under a valve-wide-open condition. Then, the calculated capability is compared with that at the reference point, say, the unit output at 1 in. Hg abs. The difference between these two is defined as the loss of generating capability. This loss is further converted into the penalty cost by the unit installation cost. Evidently, the penalty cost thus calculated is relative in nature. The procedure is entirely acceptable for system design.

Generating Energy Loss Penalty

The annual loss of energy generation is obviously dependent upon the loss of generating capability. In addition, it is related to the unit loading pattern and weather conditions. The calculational procedures are as follows:

- Calculate the maximum generating capability of turbine-generator unit at a given set of weather conditions.
- Determine the loss of generating capability by substracting the calculated value from that of the reference point.
- Obtain the loss of energy generation by multiplying the loss of generating capability with the number of operation hours.
- Repeat the above procedures for all weather condition sets.
- Determine the annual loss of energy generation by summing up all losses under various weather conditions.
- Convert the annual energy loss into penalty cost by using the unit energy cost.

It should be pointed out that the unit energy cost may vary from year to year. An escalation factor must be taken into account accordingly.

12.7 COOLING SYSTEM CONFIGURATION DESIGN

The general objective of cooling system configuration design is to determine the optimum combination of waste heat rejection equipment and steam condenser for a given turbine-generator unit. The scope of design study can vary widely. It may involve a comparative study of various heat rejection systems, leading us to select the cooling tower or any other system. However, in most cases in which the waste heat rejection system is specified, say, a mechanical-draft cooling tower, the scope will then include the application of cooling towers of various water temperature approaches to the design ambient wet-bulb temperature and various temperature cooling ranges. These towers must be matched with compatible condensers.

In system design various condenser configurations must be taken into consideration. These may include the single-pressure, two-pass condenser and the multipressure, series-flow condenser. For economic reasons, single-pressure, single-pass condensers are generally eliminated from consideration when wet towers are used. Condenser tube length, diameter, and materials are also the variables to be included in system design.

Presented in the following is the case of system design in which the most economic combination of condenser and mechanical-draft cooling tower was determined for a 800 MW turbine unit with 4 flows and 28 inches last stage blade length. The unit was planned for commercial operation in 1977. Five design parameters, cooling tower approach, condenser design pressure, number of passes in condensers, and condenser tube length and diameter were to be determined in this system design. The main purpose was to generate enough information so that the equipment specifications for condensers and towers could be prepared. The important inputs include:

> Fixed charge rate = 16%
> Levelized fuel cost = $0.5/MBtu
> Levelized auxiliary energy cost = $0.0085/kWh
> Combined penalty cost = $30/kW-yr
> Unit loading and duration in a typical year
> Summer Operation
>
> | | |
> |---|---|
> | 800 MW | 600 hr |
> | 600 MW | 900 hr |
> | 400 MW | 500 hr |
>
> Spring-Fall Operation
>
> | | |
> |---|---|
> | 800 MW | 600 hr |
> | 600 MW | 1400 hr |
> | 400 MW | 2000 hr |

Winter Operation

800 MW	300 hr
600 MW	700 hr
400 MW	1000 hr

Design wet-bulb temperature 78 F

Average wet-bulb temperature

Summer	71 F
Spring-Fall	56 F
Winter	41 F

Type of Condenser: single-pressure surface condenser

Design maximum heat load	4080×10^6 Btu/hr
Condenser tube materials	90–10 Cu-Ni

On the basis of minimum capitalized annual cost, the most economical combination of condenser and mechanical-draft cooling tower was determined. Here is the system configuration.

Condenser configuration

Number of shells = 2

Number of passes = 2

Inlet water temperature = 92 F

Temperature rise = 30.6 F

Condenser design pressure = 4.5 in. Hg abs.

Terminal temperature difference = 7.14 F

Water flow rate = 266,338 gpm

Tube diameter = 1 in. OD

Tube length = 42 ft

Tube velocity = 8 fps

Number of tubes = 31,452

Surface area = 345,837 ft^2

Mechanical-draft cooling towers

Water flow rate = 266,338 gpm

Tower approach = 14 F

Cooling ranges = 30.6 F

Fan power = 4022 bhp

For convenience, the optimum values of the five parameters are also summarized

below:

> Cooling tower approach = 14.0 F
> Condenser design pressure = 4.5 in. Hg abs.
> Number of passes in condenser = 2
> Condenser effective tube length = 42 ft
> Condenser tube diameter = 1.0 in.

In this case study, the steam turbine was specified in advance and only mechanical-draft cooling towers were considered as the waste heat rejection method. For simplicity, multipressure condensers were not considered here. The input data and other assumptions were arbitrary, but they are typical values frequently used by consulting engineers.

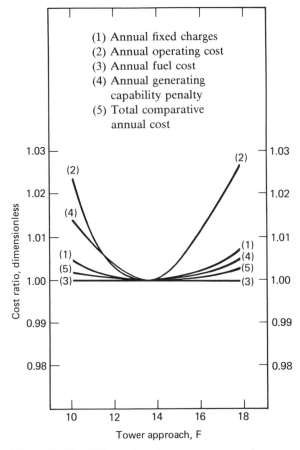

Figure 12-23.   Effects of cooling tower approach.

Equipment price information plays an important role in system design. In this case study the price of cooling towers were derived from the literature published by the Marley Company and the price of condensers came from the Westinghouse Corporation. These prices were adjusted and modified by using various escalation factors and taking into account the interest paid during the construction period. The maintenance costs were omitted in the evaluation because they would not vary significantly for all configurations under consideration.

The configuration design of a cooling system is an important task. Various computer programs have been developed to determine the economic optimum of turbine exhaust frame size, condenser configuration, and cooling tower configuration (if cooling towers are used). Engineers should be familiar with these programs and

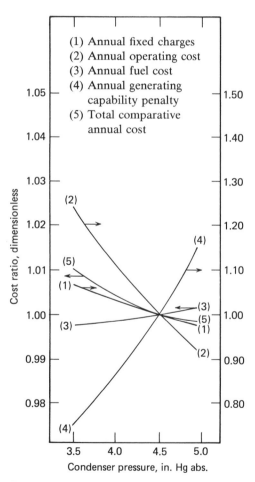

Figure 12-24. Effects of condenser design pressure.

Table 12-4
Cost Comparison of Two Design Alternatives (@ $1000)

Item for Comparison	One-Pass	Two-Pass
Fixed charges on investment	9175.27	8684.77
Operating costs	534.85	400.63
Fuel cost resulting from condenser back pressure	21195.74	21193.22
Combined generating penalty	903.29	809.96
Total comparative annual cost	31809.14	31088.57

encouraged to use them in their assignments. Unfortunately, most computer programs are the proprietary material and not readily available to the public.

Returning back to this case study, it is interesting to know how the design parameters affect the evaluated cost. Figure 12-23 shows the effects of tower approach on the cost ratio, which is simply the ratio of the evaluated cost of one particular configuration to that of the optimal configuration. The ratio becomes a unity at the optimal value of tower approach. In this case the optimal tower approach was 14 F. Also shown in Fig. 12-23 are the effects of tower approach on the cost components. In this case study the cost components included the annual fixed cost, fuel cost, operating cost, and combined generating penalty cost. Of these four cost components, the annual fuel cost was most sensitive to the change of tower approach.

Figure 12-24 shows the cost ratio in terms of condenser pressure. The cost ratio decreases as the condenser pressure increases. The pressure 4.5 in. Hg abs. was selected because it satisfied the upper limit of 5 in. Hg abs. imposed by the turbine manufacturer and also provided some margin for the time when the ambient temperature is higher than the design value.

As mentioned previously, the single-pressure, two-pass condenser is always superior to the single-pressure, one-pass condenser whenever evaporative cooling towers are used. This case study was no exception. Table 12-4 summarized the relative costs for these two arrangements.

In the cooling system optimization several limitations must be observed. These include the condenser tube diameter and tube length. Too small a diameter may result in some operational difficulties. When the condenser tube length is optimized, the minimum terminal temperature difference and condenser room space must be taken into account. In this case study, the condenser tube length and tube diameter selected were 42 ft and 1 in, respectively.

Appendix D presents an illustrative cooling system design.

SELECTED REFERENCES

1. Dynatech R/D Company, "A Survey of Alternative Methods for Cooling Condenser Discharge Water: Large-Scale Heat Rejection Equipment," Water Pollution

Control Research Series, 16130 DHS 07169, U.S. Environmental Protection Agency, 1969.

2. Dynatech R/D Company, "A Survey of Alternative Methods for Cooling Condenser Discharge Water: Operating Characteristics and Design Criteria," Water Pollution Control Research Series, 16130 DHS 08/70, U.S. Environmental Protection Agency, 1970.

3. L. G. Hauser, K. A. Oleson, and R. J. Budenholzer, "An Advanced Optimization Technique for Turbine, Condenser, Cooling System Combinations," Proceedings of the American Power Conference, Chicago, 1971.

4. D. C. Senges, H. A. Alsentzer, G. A. Englesson, M. C. Hu, and C. Murawczyk, "Close-Cycle Cooling Systems for Steam-Electric Power Plants: A State-of-the-Art Manual," EPA-600/7-79-001, U.S. Environmental Protection Agency, 1979.

5. Littleton Research and Engineering Corporation, "An Engineering-Economic Study of Cooling Pond Performance," Water Pollution Control Research Series, 16130 DFX 05/70, U.S. Environmental Protection Agency, 1970.

6. D. K. Brady, W. L. Graves, Jr., and J. C. Geyer, "Surface Heat Exchange in Power Plant Cooling Lakes," research report, Johns Hopkins University Press, 1969.

7. J. S. Neill and J. H. Gibbons, "Sizing a Cooling Pond for a Power Plant," ASME paper 75-WA/HT-63, 1975.

8. E. S. Miliares, *Power Plants with Air-Cooled Condensing Systems*, MIT Press, 1974.

9. J. P. Rossie and W. A. Williams, Jr., "The Economics of Using Conventional Nuclear Steam Turbine-Generators with Dry Cooling Systems," in the symposium volume, *Dry and Wet/Dry Cooling Towers for Power Plants*, ASME Publication, 1973.

10 M. W. Larinoff, "Dry Cooling Tower Power Plant Design Specifications and Performance Characteristics," in the symposium volume, *Dry and Wet/Dry Cooling Towers for Power Plants*, ASME publication, 1973.

11. K. W. Li, "Analytical Studies of Dry/Wet Cooling Systems for Power Plants," in the symposium volume, *Dry and Wet/Dry Cooling Towers for Power Plants*, ASME publication, 1973.

12. K. W. Li, "Hybrid Cooling Systems for Power Plants," in the symposium volume, *Cooling Systems* by BHRA Fluid Engineering, Cranfield, Bedford MK43 OAJ. England, 1975.

13. W. M. Kays and A. L. London, *Compact Heat Exchangers*, McGraw-Hill, 1964.

14. U.S. Department of Commerce, *Climatic Atlas of the United States*, Environmental Data Service, Environmental Science Services Administration, U.S. Government Printing Office, Washington, D.C., June 1968.

15. J. P. Rossie and E. A. Cecil, "Research on Dry-Type Cooling Tower for Thermal Electric Generation," Part I, Water Pollution Control Research Series, U.S. Environmental Protection Agency, November, 1970.

PROBLEMS

12-1. A once-through cooling system is planned for a 800 MW fossil-fuel generating unit. The design heat load and inlet water temperature are, respectively, 4080×10^6 Btu/hr and 74 F. Calculate the condenser surface area and circulating water pump horsepower for the following two different condenser pressures:

(a) 4.5 in. Hg abs.
(b) 3.0 in. Hg abs.

The condenser is assumed single-pressure and one-pass. The condenser tube has an effective length 40 ft and diameter 1 in OD. The total pressure drop in the intake and outfall channels is approximately equal to 30 ft of water.

12-2. Optimize the condenser tube length for the design conditions as follows:

Heat load	1000×10^6 Btu/hr
Condenser pressure	2.00 in. Hg abs.
Cooling inlet water temperature	75 F
Cooling water velocity	7.00 fps
Number of passes	Single
Tube diameter	1 in. OD
Tube gauge and material	18 BWG Admiralty
Tube cleanliness	0.85

The circulating water pipe, 1000 ft in length and 5.4 ft in diameter, delivers the water between the lake and condenser. The system is expected to operate 5000 hours per year and have an economic life of 30 years. The relative cost information used for the evaluation is summarized below:

Condenser
    Cost $= 1.25$ [4000 + 8.0 (tube surface area) + 6.4
                ×(number of tubes)]
Pump
    Cost $= 3.8$ (water flow rate in gpm)
Pump Motor
    Cost $= 72$ (rated horsepower)
Electrical Energy
    Cost $= 50$ mills/kWh

The optimization should be made on the basis of relative annual cost, which is the sum of the annual capital and operating costs.

12-3. Estimate the equilibrium temperature for the location where the weather conditions are

Wind speed	15 mph
Ambient air temperature	78 F
Relative humidity	50%
Net radiation absorption	2500 Btu/day-ft^2

12-4. A well-mixed water body with an average depth 6 ft and water surface area 10,000 ft^2 is located in the area where the average surface heat exchange coefficient K is 160 Btu/day-ft^2-F.

   a. Derive the equation predicting the water temperature change in terms of time variable. The equilibrium temperature is assumed constant.

   b. Derive the temperature prediction equation again assuming the equilibrium temperature as

$$T_e = T_{em} + \Delta T_e \sin(2\pi f\tau)$$

where

$$T_e = \text{equilibrium temperature}$$

$$T_{em} = \text{the mean value of } T_e$$

$$\Delta T_e = \text{the amplitude of the fluctuation in } T_e$$

$$f = \text{frequency of the flucuation}$$

12-5. Estimate the equilibrium temperature and surface exchange coefficient for the site conditions: ambient air temperature 90 F, relative humidity 55%, wind speed 8 mph, and net radiation absorption 3500 Btu/day-ft^2.

12-6. Repeat Problem 12-5 by using a wind speed of 12 mph.

12-7. Repeat Problem 12-5 by using the more accurate equations such as Eq. (12-20) and (12-23). Compare the results with those obtained in Problem 12-5 and indicate the difference by percentage

12-8. A cooling pond is to be designed for a 480 MW generating unit. The design heat load is 2500 × 10^6 Btu/hr. The condenser temperature rise is from 92 to 110 F. The pond site conditions are: ambient air temperature 92 F, relative humidity 50%, wind speed 7 mph, and net radiation absorption 4000 Btu/day-ft^2.

   If the cooling pond is a flow-through pond, calculate (1) the pond surface area required, (2) the evaporation loss per day, and (3) the percentage of the evaporation loss due to the plant operation.

12-9. Estimate the performance characteristics of the cooling pond designed in Problem 12-8. This can be done by calculating the water intake and discharge temperature in terms of ambient air temperature and heat load. The performance curves should be plotted in a manner similar to that for an evaporative cooling tower.

12-10. Repeat Problem 12-8 by assuming that the cooling pond is a completely mixed cooling pond.

12-11. It is of interest to compare a completely mixed cooling pond with a flow-through cooling pond. When both have the same discharge and intake temperatures, the required surface areas are related by the equation

$$r_m = e^{r_f} - 1$$

where the subscripts m and f denote, respectively, completely mixed and flow-through pond, and the parameter r is defined by Eq. (12-30). Derive the above equation.

12-12. A completely mixed cooling pond is planned for a 800 MW fossil-fuel generating unit with the condenser pressure 2.5 in. Hg abs. The design heat load is 4080×10^6 Btu/hr. Estimate the pond size based on a temperature rise in the condenser of 14 F. The plant site is near Bismark, North Dakota.

12-13. Estimate the evaporation loss for the pond designed in Problem 12-12.

12-14. Repeat Problem 12-12 and Problem 12-13 by assuming the cooling pond is of the flow-through type.

12-15. Design a mechanical-draft, dry-type cooling tower for a 800 MW fossil-fuel generating unit. The following conditions should be observed and used in the design:

Design ambient air temperature	90 F
Design initial temperature difference	75 F
Heat transfer surface	surface CF-7.34
	(see Fig. 12-15)
Cooling range	15 F
Design heat load	4600×10^6 Btu/hr

The report should include the heat transfer surface area and fan power calculations. Also included in the report are the drawings indicating the heat exchanger arrangements and relative positions to the fans.

12-16. The temperature of water leaving the tower designed for Problem 12-15 is expected to decrease as the ambient air temperature decreases. Calculate the cold water temperature at the time when the ambient air temperature is 50 F. Assume that the heat load is still equal to 4600×10^6 Btu/hr and that all fans are still in full operation.

Also estimate the values of a and b used in the performance equation $ITD = aQ^b$.

12-17. Repeat Problem 12-15 by using the finned flat tubes specified in Fig.

12-17.

12-18. The ABC power company is planning to construct a fossil-fuel electric generating unit. The plant has a nominal output of 600,000 kW at the condenser pressure 2.5 in. Hg. abs. In the previous study the plant net heat rate was determined and given by

$$PNHR = 12,932 - 8.2493(L) + 0.008389(L)^2 + 223.45(CP)$$

$$+ 0.45779(CP)(L) - 0.0010746(CP)(L)^2$$

where L is the output in MW and CP is the condenser pressure in inches of mercury. Optimize the condenser and mechanical-draft cooling tower configuration for the above steam turbine system by specifying the condenser pressure, tower approach, condenser tube water velocity, tube length, and diameter. The conditions below should be utilized in the system evaluation.

Cooling system design heat load	3,754 MBtu/hr
Design site temperature	90 F
pressure	14.7 psia
relative humidity	50%
Plant economic life	30 years
Interest rate	14%
Annual fixed charge rate	18%
Annual plant loading schedule	
Summer (WBT = 70 F)	
552,000 kW	600 hr
414,000 kW	900
276,000 kW	500
Spring-Fall (WBT = 56 F)	
552,000 kW	600 hr
414,000 kW	1400
276,000 kW	2000
Winter (WBT = 45 F)	
552,000 kW	300 hr
414,000 kW	700
276,000 kW	1000
First year fuel cost	$1.00 per MBtu
Fuel cost escalation	6%
Auxiliary power cost	12.6 mills/kwh
Auxiliary power cost escalation	6%
Condenser cost	1.25 [16,000 + 8(A) + 6.4(NT)]
Tower cost	1.25 [3.24 × 10^6 + 14(TU)]
Pump cost	3.8 (water flow rate in gpm)
Pump motor cost	72 (rated pump horsepower)
Indirect cost	25% of capital cost

where A is the condenser tube surface area in ft^2, NT is the number of condenser tubes, and TU is the tower unit as defined in Chapter 11. For simplicity the condenser is assumed to be of single pressure. The condenser heat load is given by

$$HL = 413.01 + 4.5306(L) + 0.0015321(L)^2 - 39.374(CP)$$

$$+ 0.56662(CP)(L) - 0.00065138(CP)(L)^2$$

The heat load (HL) is in MBtu/hr, condenser pressure (CP) in inches of mercury, and plant load (L) in MW.

12-19. Repeat Problem 12-18 by using the fuel cost $1.20/MBtu and an inflation rate of 10% per year.

CHAPTER THIRTEEN

Gas Turbines, Combined Cycles, and Cogeneration

13.1 GAS TURBINE PLANTS

The principle of the gas turbine consists of the channeling of hot gases from burning fuel to spin a turbine wheel. It had limited application until the twentieth century, when alloy metals were developed to withstand high temperatures. In some of the early gas turbines low-temperature steam was injected into combustion gases to temper the hot gases to a temperature of 800 F or lower, which was the limit of steel metals. Water injection also was used. Neither of these methods was significantly effective because the power produced was hardly more than enough to drive the air compressor required for the combustion gases.

Metal developments in recent years have permitted gas temperatures of 1200 F by 1950, then to 1800 F in the 1970s. At the present time advanced research and development have projected temperatures as high as 2500 F. During the 1970s conservative gas turbine manufacturers limited gas temperatures to around 1500 F, while other manufacturers designed their units for 1800 F.

Air cooling has been applied to most of the high-temperature turbines by means of the injection of air into the first rows of stationary blades. The air is emitted at the trailing edge. Steam injection and water injection are again being utilized, but now the purpose is primarily for air pollution control of the exhaust gases. The increase in firing temperature has resulted not only in greater economy, but also in greater design capability ratings in relatively smaller physical sizes. Currently the unit size ranges from 2000 to 100,000 kW.

The gas turbine was developed primarily for the aircraft industry and for internal combustion engine superchargers. Some units today are derivatives of the aircraft type units, which operate at higher speeds and are reduced to the electrical generator synchronous speeds of 3600 rpm, or 3000 rpm by reduction gears. Most of the larger power generating units are designed for either 3000 or 3600 rpm and are physically larger than the higher speed units.

For the past three decades gas turbines have played a unique role in the power industry. Because of their relatively low initial cost, gas turbines are frequently used for emergency service and handling daily peak loads on a system. In many systems gas turbines are also operated in the spinning reserve mode, saving the total system fuel cost or delaying some expensive plant construction projects.

Natural gas is the most suitable fuel for gas turbines. No fuel preparation is required. However, natural gas is expensive for power production and is not available at most sites. Distillate oil No. 2 is almost equally suitable as a fuel. This is the same fuel that is used for diesels and household heating.

529

Heavier oils are less expensive and are more available for power generation. The heavier oils, such as No. 6 Residual, have been used extensively in gas turbines. Preheating must be employed in order to reduce the viscosity for proper flow and injection. Fuel oil heating is usually accomplished by shell and tube heat exchangers served by steam supplied by an external source. Tankage must be provided for storage of the oil.

Crude oil is being used as gas turbine fuel in the Middle East and other places where it is economically available. The oil must be low in sulphur and sodium content or excessive corrosion may occur. Research and experimentation have been pursued to develop other substitute fuels such as coal gas and synthetic fuels.

The operating principle of gas turbine has been presented in Chapter 2. Basically, ambient air is drawn into a multistage compressor and is compressed to about 10 times atmospheric pressure. The compressed air then passes through the combustion chamber where fuel is injected and burned. The products of combustion enter the turbine and expand to approximately atmospheric pressure. About two-thirds of the turbine output is used to drive the compressor while the remainder is for power generation. This arrangement is called the simple cycle gas turbine, which is characterized by large exhaust energy loss. In general, the simple-cycle gas turbine has relatively low efficiency (25 to 30%) as compared with the coal-fired steam turbine system. There are other systems where the exhaust heat energy can be utilized to improve thermal efficiency. One of these is called the regenerative gas turbine. As shown in Chapter 2, the exhaust gas leaving from the turbine enters a heat exchanger where heat is transferred to the air discharged from the compressor. Because the air entering the combustion chamber has a higher temperature, less fuel

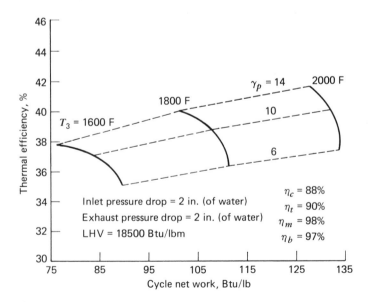

Figure 13-1. Effects of design parameters on simple-cycle gas turbine.

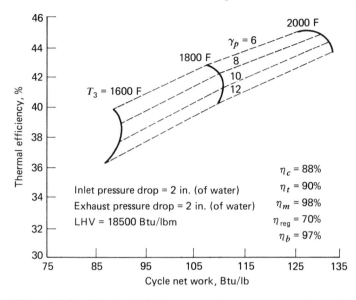

Figure 13-2. Effects of design parameters on regenerative-cycle gas turbine.

will be needed to achieve the design temperature at the turbine inlet. The regenerative gas turbine usually has a thermal efficiency as high as 36%.

In Chapter 2 it was shown that the compressor pressure ratio and the gas temperature at the turbine inlet are important design parameters. They affect not only the thermal efficiency, but also the net work produced by a unit mass of working substance. Figures 13-1 and 13-2 show these relationships, respectively, for a simple-cycle and regenerative-cycle gas turbine. It is seen that the points of maximum thermal efficiency and maximum net work output do not coincide for a given turbine inlet temperature. In a simple-cycle gas turbine the optimal value of pressure ratio for thermal efficiency is greater than the optimal pressure ratio for net work output. However, the opposite is true in the regenerative-cycle gas turbine. Since the initial cost is dependent on the value of the net work output, there are many occasions in design when the selected pressure ratio is closer to that for maximum net output than to that for maximum efficiency.

From Figs. 13-1 and 13-2 we also see that the pressure ratio for the regenerative system is usually smaller than that for the simple system. This will mean a large physical size for the regenerative system and therefore a higher cost in the initial investment.

13.2 FACTORS AFFECTING GAS TURBINE PERFORMANCE

Gas turbines can operate at full load as well as part load. As the load decreases, the turbine heat rate will increase. In addition, the turbine inlet and exhaust temperature

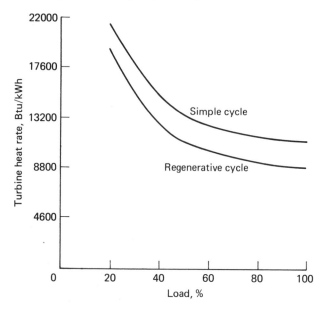

Figure 13-3. Turbine heat rates of a typical 60,000 kW unit.

will decrease with a decrease in load. Gas turbine performance at off-design conditions is important but difficult to generalize. It usually depends on specific design model. In practice engineers frequently calculate the part-load performance from the data furnished by the manufacturer or develop it from the tests of the similar or identical units. Figure 13-3 presents typical performance curves of simple-cycle and regenerative-cycle gas turbine using almost identical components. The vertical distance is an approximation of the true benefit of a regenerative system over a simple-cycle system.

Figure 13-4 shows the decrease of various turbine temperatures at part-load conditions. The turbine inlet temperatures are almost the same for both gas turbines using nearly identical components. For simplicity one curve is presented in Fig. 13-4. But the turbine exhaust temperature decreases more rapidly for the simple cycle than the regenerative cycle. While the regenerative system is more efficient, the fuel saving must be carefully evaluated in the system selection and compared with the additional cost. The example that follows presents a simplified evaluation.

EXAMPLE 13-1. A simple-cycle gas turbine plant with a base rating of 70,000 kW is to be installed on a site at sea level. The normal ambient air temperature at the site is 59 F. The unit is expected to operate on the following annual loading

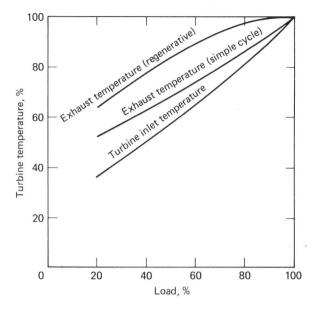

Figure 13-4. Variation of turbine temperature at various loads.

schedule.

Load (kW)	Load (%)	Hour per year
70,000	100	100
56,000	80	500
42,000	60	800
28,000	40	600

Would the addition of the regenerative air heater be justified at an increase in capital cost of $50/kW? The conditions below should be utilized for the evaluation.

Annual fixed charge rate				20%
Fuel cost				$7.00/MBtu
Turbine net heat rate differential				
Load (%)	100	80	60	40
Differential				
(Btu/kWh)	1680	1800	1450	850

Solution: The total annual fuel saving produced by the regenerative cycle is

$$\text{Fuel saving} = \left[\sum_{i=1}^{4} (\text{load})_i (\text{hour})_i (\text{HRD})_i \right] (\text{FC})$$

$$= [(70,000)(100)(1680) + (56,000)(500)(1800)$$

$$+ (42,000)(800)(1450) + (28,000)(600)(850)] 7 \times 10^{-6}$$

$$= \$876,080/\text{year}$$

The annual fixed charge on the additional investment is

$$\text{Cost} = (\text{rating})(\text{additional cost})(\text{AFCR})$$

$$= (70,000)(50)(0.20)$$

$$= \$700,000/\text{year}$$

Finally, the annual saving from the regenerative cycle, therefore, is

$$\text{Annual saving} = 876,000 - 700,000$$

$$= \$176,000/\text{year}$$

Gas turbine design ratings are usually based upon standard site conditions. The standard site conditions differ from country to country. For the purpose of comparison, Table 13-1 shows the standard conditions of various standards. One of the popular standards is that of the International Standards Organization (ISO). The site conditions in this standard are sea level altitude (i.e., 14.7 psia), 59 F and 60% relative humidity. However, some manufacturers have referenced their standard design to the conditions specified by the National Electric Manufacturers Association (NEMA). In this text the ISO standard conditions are assumed unless otherwise indicated.

Gas turbine performance varies significantly with site conditions. Of many site conditions, the ambient air temperature is the main factor affecting the performance

Table 13-1
Various Standard Site Conditions for Gas Turbines

Site conditions	CIMAC	ASME PTC-22	NEMA SM-30	ISO
Temperature	15 C	80 F	80 F	59 F
Pressure	1013 mb	14.17 psia	14.17 psia	14.7 psia
Relative humidity	60%	50%	—	60%
Cooling water temperature	15 C	—	—	—

of gas turbine. As the ambient air temperature drops, the turbine capability and thermal efficiency increase. A typical situation indicates a 23% increase in turbine capability when the air temperature drops from 59 to 0 F. In this temperature range, the thermal efficiency increases about 5%. The change of the ambient air temperature also induces a change of the specific weight of the suction air. For a constant-speed compressor, it means an increase in air mass flow as the ambient air temperature decreases. For most gas turbines about 12% increase in air mass flow will occur when the ambient air temperature changes from 59 to 0 F.

Exact effects of ambient air temperature on gas turbine performance are not easy to predict. As the ambient air temperature decreases, the component compressor and turbine efficiencies will fall off slightly. Also, the pressure ratio of the compressor will deviate from the design value. For most gas turbine used for electric power generation, the pressure ratio increases slightly as the ambient air temperature drops. All these variables have made it difficult to have a theoretical prediction on the gas turbine performance change. In practice, engineers frequently use an empirical approach and estimate the performance change by applying the correction factors.

Figure 13-5 shows the correction factor for the site temperatures different from the standard value (i.e., 59 F for the ISO standard). These correction factors should be used only for approximation or whenever the data from the manufacturer are not available. It should be noted that the effects of ambient air temperature are almost independent of gas turbine design. The effects are not altered radically by the

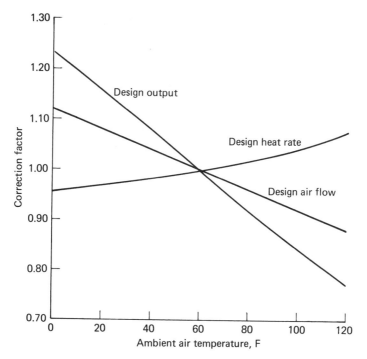

Figure 13-5.   Effects of ambient air on gas turbine performance.

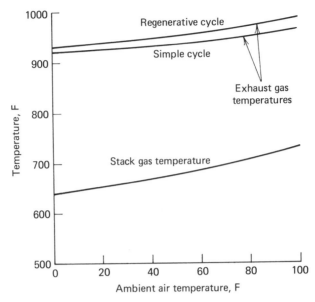

Figure 13-6. Variation of turbine temperatures with the ambient air temperature.

selection of design pressure ratio. They are also insensitive to the addition of a regenerator. Therefore, the correction factors shown in Fig. 13-5 are also valid for a regenerative gas turbine.

Figure 13-6 shows the change of turbine exhaust temperature as a function of a change in ambient air temperature. It is seen that turbine temperatures will decrease as the ambient air temperature drops. In case of a regenerative gas turbine the stack temperature is different from the turbine exhaust temperature. Again, the stack temperature will drop as the ambient air temperature decreases.

The ambient air pressure has a significant effect upon turbine capability, but no effect on turbine heat rate. As the ambient air pressure drops, the turbine capability decreases. These effects are mainly due to the reduction in the specific weight of the moist air that in turn leads to a reduction of air mass flow. Based upon past experience, the correction factor for the ambient air pressure is

$$f_p = \frac{P_s}{14.7} \tag{13-1}$$

where P_s is the ambient air pressure at the plant site and has a unit of pounds per square inch.

The ambient air humidity has a minor effect on gas turbine performance. In practice, this effect is usually ignored. However, when the ambient air temperature is above 30 C and the relative humidity is above 70% the ambient air humidity should be taken into account.

The temperature of cooling water should be considered if the gas turbine system has intercoolings on the compression side. Generally, the effects of cooling water temperature are relatively small. They become serious only when the pressure ratio for compressor is high such as those in the underground compressed air-storage plant.

In addition to the above corrections, the gas turbine output and heat rate must be adjusted for variation of frictional losses in inlet and exhaust duct. In practice, gas turbine manufacturers usually establish design performance upon the standard losses which are 2 and 4 in. of water, respectively, for the inlet and exhaust losses. When the plant system design exceed those values, the gas turbine output and heat rate should be penalized by the correction factors as shown in Figs. 13-7 and 13-8.

Often actual pressure drops are greater than those assumed by gas turbine manufacturers. For example, the gas turbine plant is frequently equipped with an air inlet filter and silencer. The filters are required for the protection of the air compressor functions, and the inlet silencer is required for environmental protection. Both items produce an air friction loss in the inlet to the air compressor, and this loss has an effect upon the unit performance. In some plants it is desirable to install also an evaporative air cooler in the inlet duct. Such an installation may be warranted in plants where the ambient temperature is normally higher than the standard temperature during the periods in which the generating unit is required to operate. It is a question of optimum economy. The cooler may cost $20/kW installed and requires water supply. There also is a performance penalty for the air

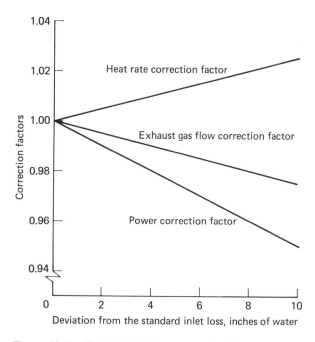

Figure 13-7.   Typical inlet loss correction factors.

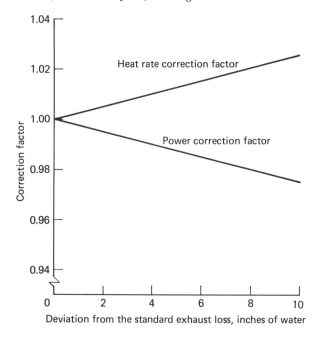

Figure 13-8.   Typical exhaust duct loss correction factors.

friction loss at the compressor inlet. However, these penalties may be more than overcome by the saving in operation costs resulting from the lowered operating air temperature.

The turbine inlet pressure drop affects not only the gas turbine performance, but also the exhaust gas flow rate. As the inlet pressure drop increases, the gas flow rate will decrease. This correction must be taken into account in a combined-cycle system design.

In the gas turbine exhaust end the standard pressure drop (4 in. of water) is provided to overcome the resistance to a short stack closely coupled to the unit. When the unit is used in a combined-cycle system, the exhaust pressure drop will increase. Depending on the fuel burned and the mode of operation, the exhaust pressure drop can vary in the range of 6 to 14 in. of water. Table 13-2 presents typical pressure drops at the turbine inlet and exhaust end.

EXAMPLE 13-2.   An industrial gas turbine-generator plant has been designed for 60,000 kW at standard ISO conditions (59 F at 14.7 psia), and with inlet and exhaust ducts each designed, respectively, for 2.0 in. and 4 in. water gauge friction losses. Its base load heat rate under these conditions is 11,000 Btu/kWh. If the unit is to be installed on a site at an altitude of 3000 ft above sea level with ambient air temperature of 120 F, what are the adjusted base capability and corresponding heat rate?

Table 13-2
Typical Pressure Drops at Turbine Inlet and Exhaust

Inlet Losses (inches of water)	
Air filter	0.5
Low level silencer	1.5
High level silencer	3.5
Evaporative cooler	0.8
Exhaust Losses (inches of water)	
Waste heat boiler for combined cycle	10–14
Bypass duct for combined cycle	6
Regenerative air heater	8
Open damper	1

Solution: The atmospheric pressure is inversely related to the altitude above sea level. In equation form the approximate relation is expressed as

$$P_s = 14.7 - 0.49(H) \qquad \text{(a)}$$

where

$$P_s = \text{atmospheric pressure (psia)}$$

$$H = \text{altitude above sea level (1000 ft)}$$

Therefore the atmospheric pressure at the altitude of 3000 ft becomes 13.23 psia. The gas turbine capability is affected by both atmospheric temperature and pressure. The site baseload capability is obtained by multiplying the design capability by the temperature and pressure correction factors. That is,

$$(\text{kW})_s = (\text{kW})_i (\text{CF})_{t1} (\text{CF})_p$$

Since

$$(\text{CF})_{t1} = 0.78 \text{ (reading from Fig. 13-5)}$$

$$(\text{CF})_p = \frac{13.23}{14.7} = 0.90$$

we have

$$(\text{kW})_s = 60,000 \times 0.78 \times 0.90$$

$$= 42,120 \text{ kW}$$

The atmospheric pressure has no effect upon the net heat rate. The site baseload heat rate is simply the design heat rate multiplied by the temperature correction factor such as

$$(HR)_s = (HR)_i (CF)_{t2} \qquad (b)$$

Inserting the numerical values into the equation gives us a site baseload heat rate of

$$(HR)_s = 11,000 \times 1.075$$

$$= 11,825 \text{ Btu/kWh}$$

EXAMPLE 13-3.   Using the same gas turbine unit as in Example 13-2 and the same site conditions. Calculate the revised capability and heat rate if the unit is to be used in a combined-cycle plant with a total pressure drop of 13 inches of water at the exhaust end.

Solution: The site capability and heat rate developed in Example 13-2 are

$$(kW)_s = 42,120 \text{ kW}$$

and

$$(HR)_s = 11,825 \text{ Btu/kWh}$$

Because of pressure drop deviation at the gas turbine exhaust end, corrections must be made to obtain the new output and heat rate. From Fig. 13-8, we have the correction factor for the turbine output as

$$\text{Correction factor for pressure drop} = 0.977$$

at the deviation 9 in. of water. Therefore the new site capability becomes

$$(kW) = 42,120 \times 0.977 = 41,151 \text{ kW}$$

Similarly, the gas turbine new heat rate is

$$(HR) = 11,825 \times 1.028$$

$$= 12,156 \text{ Btu/kWh}$$

13.3 THE COMBINED CYCLES

The combined-cycle plant consists of one or more gas turbine units generating electric power, with the hot exhaust gases discharged into waste heat recovery boilers. Steam is generated in the waste heat recovery boiler to serve a steam turbine generator, and the cooled exhaust gases are discharged to the atmosphere through a stack. In this way a greater proportion of the gas turbine fuel is utilized in net electric generation output.

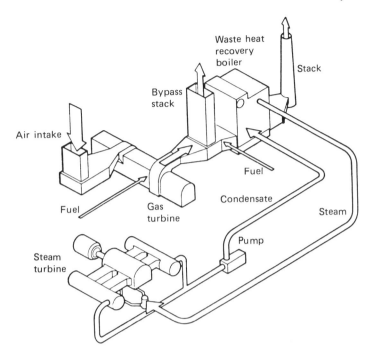

Figure 13-9. A simplified diagram for a combined-cycle plant.

Additional fuel may be burned in the waste heat recovery boiler, if required, to supplement the heat in the gas turbine exhaust. The steam produced in the waste heat recovery boiler is expanded in the steam turbine and condensed in a surface type condenser. The condensate is returned to the boiler where it is reheated by the gas turbine exhaust.

On Fig. 13-9 there is shown a simplified diagram of a combined cycle plant. Figure 13-9 shows only one gas turbine unit, although most plants utilize at least two gas turbine units and in some cases as many as five. This occurs because steam turbine-generators are commercially available in many size ratings and usually only one is required to serve several gas turbine units.

The combined-cycle plant can be formed by adding waste heat recovery boilers and steam turbine-generators to the existing gas turbine generating units. This idea may become attractive at a time when fuel cost is rising rapidly. Also, a combined-cycle system may be used in a repowering effort, in which case gas turbine units and waste heat recovery boilers are added to an existing steam turbine-generator.

An essential prerequisite to the development of a combined-cycle plant, and its economic comparison to other types of plants, is the selection of equipment and cycle arrangement. Gas turbine-generators, heat recovery boilers, and steam turbine-generators are commercially available in specific sizes. Some variations can be made, but equipment manufacturers must have prepared specific design shop drawings, and manufacturing shop tools for each model that they offer. The engineer must investigate the commercial availability of the different components. A "custom-

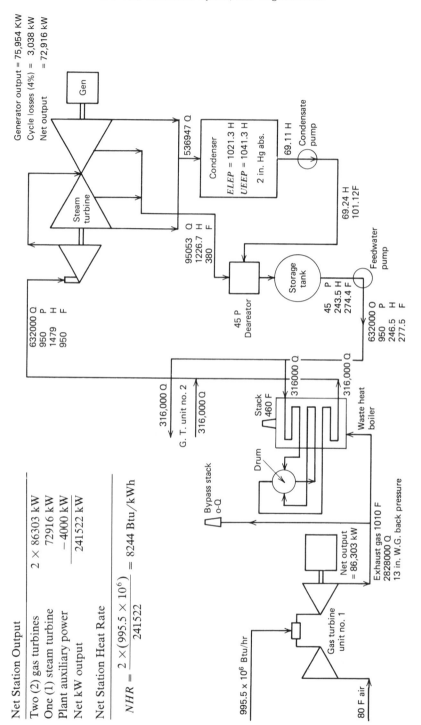

Figure 13-10. Combined-cycle heat balance (unfired boiler).

designed" component would require an exorbitant cost and could not compete economically with other modes of generation.

Gas turbines for power generation are commercially available in sizes from approximately 2500 to 100,000 kW. Their performance has been described in previous sections. Maximum exhaust temperatures are usually in the range of 800 to 1100 F and vary with the unit loading and ambient air conditions. The exact performance data are usually provided by gas turbine manufacturers.

In order to accommodate to the gas turbine exhaust temperature range, the steam turbine design conditions will be limited. Turbine throttle steam pressure and temperature are usually lower than 1250 psia and 950 F, respectively. Suitable steam turbines are commercially available in several sizes up to 100,000 kW or higher.

There is a greater range of availability in waste heat recovery boiler sizes and design parameters. Boilers may be unfired or supplementally fired and single pressure or dual pressure. Sizes are available up to about 750,000 pounds per hour by a number of manufacturers. This topic is discussed further in the next section. Figure 13-10 is a simplified energy and mass flow diagram for one version of a combined-cycle plant. In this plant the base units are two gas turbine generators, each rated at 100,000 kW at standard ISO conditions. After adjustments for the site conditions, and inlet and exhaust end pressure drops, each gas turbine unit can have an output of 86,303 kW at the full-load conditions. The exhaust gases leaving the gas turbine at 1010 F enter the waste heat recovery boiler, which produces steam at the rate of 316,000 lb/hr for each unit. The steam conditions at the turbine inlet are at a pressure of 950 psia and a temperature of 950 F. In this arrangement the steam turbine net output is approximately 72,916 kW. The heat rate developed in this combined cycle is 8244 Btu/kWh.

Many combinations of cycle arrangements are possible. There can be several feedwater heaters in the cycle. Additional feedwater heaters more than that shown on Fig. 13-10 would produce a lower heat rate, but would add to the first cost. The optimum cycle should be determined by the engineer, based upon fuel cost, initial cost, availability of equipment, and operating conditions.

13.4 WASTE HEAT RECOVERY BOILER

The waste heat recovery boiler is a key element in a combined-cycle plant. The boiler selection will affect not only the initial cost, but also the operating cost. The waste heat recovery process is the heat transfer accomplished in a series of arrangement of boiler sections, which is illustrated on a temperature profile diagram such as Fig. 13-11. This diagram is similar to that in Fig. 2-30. Water enters the boiler in form of compressed liquid. As water receives heat from the hot exhaust gases, it becomes saturated, evaporated, and eventually superheated. On the hot side, the exhaust gases leaving the gas turbine enter the steam generator and release thermal energy. The steam production on the basis of one pound of exhaust gas can be determined by applying the mass and energy conservation principles, such as shown in Section 2.6. To insure the most economical unit, it is necessary to evaluate the design

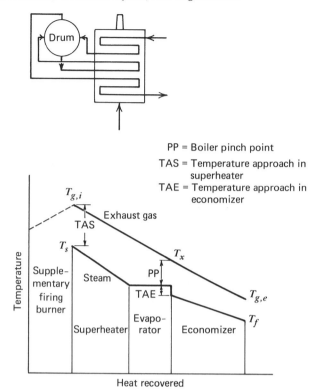

Figure 13-11. Temperature profile in waste heat recovery boiler.

parameters listed below:

1. Allowable back pressure
2. Boiler pinch point
3. Stack temperature
4. Steam pressure and temperature
5. Temperature approach for superheater and economizer

As indicated earlier, many design parameters must be evaluated, some based upon experience and others upon economic considerations. The allowable back pressure is one of important parameters. It controls the size of the free flow area for the hot gases through the boiler. As the back pressure increases, the boiler size is reduced. However, it must be remembered that an increase in back pressure will also reduce the gas turbine power typically at the rate of 0.3% for each inch of water increased. Currently, the back pressure ranges from 10 to 15 in. of water.

Another design parameter is the boiler pinch point which, as defined in Chapter 2, is the difference between the gas temperature and steam saturation temperature at the boiler economizer interface. As the boiler pinch point is reduced, the more waste heat will be utilized. Since the pinch point is the minimum temperature difference in the boiler, a reduction in pinch point means a reduction of average temperature difference in which the heat is transferred from the hot gases to the steam. Thus a reduction of pinch point will significantly increase the boiler size and therefore the initial cost. In current practice, the boiler pinch point is in the range of 50 to 80 F.

A low stack temperature is always desirable from the viewpoint of waste heat recovery. However, it does not mean that a low stack temperature should be always used. To avoid corrosion from moisture formation in economizer and stack, the minimum stack temperature should be always kept higher than the dew point temperature. Also, as the stack temperature decreases, the size of the economizer will be rapidly increased. Therefore, before a low stack temperature is used, it must be justified economically.

One device that has been used to lower the boiler exit gas temperature is a "low-level" evaporator. This evaporator, not shown in Fig. 13-11, is located right after the economizer in the gas path. It receives the feedwater ahead of the economizer and produces steam that could serve a feedwater heater.

As mentioned earlier, the conditions of steam produced must match those of exhaust gases from the gas turbine. The selection of these steam conditions affects not only the performance of steam turbine, but also the initial cost of boiler. The most common combinations of steam pressure and temperature are presented in Table 13-3. In practice, an economical evaluation must be made before the steam conditions are decided.

To insure a reasonable boiler size, the temperature approach in superheater and economizer should be carefully evaluated. For an unfired boiler, the temperature approach in superheater is frequently greater than 50 F. At a lower value, the boiler cost will increase very rapidly and, therefore, become economically unfeasible. For a fired boiler, there is no such limitation because the firing temperature can be always adjusted to maintain a reasonable temperature approach.

The temperature approach in the economizer is defined as the difference between the steam saturation temperature and the temperature of water leaving the econo-

Table 13-3
Common Steam Conditions for Waste Heat Boilers

Steam Pressure (psig)	Temperature (F)
150	450
250	550
400	650
600	750
850	825
1000	900
1250	950

mizer. When the water leaving the economizer is saturated, this temperature approach will become zero. There is considerable difference of opinion regarding heat recovery boiler operation with steaming in the economizer. In a steaming economizer, more steam will be generated from the boiler. However, some manufacturers prefer the nonsteaming economizer because the dangers of water hammer and steam blanketing in the economizer can be avoided. In the nonsteaming arrangement the typical temperature approach is around 40 F. At a lower value, the increase in economizer surface will occur and therefore results in an increase in boiler cost.

In combined-cycle system design the economics of unfired and supplementary-fired operation should be examined. The main advantages of a supplementary fired boiler are to increase the steam generating capacity and, thus, permit a large steam turbine

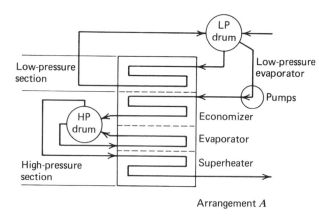

Arrangement *A*

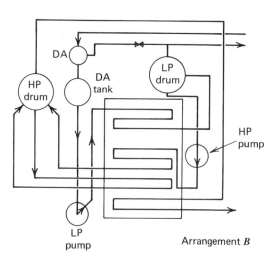

Arrangement *B*

Figure   13-12.   Flow   diagrams   for   waste heat boilers.

capability in the combined-cycle plant. It also will serve to maintain steam pressure and temperature in the boiler when its associated gas turbine load is lowered. This may be a distinct advantage if two or more gas turbines are associated with a single steam turbine generator. One gas turbine may be operated at a different load than the other (or others). Without supplemental firing, each gas turbine must share the load with sufficient closeness in order that the steam temperatures and pressures will be sufficiently compatible.

Supplemental firing adds substantial complication to plant load controls, which has influenced some plant designers and operators to prefer unfired units. Plant controls for combined cycles are beyond the scope of this book.

Mention was made previously of a low-level evaporator section in the heat recovery boiler. Because the boiler produces steam at two different pressures, it is sometimes referred to as a dual-pressure boiler. Figure 13-12 presents two dual-pressure boiler arrangements. In the first arrangement the low-pressure section is placed at the rear of the economizer and receives feedwater directly from the condensate system. The low-pressure section can be equipped to deaerate the feedwater, thus eliminating the need for any feedwater heating installation. The low-pressure drum usually serves as the deaerator storage tank, and the feedwater pumps for high-pressure section would take suction from the low-pressure drum.

The second dual-pressure boiler arrangement is relatively complicated. This boiler can produce simultaneously high-pressure superheated steam and low-pressure saturated steam. There are pumps at two locations. The low-pressure pumps are connected between the deaerator storage tank and the low-pressure drum. The high-pressure pumps take suction from the low-pressure drum and discharge water to the high-pressure section.

The dual-pressure boiler can utilize much more waste heat from the gas turbine exhaust than the single-pressure boiler. It generally has a better heat rate. However, a major drawback to the dual-pressure boiler is that it must be specially designed and built, and is not readily available.

EXAMPLE 13-4. Estimate the steam production rate for a single-pressure unfired waste heat boiler. The conditions are as follows:

Exhaust gas flow rate	2,255,670 lb/hr
Exhaust gas temperature	914.3 F
Steam conditions	1250 psia
	950 F
Water inlet temperature	326.4 F
Boiler pinch point	60 F

Solution: The procedure to determine the steam production rate is similar to that in Example 2-9. We first calculate the gas temperature at the boiler pinch point by the equation

$$T_x = T_{sat} + PP$$

since

$$T_{sat} = 572.4 \text{ F (at } P = 1250 \text{ psia)}$$

and

$$PP = 60 \text{ F}$$

$$T_x = 572.4 + 60 = 632.4 \text{ F}$$

Taking a heat balance on the superheater and evaporator, we have

$$m_g c_{pg} \left(T_{gi} - T_x \right) = m_{st} \left(h_s - h_f \right)$$

The numerical values for this example are

$$m_g = 2{,}255{,}670 \text{ lb/hr}$$

$$c_{pg} = 0.25 \text{ Btu/lb-F}$$

$$T_{gi} = 914.3 \text{ F}$$

$$h_s = 1468.6 \text{ Btu/lb (at 1250 psia and 950 F)}$$

$$h_f = 578.8 \text{ Btu/lb (saturated liquid at 1250 psia)}$$

Substituting these values into the above equation gives

$$m_{st} = 178{,}668 \text{ lb/hr}$$

As mentioned earlier, the boiler size is inversely related to the selection of boiler pinch point. Referring back to Example 13-4, but with a boiler pinch point of 40 F, we found that the steam production would increase to 191,344 lb/hr, or an increase of 7.1%. The supplemental firing in the waste boiler also increases the steam production. Again, referring to Example 13-4, a supplemental firing at a rate of 46.89×10^6 Btu/hr is added while other conditions remain unchanged. For this case the boiler will produce steam at the rate of 233,720 lb/hr or an approximately 31% increase over the unfired waste boiler.

EXAMPLE 13-5.   Estimate the steam production rate for a dual-pressure waste heat boiler. The boiler has an arrangement B shown in Fig. 13-12. The pump effects are assumed to be negligible. The input conditions for this calculation are:

High-pressure steam conditions	1250 psia
	950 F
Low-pressure steam conditions	225 psia
	saturated vapor
Deareator pressure	95 psia
Boiler pinch point	60 F
Boiler water inlet temperature	114 F
Exhaust gas temperature	914.3 F
Exhaust gas flow rate	2,255,670 lb/hr

Solution: Designating the various boiler locations as in Fig. 13-13 we note that the high-pressure section is identical to the single-pressure boiler in the previous example. For simplicity, we simply present the results as

$$T_x = 632.4 \text{ F}$$

and

$$m_{hp} = 178,668 \text{ lb/hr}$$

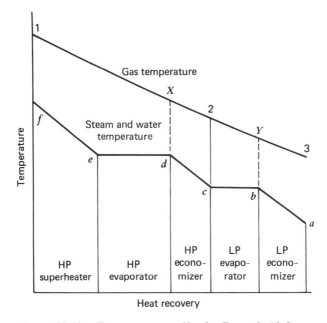

Figure 13-13.   Temperature profiles for Example 13-5.

To determine the temperature of gas exiting the high-pressure section, we take a heat balance on the high-pressure economizer, the equation is

$$m_g c_{pg}(T_x - T_2) = m_{hp}(h_d - h_{cf}) \tag{a}$$

Substituting the numerical values into the above equation gives

$$(2{,}255{,}670)(0.25)(632.4 - T_2) = (178{,}668)(578.6 - 366.3)$$

or

$$T_2 = 565.2 \text{ F}$$

Now, we calculate the amount of steam leaving the low-pressure drum, which is also equal to the sum of low-pressure steam produced and the extraction steam used for the deareator. Taking a heat balance on the low-pressure evaporator gives

$$m_g c_{pg}(T_2 - T_y) = (m_{ext} + m_{lp})(h_{cv} - h_b) \tag{b}$$

where T_y is the gas temperature at the pinch point of low-pressure evaporator. This temperature must be the steam saturation temperature plus the pinch point (i.e., $391.87 + 60 = 451.8$ F). Inserting the numerical values into Eq. (b), we have

$$(2{,}255{,}670)(0.25)(565.2 - 451.8) = (m_{ext} + m_{lp})(1200.8 - 366.3)$$

or

$$m_{ext} + m_{lp} = 76{,}630 \text{ lb/hr}$$

To determine the extraction steam flow for the deareator operation, we simply take the deareator as a control volume and apply the first law to it as below:

$$(m_{hp} + m_{lp})h_{i1} + (m_{ext})h_{i2} = (m_{hp} + m_{lp} + m_{ext})h_0 \tag{c}$$

where

$$h_{i1} = \text{water inlet enthalpy}$$

$$h_{i2} = \text{steam inlet enthalpy}$$

$$h_0 = \text{deareator discharge enthalpy}$$

These enthalpies are readily available because the states at these locations are well defined. Substituting the numerical values into Eq. (c) gives

$$(178{,}668 + 76{,}630 - m_{ext})(82.01) + (m_{ext})(1200.8)$$

$$= (178{,}668 + 76{,}630)(294.7)$$

or

$$m_{ext} = 48{,}534 \text{ lb/hr}$$

Thus the production rate of low-pressure steam is

$$m_{lp} = 76{,}630 - 48{,}534 = 28{,}096 \text{ lb/hr}$$

Finally, we want to determine the temperature of gas leaving the dual-pressure boiler. This can be accomplished by taking a heat balance around the low-pressure economizer. The equation is

$$m_g c_{pg}(T_y - T_3) = (m_{hp} + m_{lp} + m_{ext})(h_b - h_a) \tag{d}$$

where h_a and h_b are the enthalpies of saturated water, respectively, at the pressures 95 and 225 psia. With the numerical values substituted into Eq. (d), we have

$$(2{,}255{,}670)(0.25)(451.8 - T_3) = (178{,}668 + 76{,}630)(366.3 - 294.7)$$

Therefore the stack gas temperature is

$$T_3 = 419.4 \text{ F}$$

Boiler part-load performance is also important. In part-load operation the gas turbine load is reduced, and then exhaust gas temperature will fall off as previously indicated. Because of reduced available heat the waste heat boiler output must "follow." In a unfired boiler there is not only reduced steam flow, but also changes in steam conditions. The boiler drum pressure must be decreased in order to maintain a proper pinch point temperature difference. Inasmuch as the gas flow is virtually constant, the heat transfer coefficient (U) in the boiler system is approximately the same as at full load. Therefore the heat transfer rate in boiler part-load operation is proportional to the log-mean-temperature difference.

The part-load calculations for gas temperature leaving the boiler, and steam conditions (pressure, temperature and enthalpy) involve "cut and try" procedures and will not be presented in this text.

Waste heat boiler evaluation and selection cannot be isolated from other parts of combined-cycle system. Different boiler selection will affect not only the boiler performance, but also the combined-cycle system of which the boiler is only a component. System considerations to combined cycles are discussed in the next section.

13.5 COMBINED-CYCLE HEAT BALANCE AND SYSTEM CONSIDERATIONS

The combined-cycle plant presents no new problem in the heat and mass balance. It usually involves an application of the energy and mass conservation principle. Since the components of the combined-cycle system are the equipment readily available, the exact equipment performance must be taken into account. The following example illustrates a combined-cycle heat balance.

EXAMPLE 13-6. Calculate the combined-cycle plant heat balance on Fig. 13.10. There are two gas turbine-generators, each rated 100,000 kW at standard ISO conditions. After adjustments for the site conditions and additional exhaust end pressure drop, each gas turbine has a generating capability of 86,303 kW and a heat rate of 11,535 Btu/kWh. The exhaust gas has a temperature 1010 F and flows at the rate of 2,828,000 lb per hour. Other conditions are listed as below:

Boiler	
Superheater outlet pressure	950 psia
Superheater outlet temperature	950 F
Pinch point	50 F
Steam turbine	
Inlet conditions	950 psia
	950 F
Internal efficiency	80%
Condenser pressure	2 in. Hg abs.
Extraction pressure	45 psia
Turbine exhaust end loss	20 Btu/lb
Pump efficiency	75%
Miscellaneous cycle loss	4%
Combined-cycle plant	
Auxiliary power	4000 kw

Solution: First we determine the amount of steam generated in the waste heat boiler. We take the evaporator and superheater as a control volume and apply the first law to it. This gives

$$m_g c_p \left(T_g - T_x \right) = m_s \left(h_s - h_f \right) \tag{a}$$

where T_x is the gas temperature at the boiler pinch point. Since the steam saturation temperature at 950 psia is 538 F and the boiler pinch point is 50 F, the temperature T_x must be 588 F. Other conditions used for Eq. (a) are

$$m_g = 2,828,000 \text{ lb/hr, } c_p = 0.25 \text{ Btu/lb-F, } T_g = 1010 \text{ F}$$

$$h_s = 1479 \text{ Btu/lb, } h_f = 533 \text{ Btu/lb}$$

Inserting these values into Eq. (a) provides the rate of steam production for each boiler as

$$m_s = 316,000 \text{ lb/hr}$$

Now we move to the steam turbine side and determine the steam conditions at various cycle locations. Using the turbine inlet conditions (950 psia, 950 F) and turbine internal efficiency (80%), we estimate the enthalpy at the extraction as

$$h_{ext} = 1226.8 \text{ Btu/lb}$$

and the enthalpy at the turbine exhaust end as

$$h_{exh} = 1021 \text{ Btu/lb}$$

It is noted that the enthalpy h_{exh} is identical to the expansion line end point (ELEP) defined in Chapter 7. Since the turbine exhaust end loss is given as 20 Btu/lb, the used energy end point (UEEP) must be

$$\text{UEEP} = h_e = 1021 + 20$$

$$= 1041 \text{ Btu/lb}$$

Similarly, the steam enthalpies at other locations are determined and summarized as

Condenser outlet, h_{con}	69.1 Btu/lb
Deareator inlet, h_{wi}	69.3
Deareator outlet, h_{we}	243.5
Feedwater pump outlet, h_p	246.5

To calculate the steam extraction flow for the deareator, we take the energy and mass balance on the deareator and have the equations, respectively, as

$$m_{ext}h_{ext} + m_{wi}h_{wi} = m_{we}h_{we} \qquad \text{(b)}$$

$$m_{wi} + m_{ext} = m_{we} \qquad \text{(c)}$$

where m_{wi} and m_{we} are, respectively, the water flow rates at the deareator inlet and outlet. Since two boilers serve one steam turbine system, the total steam flow (or water flow rate, m_{we}) must be $2 \times 316{,}000$ lb/hr. Inserting the numerical values into the above equations and combining them give

$$m_{ext}(1226.8) + (2 \times 316{,}000 - m_{ext})(69.3) = (2 \times 316{,}000)(243.5)$$

or

$$m_{ext} = 95{,}053 \text{ lb/hr}$$

The output of steam turbine is determined by applying the first law. The equation is

$$w_{st} = m_i h_i - m_{ext} h_{ext} - m_e h_e \tag{d}$$

With the numerical values inserted, we have

$$w_{st} = (632{,}000)(1479) - (95{,}053)(1226.7) - (536{,}947)(1041.3)$$

$$= 259 \times 10^6 \text{ Btu/hr} \quad \text{or} \quad 75{,}954 \text{ kW}$$

With a miscellaneous cycle loss of 4%, the steam turbine-generator output becomes

$$w_{st} = (1 - 0.04)(75{,}954) = 72{,}916 \text{ kW}$$

The output of the combined-cycle system is the total power produced by the gas turbines and steam turbine minus the auxiliary power. For this example the net output is

$$w_{net} = w_{gt} + w_{st} - w_{aux}$$

$$= 2 \times 86303 + 72916 - 4000$$

$$= 241{,}522 \text{ kW}$$

The plant heat input is simply the product of the gas turbine output and its heat rate, and the plant heat rate is the ratio of heat input to the plant net output. That is,

$$\text{NHR} = \frac{\text{heat input}}{\text{net output}}$$

$$= \frac{2 \times (86303 \times 11535)}{2 \times 86303 + 72916 - 4000}$$

$$= 8244 \text{ Btu/kWh}$$

It is important to check the temperature of the gas leaving the boiler stack. This temperature can be determined by taking a heat balance on the economizer. The equation is

$$m_g c_p (T_x - T_{stack}) = m_s (h_f - h_p) \qquad \text{(e)}$$

where h_p is the water enthalpy at the boiler inlet and also the enthalpy at the discharge of feedwater pump. The numerical conditions for Eq. (e) are

$$m_g = 2,828,000 \text{ lb/hr}, \qquad T_x = 588 \text{ F}$$
$$m_s = 316,000 \text{ lb/hr}, \qquad h_f = 533 \text{ Btu/lb}$$
$$h_p = 246.5 \text{ Btu/lb}$$

Inserting these values into Eq. (e) gives

$$T_{stack} = 460 \text{ F}$$

It is seen from the example that the combined-cycle system has not only improved heat rate, but also increased the output as compared with the simple gas turbine. As mentioned earlier, there is always more than one boiler arrangement available. Engineers must evaluate each arrangement carefully before the boiler selection is made. For example, when a supplementary firing is added to the combined-cycle system shown in Fig. 13-10, the system would have a different performance picture. Figure 13-14 presents the heat balance of the new combined cycle with the net output and heat rate at 258,312 kW and 8247 Btu/hr respectively. The supplementary fired system has a greater output but a poorer heat rate than the unfired system. Another example involves a use of low-level evaporator in the waste heat boiler. This evaporator will replace the deareator as shown in Fig. 13-15. Since the heat balance procedure is similar to that in Example 13-6, the detailed calculations are omitted and the results are simply presented in Fig. 13-15. Compared with the single-pressure system shown in Fig. 13-10, this system shows a significant improvement in both output and heat rate. The output and heat rate are, respectively, 245,259 kW and 8118 Btu/kWh.

In addition to the cycle evaluation, the combined-cycle plant design should take into consideration component availability, efficiency, and reliability. As mentioned earlier, the component availability is extremely important. the custom-made component would cost so much that the system will not generate electric power at a competitive cost. Like other generating systems, the combined-cycle plant must be also evaluated in terms of its owning cost. The costs to own and operate any type of power plant may consist of:

1. Fixed charges on initial investment
2. Fuel cost
3. Operating and maintenance cost

These economic factors were discussed in Chapter 3. The following examples

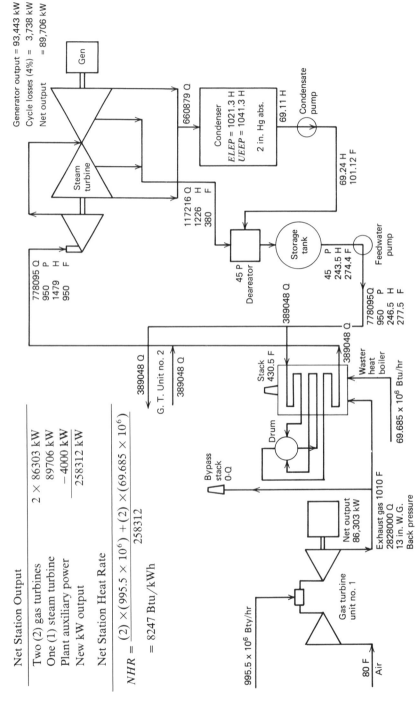

Figure 13-14.   Combined-cycle heat balance (supplementary fired boiler).

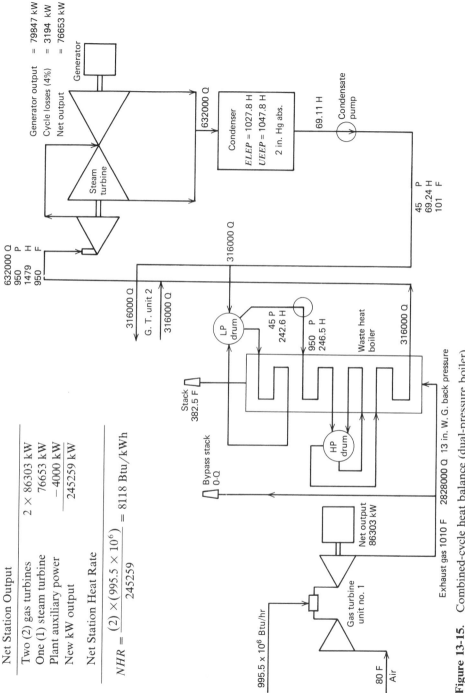

Net Station Output

Two (2) gas turbines	2 × 86303 kW
One (1) steam turbine	76653 kW
Plant auxiliary power	−4000 kW
New kW output	245259 kW

Net Station Heat Rate

$$NHR = \frac{(2) \times (995.5 \times 10^6)}{245259} = 8118 \text{ Btu/kWh}$$

Figure 13-15. Combined-cycle heat balance (dual-pressure boiler).

illustrate a simplified comparison of combined-cycle systems.

EXAMPLE 13-7. The two combined-cycle systems shown in Figs. 13-10 and 13-15 are being evaluated. Assuming that the annual fixed charge rate is 16% and the fuel cost is $7/MBtu, calculate the annual owning cost (fuel and initial cost) in terms of dollar per kilowatt and per year for the capacity factors 40, 50, and 60%.

The capital cost for the single-pressure and dual-pressure unit are, respectively, $560 and $580 per kilowatt.

Solution: The equation for the annual owning cost is

$$AOC = (\text{unit capital cost})(\text{annual fixed charge rate})$$

$$+ (8760)(\text{lifetime capacity factor})(\text{plant net heat rate})$$

$$\times (\text{levelized fuel cost}) \tag{a}$$

From Figs. 13-10 and 13-15, the plant net heat rate for these two systems are

$$(NHR)_s = 8244 \text{ Btu/kWh}$$

and

$$(NHR)_d = 8118 \text{ Btu/kWh}$$

With these heat rates and the economic factors given in this problem, the annual owning costs can be calculated by Eq. (a) and presented as below:

CAPACITY FACTOR (%)	COMBINED CYCLE SPECIFIED IN FIG. 13-10 ($/kW-yr)	COMBINED CYCLE SPECIFIED IN FIG. 13-15 ($/kW-yr)
40	291.81	291.92
50	342.36	341.70
60	392.91	391.48

Thus, the dual-pressure system is superior if the capacity factor is 50% or higher.

It would be interesting to examine the above example in terms of acceptable capital cost to the dual-pressure system. The acceptable capital cost is the cost such that the annual ownership cost for the dual-pressure system is equal to that for the single-pressure system. Still assuming a capital cost of $560/kW for the single-pressure system and making use of Eq. (a), we have

CAPACITY FACTOR (%)	ACCEPTABLE CAPITAL COST FOR THE DUAL-PRESSURE SYSTEM ($/kW)
40	579.32
50	584.14
60	588.97

It is seen that as the plant capacity factor increases, the acceptable cost for the dual-pressure system increases. Here, the acceptable cost is the maximum capital cost affordable for the dual-pressure system. If the actual capital cost is greater than this value, the dual-pressure system will be less competitive than the single-pressure system.

Combined-cycle plants have some attractive economic features. Low capital costs and low net station heat rates are a competitive advantage over other types of plants. Furthermore, the gas turbine portion of the plant could become commercially ready for operation in two or three years after project inception. The steam turbine portion of the combined-cycle plant would require an additional two to three years. Conventional coal fired plants require six to eight years, and nuclear plants require 10 to 12 years, or possibly even more.

Combined-cycle plants become more attractive as improvements in gas turbine are made. Currently, high-temperature, high-performance gas turbines are limited by the materials from which they are made. Normally, excess air must be supplied in the combustor to keep the temperature of the blades down. As mentioned earlier, one method that permits firing at high temperature is the passage of cooling air through holes in the blades. Future turbines may use film cooling and transpiration cooling. Improved blade and nozzle materials, such as ceramics, can also allow higher firing temperature. It is expected in the near future that the turbine firing temperature will be as high as 2500 F.

The one major drawback to combined-cycle plants is the fuel that is used. Currently, the combined-cycle plant uses either natural gas or oil. Both are relatively expensive. This fuel disadvantage will be significantly reduced when large coal gasifiers are economically available.

Fluidized-bed combustion systems are another alternative method by which coal energy can be utilized to run gas turbines or combined-cycle plants. In a fluidized-bed unit, combustion occurs at temperatures from 1600 to 1800 F under pressures from 4 to 10 atmospheres. The hot, high-pressure flue gas, exiting from the combustion unit is expanded through a gas turbine to generate electric power. Becuase the coal is relatively abundant and inexpensive, the development of fluidized-bed combustion system will have positive impact on use of gas turbine and combined cycle.

13.6 COGENERATION

The production of electricity from fossil fuels is a relatively inefficient process. The efficiency of modern power plants ranges from 32% for nuclear power plant to 43% for advanced, sophisticated gas/steam combined cycle plants. This means that currently between 57 and 68% of the fuel heating value is being discharged as waste heat. Some of the waste heat is unrecoverable, such as radiation and stack loss from the steam generator. But much of the waste heat can be used for industrial applications or for district heating. Through the utilization of waste heat, the overall conversion of fuel to usable energy can be as high as 80%.

Cogeneration is the simultaneous production of more than one form of useful energy. In industry, cogeneration has been used as a means of producing both electric and thermal energy. The cogeneration system is sometimes referred to as the

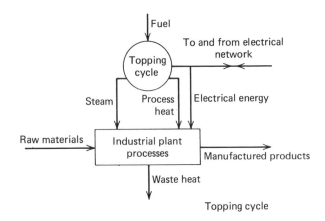

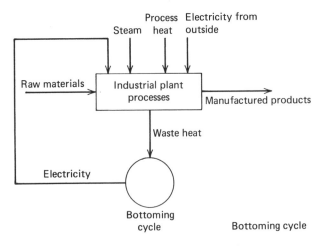

Figure 13-16. Two basic arrangements for cogeneration system.

total energy system or waste heat utilization system. Cogeneration is not a new concept. It was, in fact, implemented in the United States at the end of the nineteenth century. In 1900, most industrial plants generated their own utilities, but in 1960 they generated only 21% of electricity they consumed. This percentage continuously declined to 17% by 1968 primarily because of the availability and relatively low price of fossil fuels. The decline of cogeneration was also due to the institutional constraints and the economic distortion they produced. But in recent years the combined impacts of increasing primary fuel costs and the inflationary capital cost of energy-producing systems have made energy-saving measures economically viable. Energy conservation technology, ignored in the past, has become again important in all industries.

A cogeneration system requires equipment that burns fuel to produce shaft work (including electric power) and thermal energy. Thermal energy may be used as dry heat for an industrial process or converted to steam. The equipment for that purpose is called a prime mover. Steam and gas turbines and diesel engines have been used as prime movers in industry. There are two basic arrangements for cogeneration systems. One is topping of process steam or process heat while another is a bottoming process. Figure 13-16 shows the schematic diagrams of these two basic systems. In the topping cycle, the prime mover is used to generate electric power and the waste heat or the byproduct steam from it is used for plant processes. Topping cycles are significantly more fuel efficient than conventional systems that generate electric and thermal energy separately.

The bottoming cycle is completely different from the topping cycle. In the bottoming cycle, plant processes are run in their own fuel, and the waste heat from them is utilized for electric power generation. Bottoming cycles also save fuel when compared with conventional systems.

Below is a brief description of the prime movers used for cogeneration systems: Steam turbine, gas turbine, and diesel engine.

Steam Turbine

The steam turbine is the most common prime mover for cogeneration system. There are several types of turbogenerators that can meet the widely varying electric and thermal energy demand. One of these is the back-pressure turbine that takes steam from the boiler and produces electric power by expanding the steam through it. Upon discharge from the turbine, the steam is being directed to the steam users. While this system is simple and efficient, it has a major drawback in simultaneously meeting the electric power and thermal energy need.

One convenient approach to balance electric power and process steam demands is to connect the cogeneration system with an electric power network, as shown in Fig. 13-17. In this case the cogeneration system is operated according to the need of process steam and the steam flow determines the output of the turbine–generator. When the electric output exceeds the need, the surplus is exported to the electric power network. Conversely, the electric power is imported to cover any power deficiency.

Figure 13-17 also presents another arrangement that would simultaneously meet the electric and steam energy need. It involves use of a pressure-reducing valve (including a desuperheater) and/or a surplus steam valve. In this case the turbine is speed-governed and passes a quantity of steam corresponding to the electrical demand. The reducing valve makes up any process steam deficiency and the surplus valve takes care of any surplus steam.

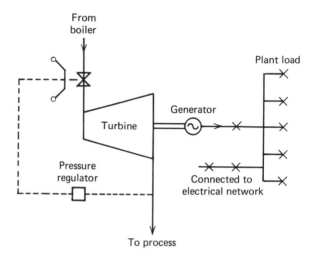

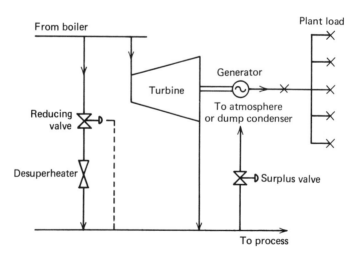

Figure 13-17. Back-pressure steam turbine in a cogeneration system.

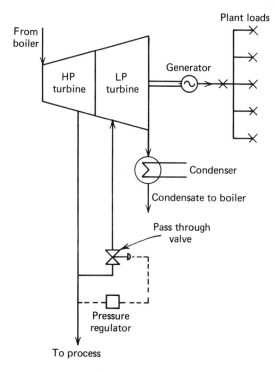

Figure 13-18. Extraction steam turbine in a cogeneration system.

When the process steam demand is relatively small as compared with the electric power need, an extraction turbine may provide the best solution. Figure 13-18 shows a schematic diagram for this arrangement. The extraction turbine receives steam from the boiler and expands it to the pressure equal to that of the process steam. At this point some of steam is extracted out as the process steam and the remainder reenters the turbine through a control valve for final expansion to the condenser pressure. The turbine is speed-governed, and the pressure regulator is used to maintain the process steam at a desirable pressure level.

The extraction turbine has one advantage that the back-pressure turbine does not have. It can easily meet the demand of both electrical and thermal energy. In turbine selection it is important that the steam inlet temperature and pressure should be as high as possible. These throttle conditions enable the turbine to generate more electricity per unit of process steam. However, high throttle steam conditions require slightly more expensive equipment.

The boiler is a key element in steam turbine cogeneration system. When the steam turbine is used as a prime mover in a topping cycle, the steam pressure and temperature are relatively high. Therefore the boiler should be carefully evaluated and selected. Based upon current energy economics, it is likely that boilers will be designed to burn coal or residual oil.

Gas Turbine

The gas turbine as described in the previous sections is a convenient and compact system for electric power generation. It usually operates at higher temperatures than the steam turbine and exhausts gas at a temperature of about 1000 F. The thermal energy of exhaust gas can be recovered in a variety of ways, as shown in Fig. 13-19. For instance, the exhaust gas can be used directly for heating and drying, which is important in the food industry. It can be also passed through a heat exchanger where heat is transferred to a process fluid such as air, water, and oil. As indicated in Fig. 13-19, it can also enter a waste heat recovery boiler to produce steam.

The gas turbine has a drawback. It has not only a low efficiency in electric power production, but also a high fuel cost. While the use of gas turbines for cogeneration is expected to increase, gas turbines will not sell well until coal-derived fuels are economically available.

When compared with the steam turbine, the gas turbine is usually suitable for small cogeneration units. Because gas turbine operates at a high temperature, it requires more maintenance. Major overhauls are required at 10,000 to 20,000 hours.

A combined gas and steam turbine system has been used for electric power generation. It has a higher efficiency than the gas turbine or steam turbine system. When used for cogeneration, it usually has the condensing steam turbine replaced by a back-pressure turbine. Figure 13-20 shows a schematic diagram for a combined-cycle cogeneration unit. (See the previous section for a discussion of the combined cycle.)

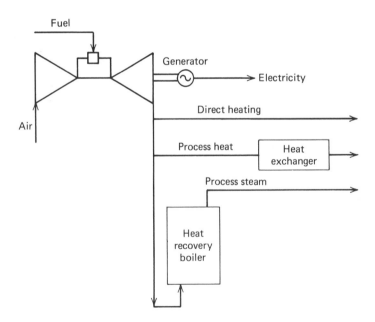

Figure 13-19.   Gas turbine in a cogeneration system.

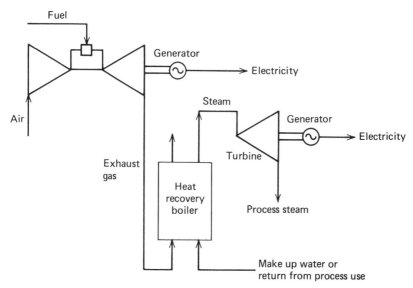

Figure 13-20. A combined-cycle cogeneration system.

Diesel Engine

Diesel engines have relatively high thermal efficiency, usually around 39%. The waste heat rejected from diesel engine is mainly in the form of exhaust gas and thermal energy carried away by the engine jacket water. At a full load the typical diesel engine would have the heat balance given below:

Shaft work	39.20%
Heat in exhaust gas	33.20
Heat in jacket water	13.84
Heat in lubricating oil	4.61
Radiation and other losses	9.15

It should be noted that not all waste heat can be recovered. For instance, only about 20% of exhaust gas energy is economically usable.

Figure 13-21 presents a diesel engine cogeneration arrangement. The diesel engine drives an electric generator and exhausts the gas into a waste heat recovery boiler for steam generation. For economic reasons, the heat removed in the engine jacket is also utilized.

The diesel engine has a small unit size and usually generates no more than 30,000 kW. Another major drawback of diesels is the expensive fuel it consumes. Unless coal-derived fuels are made more economical and feasible for use in diesels, the market development for diesel engines will not change significantly in the near future.

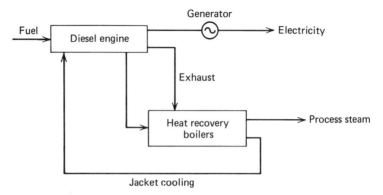

Figure 13-21.   Diesel engine in a cogeneration system.

The diesel engine cogeneration system has a relatively high electricity to thermal energy ratio. Electricity generation per unit of thermal energy produced is around 400 kWh per million Btu. Therefore, to use the diesel engine as a prime mover in cogeneration, the excess of elecric power must be put to productive use, either by selling it or storing it for later use.

The selection of the prime mover for cogeneration can be made only on a system-by-system basis. Each system has its own requirements for thermal and electrical energy. The type of fuel and space available would also affect the decision. As mentioned earlier, steam and gas turbines and diesel engines have different generating capabilities both in thermal and electrical energy. The typical ratios of electrical-to-thermal energy (E/T) are

Prime mover	E/T (kWh/MBtu)
Steam turbine	36
Gas turbine	175
Diesel engine	400

In cogeneration system design, engineers should match the energy demand with the system operating characteristics. They should also consider whether the excess electrical energy can be sold. For instance, assume that a plant has thermal and electrical energy demands, respectively, at 500 MBtu/hr and 40 MW. When the steam turbine is selected as a prime mover and designed to meet the plant thermal energy need, the steam turbine would generate only 18 MW of electric power, which is 22 MW less than the plant electric power need. However, a gas turbine would generate approximately 87 MW of electricity, producing a 47 MW surplus. When a diesel engine is selected and designed to meet the plant thermal energy demand, it would generate approximately 200 MW, yielding a surplus of 160 MW.

Generally, the steam turbine has the best ratio of electric-to-thermal energy to meet the needs of various industrial cogeneration systems. Diesel engines generate the most electricity per unit of thermal energy and have the greatest surplus of electricity for most cogeneration systems. In this respect gas turbines are somewhat between the steam turbine and diesel engine.

In cogeneration system evaluation, an incremental heat rate for power generation is frequently used. It is defined as

$$\text{IHR} = \frac{Q_{in} - Q_{th}}{W_{el}} \tag{13-2}$$

where

Q_{in} = heat supplied to cogeneration system

Q_{th} = process heat produced by cogeneration system

W_{el} = electrical energy produced by cogeneration system

The steam turbine usually has an incremental heat rate of around 4700 Btu/kWh when used as a topping of process steam (or process heat). For the gas turbine and diesel engine, the incremental heat rate is little higher. The range is 5500 to 6000 Btu/kWh for the gas turbine and 6,000 to 7,000 Btu/kWh for the diesel engine. Cogeneration systems are therefore much more efficient than conventional systems for electric power generation.

The definition of thermal efficiency for a cogeneration system is an interesting question. By tradition, the system thermal efficiency is defined as a ratio of the system output to input. In a cogeneration system the output consists of the electrical and thermal energy, while the input is simply the heating value of the fuel that the system consumes. In equation form the thermal efficiency is

$$\eta_1 = \frac{W_{el} + Q_{th}}{Q_{in}} \tag{13-3}$$

The above definition is evidently formulated on the basis of the energy concept. In other words, electrical energy is not treated any differently from thermal energy. This is of course incompatible with our commonsense notion that the electrical energy is much more valuable than the thermal energy, especially energy in low-temperature and low-pressure conditions. While Eq. (13-3) is still used, the thermal efficiency, based on the concept of thermodynamic availability, has frequently appeared in literature. This second-law thermal efficiency is defined as

$$\eta_{II} = \frac{A_{el} + A_{th}}{A_{fuel}} \tag{13-4}$$

where

A_{el} = electric work availability produced by cogeneration system

A_{th} = thermal availability produced by cogeneration system

A_{fuel} = availability of the fuel supplied to cogeneration system

Numerically, the work availability is equal to the electric work itself. The fuel availability is the product of fuel consumption and fuel unit availability. For approximation, the fuel unit availability is frequently treated as the fuel heating value. As to the availability associated with the thermal energy (either process heat or process steam), one should make use of the definition of thermodynamic availability shown in Chapter 2 and calculate the exact availability output according to the system conditions.

13.7 STEAM COSTING

When a conventional system such as a boiler generates steam for heating purposes or for electric power generation, there is no difficulty in estimating the cost of the steam. However, when a cogeneration system generates process steam, the costing of process steam is not so easy. In a cogeneration system involving a back-pressure steam turbine, steam is first used for electric power production and then used as the process steam. Engineers must determine a fair means of dividing the cost between the two cogenerated products.

Different steam costing methods have been proposed. Basically, these methods can be classified into two groups. The first group involves the costing of steam and electricity on the basis of the energy content, while the second group is on the basis of the available energy. There is no generalized method that can be used for all situations. Some methods are particularly valuable at the time when the decision for cogeneration is made, and others are more suitable for an existing cogeneration unit. Each method is biased to some degree because it distributes the economic benefits of cogeneration to users of either electricity or process steam at the expense of the users of the other product. Below are brief descriptions of the three costing methods.

The Incremental Method

This steam costing method involves the calculation of a fictitious cost for the generation of process steam in low-pressure boiler. The cost for the power generated by cogeneration is simply the incremental cost, that is, the costs that are incurred to produce power minus the above fictitious cost. In equation form, they are

$$c_s = \left(c_f \dot{m}_{fl} + (IC)_{lp} \right) / \dot{m}_{lp} \qquad (13\text{-}5)$$

and

$$c_{el} = \left[c_f(\dot{m}_{fc} - \dot{m}_{fl}) + (IC)_{tur} + \Delta(IC)_b \right] / \dot{W}_{el} \qquad (13\text{-}6)$$

where

c_s = steam unit cost

c_{el} = electrical energy unit cost

c_f = boiler fuel unit cost

$\dot{m}_{fl}$ = fuel consumption for low-pressure boiler

$\dot{m}_{fc}$ = fuel consumption for cogeneraton boiler

$(IC)_{lp}$ = initial cost for low-pressure boiler

$(IC)_{tur}$ = initial cost for turbine

$\Delta(IC)_b$ = cost differential between cogeneration boiler and low-pressure boiler

$\dot{m}_{lp}$ = low-pressure process steam output

$\dot{W}_{el}$ = electrical energy output

This method is usually applied in situations where engineers in an industrial plant decide whether or not a cogeneration facility is needed. When the cost of electricity calculated above is lower than the purchase price, the investment in a cogeneration facility would be probably worthwhile. It should be noted that this incremental method is biased toward the decision of cogeneration, because it uses all economic benefits of cogeneration to lower the electric power cost.

The Lost-Kilowatt Method

This steam costing method involves the calculation of kilowatts lost in a production of process steam. The lost kilowatt can be easily determined by comparing the output of an electric power generating system and the electrical output of the corresponding, but hypothetical cogeneration plant. Once the reduction of electrical output is determined, the market value of the electrical energy is simply the cost of the process steam. This steam costing method is suitable in situations where engineers in a power plant consider developing a steam market. When the cost of the steam is lower than the steam price available, the decision for cogeneration is probably positive.

The lost-kilowatt method is applicable to both back-pressure and extraction turbine systems. Unlike the incremental method, the lost-kilowatt method distributes all cogeneration economic benefites to the steam users. Therefore the cost of process steam by this method is usually lower than that by other methods.

The Available-Energy Method

This method utilizes the concept of available energy (also referred to as thermodynamic availability or simply availability). As presented in Chapter 2, available energy represents a potential of substance to produce useful work. This concept provides a basis for the costing of process steam and electric power produced in a cogeneration system. For simplicity, we present the available energy method as applied to a back-pressure steam turbine system in Figure 13-22. Like any energy converter, the cost of products by the back-pressure turbine must be equal to the total expenditure it requires. In equation form, it is

$$c_{el}\dot{W}_{el} + c_{lp}\dot{A}_{lp} = c_{hp}\dot{A}_{hp} + (IC)_{tur} \qquad (13\text{-}7)$$

where

$$c_{el} = \text{unit cost for electricity}$$

$$c_{lp} = \text{unit cost for low-pressure steam}$$

$$c_{hp} = \text{unit cost for high-pressure steam}$$

$$(IC)_{tur} = \text{annual capital cost for turbine}$$

$$\dot{W}_{el} = \text{annual electrical energy generation}$$

$$\dot{A}_{lp} = \text{annual availability flow in low-pressure steam}$$

$$\dot{A}_{hp} = \text{annual availability flow in high-pressure steam}$$

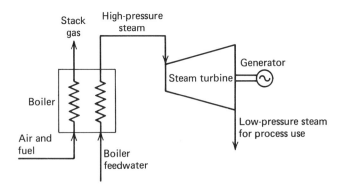

Figure 13-22. A back-pressure steam turbine for cogeneration.

To make use of the equation, the unit cost for high-pressure steam (c_{hp}) must be determined and known. This can be accomplished by applying the above money balance to the boiler. That is,

$$c_{hp}\dot{A}_{hp} = c_f\dot{A}_f + (IC)_b \qquad (13\text{-}8)$$

where $(IC)_b$ is annual capital cost for boiler, c_f is unit cost for fuel availability, and $\dot{A}_f$ is the annual fuel availability flow, which is simply the annual fuel consumption times the unit fuel availability. For hydrocarbon fuel, the unit availability can be approximated by the higher heating value of the fuel. The exact calculation of fuel availability is beyond the scope of this text.

When the cogeneration system is specified, the steam availability flow can be calculated by the equation

$$\dot{A} = \dot{m}a = \dot{m}\big[(h - h_0) - T_0(s - s_0)\big] \qquad (13\text{-}9)$$

Substituting Eq. (13-9) into Eq. (13-8) enables us to determine the unit cost for high-pressure steam. However, when we substitute Eq. (13-9) into Eq. (13-7), there are still two unknowns, c_{lp} and c_{el}. Therefore, a complementary equation is needed.

Many different methods have been proposed for establishing the complementary equation. The selection of the method depends upon the situations. For instance, we can set up the complementary equation as

$$c_{lp} = c_{el} \qquad (13\text{-}10)$$

when we consider that the electrical and thermal energy are equally important products in a cogeneration plant. Substituting Eq. (13-10) into Eq. (13-7) yields the unit costs for both products as

$$c_{lp} = c_{el} = \frac{c_{hp}\dot{A}_{hp} + (IC)_{tur}}{\dot{A}_{lp} + \dot{W}_{el}} \qquad (13\text{-}11)$$

Evidently in this method, the electrical and thermal outputs are equally charged for the cost of high-pressure steam and turbine investment.

There may arise another situation where engineers consider the back-pressure turbine as an equipment for electric power generation. In this case the electrical output must be charged for the entire cost associated with the turbine process and the thermal energy is charged only for high-pressure steam cost. Thus the complementary equation is

$$c_{lp} = c_{hp} \qquad (13\text{-}12)$$

and the unit cost for electric output is

$$c_{el} = c_{hp}\left(\frac{\dot{A}_{hp} - \dot{A}_{lp}}{\dot{W}_{el}}\right) + \frac{(IC)_{tur}}{\dot{W}_{el}} \tag{13-13}$$

At this point it should be emphasized that there are essentially two equations needed in the available energy method. The first equation is the money balance equation, such as Eq. (13-7). This equation involves a use of available energy of the cogenerated products rather than the energy content. The second equation is the so-called complementary equation, which varies case by case. Therefore, there is no generalized expression for it. To formulate the second equation, engineers must exercise their judgment and take into account the conditions under which the costings are made.

SELECTED REFERENCES

1. H. Cohen, G. F. C. Rogers, and H. I. S. Saravanamuttoo, *Gas Turbine Theory*, Longman Group Limited, 1974.

2. G. M. Dusinberre and J. C. Lester, *Gas Turbine Power*, International Textbook Company, 1962.

3. L. Ushiyama, "Theoretically Estimating the Performance of Gas Turbines Under Varying Atmospheric Conditions," ASME Transaction, Series A, *Journal of Engineering for Power*, January 1976.

4. M. P. Boyce and D. A. Hanawa, "Parametric Study of a Gas Turbine," ASME Transaction, Series A, *Journal of Engineering for Power*, July 1975.

5. G. F. Pavlenco, G. A. Englesson, and L. Denesdi, "Review and Comparison of Allocation Methods for the Separation of Electrical and Thermal Cogeneration— District Heating Costs," *Proceedings of the American Power Conference*, Volume 42, 1980.

6. W. J. Wepfer and B. G. Crutcher, "Comparison of Costing Methods for Cogenerated Process Steam and Electricity," *Proceedings of the American Power Conference*, Volume 43, 1981.

7. K. W. Li, "Allocating Steam and Electricity Cost in a Cogeneration System," *Proceedings of the American Power Conference*, Volume 44, 1982.

8. A. P. Priddy and J. J. Sullivan, "Engineering Considerations of Combined Cycles," *Proceedings of the American Power Conference*," Volume 34, 1972.

9. R. C. Sheldon and D. M. Todd, "Optimization of the Gas Turbine Exhaust Heat Recovery System," an ASME paper 71-GT-79, 1971.

10. J. C. Stewart, "Computer Techniques for Evaluating Gas Turbine Heat Recovery Application," an ASME paper 72-FT-103, 1972.

11. United Aircraft Research Laboratories, "Advanced Nonthermally Polluting Gas Turbines in Utility Applications," The U.S. Environmental Protection Agency, Water Pollution Control Research Series, 16130 DNE 03/71, 1971.

PROBLEMS

13-1. The simple gas turbine cycle has the thermal efficiency as

$$\eta_{cy} = \frac{\left(1 - \frac{1}{\rho_p}\right)(\alpha - \rho_p)}{\eta_c(k_1 - 1) - \rho_p + 1}$$

where the dimensionless terms are defined in Chapter 2. Derive an expression for the pressure ratio in which the thermal efficiency will become maximum.

13-2. The simple gas turbine has the network on the basis of unit mass as

$$w_{net} = \frac{c_p T_1}{\eta_c}\left(1 - \frac{1}{\rho_p}\right)(\alpha - \rho_p)$$

The derivation of this equation was presented in Chapter 2. Derive an expression for the pressure ratio in which the cycle network will become maximum.

13-3. A simple gas turbine-generator has been designed for 100,000 kW at standard ISO conditions. Its base load heat rate under these conditions is 11,000 Btu/kWh. What is the base capability and corresponding heat rate at the site where the ambient air temperature and pressure are, respectively, 0 F and 12.3 psia?

13-4. Repeat Problem 13-3 for the site ambient temperature of 110 F and an atmospheric pressure of 15.0 psia.

13-5. A small gas turbine-generator has been designed with an output of 24,250 kW and a heat rate of 12,400 Btu/kWh. Consulting engineers decide to use an inlet air filter and low-level silencer for this turbine. What are the penalty both in the turbine capability and heat rate?

13-6. A manufacturer provides the design performance data at the ISO conditions for one regenerative gas turbine model. The data include

Output		57,400 kW		
Heat rate		9120 Btu/kWh		
Air flow rate		1,896,000 lb/hr		
Full-load fuel consumption		523.6×10^6 Btu/hr		
(based on LHV 18,500 Btu/lb)				
Part-load fuel consumption				
Load (%)	40	60	80	100
Consumption (%)	56	68	83	100

Estimate the turbine heat rate and exhaust gas flow at various loads.

13-7. A simple-cycle gas turbine has the following performance data at the site conditions:

Output				51,100 kW
Heat rate				12,090 Btu/kWh
Air flow rate				1,850,000 lb/hr
Full-load fuel consumption				617.7×10^6 Btu/hr
(based on LHV 18,500 Btu/lb)				
Part-load fuel consumption				

Load (%)	40	60	80	100
Consumption (%)	57	71	86	100

Estimate the turbine heat rate at various turbine loads and plot the heat rate curve.

13-8. The waste heat boiler shown in Fig. 13-11 is being considered in a combined-cycle system design. The boiler is expected to provide steam at 1250 psia and 950 F at an inlet water temperature of 274 F. Assuming that the boiler pinch point is 50 F, calculate the steam flow rate and stack gas temperature for the following conditions:

Exhaust gas flow rate	1.68×10^6 lb/hr
Exhaust gas temperature	960 F
Additional firing rate	26.57×10^6 Btu/hr

13-9. Repeat Problem 13-8 with the boiler pinch point changed to 80 F.
13-10. A waste heat boiler has an arrangement and conditions shown in Fig.

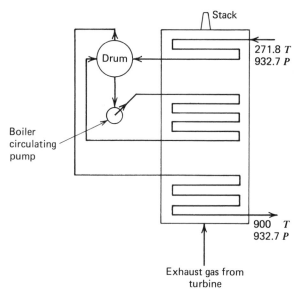

Figure 13-23. The waste heat boiler diagram for Problem 13-10.

13-23. Assume that the boiler pinch point is 50 F and the effect of the circulating pump is negligible, calculate the steam flow rate and the stack gas temperature for the turbine exhaust gas flow of 1.72×10^6 lb/hr at a temperature of 1055 F.

13-11. The combined-cycle arrangement shown in Fig. 13-10 is being considered in the plant system design. Another turbine manufacturer now offers a gas turbine model of which the adjusted performance at the site is as follows:

Gas turbine output	73,600 kW
Fuel consumption	963.7×10^6 Btu/hr
Exhaust gas flow rate	2,476,000 lb/hr
Exhaust gas temperature	1050 F

What would be the plant net output and heat rate if the original gas turbine is replaced by this model? All other conditions are identical to those in Fig. 13-10.

13-12. Calculate the plant net output and heat rate for the combined cycle system shown in Fig. 13-24. There are four gas turbines coupling with one steam turbine. The conditions listed below should be utilized for calculations.

Gas turbine	
Turbine-generator output	73,600 kW
Fuel consumption	963.7×10^6 Btu/hr
Exhaust gas flow	2,476,000 lb/hr
Exhaust gas temperature	1030 F
Boiler	
Supplementary firing	60.93×10^6 Btu/hr
Pinch point	60 F
Steam turbine	
Inlet conditions	1250 psia, 950 F
Extraction pressure	24.8 psia
Condenser pressure	3 in. Hg abs.
Exhaust end loss	20 Btu/lb
Turbine internal efficiency	88%
Pump efficiency	82%
Turbine generator efficiency	96%
Combined-cycle plant auxiliary power	7%

13-13. A simple-cycle gas turbine is used as a prime mover in cogeneration

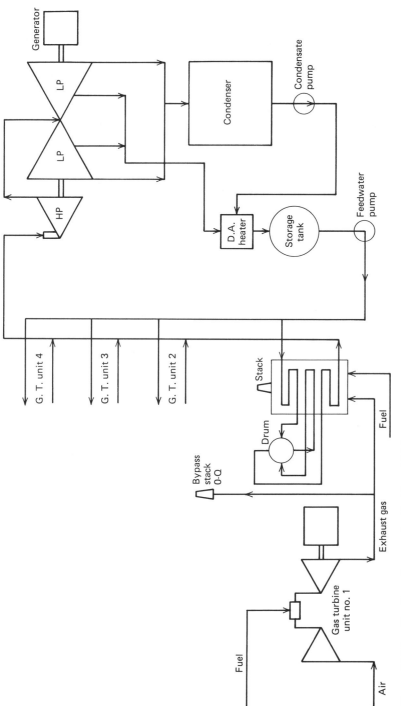

Figure 13-24. The flow diagram for Problem 13-12.

system. The turbine adjusted design performance is

Electrical output	59,000 kW
Heat rate	11,850 Btu/kWh
Fuel consumption	698.8 × 10^6 Btu/hr
(based on LHV 18,500 Btu/lb)	
Air flow rate	1,850,000 lb/hr
Exhaust gas temperature	945 F

The exhaust gas is directly used for heating and experiences a total temperature drop of 400 F in the entire process. Calculate the incremental heat rate for power generation and the second-law fuel utilization efficiency.

Assume that the fuel availability is equal to the fuel heating value.

14-14. A back-pressure steam turbine of 100,000 kW serves as a prime mover in a cogeneration system. The boiler admits the return water at a temperature of 150 F and produces the steam at 950 psia and 850 F. Steam then enters a back-pressure turbine and expands to the pressure of the process, which is 75 psia. Neglecting the effects of pumpings and the pressure drops at various locations, calculate the incremental heat rate for electric power generation and the second-law fuel utilization efficiency.

The boiler efficiency is assumed to be 80%.

13-15. By the available energy method estimate the unit cost for the thermal energy from the cogeneration system described in Problem 13-13. The economic parameters are

Gas turbine capital cost	$350/kW
Fuel cost	$7.00/MBtu
Annual fixed charge rate	18%
Annual operating hours	4000 hr

Also discuss how the thermal energy unit cost changes with the selection of the complementary equation.

13-16. By the available energy method, calculate the unit cost for the process steam for Problem 13-14. The economic parameters are

Boiler capital cost	$300/kW
Steam turbine capital cost	$250/kW
Boiler fuel cost	$1.10/MBtu
Annual fixed charge rate	22%
Annual operating hours	4500 hr

Assume that two cogenerated products are equally needed in the plant.

APPENDIX A

Heat Balance Calculations for a Steam Turbine-Cycle System

Heat balance calculations for a steam turbine system are frequently performed by consulting engineers. Here is the general procedure.

1. Use the turbine inlet conditions, internal efficiency, extraction and condenser pressures to estimate the turbine expansion line.

2. Use the pressure drops in boiler, reheater, and extraction lines and the terminal temperature difference and drain cooler approach of feedwater heaters to determine the steam properties at various locations.

3. Determine the extraction steam flow rates, starting with the feedwater heater closest to the steam generator and continuing the calculation, heater by heater, until the last one (i.e., the heater next to the condenser).

4. Calculate the outputs of all turbine cylinders. The sum of these outputs is the gross output of the steam turbine. When it is multiplied by mechanical efficiency and generator efficiency, the turbo-generator net output will be obtained.

5. Calculate the power consumption by feedwater pumps and the heat inputs supplied to the turbine-cycle system and finally estimate the turbine net heat rate.

In some cases the steam turbine output is specified. If this is the case, a trial-and-error approach should be taken. That is, the steam flow rate at the turbine inlet is first assumed. The above procedure is applied to determine the turbo-generator net output. If the calculated value is not matched with the input, the new steam flow rate must be assumed and new heat balance calculation performed. This procedure is repeated until the deviation is within the desirable range.

When the turbine exhaust end loss is available, this must be subtracted from the turbine output. Occasionally, the exhaust end loss is included in the turbine internal efficiency. In some power plant designs a certain amount of steam is extracted for an in-house use. If this is the case, this extraction must be taken into consideration. The following example illustrates the above mentioned procedure.

   1.   Inputs:

Temperature	1000 F
Pressure	3515 psia
Flow rate	4,994,457 lb/hr
Boiler pressure drop	17%

2. Reheater

       Outlet temperature   1000 F
       Pressure drop   7%

3. Steam turbine

	INLET PRESSURE	OUTLET PRESSURE	INTERNAL EFF.
HP Turbine	3515 psia	744 psia	83%
IP Turbine		153 psia	88.4%
LP Turbine		2.5 in. Hg abs.	89%

4. Feedwater heaters*

Heater number	1	2	3	4	5	6	7
Extraction steam pressure	6.82	18.70	34.50	60.60	153.0	307.0	744.0
Terminal temperature difference (F)	10	10	5	5	0	5	5
Drain cooler approach (F)	15	15	15	15	—	15	15

5. Pumps

	Inlet Pressure	Pump Efficiency	Drive
Condensate pump	2.5 in. Hg abs.	82%	Motor-driven
Drain pump		80%	Motor-driven
Feedwater pump		85%	Auxiliary turbine

6. Miscellaneous inputs

Motor efficiency	95%
Mechanical coupling efficiency	100%
Steam generator efficiency	90%
Auxiliary turbine efficiency	80%
Generator efficiency	98.5%
Auxiliary turbine exhaust pressure	2.75 in. Hg abs.

Steam extraction for the auxiliary turbine is at the extraction point no. 5. (counted from the condenser side.)

Steam extraction for in-house use is at the extraction point no. 5. The steam flow rate is 103,400 lb/hr and the return condition is 359.8 F.

The turbine exhaust end loss	11.4 Btu/lb
The pressure drop in all extraction lines	0%

    The heat balance calculation is performed in the following steps.

---

*All heaters except no. 5 are surface heaters with drain coolers. Number 5 is of the contact type. The drain from each feedwater heater will be cascaded to the next low pressure heater. The last one is directed to the condenser.

1. CONSTRUCT THE TURBINE EXPANSION LINES

To determine the expansion line for the HP turbine, we make use of its internal efficiency

$$\eta_i = \frac{h_i - h_e}{h_i - h_{es}}$$

where

h_i = enthalpy at the turbine inlet

h_e = enthalpy at the turbine outlet

h_{es} = enthalpy at the isentropic turbine outlet

For the steam enthalpy at the HP turbine outlet, substituting the numerical values into the above equation gives

$$0.83 = \frac{1421.7 - h_e}{1421.7 - 1245.4}$$

or

$$h_e = 1275.4 \ \text{Btu/lb}$$

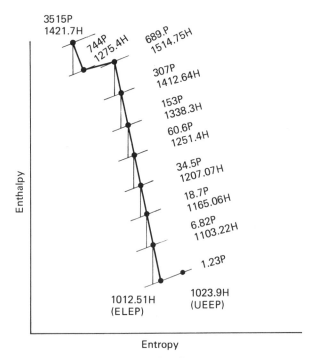

Figure A-1.   Turbine expansion line.

The turbine exhaust pressure (744 psia) and exhaust enthalpy (1275.4 Btu/lb) will determine the HP turbine exhaust end point. The straight line connecting the inlet and exhaust end points is an approximation of the HP turbine expansion curve.

Similar calculations are performed to determine the expansion lines for the IP and LP steam turbines. Figure A-1 presents the entire expansion curve including the expansion line-end point (ELEP) and the used-energy-end point (UEEP). As indicated in Chapter 7, the difference between these two points is the so-called turbine exhaust end loss. For this case the end loss is given as 11.4 Btu/lb.

The pressure drop across the reheater is given as 7%. Therefore the reheater outlet pressure becomes 689 psia.

2. DETERMINE THE STEAM CONDITIONS AT VARIOUS LOCATIONS

To determine the steam extraction conditions, we make use of the fact that turbine expansion line can be approximated by a straight line and the input information about the extraction pressure. The results are shown in Fig. A-1. Since there is no pressure drop along the extraction line, the extraction conditions are also the inlet conditions of the corresponding feedwater heater. Next, the conditions around the heater are determined by using the terminal temperature difference (*TTD*) and drain cooler approach (*DCA*). For example, the temperature of water leaving the heater is

$$T_e = T_s - TTD$$

where

$$T_s = \text{saturation temperature of extracted steam}$$

and the temperature of drain leaving the heater is

$$T_d = T_i + DCA$$

where

$$T_i = \text{the temperature of feedwater entering the heater}$$

Since the pressure drop is neglected for the heater, the inlet pressure must be equal to the outlet pressure. Table A-1 summarizes the pressure and temperature conditions for all seven heaters.

The steam properties at various locations in the turbine-cycle system are also presented in Fig. 7-7.

3. CALCULATE THE EXTRACTION FLOW RATES

Calculations must be started with the high-pressure heater and continued in the direction of the condenser. In other words, we must determine the extraction steam flow rate for the heaters in the descending order. Starting with the heater 7 shown in

Table A-1
Temperature and Pressure Conditions for All Feedwater Heaters

	Temperature (F)	Pressure (psia)
Feedwater heater 1		
Condensate inlet	108.42	153.00
Condensate outlet	165.68	153.00
Extraction steam inlet	175.68	6.82
Drain conditions	123.42	6.82
Feedwater heater 2		
	165.68	153.00
	214.41	153.00
	244.89	18.70
	180.68	18.70
Feedwater heater 3		
	214.41	153.00
	253.44	153.00
	338.47	34.50
	229.41	34.50
Feedwater heater 4		
	253.44	153.00
	288.36	153.00
	435.94	60.60
	268.44	60.60
Feedwater heater 5		
	288.36	153.00
	359.89	153.00
	624.18	153.00
	359.89	153.00
Feedwater heater 6		
	369.07	4234.94
	414.47	4234.94
	784.01	307.00
	384.07	307.00
Feedwater heater 7		
	414.47	4234.94
	504.93	4234.94
	598.15	744.00
	429.47	744.00

Fig. A-2, we have

$$\dot{m}_s h_{s,i} + \dot{m}_w h_{w,i} = \dot{m}_s h_{s,e} + \dot{m}_w h_{w,e}$$

Since the temperature and pressure conditions around this heater are available, the enthalpy terms can be easily obtained from steam table. Substituting the numerical

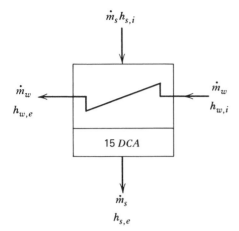

$\dot{m}_s h_{s,i}$

$\dot{m}_w$

$h_{w,e}$

$\dot{m}_w$

$h_{w,i}$

15 DCA

$\dot{m}_s$

$h_{s,e}$

Figure A-2. Schematic diagram for a feedwater heater.

values into the above equation will give the extraction flow rate for the heater 7.

$$\dot{m}_s(1275.4) + (4994{,}457)(396.12) = \dot{m}_s(407.5) + (4994{,}457)(496.12)$$

or

$$\dot{m}_s = 573{,}285 \text{ lb/hr}$$

Repeating this procedure for every heater, we can obtain the results as shown in Table A-2.
It should be pointed out that the steam flow rate at the extraction point 5 is larger than that at the heater inlet. As shown in Fig. 7-7, the in-house and auxiliary turbine steam use must be taken into consideration.
   The steam requirement for the auxiliary turbine is calculated by the equation

$$\dot{m}_{a,t} = \dot{W}_{a,t}/(h_i - h_{e,s})\eta_{a,t}$$

Table A-2
Steam Extraction Flow Rates for the Feedwater Heaters

Heater Number	Steam Flow Rate (lb / hr)
1	199,656
2	174,864
3	154,293
4	131,515
5	281,464
6	193,834
7	573,285

where

$$\dot{W}_{a,t} = \text{turbine output}$$

$$\eta_{a,t} = \text{turbine internal efficiency}$$

$$h_i = \text{steam enthalpy at the turbine inlet}$$

$$h_{e,s} = \text{steam enthalpy at the isentropic turbine outlet}$$

The output of the auxiliary turbine is evidently equal to the pump power, which can be estimated by the equation

$$\dot{W}_p = \frac{\dot{m}_w v \Delta p}{\eta_p}$$

where

$$\dot{m}_w = \text{water flow rate}$$

$$v = \text{water specific volume}$$

$$\Delta p = \text{pressure rise}$$

$$\eta_p = \text{pump overall efficiency}$$

In this numerical example the steam requirement for the auxiliary turbine is 275,565 lb/hr. Therefore, the total steam flow at the extraction point 5 is

$$= 103,400 + 275,565 + 281,464$$

$$= 660,429 \text{ lb/hr}$$

4. DETERMINE THE STEAM TURBINE OUTPUT

The steam turbine output is the sum of the outputs of all turbine cylinders. The equation used for this calculation is

$$\dot{W}_t = \sum \dot{m}_i h_i - \sum \dot{m}_e h_e$$

Applying this equation to the HP turbine cylinder, we have

$$\dot{W}_{\text{HP}} = (4,994,457)(1421.7)$$

$$- (573,285)(1275.4) - (4,421,171)(1275.4)$$

$$= 730.69 \times 10^6 \text{ Btu/hr}$$

Similiary for the IP turbine

$$\mathring{W}_{IP} = 765.72 \times 10^6 \text{ Btu/hr}$$

and for the LP turbine

$$\mathring{W}_{LP} = 1055.84 \times 10^6 \text{ Btu/hr}$$

Therefore the total turbine output is

$$\mathring{W}_t = \mathring{W}_{HP} + \mathring{W}_{IP} + \mathring{W}_{LP}$$

$$\mathring{W}_t = 2.552 \times 10^9 \text{ Btu/hr}$$

The turbine exhaust end loss is

$$(\text{loss})_{ex} = \mathring{m}_{ex} \, (\text{exhaust loss})$$

where

$$\mathring{m}_{ex} = \text{the steam flow rate at the turbine exhaust end}$$

Substituting the numerical values into the equation, we have

$$(\text{loss})_{ex} = 2{,}906{,}576(11.4)$$

$$= 33.13 \times 10^6 \text{ Btu/hr}$$

The turbogenerator net output can be obtained by subtracting the exhaust end loss from the gross output and multiplying the resultant by the mechanical coupling efficiency and generator efficiency; that is,

$$\mathring{W}_{net} = \eta_{coupling} \eta_{gen} \left(\mathring{W}_t - \text{loss}_{ex} \right)$$

$$\mathring{W}_{net} = 1.0 \times 0.985 \times (2552 - 33.13) \times 10^6$$

or

$$= 2.481 \times 10^9 \text{ Btu/hr} \quad \text{or} \quad 726{,}950 \text{ kW}$$

Compared with the computer result (726,171 kW) shown in Fig. 7-7, the difference is indeed very small, and probably due to different numerical truncations.

5. CALCULATE THE TURBINE SYSTEM NET HEAT RATE

The turbine net heat rate (NHR) is calculated by Eq. (7-12). The heat supplied to the turbine system is

$$\mathring{Q} = \mathring{m}_b (h_{b,e} - h_{b,i}) + \mathring{m}_r (h_{r,e} - h_{r,i})$$

The first term represents the amount of heat received in the boiler and the second term is that received in the reheater. Substituting the numerical values into the equation, we have

$$\mathring{Q} = 4{,}994{,}457(1421.7 - 496.1)$$

$$+ 4{,}421{,}171(1514.7 - 1275.4)$$

$$\mathring{Q} = 5680.7 \times 10^6 \text{ Btu/hr}$$

Therefore, the turbine net heat rate is

$$NHR = \frac{\text{heat input}}{\text{turbo-generator net output}}$$

$$= \frac{5680.7 \times 10^6}{726{,}950}$$

$$NHR = 7815 \text{ Btu/kWh}$$

Power Plant Availability Balance Calculations

Referring back to the turbine-cycle system shown in Fig 7-7, we estimate the coal consumption of 475,000 lb/hr assuming the steam generator efficiency (based on the first law) to be around 90%. Also, we assume the coal availability is 13,186 Btu/lb. Then, the fuel availability supplied to the steam generator is

$$\mathring{A}_{fuel} = \mathring{m}_{fuel} a_{fuel}$$

$$= 475,000 \times 13,186 = 6263.35 \times 10^6 \text{ Btu/hr}$$

Since the steam is used to heat a portion of air supply, the availability of steam must be taken into consideration. That is

$$\mathring{A}_{steam} = \mathring{m}_{steam} \times \Delta a$$

$$= 103,400(416.04 - 57.94) = 37.03 \times 10^6 \text{ Btu/hr}$$

Then the availability input to the steam generator is

$$\mathring{A}_{in} = 6263.35 \times 10^6 + 37.03 \times 10^6$$

$$= 6300.38 \times 10^6 \text{ Btu/hr}$$

The availability output of steam generator is the availability gained by steam in the boiler as well as the reheater. Numerically, it is

$$\mathring{A}_{out} = 4,994,457(633.92 - 133.0)$$

$$+ 4,421,171(604.66 - 473.04)$$

$$= 3083.74 \times 10^6 \text{ Btu/hr}$$

The availability loss in the steam generator is, by definition, the difference between the availability input and output. The loss is for this case

$$\mathring{A}_{loss} = 3216.64 \times 10^6 \text{ Btu/hr}$$

The effectiveness (or the second-law efficiency) of a steam generator is calculated by Eq. (7-20). For this case the numerical result is

$$\epsilon = 3083.74 \times 10^6 / 6300.38 \times 10^6$$

$$= 0.489 \quad \text{or} \quad 48.9\%$$

The turbine-cycle system consists of three turbine cylinders, seven feedwater heaters, condensers, pumps, auxiliary turbine, and other associated equipment. The availability input to the turbine-cycle system is the availability gained by the steam in the steam generator and the availability input to the condensate pump. As indicated in Section 7.5, the availability input to the pump is numerically equal to the pump work. Then, the availability input to the turbine-cycle ₁stem is

$$\mathring{A}_{\text{input}} = 3083.7 \times 10^6 + 2.17 \times 10^6$$

$$= 3085.8 \times 10^6 \text{ Btu/hr}$$

The output of the system is, by definition, equal to the turbine work plus the steam output for the air preheater (i.e. $2519.9 \times 10^6 + 37.03 \times 10^6 = 2556.9 \times 10^6$ Btu/hr). Using Eq. (7-20) and Eq. (7-24), we obtain the effectiveness and availability loss of the turbine system, respectively, as

$$\epsilon = 2556.9 \times 10^6 / 3085.8 \times 10^6$$

$$\epsilon = 0.829 \quad \text{or} \quad 82.9\%$$

and

$$\mathring{A}_{\text{loss}} = 3085.8 \times 10^6 - 2556.9 \times 10^6$$

$$= 528.9 \times 10^6 \text{ Btu/hr}$$

The availability loss in the turbine-cycle system occurs mainly at locations such as turbines, pumps, and heat exchangers (condensers and feedwater heaters). As indicated in Section 7.5, the availability loss of this kind is due to either the internal friction or the heat transfer through a finite temperature difference. The following equations are presented for calculations of $\mathring{A}_{\text{input}}$ and $\mathring{A}_{\text{output}}$ for these items of equipment. With these two terms, the availability loss and effectiveness can be determined by Eq. (7-20) and Eq. (7-24).

The equations for turbine $\mathring{A}_{\text{input}}$ and $\mathring{A}_{\text{output}}$ are, respectively,

$$\mathring{A}_{\text{input}} = \Sigma \mathring{m}_i a_i - \Sigma \mathring{m}_e a_e$$

and

$$\mathring{A}_{\text{output}} = \Sigma \mathring{m}_i h_i - \Sigma \mathring{m}_e h_e$$

Apparently, these two equations are identical to Eqs. (7-22) and (7-23). The numerical example is presented in Section 7.5. Other results of the turbine calculation are shown in Table 7-4. It is seen that the turbine process effectiveness is generally in the range 87 to 94%. The process is quite efficient from the second law's viewpoint.

Because the pumping process is a reverse of turbine process, the equations for pumps are identical to those for turbines. In this availability balance, the effectiveness for feedwater pumps is approximately equal to 92%. Other calculated results can be found in Table 7-4.

Figure B-1 presents a schematic diagram of typical feedwater heater. Heat is transferred from the condensing steam to the feedwater. In the heat exchanger process energy (heat) is conserved. Examining the process in terms of thermodynamic availability, however, the conservation principle is not valid. It is expected that the availability gained by the feedwater is somewhat less than that given up by the steam (or hot substances in general). The difference is defined as the availability loss of the heat exchanger. The ratio of these two is the process effectiveness.

The availability input to the heat exchanger, $\mathring{A}_{\text{input}}$ is the availability reduction in the hot substances (such as steam and hot water). The availability output, $\mathring{A}_{\text{output}}$, is the availability gain in the feedwater. The expressions for both terms are

$$\mathring{A}_{\text{input}} = \Sigma \dot{m}_{s,i} a_{s,i} - \Sigma \dot{m}_{s,e} a_{s,e}$$

and

$$\mathring{A}_{\text{output}} = \Sigma \dot{m}_{w,e} a_{w,e} - \Sigma \dot{m}_{w,i} a_{w,i}$$

Using the feedwater heater #7 as an example, we have

$$\mathring{A}_{\text{input}} = 573{,}285(473.04 - 108.16)$$

$$= 209.2 \times 10^6 \text{ Btu/hr}$$

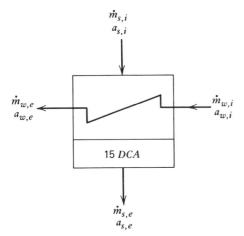

Figure B-1. Availability balance for a typical feedwater heater.

and

$$\mathring{A}_{\text{output}} = 4,994,457(133.0 - 91.65)$$

$$= 206.5 \times 10^6 \text{ Btu/hr}$$

With these terms, we get

$$\mathring{A}_{\text{loss}} = 209.2 \times 10^6 - 206.5 \times 10^6$$

$$= 2.7 \times 10^6 \text{ Btu/hr}$$

and

$$\epsilon = 206.5 \times 10^6/209.2 \times 10^6$$

$$= 0.987 \quad \text{or} \quad 98.7\%$$

The results for other feedwater heaters are presented in Table 7-4.

The calculations for condensers are similar to those for feedwater heaters. The hot substances in condensers are the condensing steam and hot water returned from the heater and other sources. The reduction of availability in these substances is equal to the total loss of availability in condensers. This approximation is entirely acceptable because the availability gained by cooling water is eventually wasted. The numerical results are shown in Table 7-4.

It may be of interest to take the power plant as a system. In this case, the plant output is the turbine net output 2519.9×10^6 Btu/hr, if we neglect the plant auxiliary power, and other losses. The plant input is simply the fuel availability and is equal to 6263.35×10^6 Btu/hr. The ratio, which is also the second-law efficiency, is

$$= 2519.9 \times 10^6/6263.35 \times 10^6$$

$$= 0.402 \quad \text{or} \quad 40.2\%$$

Tables and Charts

Notes: Tables C-1 and C-2 are reprinted with permission from *The Standards for Steam Surface Condensers*, by the Heat Exchange Institute.

Tables C-3 to Table C-5 are reprinted with permission from *The Performance Curves*, by the Cooling Tower Institute.

Table C-6 is abstracted from J. H. Keenan, F. G. Keyes, P. G. Hill, and J. G. Moore, *Steam Tables*, Wiley, New York, 1969. Reprinted by permission of John Wiley and Sons, Inc.

Table C-7 is extracted from J. H. Keenan and J. Kaye, *Gas Tables*, Wiley, New York, 1945. Reprinted by permission of John Wiley and Sons, Inc.

Table C-1
Tube Characteristics

OD of Tubing Inches	Bwg	Thickness	Inside Diameter	Surface External, Sq Ft per Linear Foot	Length in Feet for 1 sq ft Surface	Water-GPM at 1 Foot per Sec Velocity
$\frac{5}{8}$	12	0.109	.407	.1636	6.112	.406
	13	0.095	.435	.1636	6.112	.463
	14	0.083	.459	.1636	6.112	.516
	15	0.072	.481	.1636	6.112	.566
	16	0.065	.495	.1636	6.112	.600
	17	0.058	.509	.1636	6.112	.634
	18	0.049	.527	.1636	6.112	.680
	19	0.042	.541	.1636	6.112	.716
	20	0.035	.555	.1636	6.112	.754
	21	0.032	.561	.1636	6.112	.770
	22	0.028	.569	.1636	6.112	.793
	23	0.025	.575	.1636	6.112	.809
	24	0.022	.581	.1636	6.112	.826
$\frac{3}{4}$	12	0.109	.532	.1963	5.094	.693
	13	0.095	.560	.1963	5.094	.768
	14	0.083	.584	.1963	5.094	.835
	15	0.072	.606	.1963	5.094	.899
	16	0.065	.620	.1963	5.094	.941
	17	0.058	.634	.1963	5.094	.984
	18	0.049	.652	.1963	5.094	1.041
	19	0.042	.666	.1963	5.094	1.086
	20	0.035	.680	.1963	5.094	1.132
	21	0.032	.686	.1963	5.094	1.152
	22	0.028	.694	.1963	5.094	1.179
	23	0.025	.700	.1963	5.094	1.200
	24	0.022	.706	.1963	5.094	1.220
$\frac{7}{8}$	12	0.109	.657	.2291	4.367	1.057
	13	0.095	.685	.2291	4.367	1.149
	14	0.083	.709	.2291	4.367	1.231
	15	0.072	.731	.2291	4.367	1.308
	16	0.065	.745	.2291	4.367	1.359
	17	0.058	.759	.2291	4.367	1.410
	18	0.049	.777	.2291	4.367	1.478
	19	0.042	.791	.2291	4.367	1.532
	20	0.035	.805	.2291	4.367	1.586
	21	0.032	.811	.2291	4.367	1.610
	22	0.028	.819	.2291	4.367	1.642
	23	0.025	.825	.2291	4.367	1.666
	24	0.022	.831	.2291	4.367	1.690
1	12	0.109	.782	.2618	3.817	1.497
	13	0.095	.810	.2618	3.817	1.606
	14	0.083	.834	.2618	3.817	1.703

Table C-1 (continued)

OD of Tubing Inches	Bwg	Thickness	Inside Diameter	Surface External, Sq Ft per Linear Foot	Length in Feet for 1 sq ft Surface	Water-GPM at 1 Foot per Sec Velocity
	15	0.072	.856	.2618	3.817	1.794
	16	0.065	.870	.2618	3.817	1.853
	17	0.058	.884	.2618	3.817	1.992
	18	0.049	.902	.2618	3.817	1.992
	19	0.042	.916	.2618	3.817	2.054
	20	0.035	.930	.2618	3.817	2.117
	21	0.032	.936	.2618	3.817	2.145
	22	0.028	.944	.2618	3.817	2.182
	23	0.025	.950	.2618	3.817	2.209
	24	0.022	.956	.2618	3.817	2.237
$1\frac{1}{8}$	12	0.109	.907	.2944	3.397	2.014
	13	0.095	.935	.2944	3.397	2.140
	14	0.083	.959	.2944	3.397	2.251
	15	0.072	.981	.2944	3.397	2.356
	16	0.065	.995	.2944	3.397	2.424
	17	0.058	1.009	.2944	3.397	2.492
	18	0.049	1.027	.2944	3.397	2.582
	19	0.042	1.041	.2944	3.397	2.653
	20	0.035	1.055	.2944	3.397	2.725
	21	0.032	1.061	.2944	3.397	2.756
	22	0.028	1.069	.2944	3.397	2.797
	23	0.025	1.075	.2944	3.397	2.829
	24	0.022	1.081	.2944	3.397	2.861
$1\frac{1}{4}$	12	0.109	1.032	.3271	3.057	2.607
	13	0.095	1.060	.3271	3.057	2.751
	14	0.083	1.084	.3271	3.057	2.877
	15	0.072	1.106	.3271	3.057	2.994
	16	0.065	1.120	.3271	3.057	3.071
	17	0.058	1.134	.3271	3.057	3.148
	18	0.049	1.152	.3271	3.057	3.249
	19	0.042	1.166	.3271	3.057	3.328
	20	0.035	1.180	.3271	3.057	3.409
	21	0.032	1.186	.3271	3.057	3.443
	22	0.028	1.194	.3271	3.057	3.490
	23	0.025	1.200	.3271	3.057	3.525
	24	0.022	1.206	.3271	3.057	3.560

Table C-2
Saturated Steam Temperatures (Pressure in in. Hg Abs)

Absolute Pressure In. Hg	.00	.01	.02	.03	.04	.05	.06	.07	.08	.09	Absolute Pressure In. Hg
.2	34.56	35.78	36.96	38.09	39.18	40.23	41.23	42.22	43.17	44.08	.2
.3	44.96	45.83	46.67	47.48	48.28	49.05	49.80	50.53	51.25	51.96	.3
.4	52.64	53.31	53.98	54.62	55.25	55.88	56.48	57.08	57.66	58.24	.4
.5	58.80	59.35	59.90	60.43	60.96	61.48	62.00	62.49	62.99	63.47	.5
.6	63.96	64.43	64.90	65.35	65.81	66.26	66.70	67.13	67.56	67.99	.6
.7	68.40	68.82	69.23	69.63	70.03	70.43	70.81	71.20	71.58	71.96	.7
.8	72.33	72.70	73.06	73.42	73.78	74.13	74.48	74.83	75.17	75.51	.8
.9	75.85	76.18	76.51	76.83	77.15	77.47	77.79	78.11	78.42	78.73	.9
1.0	79.03	79.33	79.64	79.94	80.23	80.52	80.81	81.10	81.39	81.67	1.0
1.1	81.95	82.23	82.51	82.78	83.06	83.33	83.60	83.87	84.13	84.39	1.1
1.2	84.65	84.91	85.17	85.43	85.68	85.93	86.18	86.43	86.68	86.92	1.2
1.3	87.17	87.41	87.65	87.89	88.12	88.36	88.59	88.83	89.06	89.28	1.3
1.4	89.51	89.74	89.97	90.19	90.41	90.63	90.85	91.07	91.29	91.50	1.4
1.5	91.72	91.93	92.14	92.35	92.56	92.77	92.98	93.19	93.39	93.60	1.5
1.6	93.80	94.00	94.20	94.40	94.60	94.80	95.01	95.20	95.39	95.59	1.6
1.7	95.78	95.97	96.16	96.35	96.54	96.72	96.91	97.10	97.28	97.46	1.7
1.8	97.65	97.83	98.01	98.19	98.37	98.55	98.73	98.91	99.09	99.26	1.8
1.9	99.44	99.61	99.78	99.96	100.13	100.30	100.47	100.64	100.81	100.97	1.9
2.0	101.14	101.31	101.47	101.64	101.80	101.97	102.13	102.30	102.46	102.62	2.0
2.1	102.78	102.94	103.10	103.25	103.41	103.57	103.73	103.88	104.04	104.19	2.1
2.2	104.34	104.50	104.65	104.80	104.95	105.11	105.26	105.41	105.55	105.70	2.2
2.3	105.85	106.00	106.15	106.29	106.44	106.58	106.73	106.87	107.02	107.16	2.3
2.4	107.31	107.45	107.59	107.73	107.87	108.01	108.15	108.29	108.43	108.57	2.4
2.5	108.71	108.84	108.98	109.12	109.25	109.39	109.52	109.66	109.79	109.92	2.5
2.6	110.06	110.19	110.32	110.46	110.59	110.72	110.85	110.98	111.11	111.24	2.6
2.7	111.37	111.49	111.62	111.75	111.88	112.01	112.13	112.26	112.38	112.51	2.7
2.8	112.63	112.76	112.88	113.01	113.13	113.25	113.37	113.50	113.62	113.74	2.8
2.9	113.86	113.98	114.11	114.23	114.35	114.46	114.58	114.70	114.82	114.94	2.9
3.0	115.06	115.19	115.30	115.41	115.52	115.64	115.76	115.88	116.00	116.11	3.0
3.1	116.22	116.33	116.45	116.57	116.68	116.79	116.90	117.00	117.11	117.22	3.1
3.2	117.35	117.46	117.57	117.69	117.80	117.90	118.01	118.12	118.22	118.33	3.2
3.3	118.45	118.56	118.67	118.78	118.89	118.99	119.10	119.20	119.30	119.41	3.3
3.4	119.52	119.62	119.72	119.83	119.93	120.04	120.15	120.25	120.35	120.46	3.4
3.5	120.56	120.66	120.77	120.87	120.98	121.07	121.17	121.28	121.39	121.49	3.5
3.6	121.58	121.68	121.78	121.88	121.98	122.08	122.18	122.28	122.37	122.47	3.6
3.7	122.57	122.66	122.76	122.85	122.96	123.06	123.15	123.25	123.35	123.45	3.7
3.8	123.55	123.63	123.72	123.82	123.92	124.02	124.10	124.20	124.30	124.40	3.8
3.9	124.49	124.59	124.69	124.78	124.87	124.97	125.05	125.14	125.23	125.32	3.9
4.0	125.42	125.51	125.60	125.70	125.79	125.88	125.97	126.05	126.13	126.23	4.0
4.1	126.33	126.41	126.50	126.60	126.70	126.78	126.87	126.96	127.05	127.13	4.1
4.2	127.22	127.31	127.40	127.50	127.59	127.67	127.76	127.85	127.93	128.02	4.2
4.3	128.10	128.18	128.26	128.35	128.44	128.52	128.60	128.69	128.78	128.86	4.3
4.4	128.95	129.03	129.11	129.20	129.28	129.37	129.45	129.53	129.62	129.70	4.4
4.5	129.79	129.87	129.95	130.03	130.11	130.19	130.28	130.36	130.44	130.52	4.5
4.6	130.61	130.69	130.77	130.85	130.93	131.01	131.09	131.17	131.25	131.34	4.6
4.7	131.42	131.50	131.58	131.66	131.74	131.82	131.90	131.98	132.06	132.14	4.7
4.8	132.22	132.30	132.38	132.46	132.53	132.61	132.69	132.77	132.85	132.93	4.8
4.9	133.00	133.09	133.17	133.25	133.33	133.41	133.49	133.57	133.65	133.73	4.9

Table C-3
Enthalpy of Saturated Air-Water Vapor Mixtures at 29.921 in. Hg
(Btu per Pound of Dry Air)

F	0.0	0.1	0.2	0.3	0.4	0.5	0.6	0.7	0.8	0.9	
0	0.835	0.863	0.892	0.920	0.949	0.977	1.006	1.034	1.063	1.091	0
1	1.120	1.149	1.777	1.206	1.235	1.264	1.293	1.321	1.350	1.379	1
2	1.408	1.437	1.466	1.495	1.524	1.553	1.582	1.611	1.640	1.669	2
3	1.698	1.727	1.756	1.786	1.815	1.844	1.874	1.903	1.932	1.962	3
4	1.991	2.020	2.050	2.079	2.109	2.138	2.168	2.197	2.227	2.256	4
5	2.286	2.316	2.345	2.375	2.405	2.434	2.464	2.494	2.523	2.553	5
6	2.583	2.613	2.643	2.673	2.703	2.732	2.763	2.793	2.823	2.853	6
7	2.883	2.913	2.944	2.974	3.005	3.035	3.066	3.096	3.127	3.157	7
8	3.188	3.219	3.249	3.280	3.310	3.341	3.371	3.402	3.433	3.463	8
9	3.494	3.525	3.556	3.586	3.617	3.648	3.679	3.710	3.741	3.772	9
10	3.803	3.834	3.865	3.897	3.928	3.959	3.990	4.022	4.053	4.085	10
11	4.116	4.147	4.179	4.210	4.242	4.273	4.305	4.337	4.368	4.400	11
12	4.432	4.464	4.496	4.528	4.560	4.592	4.624	4.656	4.689	4.721	12
13	4.753	4.785	4.817	4.850	4.882	4.914	4.946	4.979	5.011	5.044	13
14	5.076	5.109	5.141	5.174	5.206	5.239	5.272	5.304	5.337	5.370	14
15	5.403	5.436	5.469	5.502	5.535	5.568	5.602	5.635	5.668	5.702	15
16	5.735	5.768	5.802	5.835	5.869	5.902	5.936	5.970	6.003	6.037	16
17	6.071	6.105	6.139	6.173	6.207	6.241	6.275	6.309	6.343	6.378	17
18	6.412	6.446	6.480	6.515	6.549	6.583	6.618	6.652	6.687	6.721	18
19	6.756	6.791	6.826	6.860	6.895	6.930	6.965	7.000	7.036	7.071	19
20	7.106	7.141	7.176	7.212	7.247	7.282	7.318	7.353	7.389	7.424	20
21	7.460	7.496	7.532	7.567	7.603	7.639	7.675	7.711	7.748	7.784	21
22	7.820	7.856	7.893	7.929	7.966	8.002	8.039	8.076	8.112	8.149	22
23	8.186	8.223	8.260	8.297	8.334	8.371	8.408	8.445	8.482	8.520	23
24	8.557	8.594	8.632	8.669	8.707	8.745	8.782	8.820	8.858	8.896	24
25	8.934	8.972	9.010	9.048	9.086	9.125	9.163	9.201	9.240	9.278	25
26	9.317	9.356	9.394	9.433	9.472	9.511	9.550	9.589	9.628	9.667	26
27	9.706	9.745	9.785	9.824	9.864	9.904	9.943	9.983	10.023	10.063	27
28	10.103	10.143	10.183	10.223	10.263	10.304	10.344	10.384	10.425	10.465	28
29	10.506	10.547	10.587	10.628	10.669	10.710	10.750	10.791	10.833	10.874	29
30	10.915	10.956	10.998	11.040	11.081	11.123	11.165	11.207	11.249	11.291	30
31	11.333	11.376	11.418	11.461	11.503	11.546	11.589	11.631	11.673	11.716	31
32	11.758	11.799	11.841	11.882	11.923	11.964	12.005	12.046	12.087	12.128	32
33	12.169	12.210	12.252	12.293	12.335	12.376	12.418	12.460	12.501	12.543	33
34	12.585	12.627	12.669	12.711	12.753	12.796	12.838	12.880	12.923	12.965	34
35	13.008	13.051	13.093	13.136	13.179	13.222	13.265	13.308	13.351	13.395	35
36	13.438	13.481	13.525	13.568	13.612	13.655	13.699	13.742	13.786	13.830	36
37	13.874	13.918	13.962	14.007	14.051	14.095	14.140	14.185	14.229	14.274	37
38	14.319	14.364	14.409	14.454	14.499	14.544	14.589	14.635	14.680	14.725	38
39	14.771	14.817	14.862	14.908	14.954	15.000	15.045	15.092	15.138	15.184	39
40	15.230	15.276	15.323	15.369	15.416	15.462	15.509	15.556	15.603	15.650	40
41	15.697	15.744	15.791	15.839	15.886	15.933	15.981	16.029	16.076	16.124	41
42	16.172	16.220	16.268	16.317	16.365	16.413	16.462	16.511	16.559	16.608	42
43	16.657	16.706	16.755	16.804	16.853	16.902	16.951	17.001	17.050	17.099	43
44	17.149	17.199	17.248	17.298	17.348	17.398	17.448	17.499	17.549	17.599	44
45	17.650	17.701	17.751	17.802	17.853	17.904	17.956	18.007	18.058	18.110	45
46	18.161	18.212	18.264	18.316	18.367	18.419	18.471	18.523	18.575	18.628	46
47	18.680	18.733	18.785	18.838	18.891	18.944	18.997	19.051	19.104	19.157	47
48	19.211	19.265	19.318	19.372	19.426	19.480	19.534	19.588	19.642	19.697	48
49	19.751	19.806	19.860	19.915	19.970	20.025	20.080	20.135	20.190	20.246	49
50	20.30	20.36	20.41	20.47	20.53	20.58	20.64	20.70	20.75	20.81	50
51	20.86	20.92	20.97	21.03	21.09	21.14	21.20	21.26	21.32	21.37	51
52	21.43	21.49	21.55	21.60	21.66	21.72	21.78	21.84	21.90	21.96	52

Table C-3 (continued)

F	0.0	0.1	0.2	0.3	0.4	0.5	0.6	0.7	0.8	0.9	
53	22.02	22.07	22.13	22.19	22.25	22.31	22.37	22.43	22.49	22.55	53
54	22.61	22.67	22.73	22.79	22.85	22.91	22.97	23.03	23.09	23.16	54
55	23.22	23.28	23.34	23.40	23.46	23.53	23.59	23.65	23.71	23.78	55
56	23.84	23.90	23.96	24.03	24.09	24.16	24.22	24.28	24.35	24.41	56
57	24.48	24.54	24.60	24.67	24.73	24.80	24.86	24.92	24.99	25.05	57
58	25.12	25.18	25.25	25.31	25.38	25.45	25.51	25.58	25.65	25.71	58
59	25.78	25.85	25.91	25.98	26.05	26.12	26.18	26.25	26.32	26.39	59
60	26.46	26.53	26.60	26.67	26.74	26.80	26.87	26.94	27.01	27.08	60
61	27.15	27.22	27.29	27.36	27.43	27.50	27.57	27.64	27.71	27.78	61
62	27.85	27.92	27.99	28.06	28.13	28.21	28.28	28.35	28.42	28.50	62
63	28.57	28.64	28.72	28.79	28.86	28.94	29.01	29.09	29.16	29.24	63
64	29.31	29.38	29.46	29.53	29.61	29.68	29.76	29.83	29.91	29.98	64
65	30.06	30.13	30.20	30.28	30.36	30.44	30.51	30.59	30.67	30.75	65
66	30.83	30.90	30.98	31.06	31.14	31.22	31.30	31.38	31.46	31.54	66
67	31.62	31.70	31.78	31.86	31.94	32.02	32.10	32.18	32.26	32.34	67
68	32.42	32.50	32.59	32.67	32.75	32.83	32.92	33.00	33.08	33.17	68
69	33.25	33.33	33.41	33.50	33.58	33.66	33.74	33.83	33.91	34.00	69
70	34.09	34.17	34.25	34.34	34.43	34.51	34.60	34.69	34.78	34.87	70
71	34.95	35.04	35.13	35.22	35.30	35.39	35.48	35.57	35.66	35.75	71
72	35.83	35.92	36.01	36.10	36.19	36.28	36.37	36.46	36.56	36.65	72
73	36.74	36.83	36.92	37.01	37.10	37.20	37.29	37.38	37.48	37.57	73
74	37.66	37.76	37.85	37.94	38.04	38.13	38.23	38.32	38.41	38.51	74
75	38.61	38.70	38.80	38.89	38.98	39.08	39.18	39.27	39.37	39.47	75
76	39.57	39.67	39.77	39.87	39.97	40.07	40.17	40.27	40.37	40.47	76
77	40.57	40.67	40.77	40.87	40.97	41.07	41.17	41.27	41.37	41.48	77
78	41.58	41.68	41.78	41.88	41.99	42.10	42.20	42.30	42.41	42.52	78
79	42.62	42.73	42.84	42.94	43.05	43.16	43.26	43.37	43.47	43.58	79
80	43.69	43.80	43.91	44.01	44.12	44.23	44.34	44.45	44.56	44.67	80
81	44.78	44.89	45.00	45.11	45.22	45.34	45.45	45.56	45.67	45.78	81
82	45.90	46.01	46.12	46.23	46.35	46.46	46.58	46.70	46.81	46.92	82
83	47.04	47.15	47.27	47.39	47.51	47.63	47.74	47.86	47.98	48.10	83
84	48.22	48.34	48.46	48.58	48.70	48.82	48.94	49.06	49.18	49.30	84
85	49.43	49.55	49.67	49.79	49.91	50.03	50.15	50.28	50.40	50.53	85
86	50.66	50.78	50.90	51.03	51.16	51.28	51.41	51.54	51.67	51.80	86
87	51.93	52.06	52.19	52.32	52.45	52.58	52.71	52.84	52.97	53.10	87
88	53.23	53.36	53.49	53.62	53.75	53.88	54.02	54.15	54.28	54.42	88
89	54.56	54.69	54.82	54.96	55.09	55.23	55.37	55.51	55.65	55.79	89
90	55.93	56.07	56.21	56.35	56.49	56.63	56.77	56.91	57.05	57.19	90
91	57.33	57.47	57.61	57.76	57.90	58.05	58.19	58.34	58.48	58.63	91
92	58.78	58.92	59.07	59.21	59.36	59.50	59.65	59.80	59.95	60.10	92
93	60.25	60.40	60.55	60.70	60.85	61.00	61.15	61.31	61.46	61.61	93
94	61.77	61.92	62.07	62.23	62.38	62.54	62.69	62.85	63.00	63.16	94
95	63.32	63.48	63.63	63.79	63.95	64.11	64.27	64.44	64.60	64.76	95
96	64.92	65.08	65.25	65.41	65.58	65.74	65.90	66.06	66.23	66.39	96
97	66.55	66.72	66.88	67.05	67.22	67.39	67.56	67.73	67.90	68.07	97
98	68.23	68.40	68.57	68.74	68.91	69.08	69.26	69.43	69.61	69.78	98
99	69.96	70.14	70.32	70.50	70.67	70.85	71.02	71.20	71.38	71.55	99
100	71.73	71.91	72.09	72.27	72.45	72.63	72.82	73.00	73.19	73.37	100
101	73.55	73.73	73.92	74.11	74.29	74.48	74.67	74.86	75.04	75.23	101
102	75.42	75.62	75.82	76.01	76.20	76.39	76.58	76.77	76.96	77.15	102
103	77.34	77.54	77.73	77.93	78.12	78.32	78.52	78.72	78.92	79.12	103
104	79.32	79.52	79.72	79.92	80.12	80.32	80.52	80.72	80.93	81.13	104
105	81.34	81.54	81.75	81.95	82.16	82.37	82.58	82.79	83.00	83.21	105
106	83.42	83.63	83.84	84.05	84.26	84.48	84.69	84.91	85.12	85.34	106
107	85.56	85.77	85.99	86.21	86.43	86.65	86.87	87.10	87.32	87.54	107

Table C-3 (continued)

F	0.0	0.1	0.2	0.3	0.4	0.5	0.6	0.7	0.8	0.9	
108	87.76	87.99	88.22	88.44	88.67	88.89	89.11	89.34	89.57	89.80	108
109	90.03	90.25	90.48	90.71	90.94	91.17	91.40	91.64	91.87	92.10	109
110	92.34	92.57	92.81	93.05	93.29	93.52	93.76	94.00	94.24	94.48	110
111	94.72	94.96	95.21	95.45	95.70	95.94	96.19	96.44	96.68	96.93	111
112	97.18	97.43	97.68	97.93	98.18	98.43	98.68	98.94	99.19	99.45	112
113	99.71	99.96	100.22	100.48	100.74	101.00	101.26	101.52	101.78	102.05	113
114	102.31	102.58	102.84	103.10	103.37	103.63	103.90	104.17	104.44	104.71	114
115	104.98	105.25	105.52	105.79	106.06	106.34	106.61	106.89	107.17	107.45	115
116	107.73	108.01	108.29	108.57	108.85	109.13	109.41	109.70	109.98	110.27	116
117	110.55	110.84	111.13	111.42	111.71	112.00	112.29	112.58	112.87	113.16	117
118	113.46	113.75	114.05	114.35	114.65	114.95	115.25	115.55	115.86	116.16	118
119	116.46	116.77	117.07	117.38	117.69	118.00	118.30	118.61	118.92	119.23	119
120	119.54	119.85	120.17	120.48	120.80	121.12	121.44	121.76	122.08	122.40	120
121	122.72	123.04	123.36	123.68	124.01	124.34	124.67	125.00	125.33	125.65	121
122	125.98	126.31	126.64	126.98	127.31	127.65	127.99	128.33	128.67	129.01	122
123	129.35	129.69	130.03	130.37	130.72	131.06	131.41	131.75	132.10	132.45	123
124	132.80	133.15	133.50	133.85	134.21	134.57	134.93	135.29	135.66	136.03	124
125	136.4	136.7	137.1	137.5	137.8	138.2	138.6	139.0	139.3	139.7	125
126	140.1	140.5	140.8	141.2	141.6	142.0	142.3	142.7	143.1	143.5	126
127	143.9	144.3	144.7	145.1	145.5	145.9	146.3	146.7	147.1	147.4	127
128	147.8	148.2	148.6	149.0	149.4	149.8	150.2	150.6	151.0	151.4	128
129	151.8	152.2	152.6	153.0	153.4	153.8	154.2	154.6	155.1	155.5	129
130	155.9	156.3	156.8	157.2	157.6	158.0	158.5	158.9	159.4	159.8	130
131	160.3	160.7	161.2	161.6	162.0	162.5	162.9	163.4	163.8	164.2	131
132	164.7	165.1	165.6	166.0	166.5	167.0	167.4	167.9	168.3	168.8	132
133	169.3	169.7	170.2	170.7	171.1	171.6	172.1	172.6	173.0	173.5	133
134	174.0	174.5	175.0	175.4	175.9	176.4	176.9	177.4	177.9	178.4	134
135	178.9	179.4	179.9	180.4	180.9	181.4	181.9	182.4	182.9	183.4	135
136	183.9	184.4	184.9	185.4	185.9	186.4	186.9	187.4	188.0	188.5	136
137	189.0	189.5	190.0	190.6	191.1	191.6	192.2	192.7	193.3	193.8	137
138	194.4	194.9	195.5	196.0	196.6	197.1	197.6	198.2	198.8	199.4	138
139	199.9	200.5	201.1	201.7	202.2	202.8	203.4	204.0	204.5	205.1	139
140	205.7	206.3	206.9	207.5	208.1	208.7	209.3	209.9	210.4	211.0	140
141	211.6	212.2	212.8	213.4	214.0	214.6	215.2	215.8	216.4	217.0	141
142	217.7	218.3	218.9	219.5	220.2	220.8	221.5	222.1	222.7	223.4	142
143	224.1	224.7	225.3	226.0	226.6	227.3	228.0	228.6	229.3	229.9	143
144	230.6	231.3	232.0	232.6	233.3	234.0	234.7	235.3	236.0	236.7	144
145	237.4	238.1	238.8	239.5	240.2	240.9	241.6	242.3	243.0	243.7	145
146	244.4	245.1	245.8	246.5	247.2	248.0	248.7	249.5	250.2	250.9	146
147	251.7	252.4	253.2	254.0	254.7	255.5	256.2	257.0	257.7	258.5	147
148	259.3	260.0	260.8	261.6	262.4	263.2	263.9	264.7	265.5	266.3	148
149	267.1	267.9	268.7	269.5	270.3	271.1	271.9	272.7	273.5	274.4	149
150	275.3	276.1	276.9	277.7	278.5	279.4	280.2	281.1	281.9	282.8	150
151	283.6	284.5	285.3	286.2	287.1	287.9	288.8	289.7	290.6	291.5	151
152	292.4	293.3	294.2	295.1	296.0	296.9	297.8	298.7	299.7	300.6	152
153	301.5	302.4	303.3	304.3	305.2	306.1	307.1	308.0	309.0	309.9	153
154	310.9	311.9	312.8	313.8	314.8	315.8	316.8	317.8	318.8	319.8	154
155	320.8	321.8	322.8	323.8	324.8	325.8	326.9	327.9	328.9	330.0	155
156	331.0	332.1	333.1	334.2	335.2	336.3	337.4	338.4	339.5	340.6	156
157	341.7	342.8	343.9	345.0	346.1	347.1	348.3	349.4	350.5	351.6	157
158	352.7	353.8	355.0	356.1	357.2	358.4	359.5	360.7	361.9	363.0	158
159	364.2	365.4	366.6	367.8	369.0	370.2	371.4	372.6	373.8	375.1	159
160	376.3	377.5	378.8	380.0	381.2	382.5	383.7	385.0	386.3	387.5	160
161	388.8	390.1	391.4	392.7	394.0	395.3	396.6	398.0	399.3	400.7	161
162	402.0	403.3	404.7	406.1	407.4	408.8	410.2	411.5	412.9	414.3	162

Table C-3 (continued)

F	0.0	0.1	0.2	0.3	0.4	0.5	0.6	0.7	0.8	0.9	
163	415.7	417.1	418.5	419.9	421.3	422.7	424.1	425.6	427.0	428.4	163
164	429.9	431.4	432.9	434.3	435.8	437.4	438.9	440.4	441.9	443.5	164
165	445.0	446.5	448.1	449.6	451.2	452.8	454.3	455.9	457.5	459.1	165
166	460.7	462.3	463.9	465.6	467.2	468.9	470.5	472.2	473.8	475.5	166
167	477.2	478.9	480.6	482.3	484.0	485.7	487.4	489.2	490.9	492.6	167
168	494.4	496.2	497.9	499.7	501.5	503.3	505.1	506.9	508.7	510.6	168
169	512.4	514.3	516.1	518.0	519.9	521.8	523.7	525.7	527.6	529.5	169
170	531.5	533.5	535.4	537.4	539.4	541.4	543.4	545.4	547.4	549.5	170
171	551.5	553.6	555.6	557.7	559.8	562.0	564.1	566.2	568.4	570.5	171
172	572.7	574.9	577.1	579.2	581.4	583.7	585.9	588.1	590.4	592.6	172
173	594.9	597.2	599.5	601.8	604.1	606.4	608.8	611.1	613.5	615.9	173
174	618.3	620.7	623.2	625.6	628.1	630.6	633.1	635.6	638.1	640.6	174
175	643.2	645.8	648.3	650.9	653.5	656.1	658.7	661.4	664.0	666.7	175
176	669.4	672.1	674.8	677.6	680.4	683.1	685.9	688.8	691.6	694.4	176
177	697.3	700.2	703.1	706.0	708.9	711.9	714.8	717.8	720.8	723.9	177
178	726.9	730.0	733.0	736.1	739.2	742.4	745.5	748.7	751.9	755.1	178
179	758.3	761.6	764.8	768.1	771.4	774.8	778.1	781.5	784.9	788.4	179
180	791.8	795.3	798.7	802.2	805.8	809.3	812.9	816.5	820.1	823.7	180
181	827.4	831.1	834.8	838.6	842.4	846.2	850.1	853.9	857.8	861.8	181
182	865.7	869.7	873.6	877.6	881.7	885.7	889.8	894.0	898.1	902.3	182
183	906.5	910.8	915.0	919.4	923.7	928.1	932.5	937.0	941.4	946.0	183
184	950.5	955.1	959.7	964.3	968.9	973.6	978.4	983.1	988.0	992.8	184
185	998.	1003.	1008.	1013.	1018.	1023.	1028.	1033.	1038.	1044.	185
186	1049.	1054.	1060.	1065.	1070.	1076.	1081.	1087.	1093.	1098.	186
187	1104.	1110.	1116.	1121.	1127.	1133.	1139.	1145.	1152.	1158.	187
188	1164.	1170.	1177.	1183.	1189.	1196.	1202.	1209.	1216.	1222.	188
189	1229.	1236.	1243.	1250.	1257.	1264.	1271.	1279.	1286.	1294.	189
190	1301.	1308.	1316.	1323.	1331.	1339.	1346.	1354.	1362.	1370.	190
191	1378.	1386.	1394.	1403.	1411.	1420.	1429.	1437.	1446.	1455.	191
192	1464.	1473.	1482.	1491.	1501.	1510.	1520.	1529.	1539.	1549.	192
193	1559.	1569.	1579.	1590.	1600.	1611.	1622.	1633.	1644.	1655.	193
194	1666.	1677.	1689.	1700.	1712.	1723.	1735.	1747.	1759.	1772.	194
195	1784.	1797.	1809.	1822.	1836.	1849.	1862.	1876.	1890.	1904.	195
196	1918.	1932.	1947.	1961.	1976.	1991.	2006.	2022.	2037.	2053.	196
197	2069.	2085.	2102.	2119.	2136.	2153.	2170.	2188.	2206.	2224.	197
198	2243.	2262.	2281.	2300.	2319.	2339.	2359.	2380.	2401.	2422.	198
199	2443.	2465.	2487.	2509.	2532.	2555.	2579.	2603.	2627.	2652.	199
200	2677.	2702.	2728.	2755.	2781.	2809.	2836.	2864.	2893.	2922.	200

Table C.4

Density of Saturated Air-Water Vapor Mixtures at 29.921 in. Hg (Pounds of Mixture per Cubic Foot of Mixture)

F	.0	.1	.2	.3	.4	.5	.6	.7	.8	.9	F
60	.07586	.07584	.07582	.07581	.07579	.07577	.07576	.07574	.07573	.07571	60
61	.07569	.07568	.07566	.07564	.07563	.07561	.07559	.07558	.07556	.07554	61
62	.07553	.07551	.07549	.07548	.07546	.07545	.07543	.07541	.07540	.07538	62
63	.07536	.07535	.07533	.07531	.07530	.07528	.07527	.07525	.07523	.07522	63
64	.07520	.07518	.07517	.07515	.07513	.07512	.07510	.07508	.07507	.07505	64
65	.07503	.07502	.07500	.07499	.07497	.07496	.07494	.07492	.07491	.07489	65
66	.07488	.07486	.07484	.07483	.07481	.07479	.07478	.07476	.07474	.07472	66
67	.07471	.07469	.07468	.07466	.07464	.07463	.07461	.07459	.07458	.07456	67
68	.07454	.07453	.07451	.07449	.07448	.07446	.07444	.07443	.07441	.07440	68
69	.07438	.07436	.07435	.07433	.07431	.07430	.07428	.07427	.07425	.07423	69
70	.07422	.07420	.07419	.07417	.07415	.07414	.07412	.07411	.07409	.07407	70
71	.07406	.07404	.07403	.07401	.07399	.07398	.07396	.07395	.07393	.07391	71
72	.07390	.07388	.07386	.07385	.07383	.07381	.07380	.07378	.07377	.07375	72
73	.07373	.07372	.07370	.07368	.07366	.07365	.07363	.07361	.07360	.07358	73
74	.07356	.07355	.07353	.07351	.07350	.07348	.07346	.07345	.07343	.07341	74
75	.07340	.07338	.07336	.07335	.07333	.07332	.07330	.07328	.07327	.07325	75
76	.07323	.07322	.07320	.07318	.07317	.07315	.07313	.07312	.07310	.07308	76
77	.07307	.07305	.07303	.07302	.07300	.07299	.07297	.07295	.07294	.07292	77
78	.07290	.07289	.07287	.07285	.07284	.07282	.07280	.07279	.07277	.07275	78
79	.07274	.07272	.07270	.07269	.07267	.07265	.07264	.07262	.07261	.07259	79
80	.07257	.07256	.07254	.07252	.07251	.07249	.07247	.07246	.07244	.07243	80
81	.07241	.07239	.07237	.07236	.07234	.07232	.07230	.07229	.07227	.07225	81
82	.07224	.07222	.07221	.07219	.07217	.07215	.07214	.07212	.07211	.07209	82
83	.07207	.07206	.07204	.07202	.07200	.07199	.07197	.07195	.07194	.07192	83
84	.07191	.07189	.07187	.07185	.07184	.07182	.07180	.07179	.07177	.07176	84
85	.07174	.07172	.07170	.07169	.07167	.07165	.07164	.07162	.07160	.07158	85
86	.07157	.07155	.07153	.07152	.07150	.07149	.07147	.07145	.07143	.07142	86
87	.07140	.07138	.07137	.07135	.07133	.07132	.07130	.07128	.07127	.07125	87
88	.07123	.07122	.07120	.07118	.07116	.07115	.07113	.07111	.07110	.07108	88
89	.07106	.07105	.07103	.07101	.07100	.07098	.07096	.07095	.07093	.07091	89
90	.07090	.07088	.07086	.07084	.07083	.07081	.07079	.07078	.07076	.07074	90
91	.07073	.07071	.07069	.07067	.07066	.07064	.07062	.07061	.07059	.07057	91
92	.07056	.07054	.07052	.07050	.07049	.07047	.07045	.07043	.07042	.07040	92

Table C-4 (continued)

F	.0	.1	.2	.3	.4	.5	.6	.7	.8	.9
93	.07038	.07037	.07035	.07033	.07031	.07030	.07028	.07026	.07025	.07023
94	.07021	.07019	.07018	.07016	.07014	.07013	.07011	.07009	.07007	.07006
95	.07004	.07002	.07001	.06999	.06997	.06995	.06994	.06992	.06990	.06988
96	.06985	.06983	.06982	.06980	.06978	.06976	.06975	.06973	.06971	.06970
97	.06969	.06968	.06966	.06964	.06962	.06961	.06959	.06957	.06955	.06954
98	.06952	.06950	.06948	.06947	.06945	.06943	.06942	.06940	.06938	.06937
99	.06935	.06933	.06931	.06930	.06928	.06926	.06924	.06922	.06921	.06919
100	.06917	.06915	.06914	.06912	.06910	.06908	.06907	.06905	.06903	.06901
101	.06900	.06898	.06896	.06894	.06893	.06891	.06889	.06887	.06886	.06884
102	.06882	.06880	.06878	.06877	.06875	.06873	.06871	.06870	.06868	.06866
103	.06864	.06862	.06861	.06859	.06857	.06855	.06853	.06852	.06850	.06848
104	.06846	.06844	.06843	.06841	.06839	.06837	.06836	.06834	.06832	.06830
105	.06829	.06827	.06825	.06823	.06821	.06819	.06818	.06816	.06814	.06812
106	.06810	.06809	.06807	.06805	.06803	.06801	.06800	.06798	.06796	.06794
107	.06792	.06791	.06789	.06787	.06785	.06783	.06781	.06779	.06778	.06776
108	.06774	.06772	.06770	.06769	.06767	.06765	.06763	.06761	.06760	.06758
109	.06756	.06754	.06752	.06751	.06749	.06747	.06745	.06743	.06741	.06740
110	.06738	.06736	.06734	.06732	.06730	.06728	.06727	.06725	.06723	.06721
111	.06719	.06717	.06715	.06714	.06712	.06710	.06708	.06706	.06704	.06703
112	.06701	.06699	.06697	.06695	.06693	.06691	.06689	.06688	.06686	.06684
113	.06682	.06680	.06678	.06676	.06674	.06673	.06671	.06669	.06667	.06665
114	.06663	.06661	.06660	.06658	.06656	.06654	.06652	.06650	.06648	.06646
115	.06644	.06643	.06641	.06639	.06637	.06635	.06633	.06631	.06629	.06627
116	.06625	.06624	.06622	.06620	.06618	.06616	.06614	.06612	.06610	.06608
117	.06607	.06605	.06603	.06601	.06599	.06597	.06595	.06593	.06591	.06589
118	.06587	.06585	.06583	.06581	.06579	.06577	.06576	.06574	.06572	.06570
119	.06568	.06566	.06564	.06562	.06560	.06558	.06556	.06554	.06552	.06550
120	.06548	.06546	.06544	.06542	.06540	.06538	.06537	.06535	.06533	.06531
121	.06529	.06527	.06525	.06523	.06521	.06519	.06517	.06515	.06513	.06511
122	.06509	.06507	.06505	.06503	.06501	.06499	.06497	.06495	.06493	.06491
123	.06489	.06487	.06485	.06483	.06481	.06479	.06477	.06475	.06473	.06471
124	.06469	.06467	.06465	.06463	.06461	.06459	.06457	.06455	.06453	.06451

Table C-5
Volume of Saturated Air-Water Vapor Mixtures at 29.921 in. Hg (Cubic Feet of Mixture per Pound of Dry Air)

F	.0	.1	.2	.3	.4	.5	.6	.7	.8	.9	F
60	13.329	13.332	13.336	13.339	13.343	13.346	13.349	13.353	13.356	13.360	60
61	13.363	13.367	13.370	13.374	13.377	13.381	13.384	13.388	13.391	13.395	61
62	13.398	13.402	13.405	13.409	13.412	13.416	13.419	13.423	13.426	13.430	62
63	13.433	13.437	13.440	13.444	13.447	13.451	13.454	13.458	13.461	13.465	63
64	13.468	13.472	13.475	13.479	13.482	13.486	13.490	13.493	13.497	13.500	64
65	13.504	13.508	13.511	13.515	13.518	13.522	13.525	13.529	13.532	13.536	65
66	13.539	13.543	13.546	13.550	13.554	13.558	13.561	13.565	13.569	13.572	66
67	13.576	13.580	13.583	13.587	13.591	13.595	13.598	13.602	13.606	13.609	67
68	13.613	13.617	13.620	13.624	13.628	13.632	13.635	13.639	13.643	13.646	58
69	13.650	13.654	13.657	13.661	13.665	13.669	13.672	13.676	13.680	13.683	69
70	13.687	13.691	13.694	13.698	13.702	13.706	13.709	13.713	13.717	13.720	70
71	13.724	13.728	13.732	13.735	13.739	13.743	13.747	13.751	13.754	13.758	71
72	13.762	13.766	13.770	13.774	13.778	13.782	13.785	13.789	13.793	13.797	72
73	13.801	13.805	13.809	13.813	13.817	13.821	13.825	13.829	13.833	13.837	73
74	13.841	13.845	13.849	13.853	13.857	13.861	13.865	13.869	13.873	13.877	74
75	13.881	13.885	13.889	13.893	13.897	13.901	13.905	13.909	13.913	13.917	75
76	13.921	13.925	13.929	13.933	13.937	13.942	13.946	13.950	13.954	13.958	76
77	13.962	13.966	13.970	13.974	13.978	13.983	13.987	13.991	13.995	13.999	77
78	14.003	14.007	14.011	14.016	14.020	14.024	14.028	14.032	14.037	14.041	78
79	14.045	14.049	14.053	14.058	14.062	14.066	14.070	14.074	14.079	14.083	79
80	14.087	14.091	14.096	14.100	14.104	14.109	14.113	14.117	14.121	14.126	80
81	14.130	14.135	14.139	14.143	14.148	14.152	14.156	14.161	14.165	14.170	81
82	14.174	14.178	14.183	14.187	14.192	14.196	14.200	14.205	14.209	14.214	82
83	14.218	14.222	14.227	14.231	14.236	14.240	14.245	14.249	14.254	14.258	83
84	14.263	14.267	14.272	14.276	14.281	14.285	14.290	14.294	14.299	14.303	84
85	14.308	14.313	14.317	14.322	14.326	14.331	14.336	14.340	14.345	14.349	85
86	14.354	14.359	14.363	14.368	14.373	14.377	14.382	14.387	14.391	14.396	86
87	14.401	14.405	14.410	14.415	14.420	14.424	14.429	14.434	14.439	14.443	87
88	14.448	14.453	14.458	14.462	14.467	14.472	14.477	14.482	14.486	14.491	88
89	14.496	14.501	14.506	14.511	14.516	14.521	14.526	14.530	14.535	14.540	89

Table C-5 (continued)

F	.0	.1	.2	.3	.4	.5	.6	.7	.8	.9	F
90	14.545	14.550	14.555	14.560	14.565	14.570	14.575	14.580	14.585	14.590	90
91	14.595	14.600	14.605	14.610	14.615	14.620	14.625	14.630	14.635	14.640	91
92	14.645	14.650	14.656	14.661	14.666	14.671	14.676	14.681	14.686	14.692	92
93	14.697	14.702	14.707	14.712	14.718	14.723	14.728	14.733	14.738	14.744	93
94	14.749	14.754	14.760	14.765	14.770	14.776	14.781	14.786	14.791	14.797	94
95	14.802	14.808	14.813	14.818	14.824	14.829	14.835	14.840	14.845	14.851	95
96	14.856	14.862	14.867	14.873	14.878	14.884	14.889	14.895	14.900	14.906	96
97	14.911	14.917	14.922	14.928	14.933	14.939	14.944	14.950	14.956	14.961	97
98	14.967	14.972	14.978	14.984	14.989	14.995	15.001	15.007	15.012	15.018	98
99	15.024	15.029	15.035	15.041	15.047	15.052	15.058	15.064	15.070	15.076	99
100	15.081	15.087	15.093	15.099	15.105	15.111	15.117	15.123	15.128	15.134	100
101	15.140	15.146	15.152	15.158	15.164	15.170	15.176	15.182	15.188	15.194	101
102	15.200	15.206	15.212	15.218	15.224	15.231	15.237	15.243	15.249	15.255	102
103	15.261	15.267	15.273	15.280	15.286	15.292	15.298	15.304	15.311	15.317	103
104	15.323	15.330	15.336	15.342	15.348	15.355	15.361	15.368	15.374	15.380	104
105	15.387	15.393	15.400	15.406	15.413	15.419	15.426	15.432	15.439	15.445	105
106	15.452	15.458	15.465	15.471	15.478	15.485	15.491	15.498	15.504	15.511	106
107	15.518	15.524	15.531	15.538	15.544	15.551	15.558	15.565	15.572	15.578	107
108	15.585	15.592	15.599	15.606	15.613	15.620	15.626	15.633	15.640	15.647	108
109	15.654	15.661	15.668	15.675	15.682	15.689	15.696	15.703	15.710	15.717	109
110	15.724	15.731	15.739	15.746	15.753	15.760	15.767	15.775	15.782	15.789	110
111	15.796	15.803	15.811	15.818	15.825	15.833	15.840	15.847	15.855	15.862	111
112	15.869	15.877	15.884	15.892	15.899	15.907	15.914	15.922	15.929	15.937	112
113	15.944	15.952	15.959	15.967	15.974	15.982	15.990	15.997	16.005	16.013	113
114	16.020	16.028	16.036	16.044	16.051	16.059	16.067	16.075	16.083	16.090	114
115	16.098	16.106	16.114	16.122	16.130	16.138	16.146	16.154	16.162	16.170	115
116	16.178	16.186	16.194	16.202	16.210	16.218	16.227	16.235	16.243	16.251	116
117	16.259	16.268	16.276	16.284	16.293	16.301	16.309	16.318	16.326	16.334	117
118	16.343	16.351	16.360	16.368	16.377	16.385	16.394	16.402	16.411	16.420	118
119	16.428	16.437	16.446	16.454	16.463	16.472	16.480	16.489	16.498	16.507	119
120	16.515	16.524	16.533	16.542	16.551	16.560	16.569	16.578	16.587	16.596	120
121	16.605	16.614	16.623	16.632	16.641	16.650	16.659	16.669	16.678	16.687	121
122	16.696	16.705	16.715	16.724	16.734	16.743	16.752	16.762	16.771	16.781	122
123	16.790	16.800	16.809	16.819	16.828	16.838	16.848	16.857	16.867	16.876	123
124	16.886	16.896	16.906	16.916	16.926	16.936	16.945	16.955	16.965	16.975	124

Table C-6
Thermodynamic Properties of Steam and Water

Table C-6a Properties of Saturated Water: Temperature Table v, ft^3/lb; h, Btu/lb; s, Btu/lb·R

Temperature (F) T	Pressure (lbf/in.²) p	Specific Volume		Internal Energy		Enthalpy			Entropy			Temperature (F) T
		Saturated Liquid v_f	Saturated Vapor v_g	Saturated Liquid u_f	Saturated Vapor u_g	Saturated Liquid h_f	h_{fg}	Saturated Vapor h_g	Saturated Liquid s_f	s_{fg}	Saturated Vapor s_g	
32	0.0886	0.01602	3305	−0.01	1021.2	−0.01	1075.4	1075.4	−0.00003	2.1870	2.1870	32
35	0.0999	0.01602	2948	2.99	1022.2	3.00	1073.7	1076.7	0.00607	2.1704	2.1764	35
40	0.1217	0.01602	2445	8.02	1023.9	8.02	1070.9	1078.9	0.01617	2.1430	2.1592	40
50	0.1780	0.01602	1704	18.06	1027.2	18.06	1065.2	1083.3	0.03607	2.0899	2.1259	50
60	0.2563	0.01604	1207	28.08	1030.4	28.08	1059.6	1087.7	0.05555	2.0388	2.0943	60
70	0.3632	0.01605	867.7	38.09	1033.7	38.09	1054.0	1092.0	0.07463	1.9896	2.0642	70
80	0.5073	0.01607	632.8	48.08	1037.0	48.09	1048.3	1096.4	0.09332	1.9423	2.0356	80
90	0.6988	0.01610	467.7	58.07	1040.2	58.07	1042.7	1100.7	0.11165	1.8966	2.0083	90
100	0.9503	0.01613	350.0	68.04	1043.5	68.05	1037.0	1105.0	0.12963	1.8526	1.9822	100
110	1.276	0.01617	265.1	78.02	1046.7	78.02	1031.3	1109.3	0.14730	1.8101	1.9574	110
120	1.695	0.01621	203.0	87.99	1049.9	88.00	1025.5	1113.5	0.1647	1.7690	1.9336	120
130	2.225	0.01625	157.2	97.97	1053.0	97.98	1019.8	1117.8	0.1817	1.7292	1.9109	130
140	2.892	0.01629	122.9	107.95	1056.2	107.96	1014.0	1121.9	0.1985	1.6907	1.8892	140
150	3.722	0.01634	97.0	117.95	1059.3	117.96	1008.1	1126.1	0.2150	1.6533	1.8684	150
160	4.745	0.01640	77.2	127.94	1062.3	127.96	1002.2	1130.1	0.2313	1.6171	1.8484	160
170	5.996	0.01645	62.0	137.95	1065.4	137.97	996.2	1134.2	0.2473	1.5819	1.8293	170
180	7.515	0.01651	50.2	147.97	1068.3	147.99	990.2	1138.2	0.2631	1.5478	1.8109	180
190	9.343	0.01657	41.0	158.00	1071.3	158.03	984.1	1142.1	0.2787	1.5146	1.7932	190
200	11.529	0.01663	33.6	168.04	1074.2	168.07	977.9	1145.9	0.2940	1.4822	1.7762	200
210	14.13	0.01670	27.82	178.10	1077.0	178.14	971.6	1149.7	0.3091	1.4508	1.7599	210
212	14.70	0.01672	26.80	180.1	1077.6	180.2	970.3	1150.5	0.3121	1.4446	1.7567	212
220	17.19	0.01677	23.15	188.2	1079.8	188.2	965.3	1153.5	0.3241	1.4201	1.7441	220
230	20.78	0.01685	19.39	198.3	1082.6	198.3	958.8	1157.1	0.3388	1.3901	1.7289	230
240	24.97	0.01692	16.33	208.4	1085.3	208.4	952.3	1160.7	0.3534	1.3609	1.7143	240
250	29.82	0.01700	13.83	218.5	1087.9	218.6	945.6	1164.2	0.3677	1.3324	1.7001	250

Table C-6 (continued)

Table C-6a Properties of Saturated Water: Temperature Table v, ft^3/lb; h, Btu/lb; s, Btu/lb·R

Temperature (F) T	Pressure (lbf/in.^{2}) p	Specific Volume		Internal Energy		Enthalpy			Entropy			Temperature (F) T
		Saturated Liquid v_f	Saturated Vapor v_g	Saturated Liquid u_f	Saturated Vapor u_g	Saturated Liquid h_f	h_{fg}	Saturated Vapor h_g	Saturated Liquid s_f	s_{fg}	Saturated Vapor s_g	
260	35.42	0.01708	11.77	228.6	1090.5	228.8	938.8	1167.6	0.3819	1.3044	1.6864	260
270	41.85	0.01717	10.07	238.8	1093.0	239.0	932.0	1170.9	0.3960	1.2771	1.6731	270
280	49.18	0.01726	8.65	249.0	1095.4	249.2	924.9	1174.1	0.4099	1.2504	1.6602	280
290	57.53	0.01735	7.47	259.3	1097.7	259.4	917.8	1177.2	0.4236	1.2241	1.6477	290
300	66.98	0.01745	6.47	269.5	1100.0	269.7	910.4	1180.2	0.4372	1.1984	1.6356	300
310	77.64	0.01755	5.63	279.8	1102.1	280.1	903.0	1183.0	0.4507	1.1731	1.6238	310
320	89.60	0.01765	4.92	290.1	1104.2	290.4	895.3	1185.8	0.4640	1.1483	1.6123	320
330	103.00	0.01776	4.31	300.5	1106.2	300.8	887.5	1188.4	0.4772	1.1238	1.6010	330
340	117.93	0.01787	3.79	310.9	1108.0	311.3	879.5	1190.8	0.4903	1.0997	1.5901	340
350	134.53	0.01799	3.35	321.4	1109.8	321.8	871.3	1193.1	0.5033	1.0760	1.5793	350
360	152.92	0.01811	2.96	331.8	1111.4	332.4	862.9	1195.2	0.5162	1.0526	1.5688	360
370	173.23	0.01823	2.63	342.4	1112.9	343.0	854.2	1197.2	0.5289	1.0295	1.5585	370
380	195.60	0.01836	2.34	353.0	1114.3	353.6	845.4	1199.0	0.5416	1.0067	1.5483	380
390	220.2	0.01850	2.09	363.6	1115.6	364.3	836.2	1200.6	0.5542	0.9841	1.5383	390
400	247.1	0.01864	1.87	374.3	1116.6	375.1	826.8	1202.0	0.5667	0.9617	1.5284	400
420	308.5	0.01894	1.50	395.8	1118.3	396.9	807.2	1204.1	0.5915	0.9175	1.5091	420
440	381.2	0.01926	1.22	417.6	1119.3	419.0	786.3	1205.3	0.6161	0.8740	1.4900	440
460	466.3	0.01961	1.00	439.7	1119.6	441.4	764.1	1205.5	0.6404	0.8308	1.4712	460
480	565.5	0.02000	0.82	462.2	1118.9	464.3	740.3	1204.6	0.6646	0.7878	1.4524	480
500	680.0	0.02043	0.68	485.1	1117.4	487.7	714.8	1202.5	0.6888	0.7448	1.4335	500
540	961.5	0.02145	0.47	532.6	1111.0	536.4	657.5	1193.8	0.7374	0.6576	1.3950	540
580	1324.3	0.02278	0.32	583.1	1098.9	588.6	589.3	1178.0	0.7872	0.5668	1.3540	580
600	1541.0	0.02363	0.27	609.9	1090.0	616.7	549.7	1166.4	0.8130	0.5187	1.3317	600
640	2057.1	0.02593	0.18	668.7	1063.2	678.6	453.4	1131.9	0.8681	0.4122	1.2803	640
680	2705	0.03032	0.11	741.7	1011.0	756.9	309.8	1066.7	0.9350	0.2718	1.2068	680
700	3090	0.03666	0.07	801.7	947.7	822.7	167.5	990.2	0.9902	0.1444	1.1346	700
705.4	3204	0.05053	0.05	872.6	872.6	902.5	0	902.5	1.0580	0	1.0580	705.4

Source: Abstracted from J. H. Keenan, F. G. Keyes, P. G. Hill, and J. G. Moore, *Steam Tables*. Wiley, New York, 1969. Reprinted by permission of John Wiley & Sons, Inc.

Table C-6 (continued)

Table C-6b Properties of Saturated Water: Pressure Table v, ft^3 / lb; h, Btu / lb; s, Btu / lb-R

Absolute Pressure (lbf / in.²) p	Tempera-ture (F) T	Specific Volume		Internal Energy		Enthalpy			Entropy			Absolute Pressure (lbf / in.²) p
		Saturated Liquid v_f	Saturated Vapor v_g	Saturated Liquid u_f	Saturated Vapor u_g	Saturated Liquid h_f	h_{fg}	Saturated Vapor h_g	Saturated Liquid s_f	s_{fg}	Saturated Vapor s_g	
0.4	72.84	0.01606	792.0	40.94	1034.7	40.94	1052.3	1093.3	0.0800	1.9760	2.0559	0.4
1	101.70	0.01614	333.6	69.74	1044.0	69.74	1036.0	1105.8	0.1327	1.8453	1.9779	1
2	126.04	0.01623	173.75	94.02	1051.8	94.02	1022.1	1116.1	0.1750	1.7448	1.9198	2
4	152.93	0.01636	90.64	120.88	1060.2	120.89	1006.4	1127.3	0.2198	1.6426	1.8624	4
6	170.03	0.01645	61.98	137.98	1065.4	138.00	996.2	1134.2	0.2474	1.5819	1.8292	6
8	182.84	0.01653	47.35	150.81	1069.2	150.84	988.4	1139.3	0.2675	1.5383	1.8058	8
10	193.19	0.01659	38.42	161.20	1072.2	161.23	982.1	1143.3	0.2836	1.5041	1.7887	10
14.696	211.99	0.01672	26.80	180.10	1077.6	180.15	970.4	1150.5	0.3121	1.4446	1.7567	14.696
15	213.03	0.01672	26.29	181.14	1077.9	181.19	969.7	1150.9	0.3137	1.4414	1.7551	15
20	227.96	0.01683	20.09	196.19	1082.0	196.26	960.1	1156.4	0.3358	1.3962	1.7320	20
30	250.34	0.01700	13.75	218.84	1088.0	218.93	945.4	1164.3	0.3682	1.3314	1.6996	30
40	267.26	0.01715	10.50	236.03	1092.3	236.16	933.8	1170.0	0.3921	1.2845	1.6767	40
50	281.03	0.01727	8.52	250.08	1095.6	250.24	924.2	1174.4	0.4113	1.2476	1.6589	50
60	292.73	0.01738	7.18	262.1	1098.3	262.3	915.8	1178.0	0.4273	1.2170	1.6444	60
70	302.96	0.01748	6.21	272.6	1100.6	272.8	908.3	1181.0	0.4412	1.1909	1.6321	70
80	312.07	0.01757	5.47	282.0	1102.6	282.2	901.4	1183.6	0.4534	1.1679	1.6214	80
90	320.31	0.01766	4.90	290.5	1104.3	290.8	895.1	1185.9	0.4644	1.1475	1.6119	90
100	327.86	0.01774	4.43	298.3	1105.8	298.6	889.2	1187.8	0.4744	1.1290	1.6034	100
120	341.30	0.01789	3.73	312.3	1108.3	312.7	878.5	1191.1	0.4920	1.0966	1.5886	120
140	353.08	0.01802	3.22	324.6	1110.3	325.1	868.7	1193.8	0.5073	1.0688	1.5761	140
160	363.60	0.01815	2.84	335.6	1112.0	336.2	859.8	1196.0	0.5208	1.0443	1.5651	160
180	373.13	0.01827	2.53	345.7	1113.4	346.3	851.5	1197.8	0.5329	1.0223	1.5553	180
200	381.86	0.01839	2.29	354.9	1114.6	355.6	843.7	1199.3	0.5440	1.0025	1.5464	200
250	401.04	0.01865	1.84	375.4	1116.7	376.2	825.8	1202.1	0.5680	0.9594	1.5274	250
300	417.43	0.01890	1.54	393.0	1118.2	394.1	809.8	1203.9	0.5883	0.9232	1.5115	300
350	431.82	0.01912	1.33	408.7	1119.0	409.9	795.0	1204.9	0.6060	0.8917	1.4978	350
400	444.70	0.01934	1.16	422.8	1119.5	424.2	781.2	1205.5	0.6218	0.8638	1.4856	400

Table C-6 (continued)

Table C-6b Properties of Saturated Water: Pressure Table v, ft^3/lb; h, Btu/lb; s, Btu/lb-R

Absolute Pressure (lbf/in.²) p	Temperature (F) T	Specific Volume		Internal Energy		Enthalpy			Entropy			Absolute Pressure (lbf/in.²) p
		Saturated Liquid v_f	Saturated Vapor v_g	Saturated Liquid u_f	Saturated Vapor u_g	Saturated Liquid h_f	h_{fg}	Saturated Vapor h_g	Saturated Liquid s_f	s_{fg}	Saturated Vapor s_g	
450	456.39	0.01955	1.03	435.7	1119.6	437.4	768.2	1205.6	0.6360	0.8385	1.4746	450
500	467.13	0.01975	0.93	447.7	1119.4	449.5	755.8	1205.3	0.6490	0.8154	1.4615	500
550	477.07	0.01994	0.84	458.9	1119.1	460.9	743.9	1204.8	0.6611	0.7941	1.4551	550
600	486.33	0.02013	0.77	469.4	1118.6	471.7	732.4	1204.1	0.6723	0.7742	1.4464	600
700	503.23	0.02051	0.66	488.9	1117.0	491.5	710.5	1202.0	0.6927	0.7378	1.4305	700
800	518.36	0.02087	0.57	506.6	1115.0	509.7	689.6	1199.3	0.7110	0.7050	1.4160	800
900	532.12	0.02123	0.50	523.0	1112.6	526.6	669.5	1196.0	0.7277	0.6750	1.4027	900
1000	544.75	0.02159	0.45	538.4	1109.9	542.4	650.0	1192.4	0.7432	0.6471	1.3903	1000
1200	567.37	0.02232	0.36	566.7	1103.5	571.7	612.3	1183.9	0.7712	0.5961	1.3673	1200
1400	587.25	0.02307	0.30	592.7	1096.0	598.6	575.5	1174.1	0.7964	0.5497	1.3461	1400
1600	605.06	0.02386	0.26	616.9	1087.4	624.0	538.9	1162.9	0.8196	0.5062	1.3258	1600
1800	621.21	0.02472	0.22	640.0	1077.7	648.3	502.1	1150.4	0.8414	0.4654	1.3060	1800
2000	636.00	0.02565	0.19	662.4	1066.6	671.9	464.4	1136.3	0.8623	0.4238	1.2861	2000
2500	668.31	0.02860	0.13	717.7	1031.0	730.9	360.5	1091.4	0.9131	0.3169	1.2327	2500
3000	695.52	0.03431	0.08	783.4	968.8	802.5	213.0	1015.5	0.9732	0.1843	1.1575	3000
3203.6	705.44	0.05053	0.05	872.6	872.6	902.5	0	902.5	1.0580	0	1.0580	3203.6

Table C-6 (continued)

Table C-6c Properties of Water: Superheated Vapor Table v, ft^3 / lb; h, Btu / lb; s, Btu / lb-R

Temperature (F)	v	u	h	v	u	h	s
	10 psia (193.2 F)			14.7 psia (212.0 F)			
Saturated	38.42	1072.2	1143.3	1.7877	26.80	1077.6	1150.5	1.7567
200	38.85	1074.7	1146.6	1.7927				
250	41.93	1092.6	1170.2	1.8272	28.42	1091.5	1168.8	1.7832
300	44.99	1110.4	1193.7	1.8592	30.52	1109.6	1192.6	1.8157
400	51.30	1146.1	1240.5	1.9171	34.67	1145.6	1239.9	1.8741
500	57.04	1182.2	1287.7	1.9690	38.77	1181.8	1287.3	1.9263
600	63.03	1218.9	1335.5	2.0164	42.86	1218.6	1335.2	1.9737
700	69.01	1256.3	1384.0	2.0601	46.93	1256.1	1383.8	2.0175
800	74.98	1294.6	1433.3	2.1009	51.00	1294.4	1433.1	2.0584
900	80.95	1333.7	1483.5	2.1393	55.07	1333.6	1483.4	2.0967
1000	86.91	1373.8	1534.6	2.1755	59.13	1373.7	1534.5	2.1330
1100	92.88	1414.7	1586.6	2.2099	63.19	1414.6	1586.4	2.1674

Temperature (F)	v	u	h	v	u	h	s
	20 psia (228.0 F)			40 psia (267.3 F)			
Saturated	20.09	1082.0	1156.4	1.7320	10.50	1092.3	1170.0	1.6767
250	20.79	1090.3	1167.2	1.7475				
300	22.36	1108.7	1191.5	1.7805	11.04	1105.1	1186.8	1.6993
350	23.90	1126.9	1215.4	1.8110	11.84	1124.2	1211.8	1.7312
400	25.43	1145.1	1239.2	1.8395	12.62	1143.0	1236.4	1.7606
500	28.46	1181.5	1286.8	1.8919	14.16	1180.1	1284.9	1.8140
600	31.47	1218.4	1334.8	1.9395	15.69	1217.3	1333.4	1.8621
700	34.47	1255.9	1383.5	1.9834	17.20	1255.1	1382.4	1.9063
800	37.46	1294.3	1432.9	2.0243	18.70	1293.7	1432.1	1.9474
900	40.45	1333.5	1483.2	2.0627	20.20	1333.0	1482.5	1.9859
1000	43.44	1373.5	1534.3	2.0989	21.70	1373.1	1533.8	2.0223
1100	46.42	1414.5	1586.3	2.1334	23.20	1414.2	1585.9	2.0568

Temperature (F)	v	u	h	v	u	h	s
	60 psia (292.7 F)			80 psia (312.1 F)			
Saturated	7.18	1098.3	1178.0	1.6444	5.47	1102.6	1183.6	1.6214
300	7.26	1101.3	1181.9	1.6496				
350	7.82	1121.4	1208.2	1.6830	5.80	1118.5	1204.3	1.6476
400	8.35	1140.8	1233.5	1.7134	6.22	1138.5	1230.6	1.6790
500	9.40	1178.6	1283.0	1.7678	7.02	1177.2	1281.1	1.7346
600	10.43	1216.3	1332.1	1.8165	7.79	1215.3	1330.7	1.7838
700	11.44	1254.4	1381.4	1.8609	8.56	1253.6	1380.3	1.8285
800	12.45	1293.0	1431.2	1.9022	9.32	1292.4	1430.4	1.8700
900	13.45	1332.5	1481.8	1.9408	10.08	1332.0	1481.2	1.9087
1000	14.45	1372.7	1533.2	1.9773	10.83	1372.3	1532.6	1.9453
1100	15.45	1413.8	1585.4	2.0119	11.58	1413.5	1584.9	1.9799
1200	16.45	1455.8	1638.5	2.0448	12.33	1455.5	1638.1	2.0130

Table C-6c (continued)

Temperature (F)	v	u	h	s	v	u	h	s
	100 psia (327.9 F)				120 psia (341.3 F)			
Saturated	4.434	1105.8	1187.8	1.6034	3.730	1108.3	1191.1	1.5886
350	4.592	1115.4	1200.4	1.6191	3.783	1112.2	1196.2	1.5950
400	4.934	1136.2	1227.5	1.6517	4.079	1133.8	1224.4	1.6288
450	5.265	1156.2	1253.6	1.6812	4.360	1154.3	1251.2	1.6590
500	5.587	1175.7	1279.1	1.7085	4.633	1174.2	1277.1	1.6868
600	6.216	1214.2	1329.3	1.7582	5.164	1213.2	1327.8	1.7371
700	6.834	1252.8	1379.2	1.8033	5.682	1252.0	1378.2	1.7825
800	7.445	1291.8	1429.6	1.8449	6.195	1291.2	1428.7	1.8243
900	8.053	1331.5	1480.5	1.8838	6.703	1330.9	1479.8	1.8633
1000	8.657	1371.9	1532.1	1.9204	7.208	1371.5	1531.5	1.9000
1100	9.260	1413.1	1584.5	1.9551	7.711	1412.8	1584.0	1.9348
1200	9.861	1455.2	1637.7	1.9882	8.213	1454.9	1637.3	1.9679
	140 psia (353.1 F)				160 Psia (363.6 F)			
Saturated	3.221	1110.3	1193.8	1.5761	2.836	1112.0	1196.0	1.5651
400	3.466	1131.4	1221.2	1.6088	3.007	1128.8	1217.8	1.5911
450	3.713	1152.4	1248.6	1.6399	3.228	1150.5	1246.1	1.6230
500	3.952	1172.7	1275.1	1.6682	3.440	1171.2	1273.0	1.6518
550	4.184	1192.6	1300.9	1.6944	3.646	1191.3	1299.2	1.6784
600	4.412	1212.1	1326.4	1.7191	3.848	1211.1	1325.0	1.7034
700	4.860	1251.2	1377.1	1.7648	4.243	1250.4	1376.0	1.7494
800	5.301	1290.5	1427.9	1.8068	4.631	1289.9	1427.0	1.7916
900	5.739	1330.4	1479.1	1.8459	5.015	1329.9	1478.4	1.8308
1000	6.173	1371.0	1531.0	1.8827	5.397	1370.6	1530.4	1.8677
1100	6.605	1412.4	1583.6	1.9176	5.776	1412.1	1583.1	1.9026
1200	7.036	1454.6	1636.9	1.9507	6.154	1454.3	1636.5	1.9358
	180 psia (373.1 F)				200 psia (381.9 F)			
Saturated	2.533	1113.4	1197.8	1.5553	2.289	1114.6	1199.3	1.5464
400	2.648	1126.2	1214.4	1.5749	2.361	1123.5	1210.8	1.5600
450	2.850	1148.5	1243.4	1.6078	2.548	1146.4	1240.7	1.5938
500	3.042	1169.6	1270.9	1.6372	2.724	1168.0	1268.8	1.6239
550	3.228	1190.0	1297.5	1.6642	2.893	1188.7	1295.7	1.6512
600	3.409	1210.0	1323.5	1.6893	3.058	1208.9	1322.1	1.6767
700	3.763	1249.6	1374.9	1.7357	3.379	1248.8	1373.8	1.7234
800	4.110	1289.3	1426.2	1.7781	3.693	1288.6	1425.3	1.7660
900	4.453	1329.4	1477.7	1.8175	4.003	1328.9	1477.1	1.8055
1000	4.793	1370.2	1529.8	1.8545	4.310	1369.8	1529.3	1.8425
1100	5.131	1411.7	1582.6	1.8894	4.615	1411.4	1582.2	1.8776
1200	5.467	1454.0	1636.1	1.9227	4.918	1453.7	1635.7	1.9109

Table C-6c (continued)

Temperature (F)	t	u	h	s	v	u	h	s
	250 psia (401.0 F)				300 psia (417.4 F)			
Saturated	1.845	1116.7	1202.1	1.5274	1.544	1118.2	1203.9	1.5115
450	2.002	1141.1	1233.7	1.5632	1.636	1135.4	1226.2	1.5365
500	2.150	1163.8	1263.3	1.5948	1.766	1159.5	1257.5	1.5701
550	2.290	1185.3	1291.3	1.6233	1.888	1181.9	1286.7	1.5997
600	2.426	1206.1	1318.3	1.6494	2.004	1203.2	1314.5	1.6266
700	2.688	1246.7	1371.1	1.6970	2.227	1244.6	1368.3	1.6751
800	2.943	1287.0	1423.2	1.7401	2.442	1285.4	1421.0	1.7187
900	3.193	1327.6	1475.3	1.7799	2.653	1326.3	1473.6	1.7589
1000	3.440	1368.7	1527.9	1.8172	2.860	1367.7	1526.5	1.7964
1100	3.685	1410.5	1581.0	1.8524	3.066	1409.6	1579.8	1.8317
1200	3.929	1453.0	1634.8	1.8858	3.270	1452.2	1633.8	1.8653
1300	4.172	1496.3	1698.3	1.9177	3.473	1495.6	1688.4	1.8973
	350 psia (431.8 F)				400 psia (444.7 F)			
Saturated	1.327	1119.0	1204.9	1.4978	1.162	1119.5	1205.5	1.4856
450	1.373	1129.2	1218.2	1.5125	1.175	1122.6	1209.6	1.4901
500	1.491	1154.9	1251.5	1.5482	1.284	1150.1	1245.2	1.5282
550	1.600	1178.3	1281.9	1.5790	1.383	1174.6	1277.0	1.5605
600	1.703	1200.3	1310.6	1.6068	1.476	1197.3	1306.6	1.5892
700	1.898	1242.5	1365.4	1.6562	1.650	1240.4	1362.5	1.6397
800	2.085	1283.8	1418.8	1.7004	1.816	1282.1	1416.6	1.6844
900	2.267	1325.0	1471.8	1.7409	1.978	1323.7	1470.1	1.7252
1000	2.446	1366.6	1525.0	1.7787	2.136	1365.5	1523.6	1.7632
1100	2.624	1408.7	1578.6	1.8142	2.292	1407.8	1577.4	1.7989
1200	2.799	1451.5	1632.8	1.8478	2.446	1450.7	1631.8	1.8327
1300	2.974	1495.0	1687.6	1.8799	2.599	1494.3	1686.8	1.8648
	450 psia (456.4 F)				500 psia (467.1 F)			
Saturated	1.033	1119.6	1205.6	1.4746	0.928	1119.4	1205.3	1.4645
500	1.123	1145.1	1238.5	1.5097	0.992	1139.7	1231.5	1.4923
550	1.215	1170.7	1271.9	1.5436	1.079	1166.7	1266.6	1.5279
600	1.300	1194.3	1302.5	1.5732	1.158	1191.1	1298.3	1.5585
700	1.458	1238.2	1359.6	1.6248	1.304	1236.0	1356.7	1.6112
800	1.608	1280.5	1414.4	1.6701	1.441	1278.8	1412.1	1.6571
900	1.752	1322.4	1468.3	1.7113	1.572	1321.0	1466.5	1.6987
1000	1.894	1364.4	1522.2	1.7495	1.701	1363.3	1520.7	1.7471
1100	2.034	1406.9	1576.3	1.7853	1.827	1406.0	1575.1	1.7731
1200	2.172	1450.0	1630.8	1.8192	1.952	1449.2	1629.8	1.8072
1300	2.308	1493.7	1685.9	1.8515	2.075	1493.1	1685.1	1.8395
1400	2.444	1538.1	1741.7	1.8823	2.198	1537.6	1741.0	1.8704

Table C-6c (continued)

Temperature (F)	v	u	h	s	v	u	h	s
	600 psia (486.3 F)				700 psia (503.2 F)			
Saturated	0.770	1118.6	1204.1	1.4464	0.656	1117.0	1202.0	1.4305
500	0.795	1128.0	1216.2	1.4592				
550	0.875	1158.2	1255.4	1.4990	0.728	1149.0	1243.2	1.4723
600	0.946	1184.5	1289.5	1.5320	0.793	1177.5	1280.2	1.5081
700	1.073	1231.5	1350.6	1.5872	0.907	1226.9	1344.4	1.5661
800	1.190	1275.4	1407.6	1.6343	1.011	1272.0	1402.9	1.6145
900	1.302	1318.4	1462.9	1.6766	1.109	1315.6	1459.3	1.6576
1000	1.411	1361.2	1517.8	1.7155	1.204	1358.9	1514.9	1.6970
1100	1.517	1404.2	1572.7	1.7519	1.296	1402.4	1570.2	1.7337
1200	1.622	1447.7	1627.8	1.7861	1.387	1446.2	1625.8	1.7682
1300	1.726	1491.7	1683.4	1.8186	1.476	1490.4	1681.7	1.8009
1400	1.829	1536.5	1739.5	1.8479	1.565	1535.3	1738.1	1.8321
	800 psia (518.4 F)				900 psia (532.1 F)			
Saturated	0.569	1115.0	1199.3	1.4160	0.501	1112.6	1196.0	1.4027
550	0.615	1138.8	1229.9	1.4469	0.527	1127.5	1215.2	1.4219
600	0.677	1170.1	1270.4	1.4861	0.587	1162.2	1260.0	1.4652
650	0.732	1197.2	1305.6	1.5186	0.639	1191.1	1297.5	1.4999
700	0.783	1222.1	1338.0	1.5471	0.686	1217.1	1331.4	1.5297
800	0.876	1268.5	1398.2	1.5969	0.772	1264.9	1393.4	1.5810
900	0.964	1312.9	1455.6	1.6408	0.851	1310.1	1451.9	1.6257
1000	1.048	1356.7	1511.9	1.6807	0.927	1354.5	1508.9	1.6662
1100	1.130	1400.5	1567.8	1.7178	1.001	1398.7	1565.4	1.7036
1200	1.210	1444.6	1623.8	1.7526	1.073	1443.0	1621.7	1.7386
1300	1.289	1489.1	1680.0	1.7854	1.144	1487.8	1687.3	1.7717
1400	1.367	1534.2	1736.6	1.8167	1.214	1533.0	1735.1	1.8031
	1000 psia (544.8 F)				1200 psia (567.4 F)			
Saturated	0.446	1109.0	1192.4	1.3903	0.362	1103.5	1183.9	1.3673
600	0.514	1153.7	1248.8	1.4450	0.402	1134.4	1223.6	1.4054
650	0.564	1184.7	1289.1	1.4822	0.450	1170.9	1270.8	1.4490
700	0.608	1212.0	1324.6	1.5135	0.491	1201.3	1310.2	1.4837
800	0.688	1261.2	1388.5	1.5664	0.562	1253.7	1378.4	1.5402
900	0.761	1307.3	1448.1	1.6120	0.626	1301.5	1440.4	1.5876
1000	0.831	1352.2	1505.9	1.6530	0.685	1347.5	1499.7	1.6297
1100	0.898	1396.8	1562.9	1.6908	0.743	1393.0	1557.9	1.6682
1200	0.963	1441.5	1619.7	1.7261	0.798	1438.3	1615.5	1.7040
1300	1.027	1486.5	1676.5	1.7593	0.853	1483.8	1673.1	1.7377
1400	1.091	1531.9	1733.7	1.7909	0.906	1529.6	1730.7	1.7697
1600	1.215	1624.4	1849.3	1.8499	1.011	1622.6	1847.1	1.8290

Table C-6 (continued)

Table C-6d Properties of Water: Compressed Liquid Table v, ft^3 / lb; h, Btu / lb; s Btu / lb-R

Temperature (F)	500 psia (T_{sat} = 467.1 F)				1000 psia (T_{sat} = 544.8 F)			
	v	u	h	s	v	u	h	s
32	0.015994	0.00	1.49	0.00000	0.015967	0.03	2.99	0.00005
50	0.015998	18.02	19.50	0.03599	0.015972	17.99	20.94	0.03592
100	0.016106	67.87	69.36	0.12932	0.016982	67.70	70.68	0.12901
150	0.016318	117.66	119.17	0.21457	0.016293	117.38	120.40	0.21410
200	0.016608	167.65	169.19	0.28341	0.016580	167.26	170.32	0.29281
300	0.017416	268.92	270.53	0.43641	0.017379	268.24	271.46	0.43552
400	0.018608	373.68	375.40	0.56604	0.018550	372.55	375.98	0.56472
Saturated	0.019748	447.70	449.53	0.64904	0.021591	538.39	542.38	0.74320
	1500 psia (T_{sat} = 596.4 F)				2000 psia (T_{sat} = 636.0 F)			
32	0.015939	0.05	4.47	0.00007	0.015912	0.06	5.95	0.00008
50	0.015946	17.59	22.38	0.03584	0.015920	17.91	23.81	0.03575
100	0.016058	67.53	71.99	0.12870	0.016034	67.37	73.30	0.12839
150	0.016268	117.10	121.62	0.21364	0.016244	116.83	122.84	0.21318
200	0.016554	166.87	171.46	0.29221	0.016527	166.49	172.60	0.29162
300	0.017343	267.58	272.39	0.43463	0.017308	266.93	272.33	0.43376
400	0.018493	371.45	376.59	0.56343	0.018439	370.38	377.21	0.56216
500	0.20204	481.1	487.4	0.6853	0.02014	479.8	487.3	0.6832
Saturated	0.02346	605.0	611.5	0.8082	0.02565	662.4	671.9	0.8623
	3000 psia (T_{sat} = 695.5 F)				4000 psia			
32	0.015859	0.09	8.90	0.00009	0.015807	0.10	11.80	0.00005
50	0.015870	17.84	26.65	0.03555	0.015821	17.76	29.47	0.03534
100	0.015987	67.04	75.91	0.12777	0.015942	66.72	78.52	0.12714
150	0.016196	116.30	125.29	0.21226	0.016150	115.77	127.73	0.21136
200	0.016476	165.74	174.89	0.29046	0.016425	165.02	177.18	0.28931
300	0.017240	265.66	275.23	0.43205	0.017174	264.43	277.15	0.43038
400	0.018334	368.32	378.50	0.55970	0.018235	366.35	379.85	0.55734
500	0.019944	476.2	487.3	0.6794	0.019766	472.9	487.5	0.6758
Saturated	0.034310	783.5	802.5	0.9732				

Table C-7
Thermodynamic Properties of Air at Low Pressure

Ideal Gas Properties of Air h and u, Btu / lb, s' Btu / lb-R

$T(R)$	h	u	s'	$T(R)$	h	u	s'
300	71.61	51.04	0.46007	1000	240.98	173.43	0.75042
320	76.40	54.46	0.47550	1100	265.99	190.58	0.77426
340	81.18	57.87	0.49002	1200	291.30	209.05	0.79628
360	85.97	61.29	0.50369	1300	316.94	227.83	0.81680
380	90.75	64.70	0.51663	1400	342.90	246.93	0.83604
400	95.53	68.11	0.52890	1500	369.17	266.34	0.85416
420	100.32	71.52	0.54058	1600	395.74	286.06	0.87130
440	105.11	74.93	0.55172	1700	422.59	306.06	0.88758
460	109.90	78.36	0.56235	1800	449.71	326.32	0.90308
480	114.69	81.77	0.57255	1900	477.09	346.85	0.91788
500	119.48	85.20	0.58233	2000	504.71	367.61	0.93205
520	124.27	88.62	0.59173	2100	532.55	388.60	0.94564
537	128.34	91.53	0.59945	2200	560.59	409.78	0.95868
540	129.06	92.04	0.60078	2300	588.82	431.16	0.97123
560	133.86	95.47	0.60950	2400	617.22	452.70	0.98331
580	138.66	98.90	0.61793	2500	645.78	474.40	0.99497
600	143.47	102.34	0.62607	2600	674.49	496.26	1.00623
620	148.28	105.78	0.63395	2700	703.35	518.26	1.01712
640	153.09	109.21	0.64159	2800	732.33	540.40	1.02767
660	157.92	112.67	0.64902	2900	761.45	562.66	1.03788
680	162.73	116.12	0.65621	3000	790.68	585.04	1.04779
700	167.56	119.58	0.66321	3100	820.03	607.53	1.05741
720	172.39	123.04	0.67002	3200	849.48	630.12	1.06676
740	177.23	126.51	0.67665	3300	879.02	652.81	1.07585
760	182.08	129.99	0.68321	3400	908.66	675.60	1.08470
780	186.94	133.47	0.68942	3500	938.40	698.48	1.09332
800	191.81	136.97	0.69558	3600	968.21	721.44	1.10172
820	196.69	140.47	0.70160	3700	998.11	744.48	1.10991
840	201.56	143.98	0.70747	3800	1028.09	767.60	1.11791
860	206.46	147.50	0.71323	3900	1058.14	790.80	1.12571
880	211.35	151.02	0.71886	4000	1088.26	814.06	1.13334
900	216.26	154.57	0.72438	4100	1118.5	837.4	1.14079
920	221.18	158.12	0.72979	4200	1148.7	860.8	1.14809
940	226.11	161.68	0.73509	4300	1179.0	884.3	1.15522
960	231.06	165.26	0.74030	4400	1209.4	907.8	1.16221
980	236.02	168.83	0.74540	4500	1239.9	931.4	1.16905
				4600	1270.4	955.0	1.17575
				4700	1300.9	978.7	1.18232
				4800	1331.5	1002.5	1.18876
				4900	1362.2	1026.3	1.19508
				5000	1392.9	1050.1	1.20129
				5100	1423.6	1074.0	1.20738
				5200	1454.4	1098.0	1.21336
				5300	1485.3	1122.0	1.21923

Source: Data extracted from J. H. Keenan and J. Kaye, *Gas Tables*, Wiley, New York, 1945. Reprinted by permission of John Wiley & Sons, Inc.

Illustrative Example for Cooling System Optimization

The purpose of this study was to determine the optimum combination of a mechanical-draft cooling tower and steam condenser for a previously selected turbine generator system. This generating system had a nominal output of 700,000 kW with supercritical steam 3515 psia and 1000 F at turbine inlet. The system was similar to that in Fig. 7-7. The previous study determined the plant net heat rate and condenser heat load as shown in Figs. D-1 and D-2. Other inputs to this system design were:

A. General Information
   Rate of return   = 13%
   Annual fixed charge rate (AFCR) = 18%
   Economic life   = 35 yr

B. Operation Costs
   First year fuel cost   = $1.00/MBtu
   Fuel cost escalation   = 6%/yr
   levelized replacement energy cost   = 27.3 mills/kWh
   levelized power demand costs   = $170.4/kW-yr
   levelized auxiliary power cost   15.8 mills/kWh

C. Capital Costs Expressed in 1983 Dollars (1)
   Condenser
       Cost = 1.25[16,000 + 8.0 (tube surface area) + 6.4 (number of tubes)]
   Cooling tower
       Cost = 1.25(29) (number of tower units)
   Circulating water pump
       Cost = 3.8 (circulating water flow rate in gallons per minute)
   Circulating water pump motor
       Cost = 72 (rated power in horsepower)

D. Operating Pattern

613

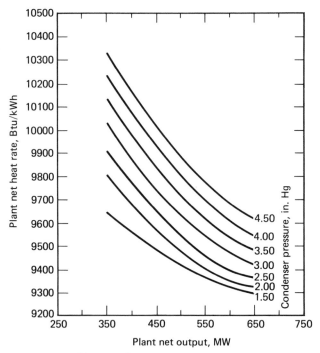

Figure D-1. Plant net heat rate.

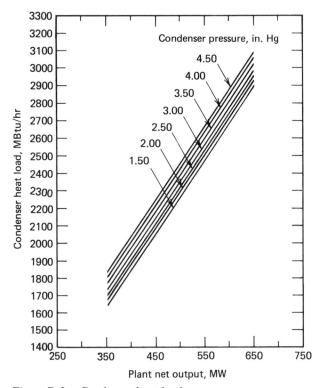

Figure D-2. Condenser heat load.

Season	Load (MW)	Operation Hours in Years		
		1–5	6–15	16–35
	VWO(2)	50.	50.	0.
Summer	650.	920.	750.	400.
$WBT = 70$ F	525.	590.	580.	600.
	350.	310.	580.	900.
Spring and	650.	1850.	1500.	400.
fall	525.	1200.	1200.	1100.
$WBT = 60$ F	350.	600.	1000.	1600.
Winter	650.	920.	750.	400.
$WBT = 45$ F	525.	590.	580.	600.
	350.	310.	580.	900.

Note: (1) represents the installed cost. (2) represents the valve wide open conditions. The throttle steam flow is 105% of the design value, and steam conditions are identical to the design values.

Table D-1 summarizes other system parameters used for this study.

The optimization was based on the minimum present worth of lifetime evaluated costs. In this study the evaluated costs consist of (1) relative initial investment, (2) plant fuel cost, (3) cooling system operating cost, (4) power demand cost and (5) replacement energy cost. To determine the best combination of mechanical-draft cooling tower and steam condenser, the study was to determine the optimal values for these four variables:

Condenser pressure
Tower approach
Condenser tube length
Water velocity in condenser tubes

The steps used in this study were (1) identify some possible system configurations, (2) simulate these systems under various operating conditions, and (3) calculate the present worth of lifetime evaluated costs. In the following was the presentation of calculations and results in each step.

STEP 1 IDENTIFY SOME POSSIBLE SYSTEM CONFIGURATIONS

In spite of so many system parameters in this system design, there were four parameters to be optimized. If five values were considered for each parameter, there would be many different combinations. Obviously, it would be very time-consuming to investigate each of these system configurations. Instead of doing it, we optimized each parameter at a time as suggested in Chapter 11. In this design, we first kept the condenser tube length and water velocity constant (36 ft and 8 fps) and examined

Table D-1
Summary of All Fixed System Parameters

A.	Design weather conditions		
	Dry-bulb temperature	=	90 F
	Relative humidity	=	48%
	Wet-bulb temperature	=	74 F
B.	Average wet-bulb temperature		
	Summer	=	70 F
	Spring and fall	=	60 F
	Winter	=	45 F
C.	Condenser		
	Design heat load	=	3286.49 MBtu/hr
	Number of passes	=	2
	Tube material	=	Admiralty metal
	Tube outside diameter	=	1.00 in.
	Tube wall gauge	=	18 BWG
	Tube cleanliness factor	=	85%
D.	Circulating water system		
	Pipe length	=	4000 ft
	Pipe diameter	=	8.5 ft
	Pipe material	=	Welded steel
	Pump efficiency	=	85%
	Pump motor efficiency	=	90%
	Increase in elevation of circulating		
	water piping (including tower height)	=	50 ft
E.	Cooling tower		
	Total water to air flow ratio	=	1.00
	Fan horsepower	=	(0.011)(tower units)
	Fan efficiency	=	80%
	Fan motor efficiency	=	90%

three different tower approaches, for three different condenser pressures. The nine system configurations were:

Configuration number	1	2	3	4	5	6	7	8	9
Condenser pressure (in. Hg Abs.)	3.0	3.0	3.0	4.0	4.0	4.0	5.0	5.0	5.0
Tower approach (F)	10	13	16	10	13	16	10	13	16

In each configuration, condenser and cooling tower were sized. Enough information was generated so that the relative cost could be estimated. Table D-2 summarized the results for the configurations 4, 5, and 6. Other six configurations were not included here for simplicity. These six configurations were more expensive than the optimal combination, as shown later.

As indicated in Table D-2, condenser calculations included water flow rate, tube surface area, number of tubes, condenser temperature rise, and terminal temperature difference. These were done by using the procedures presented in Chapter 9. The tower calculations included the tower characteristics, tower rating factor, number of tower units, and fan horsepower. In this study the water-to-air flow ratio in the tower was assumed to be a unity. This was entirely acceptable for the comparative study. To estimate the relative tower cost, the rating factor was determined by using either Fig. 11-15 or the equations in Table 11-11. The number of tower units was the product of the rating factor and water flow rate in gallons per minute.

Table D-2
Configuration Design for Alternatives 4, 5, and 6

		Design Configuration		
		4	5	6
Design condenser pressure		4.00 in. Hg	4.00 in. Hg	4.00 in. Hg
Design tower approach		10 F	13 F	16 F
Condenser				
Heat load	(MBtu/hr)	3286.49	3286.49	3286.49
Wet-bulb temperature	(F)	74	74	74
Water flow rate	(gpm)	200,105	215,037	232,584
Tube surface area	(ft^2)	236,431	254,074	274,808
Number of tubes	(1)	25,086	26,958	29,158
Temperature rise	(F)	32.9	30.6	28.3
Terminal temperature difference	(F)	8.57	7.85	7.16
Tower				
Heat load	(MBtu/hr)	3286.49	3286.49	3286.49
Cooling range	(F)	32.9	30.6	28.3
Tower water-to-air flow ratio	(1)	1.00	1.00	1.00
Tower characteristics	(1)	2.80	1.81	1.30
Marley company rating factor	(1)	1.539	1.138	0.9131
Number of tower units	(1)	307,900	244,700	212,400
Fan power	(hp)	3387	2692	2336
Fan power (fan motor efficiency = 90%)	(kW)	2807	2231	1936
Circulating water pump				
Pump power (pump efficiency = 85%)	(hp)	4434	4849	5355
Pump power (pump motor efficiency = 90%)	(kW)	3675	4019	4439

Table D-3 Summary of Circulating Pump Calculations

		Design Configuration		
		4	5	6
Design condenser pressure		4.00 in. Hg	4.00 in. Hg	4.00 in. Hg
Design tower approach		10 F	13 F	16 F
Water flow rate	(gpm)	200,105	215,037	232,584
Pipe diameter	(ft)	8.5	8.5	8.5
Pipe length	(ft)	4000	4000	4000
Water velocity	(fps)	7.86	8.44	9.13
Pumping pressure head developed for:				
Condenser	(ft)[a]	12.1	12.1	12.1
Elevation change				
(including tower)	(ft)	50.0	50.0	50.0
Pipe resistance	(ft)	9.0	10.3	11.9
Pump suction and				
velocity head	(ft)	1.5	1.5	1.5
Miscellaneous fittings	(ft)	2.0	2.0	2.0
Total	(ft)	74.6	75.9	77.5
Pump power				
(Pump efficiency = 85%)	(hp)	4434	4849	5355
Pump power				
(Pump motor efficiency = 90%)	(kw)	3675	4019	4439

[a] Expressed in feet of water.

In this optimization the tower fan power was estimated by using the empirical equation.

$$\text{Fan horsepower} = (0.0110)(\text{number of tower units})$$

This approach eliminated the detailed tower design including a selection of tower fills.

Table D-3 presented the summary of circulating pump calculations. It included the pressure drops in various locations of the water circuit. The circulating water pipe between the condenser and tower was also taken into account.

STEP 2   SIMULATE THE PROPOSED SYSTEM UNDER VARIOUS CONDITIONS

To determine the plant fuel consumption, the plant operation must be simulated under various conditions. These included different plant loadings under different weather conditions. In this system design the plant loadings to be considered were the maximum, 650 MW, 525 MW, and 350 MW, and the weather conditions were the summer, spring-fall, and winter (the spring conditions were assumed identical to

ble D-4

eration Simulation of Cooling System Configurations

sign Configuration 4

Season Load (MW)	Summer WBT = 70 F				Fall and Spring WBT = 60 F			Winter WBT = 45 F		
	VWO	650	525	350	650	525	350	650	525	350
Ieat load (MBtu/hr)	3238	2993	2424	1650	2955	2393	1622	2922	2367	1602
'ower characteristics	2.80	2.80	2.80	2.80	2.80	2.80	2.80	2.80	2.80	2.80
:ooling range (F)	32.2	29.9	24.2	16.5	29.6	23.9	16.2	29.3	23.7	16.0
'ower approach (F)	11.5	11.0	9.75	7.63	15.1	13.4	10.6	22.3	20.1	16.0
:old water temperature (F)	81.5	81.0	79.8	77.6	75.1	73.4	70.6	67.3	65.1	61.0
:ondenser pressure (in. Hg)	3.73	3.38	2.65	1.90	2.85	2.20	1.51	2.32	1.72	1.13
'lant net heat rate (Btu/kWh)	9449	9469	9525	9679	9405	9460	9652	9354	9409	9608

esign Configuration 5

Season Load (MW)	Summer WBT = 70 F				Fall and Spring WBT = 60 F			Winter WBT = 45 F		
	VWO	650	525	350	650	525	350	650	525	350
Ieat load (MBtu/hr)	3233	2993	2427	1653	2958	2396	1623	2924	2365	1603
Tower characteristics	1.81	1.81	1.81	1.81	1.81	1.81	1.81	1.81	1.81	1.81
:ooling range (F)	30.1	27.8	22.6	15.4	27.5	22.3	15.1	27.2	22.0	14.9
Tower approach (F)	14.6	14.0	12.3	9.53	18.6	16.3	12.7	26.1	23.3	18.3
:old water temperature (F)	84.6	84.0	82.3	79.5	78.6	76.3	72.7	71.1	68.3	63.3
:ondenser pressure (in. Hg)	3.73	3.38	2.68	1.93	2.89	2.24	1.55	2.32	1.76	1.15
'lant net heat rate (Btu/kWh)	9449	9469	9530	9784	9410	9466	9666	9355	9412	9610

esign Configuration 6

Season Load (MW)	Summer WBT = 70 F				Fall and Spring WBT = 60 F			Winter WBT = 45 F		
	VWO	650	525	350	650	525	350	650	525	350
Heat load (MBtu/hr)	3241	2996	2429	1656	2961	2397	1628	2926	2372	1605
Tower characteristics	1.30	1.30	1.30	1.30	1.30	1.30	1.30	1.30	1.30	1.30
Cooling range (F)	27.9	25.8	20.9	14.2	25.5	20.6	14.0	25.16	20.4	13.8
Tower Approach (F)	17.8	17.0	14.9	11.4	21.7	18.9	14.8	29.4	26.4	20.6
Cold water temperature (F)	87.75	87.0	84.9	81.4	81.7	78.9	74.8	74.4	71.4	65.6
Condenser pressure (in. Hg)	3.75	3.42	2.70	1.98	2.93	2.26	1.60	2.37	1.81	1.19
Plant net heat rate (Btu/kWh)	9453	9474	9533	9797	9414	9468	9769	9358	9417	9615

those in the fall). The weather conditions were represented by the average value of ambient wet-bulb temperature. Table D-4 presented the simulation results for the configurations 4, 5, and 6. All of these calculations were based on the assumption that the circulating water flow rate was kept constant and equal to the design value.

It is seen in Table D-4 that a different system configuration will result in a different plant net heat rate and thus result in a different plant fuel cost. As the condenser pressure increases, the plant net heat rate will increase. At a given condenser pressure the plant net heat rate will increase with an increase in the design tower approach.

The simulation technique involves an iterative approach. First assume the condenser pressure and estimate the system heat load by using the turbine performance information such as that contained in Figs. D-1 and D-2. Secondly, estimate the temperature of the water leaving the evaporative cooling tower. The method for this temperature prediction was presented in Chapter 8. Since the circulating water flow is constant, the condenser pressure will be determined by the system heat load and the water inlet temperature. This calculated condenser pressure must be checked and compared with the assumed value. If the agreement is not within a reasonable range, the above procedure should be repeated with a new assumption of condenser pressure.

STEP 3 CALCULATE THE PRESENT WORTH OF LIFE-TIME EVALUATED COSTS

The initial costs were estimated by using the empirical equations presented as the study inputs. It should be pointed out that these were not the actual cost. The costs that were essentially constant with various design configurations were omitted for simplicity. The calculated results were presented in Table D-5.

The second evaluated cost was the plant fuel cost resulting from a different selection of system configurations. The annual fuel consumption was first calculated by using the plant net heat rate and loading pattern. With the unit fuel cost and other given economic factors, the present worth of lifetime fuel cost was determined and presented in Table D-5.

Table D-5 also included the costs for cooling system operation, generating capability, and energy loss. The operation cost consisted of energy cost for operating circulating water pumps and tower fans. In this study the levelized plant capacity factor was first calculated and found to be 0.602. Then the equation below was used to estimate the present worth of lifetime operation cost.

$$PW \text{ (operation cost)} = \text{(levelized unit fuel cost)(pump power + fan power)}$$

$$\times \text{(plant capacity factor)}(8760)(\text{SPWF})$$

To determine the penalty costs for generating capability and energy loss, the generation capability loss for each system configuration was determined. As shown in Table D-6, the plant maximum output was first estimated, assuming a throttle

Table D-5
Cost Comparison

	Design Configuration		
	4	5	6
Installed capital costs in 1983 dollars			
Condenser	2,585,000	2,776,400	3,001,300
Cooling tower	11.161,400	8,870,400	7,698,500
Circulating water pump	760,400	817,100	883,800
Circulating water pump motor	319,200	349,100	385,600
Subtotal	14,826,000	12,813,000	11,969,200
Total cost (indirect and overhead costs = 25%)	18,532,500	16,016,300	14,961,500
Present worth of lifetime costs in 1983 dollars[a]			
Capital investment	25,304,300	21,868,600	20,428,500
Fuel	456,129,700	456,618,200	457,994,700
Fan and pump operation (15.8 mills/kWh)[b]	4,099,500	3,952,800	4,031,800
Capability loss penalty ($170.4/kW-yr)[b]	11,031,000	10,775,000	11,280,000
Energy loss penalty (27.3 mills/kWh)[b]	88,200	86,200	90,200
Total	496,652,700	493,300,800	493,825,200

[a]Rate of return = 13%.
 Annual fixed charge rate = 18%.
 [b]Levelized value.

steam flow of 105% of the design value. In these calculations the steam conditions were assumed to be identical to the rated conditions, and the plant auxiliary power consumption excluding those of circulating water pumps and tower fans was assumed to be 5%. The actual maximum plant net output was then determined by subtracting the cooling system auxiliary power from the previously mentioned output. For convenience, this actual maximum output was compared with a reference point (say 720 MW), and the difference was treated as the relative generating

Table D-6
Calculation of Generating Capability Loss

		Design Configuration		
		4	5	6
Plant miximum output (including cooling system auxiliary power)	(kW)	716,467	716,467	716,134
Cooling tower fans	(kW)	2,807	2,231	1,936
Circulating water pumps	(kW)	3,675	4,019	4,439
Plant net output	(kW)	709,985	710,217	709,759
Generating capability loss (based on 720 MW)	(kW)	10,015	9,783	10,241

capability loss. The present worth of the lifetime capability loss penalty was then expressed in terms of this relative generating capability loss as

PW (capability loss penalty) = (relative generating capability loss)

$$\times \text{(present worth of lifetime unit penalty cost)}$$

Similarly, the present worth of lifetime energy loss penalty was determined by

PW (energy loss penalty) = (relative generating capability loss)

$$\times \text{(operation hour per year)}$$

$$\times \text{(present worth of lifetime unit energy replacement cost)}$$

Table D-5 summarized the present worth of lifetime evaluated costs. On the basis of minimum cost the system configuration 5 was the best. Figure D-3 presented the present worth for all nine configurations. It must be remembered that the above optimization was based on the assumption that the condenser tube length was 36 ft

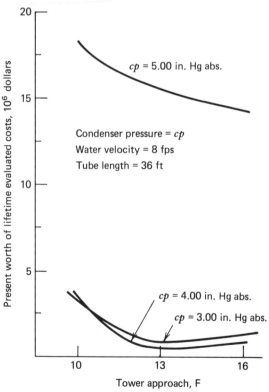

Figure D-3. Present worth of lifetime evaluated costs (relative to $\$493 \times 10^6$).

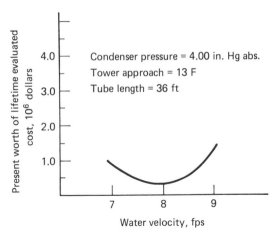

Figure D-4. Present worth of lifetime evaluated costs (relative to 493×10^6).

and the water velocity in condenser tubes was 8 fps; therefore, further optimization was necessary. The results were presented in Figs. D-4 and D-5, indicating that the above condenser tube length and water velocity were indeed the optimal value.

In summary, the cooling system under consideration should have the following:

Condenser pressure	4.00 in. Hg abs.
Tower approach	13 F
Condenser tube length	36 ft
Water velocity in condenser tubes	8 fps

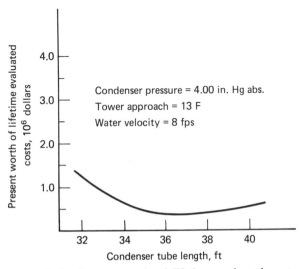

Figure D-5. Present worth of lifetime evaluated costs (relative to 493×10^6).

Cooling Tower Performance Curves and Tower Fill Data

Source: Figures E-1 to E-3 are reprinted with permission from *The Performance Curves* by the Cooling Tower Institute. Figures E-4 to E-11 are reprinted with permission from *The Kelly's Handbook of Cross-Flow Performance*.

69 WET BULB (°F)
26 RANGE (°F)

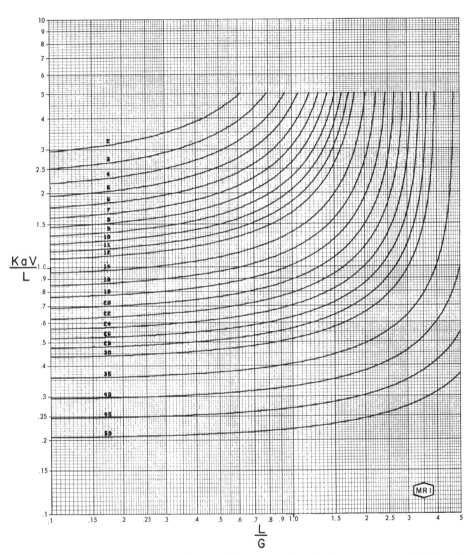

Figure E-1. Counterflow tower characteristic curves (wet-bulb temperature 69 F and cooling range 26 F).

69 WET BULB (°F)
22 RANGE (°F)

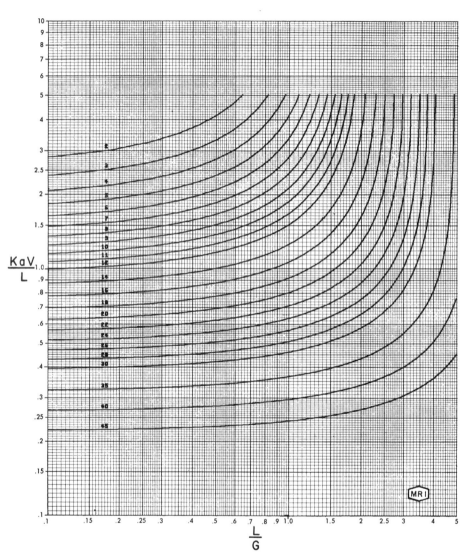

Figure E-2. Counterflow tower characteristic curves (wet-bulb temperature 69 F and cooling range 22 F).

69 WET BULB (°F)
18 RANGE (°F)

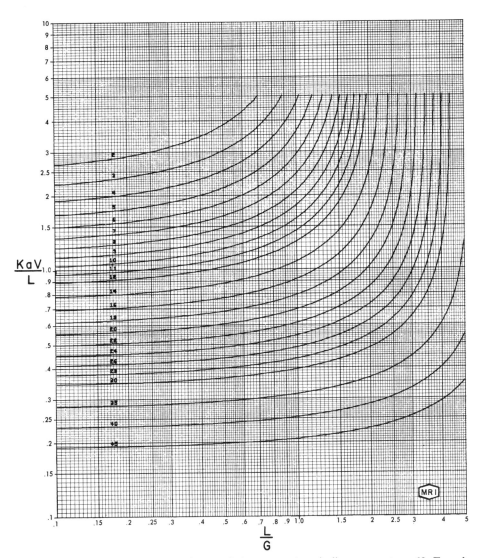

Figure E-3. Counterflow tower characteristic curves (wet-bulb temperature 69 F and cooling range 18 F).

75 °F, WET BULB
20 °F, RANGE

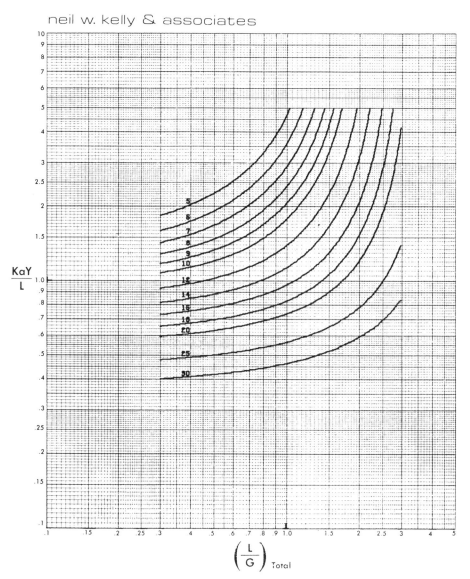

Figure E-4. Crossflow tower characteristic curves (wet-bulb temperature 75 F and cooling range 20 F).

75 °F, WET BULB
18 °F, RANGE

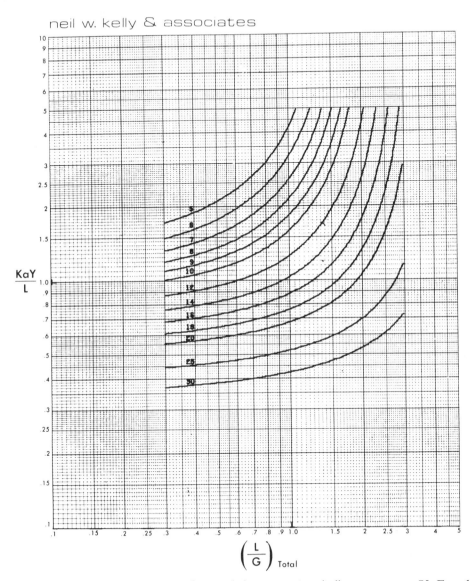

Figure E-5. Crossflow tower characteristic curves (wet-bulb temperature 75 F and cooling range 18 F).

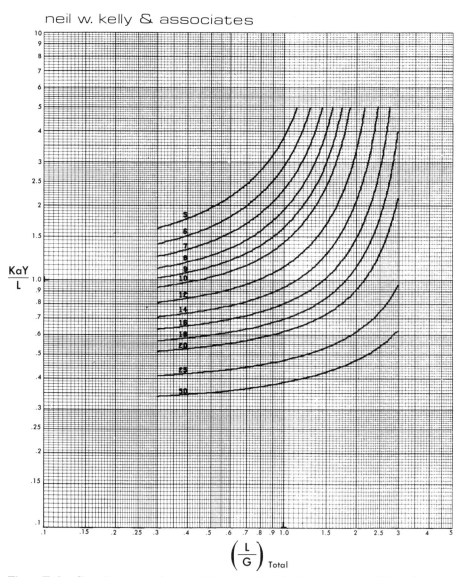

75 °F, WET BULB
16 °F, RANGE

Figure E-6.   Crossflow tower characteristic curves (wet-bulb temperature 75 F and cooling range 16 F).

FILL F

AIR TRAVEL, X=12 FT.

neil w. kelly & associates

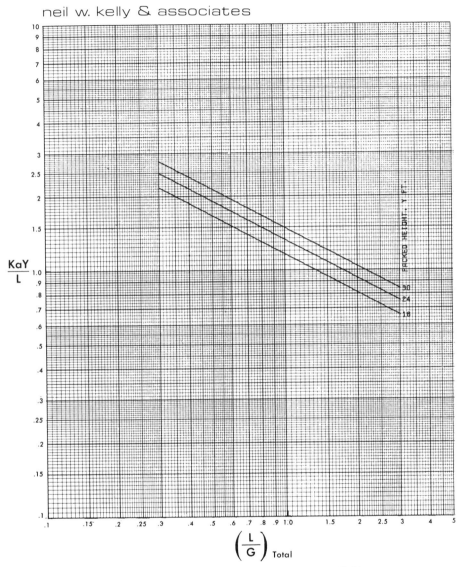

$$\frac{KaY}{L}$$

$$\left(\frac{L}{G}\right)_{Total}$$

Figure E-7. Fill characteristic curves for Kelly's fill type F ($x = 12$ ft).

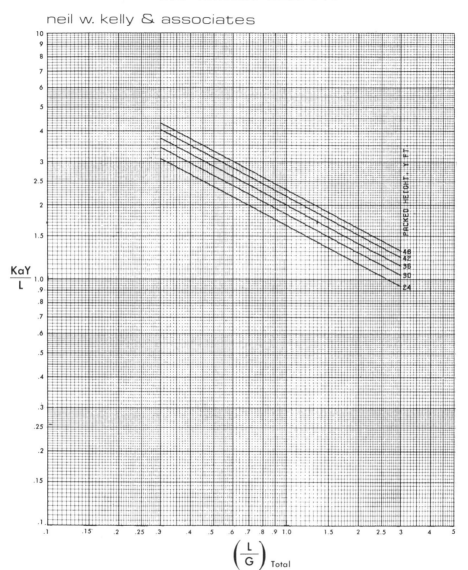

Figure E-8. Fill characteristic curves for Kelly's fill type F (x = 18 ft).

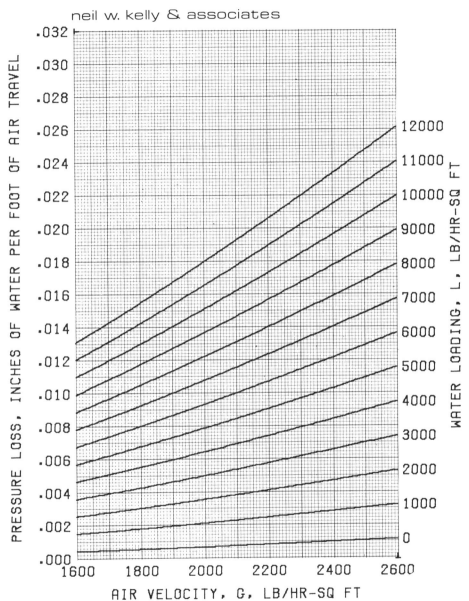

Figure E-9.   Pressure drop for Kelly's fill type F.

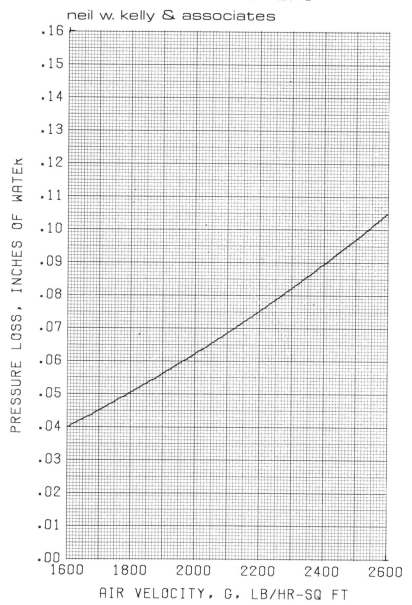

Figure E-10. Pressure drop across inlet louvers.

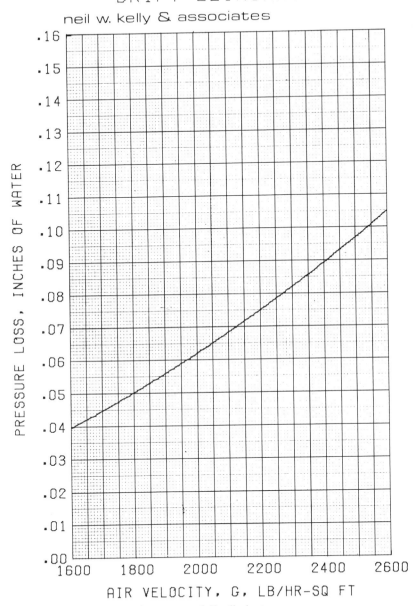

Figure E-11. Pressure drop across drift eliminator.

Conversion Factors

The following are conversion factors from English units to SI:

1 in.	$= 0.0254$ m
1 ft	$= 0.3048$ m
1 yd	$= 0.9144$ m
1 in.^{2}	$= 645.16$ mm^2
1 ft^2	$= 0.0929$ m^2
1 yd^2	$= 0.836$ m^2
1 in.^{3}	$= 1.638 \times 10^{-5}$ m^3
1 ft^3	$= 0.0283$ m^3
1 lb	$= 0.4536$ kg
1 lb/ft^3	$= 16.019$ kg/m^3
1 lb/in.^{2}	$= 6.8948$ kNm$^{-2} = 0.068948$ bar
1 psig	$= 0.068948$ bar $(g) = 6.8948$ kNm$^{-2}(g)$
1 Btu	$= 1.055$ kJ
1 Btu/hr	$= 0.293$ W
1 Btu/hr-ft^2	$= 3.15$ W/m^2
1 Btu/lb	$= 2.33$ kJ/kg
1 therm	$= 100{,}000$ Btu $= 105.5$ MJ
1 hp	$= 745.7$ W
1 R	$= \frac{5}{9}$ K
t F	$= \frac{5}{9}(t - 32)$ C
1 Btu/ft^2-hr-F	$= 5.678$ W/m^2-C
1 Btu/ft-hr-F	$= 1.7307$ W/m-C
1 Btu in/ft^2-hr-F	$= 0.1442$ W/m-C
1 Btu/ft^3	$= 37.258$ kJ/m^3
1 ton of refrigeration	$= 3.5169$ kW

Index